The Species–Area Relationship

The species–area relationship (SAR) describes a range of related phenomena that are fundamental to the study of biogeography, macroecology and community ecology. While the subject of ongoing debate for a century, surprisingly no previous book has focused specifically on the SAR. This volume addresses this shortfall by providing a synthesis of the development of SAR typologies and theory, as well as empirical research and application to biodiversity conservation problems. It also includes a compilation of recent advances in SAR research, comprising novel SAR-related theories and findings from the leading authors in the field. The chapters feature specific knowledge relating to terrestrial, marine and freshwater realms, ensuring a comprehensive volume relevant to a wide range of fields, with a mix of review and novel material and with clear recommendations for further research and application.

THOMAS J. MATTHEWS is a senior research fellow at the University of Birmingham; a research member of the Azorean Biodiversity Group, Portugal; and a visiting researcher at the University of the Ryukyus, Japan. He is a leading researcher in the fields of macroecology and biogeography and much of his research involves the species–area relationship in some form. He is also the lead-author of a key SAR software resource and an associate editor of *Frontiers of Biogeography*, the scientific journal of the International Biogeography Society, and the *Journal of Biogeography*.

KOSTAS A. TRIANTIS is an assistant professor at the National and Kapodistrian University of Athens. He is an associate editor of the *Journal of Biogeography* and, since 2015, has been Director-at-Large of the International Biogeography Society. He is a biogeographer with broad interests in island biogeography, macroecology and conservation biology and has a long-term fascination with the species–area relationship.

ROBERT J. WHITTAKER is an expert in island biogeography and conservation biogeography. He has been a professor of biogeography at the University of Oxford since 2004, and a part-time professor at University of Copenhagen, in the Centre for Macroecology, Evolution and Climate, since 2015. He has published more than 150 articles and three previous books. He is the editor-in-chief of *Frontiers of Biogeography*.

ECOLOGY, BIODIVERSITY AND CONSERVATION

The world's biological diversity faces unprecedented threats. The urgent challenge facing the concerned biologist is to understand ecological processes well enough to maintain their functioning in the face of the pressures resulting from human population growth. Those concerned with the conservation of biodiversity and with restoration also need to be acquainted with the political, social, historical, economic and legal frameworks within which ecological and conservation practice must be developed. The new Ecology, Biodiversity, and Conservation series will present balanced, comprehensive, up-to-date, and critical reviews of selected topics within the sciences of ecology and conservation biology, both botanical and zoological, and both 'pure' and 'applied'. It is aimed at advanced final-year undergraduates, graduate students, researchers, and university teachers, as well as ecologists and conservationists in industry, government and the voluntary sectors. The series encompasses a wide range of approaches and scales (spatial, temporal, and taxonomic), including quantitative, theoretical, population, community, ecosystem, landscape, historical, experimental, behavioural and evolutionary studies. The emphasis is on science related to the real world of plants and animals rather than on purely theoretical abstractions and mathematical models. Books in this series will, wherever possible, consider issues from a broad perspective. Some books will challenge existing paradigms and present new ecological concepts, empirical or theoretical models, and testable hypotheses. Other books will explore new approaches and present syntheses on topics of ecological importance.

Ecology and Control of Introduced Plants
Judith H. Myers and Dawn Bazely

Invertebrate Conservation and Agricultural Ecosystems
T. R. New

The Species–Area Relationship

Theory and Application

Edited by

THOMAS J. MATTHEWS
University of Birmingham

KOSTAS A. TRIANTIS
National and Kapodistrian University of Athens

ROBERT J. WHITTAKER
University of Oxford

CAMBRIDGE
UNIVERSITY PRESS

CAMBRIDGE
UNIVERSITY PRESS

University Printing House, Cambridge CB2 8BS, United Kingdom

One Liberty Plaza, 20th Floor, New York, NY 10006, USA

477 Williamstown Road, Port Melbourne, VIC 3207, Australia

314–321, 3rd Floor, Plot 3, Splendor Forum, Jasola District Centre, New Delhi – 110025, India

79 Anson Road, #06–04/06, Singapore 079906

Cambridge University Press is part of the University of Cambridge.

It furthers the University's mission by disseminating knowledge in the pursuit of
education, learning, and research at the highest international levels of excellence.

www.cambridge.org
Information on this title: www.cambridge.org/9781108477079
DOI: 10.1017/9781108569422

© Cambridge University Press 2021

First published 2021

Printed in the United Kingdom by TJ Books Ltd, Padstow Cornwall

A catalogue record for this publication is available from the British Library.

Library of Congress Cataloging-in-Publication Data
Names: Matthews, Thomas J. (Thomas James), 1987– editor. | Triantis, Kostas A., 1976– editor. |
    Whittaker, Robert J., 1959– editor.
Title: The species-area relationship : theory and application / edited by Thomas J. Matthews,
    University of Birmingham, Kostas A. Triantis, National and Kapodistrian University of Athens,
    Robert J. Whittaker, University of Oxford.
Description: Cambridge, UK ; New York, NY : Cambridge University Press, 2021. |
    Series: Ecology, biodiversity and conservation | Includes bibliographical references and index.
Identifiers: LCCN 2020009499 (print) | LCCN 2020009500 (ebook) | ISBN 9781108477079
    (hardback) | ISBN 9781108701877 (paperback) | ISBN 9781108569422 (epub)
Subjects: LCSH: Species diversity–Statistical methods. | Biogeography. | Macroecology. |
    Numbers of species. | Spatial ecology.
Classification: LCC QH541.15.S64 S65 2021 (print) | LCC QH541.15.S64 (ebook) |
    DDC 581.7–dc23
LC record available at https://lccn.loc.gov/2020009499
LC ebook record available at https://lccn.loc.gov/2020009500

ISBN 978-1-108-47707-9 Hardback
ISBN 978-1-108-70187-7 Paperback

Contents

Contributors

Saeid Alirezazadeh
CIBIO/InBio, Centro de Investigação em Biodiversidade e Recursos Genéticos, Laboratório Associado, Universidade do Porto, Campus Agrário de Vairão, Portugal

Maíra Benchimol
Applied Ecology and Conservation Lab, Universidade Estadual de Santa Cruz, Brazil

Tim M. Blackburn
Centre for Biodiversity and Environment Research, University College London, UK

Luís Borda-de-Água
CIBIO/InBio, Centro de Investigação em Biodiversidade e Recursos Genéticos, Laboratório Associado, Universidade do Porto, Campus Agrário de Vairão, Portugal

Paulo A. V. Borges
cE3c – Centre for Ecology, Evolution and Environmental Changes/Azorean Biodiversity Group and Universidade dos Açores, Portugal

Anderson Saldanha Bueno
Instituto Federal de Educação, Ciência e Tecnologia Farroupilha, Brazil

Pedro Cardoso
LIBRe – Laboratory for Integrative Biodiversity Research, University of Helsinki, Finland

Phillip Cassey
Centre for Applied Conservation Science, and School of Biological Sciences, University of Adelaide, Australia

Ryan A. Chisholm
Department of Biological Sciences, National University of Singapore, Singapore

Francisco Dionísio
cE3c – Centre for Ecology, Evolution and Environmental Changes, Universidade de Lisboa, Portugal

Simone Fattorini
Department of Life, Health and Environmental Sciences, University of L'Aquila, Italy

Dominique Gravel
Département de Biologie, Université de Sherbrooke, Canada

John M. Halley
Department of Biological Applications and Technology, University of Ioannina, Greece

John Harte
Energy and Resources Group and Department of Environmental Science, Policy and Management, University of California, USA

Robert D. Holt
Department of Biology, University of Florida, USA

Stephen P. Hubbell
Department of Ecology and Evolutionary Biology, University of California, USA

Isabel L. Jones
Faculty of Natural Sciences, University of Stirling, UK

Athanasios S. Kallimanis
Department of Ecology, Aristotle University of Thessaloniki, Greece

Alexandra Kraberg
Alfred-Wegener Institute Helmholtz Centre for Polar Research, Section of Polar Biological Oceanography, Germany

Mark V. Lomolino
Department of Environmental and Forestry Biology, College of Environmental Science and Forestry, USA

Thomas J. Matthews
School of Geography, Earth and Environmental Sciences, and Birmingham Institute of Forest Research, University of Birmingham, UK

Florent Mazel
Department of Botany and Biodiversity Research Centre, University of British Columbia, Canada

Manuela Neves
CEAUL, and Instituto Superior de Agronomia, Universidade de Lisboa, Portugal

Ana Filipa Palmeirim
School of Environmental Sciences, University of East Anglia, UK

Henrique M. Pereira
German Centre for Integrative Biodiversity Research (iDiv) Halle-Jena-Leipzig, Germany

Carlos A. Peres
School of Environmental Sciences, University of East Anglia, UK

Konstantinos Proios
Department of Ecology and Taxonomy, National and Kapodistrian University of Athens, Greece

Petr Pyšek
Department of Invasion Ecology, Czech Academy of Sciences, Czech Republic

François Rigal
Université de Pau et des Pays de l'Adour, Collège STEE, Sciences et technologies pour l'énergie et l'environnement, France

James Rosindell
Department of Life Sciences, Imperial College London, UK

Arnošt L. Šizling
Center of Theoretical Study, Charles University, Czech Republic

Eva Šizlingová
Center of Theoretical Study, Charles University, Czech Republic

Adrian Stier
Department of Ecology, Evolution, and Marine Biology, University of California, USA

David Storch
Department of Ecology and the Center for Theoretical Study, Charles University, Czech Republic

Danielle Storck-Tonon
Programa de Pós-Graduação em Ambiente e Sistemas de Produção Agrícola, Universidade do Estado de Mato Grosso, Brazil

Giovanni Strona
Research Centre for Ecological Change, University of Helsinki, Finland

Wilfried Thuiller
Université Grenoble Alpes, Université Savoie Mont Blanc, CNRS, LECA – Laboratoire d'Écologie Alpine, France

Even Tjørve
Inland University of Applied Sciences, Norway

Kathleen M. C. Tjørve
Inland University of Applied Sciences, Norway

Kostas A. Triantis
Department of Ecology and Taxonomy, National and Kapodistrian University of Athens, Greece

Karl Inne Ugland
Faculty of Mathematics and Natural Sciences Department of Biosciences, University of Oslo, Norway

Werner Ulrich
Faculty of Biology and Environment Protection, Nicolaus Copernicus University, Poland

Joseph A. Veech
Department of Biology, Texas State University, USA

Robert J. Whittaker
School of Geography and the Environment, University of Oxford, UK

Foreword

MARK V. LOMOLINO

'Size matters'. At face value, this appears to be an overly simplistic and vacuous statement. In reality, however, it is a surprisingly insightful declaration of the influence of size on nearly all natural phenomena, including all those involving life throughout its 3.6-billion-year history on this planet.

My first introduction to the pervasive relevance of size was in a characteristically captivating lecture by the esteemed physiological ecologist Brian McNab. 'Size matters,' he told us. 'The most important factor influencing all physiological processes – thermoregulation, metabolism, digestion, locomotion, etc., is body size.' Professor McNab would go on to say (while we struggled to keep up with the frenetic pace of his renderings of graphs on the chalkboard and his singular stream of insights), 'If a picture is worth a thousand words, then a graph must be worth millions!' The connection back to his original statement on size was that, for the preponderance of graphs we hurriedly scratched down in our notes, the x-axis was body size.

Turn the pages some ten years later to lectures in what would seem to be a starkly unrelated course – Biogeography, and James Hemphill Brown would time and again return to the fundamental importance of size. Perhaps more than any other scientist of his time, Professor Brown understands that some very fundamental principles of physics govern the marvellous myriad of natural phenomena, from chemical activity and fluid flow among cells to the properties and processes regulating the activity, longevity and diversity of individuals, species populations, biological communities and, indeed, the entire biosphere.

Given this background and, in particular, the pervasive relevance of size as a property influencing fundamental properties of the natural world, the long and distinguished history of research on the species–area relationship can be recognized as a scaled-up restatement of McNab's axiom – in this case, *the most important factor influencing the characteristic properties and processes of all species, ecological communities and ecosystems is their size, or 'area'*. Area dictates the total amount of energy from solar radiation reaching the primary producers and subsequently transferred to all consumers and decomposers. Area thus also limits the total number (or mass) of individuals that can be supported at any particular time and, through a complex set of processes and rules governing division of energy, space and other key resources, it ultimately determines the diversity of species.

Accordingly, the species–area relationship has attracted the attention of scientists throughout the histories of the fields of ecology, evolution and biogeography, and likely long before the earliest articulation of these disciplines. Indeed, in their struggles for survival, the earliest hunters and gatherers must have understood that their likelihood of encountering more and a greater variety of food plants and game animals increased with the size of their search area. The first scientifically rigorous descriptions and postulations of underlying causes for the species–area relationship, however, would await many generations of advancements, as it required an upscaling of our collective knowledge of the natural world from local to regional and eventually a global scale.

With the eventual burgeoning accumulation of empirical descriptions of how diversity of an ever-expanding menagerie of species and taxa varied across the globe came an increased sophistication in the tools to analyse and theories to explain this very fundamental pattern of nature. Within the past two decades, our empirical knowledge, theoretical constructs and analytical powers have advanced to what appears to be a watershed moment in our quest to understand the nature of biological diversity and its dependence on area. The co-editors of this volume have seized on this opportunity and produced what may well serve as a landmark contribution in modern science.

The collected contributions by distinguished biogeographers and ecologists have been organized into five sections. In Part I, a summary of the history of research on the species–area relationship (SAR) is provided, including a description of the dichotomy of island SARs versus species accumulation curves. The heuristic value of these patterns in terms of their relevance to developments in a variety of fields, including island biogeography and conservation biology, is then summarized.

Part II includes a number of interesting chapters on diversity–area relationships, including their alternative forms, underlying factors and processes, and comparisons between emergent patterns for natives versus alien species, and for alternative measures of biological diversity, including functional and phylogenetic diversity.

Part III focuses on theory and analytical approaches for describing and extrapolating from empirical patterns. Included in this section are chapters explicitly identifying assumptions behind alternative renderings of species–area curves, their relevance to related patterns in community structure such as beta diversity (community dissimilarity), the geometrical underpinnings of nested SARs and the implications of maximum entropy theory, extreme value theory, neutral theory and food web theory.

Part IV addresses the applied relevance of species–area research, including a variety of innovative approaches for identifying such important properties and phenomena as biodiversity hotspots, susceptibility of particular types of species to anthropogenic extinctions, temporal dynamics in species–area curves, the importance of shape and other geometric moments beside area, the use of 'relict' SARs to assess conservation value and applications to marine systems.

Part V comprises a single chapter that provides an especially important and provocative synthesis of the salient lessons from this collective body of contributions. The chapter author goes on to provide an outline of how to develop a more integrative understanding of this all-pervasive pattern in the geography of nature – outlining a unifying theory of SARs and how this may be developed to provide a genuinely holistic understanding of how and why diversity varies across space and time.

It is indeed an honour to have been invited to contribute to this hallmark contribution to biogeography, if only in a small way by adding a few thoughts from research my colleagues and I have conducted on the SAR and related patterns in the geography of nature. First, and at a most general level and likely relevant to all issues discussed in these collected works, we assert that in order to develop a more holistic understanding of this or any other pervasive pattern in biogeography, our conceptual constructs and empirical inferences should be based on the fundamental, unifying principles of the discipline (see Lomolino, 2016; Lomolino et al., 2017, p. 144). In short, these principles assert that all patterns in the geography of life result from: (1) non-random patterns in variation of environmental conditions across the geographic template; (2) the influence of this variation on the fundamental biogeographic processes of immigration, extinction and evolution; (3) the influence of these fundamental processes on each other (e.g. the arrival of new species altering the likelihood of extinction or evolution of resident species) and (4) system feedback in the form of ecological interactions among species, which influence the fundamental capacities of other species to immigrate, survive and evolve.

The second point follows from the first and in particular its emphasis on the fundamental processes, which exhibit scale dependence. That is, with respect to island species–area relationships specifically, immigration, extinction and evolution are strongly influenced by the fundamental dimensions of islands – that is, their isolation and size (area). One salient but easily overlooked product of this spatial scale dependence is that even such a long-studied and apparently simple pattern as the SAR may exhibit a protean persona – appearing to take on different forms depending on the spatial scale (extent) and the underlying processes that tend to dominate at that scale. As I postulated in previous papers, this protean nature includes the tendency for the SAR to be relatively flat on small islands (where colonizations and extinctions tend to be stochastic), assume the expected or 'canonical' curve for islands of intermediate size (where immigration and extinction may approximate an equilibrial condition consistent with MacArthur and Wilson's theory) and then the slope of the relationship may increase again on the largest islands (where in situ evolution supplements the species pool) (see Lomolino, 2000, 2001).

The third point from our research on these patterns touches on an admittedly darker but equally pervasive concern over how we approach research on the geography of the natural world. As an increasing number of other scientists have observed, there doesn't exist one ecosystem anywhere across our planet that hasn't been touched, if not fundamentally transformed, by humanity. Recently, this has been clearly demonstrated by a series of elegant and innovative studies conducted by S. Faurby and J. C. Svenning (2015), who were able to recreate the natural (pre-humanity) geographic gradients in native communities at regional to global scales – patterns that often differ substantially from contemporary patterns perceived to represent the 'natural' state.

In our own research in reconstructive biogeography, Alexandra van der Geer, Georgios Lyras and I discovered that human activities have qualitatively altered the SAR for mammals on oceanic islands (van der Geer et al., 2017; see also Helmus et al., 2014). To understand why this occurred, we need only to consider the impacts of human activities in light of the fundamental unifying processes. Anthropogenic transformations of the geographic signatures of nature are to be expected given that our effects on extinctions, immigrations and evolution of native wildlife and plants are species-selective. That is, human activities differentially impact species as a function of a variety of traits – the most important of which is their body size. For example, anthropogenic extinctions of terrestrial vertebrates across the globe during the late Pleistocene and early Holocene decimated the native megafauna – the largest species, which required the largest areas to maintain their populations. Species introductions appear to have further compounded this bias against larger, area intensive species (i.e. those requiring large spaces to persist as viable populations) – the most common species listed on our tallies of those introduced to islands are typically dominated by relatively small species (mice, rats and shrews). The overall impact across the world's archipelagos was a wholesale distortion of the natural, underlying patterns in species diversity – an affect that may have cascaded through species assemblages on islands as the overabundance of exotic mammals impacted native wildlife and plants across the world's archipelagos.

This leaves us with a haunting question for each of the contributors to this volume and indeed for anyone interested in understanding and conserving biological diversity of native biotas: *To what degree are the patterns we have been studying throughout the history of biogeography, evolution and ecology artifacts of the power of our species to transform the primordial signals of nature across the planet?* We have only recently amassed the databases and developed the analytical tools required to answer this question, but, given the ominous global changes on the horizon, it seems imperative we address this challenge before all too many more species join the lost legions of the ghosts of biotas past.

References

Faurby, S. & Svenning, J.-C. (2015) Historic and prehistoric human-driven extinctions have reshaped global mammal diversity patterns. *Diversity and Distributions*, **21**, 1155–1166.

Helmus, M. R., Mahler, D. L. & Losos, J. B. (2014) Island biogeography of the Anthropocene. *Nature*, **513**, 543–546.

Lomolino, M. V. (2000) Ecology's most general, yet protean pattern: The species–area relationship. *Journal of Biogeography*, **27**, 17–26.

Lomolino, M. V. (2001) The species–area relationship: New challenges for an old pattern. *Progress in Physical Geography*, **25**, 1–21.

Lomolino, M. V. (2016) The fundamental, unifying principles of biogeography. *Frontiers of Biogeography*, **8**, e29920.

Lomolino, M. V., Riddle, B. R. & Whittaker, R. J. (2017) *Biogeography*, 5th ed. Sunderland: Sinauer Associates.

van der Geer, A. A., Lomolino, M. V. & Lyras, G. (2017) Island life before man. *Journal of Biogeography*, **44**, 995–1006.

Preface

The species–area relationship is one of the longest known general patterns in ecology and biogeography: a simple enough emergent macroecological property on first glance, it remains in some ways poorly understood, beguilingly varied in form and enigmatic. Wider and contemporary interest in the species–area relationship (SAR) is evidenced by the results of a Google Scholar search undertaken in September 2019 using 'species–area relationship' as the topic field and all available years, which yielded 12,800 articles (almost 2,000 of which were published in the previous two years), distributed across a diverse array of fields, including ecology, zoology, forestry, microbiology, anthropology, biophysics and virology. All the more surprising then that, to our knowledge, no book has previously been published that focuses specifically on the SAR. When the idea of a SAR focused book was first proposed to us by Michael Usher, we contemplated writing a co-authored text. However, we subsequently decided that the field would be better served by an edited volume that provided both a synthesis of SAR material and a compilation of some of the most interesting recent advances in SAR research; topics that, due to their recent emergence in the literature, have yet to be brought together in any book or review.

The book which has resulted is thus broad in scope, with each chapter providing an in-depth evaluation of a particular area of SAR research, in cases largely empirical and in cases via review. It also provides an overview of recent advances in community ecology, macroecology and biogeography more generally: we hope it will therefore be of interest even to those who begin with no specific interest in the SAR.

We thank all the contributing authors for their willingness first to participate in this project and second to engage in article review and revision. We are also grateful to the individuals who did not contribute chapters but who helped with chapter review. We would also like to thank the following individuals at, or connected to, Cambridge

University Press, for their contribution to the development of this book: Michael Usher and Dominic Lewis for inviting us to contribute a book on SARs as part of the Ecology, Biodiversity and Conservation series, Liz Steel for proof reading, and Aleksandra Serocka for guiding us through the process and answering our numerous questions along the way. Finally, Tom would like to thank his parents, whose long-term support and guidance has allowed him to pursue a career in research, and Emilie, for her unwavering love and encouragement. Kostas would like to thank his parents, Eleni and Aristotelis, for their unparalleled support, and Anthi, for her patience and understanding.

Part I

Introduction and History

1 · *The Species–Area Relationship: Both General and Protean?*

THOMAS J. MATTHEWS, KOSTAS A.
TRIANTIS AND ROBERT J. WHITTAKER

> The history of the species–area field is long and rife with debate. Workers
> have argued about the form of the relationship, its interpretation, and the
> reasons for its existence. This argument is not trivial and without
> consequence.
>
> (McGuinness, 1984, p. 424)

An introduction to a book on the species–area relationship (SAR) would
be incomplete without the oft-repeated statement that the SAR, which
describes the increase in richness observed with increasing sample area
(Figure 1.1A), is the closest thing to a general law in ecology (Schoener,
1976; Rosenzweig, 1995; Lawton, 1999; Lomolino, 2000; Tjørve &
Tjørve, 2017). However, while its characterization as a 'law' can be
debated, there is no disputing the fact that the SAR is an almost universally observed phenomenon. It has been described for practically all taxa,
across multiple spatial and temporal scales and in a range of systems and
landscape types (MacArthur & Wilson, 1967; Connor & McCoy, 1979;
Rosenzweig, 1995; Lomolino, 2000; Drakare et al., 2006; Triantis et al.,
2012; Bolgovics et al., 2015; Matthews et al., 2016a; Dengler et al.,
2020). The few cases where the expected relationship is not observed are
where other variables exert a much stronger influence on richness than
area and either negatively co-vary with area (e.g. likelihood of wildfire;
Wardle et al., 1997) or vary independently of area. As a pattern, it has
intrigued ecologists and biogeographers for over 200 years (Chapter 2).
Indeed, the SAR has formed the focus of much of our own research (e.g.
Triantis et al., 2008, 2012; Whittaker et al., 2014, 2017; Matthews et al.,
2016a, b). It represents a fundamental component of numerous ecological and biogeographical theories, including the equilibrium theory of
island biogeography (MacArthur & Wilson, 1967) and the unified

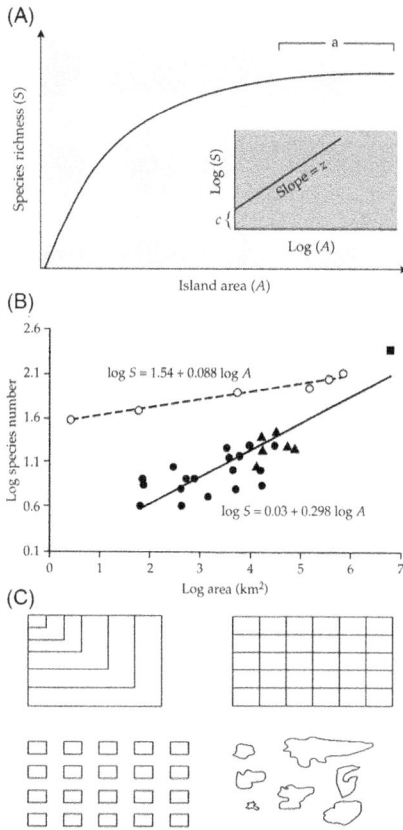

Figure 1.1 Three key representations of the Species–Area Relationship. (A) An idealized power model SAR, which describes a curved relationship in arithmetic space and a straight-line relationship in its log–log form. (B) E. O. Wilson's (1961) figure of SARs for ponerine and cerapachyine ants in Melanesia. Solid dots represent islands (ISARs); open circles, cumulative areas of New Guinea up to and including the whole island (SACs); triangles, archipelagos (not used in the regression); and the square, the whole of South-East Asia. (C) Some alternative configurations of area that might be involved in different studies: the bottom right represents a group of isolates that could be the basis for an ISAR, whereas the other three sampling scenarios would be used to construct a SAC. (A) From Lomolino et al. (2017; figure 13.8); (B) and (C) from Whittaker and Fernández-Palacios (2007; figure 4.2 and box 4.2, respectively; part (B) was originally adapted from Wilson, 1961)

neutral theory of biodiversity and biogeography (Hubbell, 2001; Chapter 11). In addition, SAR models have been widely used in applied ecology and conservation science and represent one of the most important tools in the conservation biogeographer's toolkit (Rosenzweig, 2004;

Whittaker et al., 2005; Chapters 13–17). For example, the SAR is a cornerstone of the applied 'reconciliation ecology' research agenda (Rosenzweig, 2003, 2004).

1.1 The Many Types of Species–Area Relationship

At this stage it is necessary to define what exactly is meant by the term 'species–area relationship'. While the SAR may appear to be a relatively uncomplicated concept, its application within the ecological literature is often somewhat ambiguous. Most of this ambiguity concerns the fact that there is not a single type of SAR, but instead there is a suite of relationship types with more or less distinct data structures. While the search for an agreed classification and terminology has generated much debate (e.g. Scheiner, 2003; Gray et al., 2004; Dengler, 2009), there are only two variables involved in each case: area and species number, and so the problem can be quickly described. First, the areal units analysed may be geographically separated or contiguous, permitting their analysis either as separate entities or as a nested sequence. Second, each species can be counted once as area is accumulated (whether from a nested sequence of contiguous areas or not) or separately in each (sub-)area. Third, for the accumulation curve structure, the value of richness entered for a given size of area can represent a single data point or the average richness value of multiple samples of that size (Dengler et al., 2020).

The upshot is that several forms of species accumulation curve (SAC) can be identified as types of species–area relationship (Type I–III curves in Scheiner's 2003 typology) but just one type of data structure of 'isolate' area versus species number per isolate (Gray et al., 2004), which some term island species–area relationships (ISARs; Type IV SARs in Scheiner, 2003) (e.g. Triantis et al., 2012; Matthews et al., 2016a, b). Further, SACs are often constructed using non-area-based measures of sampling effort (e.g. trap hours) and so sample area-based SACs can be distinguished as a subset of SACs.

At its most basic, SACs are constrained by their mode of construction to display a monotonic increase in species richness. A SAC must either remain constant or increase with each increment in area (Figure 1.2), such that the total richness of the study system is described by the final data point. By contrast, sometimes a larger island can have fewer species than a smaller one and, in certain rare circumstances, there can even be a negative overall ISAR or the relationship may be negative over a limited span of the area range of a study system (Figure 1.2).

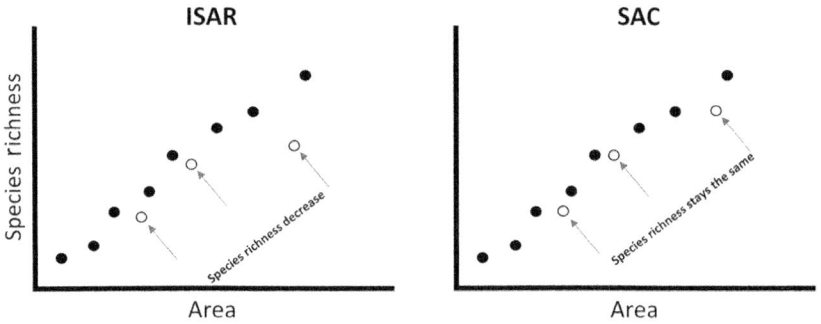

Figure 1.2 Two hypothetical species−area relationships. To the left, an island (isolate) species−area relationship (ISAR) in which there is scatter around the mean trend, to the extent that some islands are less rich than one or more smaller island; to the right, a species accumulation curve (SAC). The SAC is constructed by adding the new species encountered per each extension of area, such that the SAC is constrained never to decrease. As each island may contain different subsets of species, the ISAR only crudely constrains the possible overall system richness, whereas the final data point for the SAC describes the overall system richness.

The lack of a generally agreed terminology distinguishing SAR types has presented a real hindrance to the development of a shared understanding. Studies comparing SARs often combine ISARs and SACs without acknowledging that that the two types have different properties (but see e.g. Drakare et al., 2006; Matthews et al., 2016b). For example, the seminal SAR figure of Wilson (1961) for ponerine and cerapachyine ants in Melanesia compares an ISAR (island data) with a SAC ('mainland' New Guinea), a feature that is often neglected in work citing or reproducing it (see Figure 1.1B). Indeed, some have argued that, to fairly compare SARs from different systems, the whole sampling design must be identical, including range of area, type of SAR (SAC or ISAR) and if SACs are the focus of study, the accumulation order, sampling intensity, plot shape, use of continuous or discontinuous plots and, if the latter, the locations of the discontinuous plots within the study extent (see Chapter 7; see Figure 1.1C for some examples of SAR construction approaches).

In this volume, we have encouraged authors to make explicit reference to the type(s) of SAR (e.g. ISAR, SAC or both) that their chapter is focused on. However, and as with most things in science and life in general, different people have different opinions on the correct terminology to use and we have not imposed a standard usage. For instance, in certain chapters SACs are referred to as nested SARs, and the term saSAR is used in some chapters to refer to sample-area based SACs. We acknowledge that our own use of the term 'island SAR' (or Isolate SAR) may not be the best solution as ISARs might be constructed using data not from isolates but for other irregularly sized geographical units of varying degrees of geographical isolation (e.g. countries or biogeographic provinces).

1.2 A Flexible Biogeographical Law

Our work has generally focused on the ISAR, whether that be for true islands (e.g. Triantis et al., 2012) or habitat islands (Matthews et al., 2016a). What has struck us during this work (and in reading the work of many others) is that the idea that the ISAR is universal is only really true in the sense that larger areas tend to have more species than smaller areas. Beyond that, many characteristics of ISARs have been found to differ (often quite considerably) between datasets. Lomolino (2000; see also Whittaker & Triantis, 2012) recognized this when he described the ISAR as 'protean'. The adjective 'protean', meaning 'versatile', 'flexible' and 'able to change easily', is derived from the Ancient Greek God 'Proteus'. According to Greek mythology, Proteus was an early prophetic sea-god who had the ability to foresee the future, but would frequently change his shape to avoid those who asked him to share his prophetic knowledge. The name Proteus suggests the 'first', and the ISAR is considered to be one of the first discussed general patterns related to the diversity of life. One other obvious way that the ISAR can be considered a protean pattern is the fact that the mathematical form of the relationship often varies between datasets (Triantis et al., 2012; Matthews et al., 2016a; Leveau et al., 2019). In practice, we can often fit multiple models successfully to the same dataset. We then have the challenge of trying to work out which is the best model, that is, the closest approximation to the true form of the relationship. This may be attempted as a statistical exercise, that is, as a black box type of approach. Alternatively, we may regard the process of model fitting rather differently, as an exercise in hypothesis testing, that is, we may have theoretical grounds for predicting a particular form or a particular set of alternative forms that are linked to distinct mechanisms or processes. The step of model fitting is then used to determine the plausibility of the initial hypothesis or of selecting from a set of multiple-working hypotheses those which remain standing and then, of these, which has the greatest verisimilitude. Hence, SARs have what we might think of as phenomenological flexibility, as we are describing a suite of slightly different aspects of how diversity and area are related and then, for these different phenomena, there are advocates for particular mathematical models to represent the SAR. Indeed, over thirty different functions have been proposed (Chapter 7).

While the variation in ISAR form between datasets has (so far) precluded the identification of a universal model that provides the best fit to all datasets, the discovery of generalities regarding how and why ISARs vary has improved our understanding of the processes shaping diversity patterns more generally (see the Foreword). We know for example that the factors that influence species richness vary across spatial scale (Shmida & Wilson, 1985; Rosenzweig, 1995; Whittaker et al., 2001; Turner & Tjørve, 2005) and thus the ISAR should be scale-dependent (see Figure 1.3A). These different factors should also affect the nature of the relationship of species richness with area. Recent meta-analyses have shown that ISAR slope (z) increases from habitat, to continental shelf, to

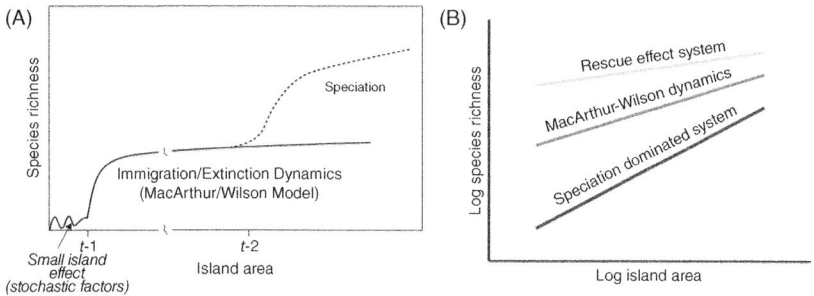

Figure 1.3 Two contrasting models of ISAR form. (A) The theoretical relationship of species richness with area in arithmetic space according to Lomolino (2000). Starting at the left side, there is little change in species number until a critical threshold is reached, but beyond threshold t1, species number increases rapidly, as a function of the immigration/extinction dynamics of the equilibrium model of island biogeography. With islands larger than t2, species number shows a further increase, as it is afforced by in situ speciation. The 'small-island effect' is shown by many island datasets (see also Chapter 19), but the generality (and even existence) of the second threshold and the upward curve towards the right side of the plot is contested. (B) As archipelago isolation increases, the ISAR slope (z) generally increases. Rescue effect systems are the least isolated and hence species populations on smaller islands are continually rescued from extinction by supplementary immigration, resulting in high intercepts and low slopes. Islands of intermediate isolation experience higher rates of species turnover and therefore feature steeper slopes and lower intercepts. Speciation dominated systems refer to remote archipelagos in which larger islands gain species through in situ diversification, generating the steepest slopes and lowest intercepts. This panel is only approximated by analyses of empirical datasets (e.g. Matthews et al. 2016a). (A) From Lomolino (2000; figure 6) under license from John Wiley and Sons; (B) from Whittaker et al. (2017; figure 4A) and reprinted with permission from AAAS

oceanic islands (Triantis et al., 2012; Matthews et al., 2016a). The shallowest slopes and higher intercepts characterize systems with minimal isolation, in which island-extirpation events are typically rapidly reversed by recolonization ('rescue effects') and which therefore feature comparatively high richness even on the smaller islands. By contrast, remote oceanic islands receive such low rates of immigration that colonizing lineages can diversify in isolation. In these systems, in theory, the smallest islands have low species richness because their small, unreliable resource bases cannot sustain marginal populations of small size or permit the origin and persistence of newly formed endemics. In practice, as we argue in Chapter 3, the more remote oceanic archipelagos can be so isolated that the configuration of the archipelago itself exerts a strong influence on ISAR form, in part overriding the effect on slope and intercept exerted by distance from the mainland (and see Matthews et al., 2019).

Thus, different processes are expected to result in different slopes; the difference is expected to be most pronounced when speciation-dominated systems are compared to systems where MacArthur–Wilson dynamics prevail and speciation has a secondary or no role (Figures 1.3 and 1.4). The cichlid

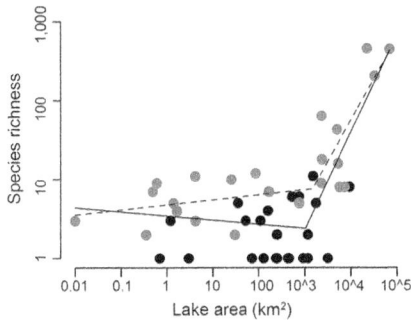

Figure 1.4 The effect of adaptive radiation on the ISAR (lighter points = lakes with speciation; darker points = lakes without speciation). The fit of two-slope ISAR models are shown (solid line = whole dataset; dashed line = lakes with speciation only). For the whole dataset model fit, the pre-breakpoint slope does not significantly differ from zero, the breakpoint occurs at 1,030 km^2 and the post-breakpoint slope (1.29) is significant and positive. For the model fit to the lakes with speciation, the pre-breakpoint slope does not significantly differ from zero, the breakpoint occurs at 1,470 km^2 and the post-breakpoint slope (0.99) is significant and positive. In both cases the two-slope model provides a better fit to the data than a one-slope regression model according to AIC$_c$. The data are of cichlid fishes in forty-six African lakes. From Wagner et al. (2014; figure 3b) under license from John Wiley and Sons

species of the Great African Lakes provide an illustration of these differences, but also of a clear threshold effect where there appears to be no relationship between species and area until a minimum lake area has been reached. Above this threshold, those lakes in which in situ diversification has occurred demonstrate a clear positive relationship. The *Anolis* lizards of the Caribbean provide another pertinent example of a threshold in the speciation–area and species–area relationships (Losos & Parent, 2010; figures 15.3 and 15.5).

Elements of the above arguments were codified in Rosenzweig's (1995) seminal scale-structured model of species–area relationships, which, however, included both ISARs and SACs (see Scheiner, 2003; Whittaker & Fernández-Palacios, 2007). His interprovincial curve represents an ISAR fitted between distinct biogeographical provinces, which he argued should exhibit a slope always >0.6 and generally close to unity (see Rosenzweig, 1995, 1998, 2001) (Figure 1.5A). SACs within continents, his intraprovincial curves, should be flatter and with higher intercepts than regular ISARs found within the province, which should display a range of values depending on their degree of isolation. Figure 1.5B was inspired by Rosenzweig's model but differs in showing exclusively ISARs. Given the propensity for remote oceanic archipelagos to generate endemics in situ, the expectation is for the between-archipelago or archipelagic species–area relationship (ASAR) to provide a steep relationship, where the relevant within-archipelago ISAR only crudely predicts the archipelagic value and, again, there is variation in the overall richness of different archipelagos (Figure 1.5A and B) in perhaps widely different locations. This reasoning is supported by analyses showing that i) for oceanic archipelagos, inter-archipelago species–area relationships (ASARs) are systematically steeper than the constituent ISARs (e.g. Triantis et al., 2015) and ii) ISARs estimated for endemic species are typically steeper with lower intercepts than those for non-endemic native species for the same archipelago (Figure 1.5C; Triantis et al., 2008). The final panel (Figure 1.5D) presents another generalization, which is that, for forest habitat island systems, ISARs calculated for generalist bird species are flatter and with higher intercepts than those for forest specialist bird species (Matthews et al., 2014). That i) specialist species, more reliant upon and restricted to the focal habitat, have the steeper slopes and ii) island endemic species have steeper slopes than non-endemic species (which may have populations on other islands or landmasses exchanging propagules with the islands in question) appears to be part of the same pattern, relating to the extent to which, as Rosenzweig put it, the islands in question are acting as sources or as sinks.

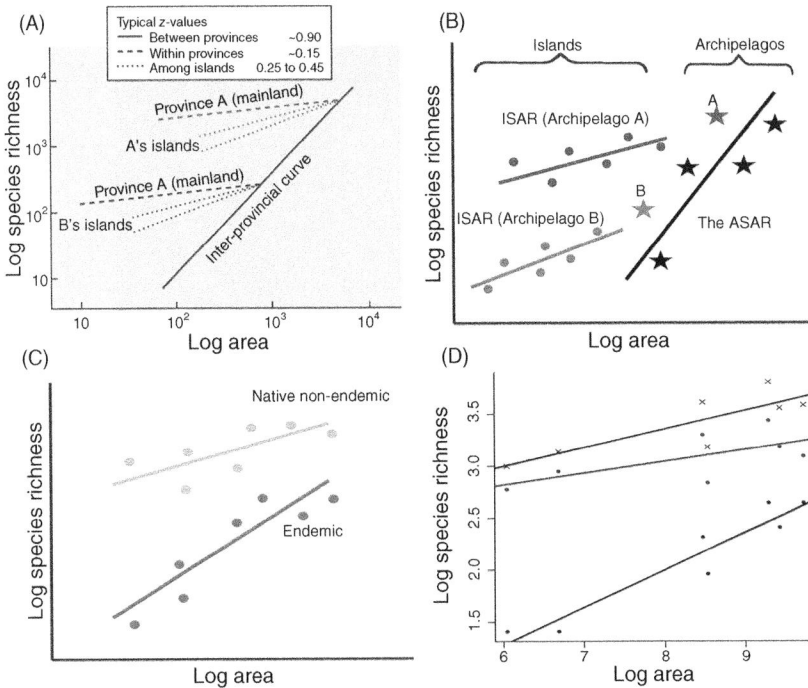

Figure 1.5 Generalized species–area relationships as a function of scale and isolation. (A) Rosenzweig's scale-structured model of species–area relationships, comprising a mix of SAC and ISAR types. (B) The archipelago species–area relationship (ASAR) and two of its constituent archipelago ISARs. Points A and B on the ASAR represent the archipelago diversity for archipelagos A and B, respectively. (C) ISARs constructed using endemic species and native non-endemic species. (D) Three ISARs for the same bird habitat island (seven islands) dataset: the bottom curve is for habitat specialist bird species, the middle curve is for habitat generalists and the top curve (and the crosses) for all bird species combined. The lines are the fit of the power (log–log) ISAR model. See text for further details. (A) From Rosenzweig (1995; figure 9.11; modified by Lomolino et al., 2017) and reprinted with permission from Cambridge University Press; (B) and (C) from Whittaker et al. (2017; figure 4c and figure 4b, respectively) and reprinted with permission from AAAS; and (D) from Matthews et al. (2014; figure 1e) under license from John Wiley and Sons

1.3 From Species–Area Relationships to Diversity–Area Relationships

Recently, and echoing the shift in macroecology and biogeography more generally, ISAR research has expanded to include other facets of diversity, such as functional (FD) and phylogenetic (PD) diversity

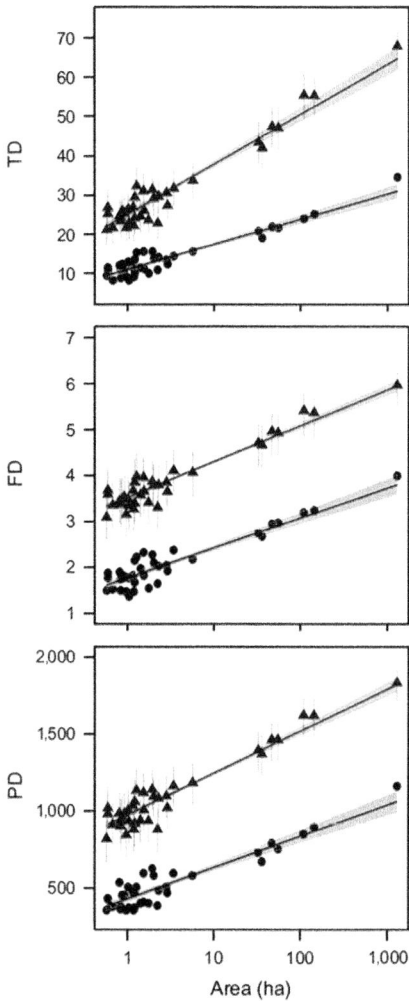

Figure 1.6 The relationship between island area and taxonomic diversity (top), functional diversity (middle) and phylogenetic diversity (bottom) for seventy-seven breeding bird species recorded on thirty-six islands in the Thousand Island Lake, China (surveyed between 2007 and 2016). The circles are observed diversity and the triangles diversity estimated using multi-species occupancy models. The lines are linear regression model fits (with the 95 per cent confidence intervals around each point in the case of the occupancy models) to the two sets of data after accounting for island isolation. Functional diversity was measured using a dendrogram-based metric and phylogenetic diversity using Faith's index. The grey bars represent 95 per cent Bayesian credible intervals of posterior mean estimates of TD, FD and PD. From Si et al. (2018; figures 1a, c and e) under license from John Wiley and Sons

(Whittaker et al., 2014; Mazel et al., 2015; Chapter 5). Figure 1.6 provides an example of an ISAR and the corresponding functional diversity–area and phylogenetic diversity–area relationships for a dataset of breeding birds in a lake island archipelago in China (Si et al., 2018). As has been argued for community ecology in general (e.g. Webb et al., 2002; Kraft et al., 2008), there is a hope that a focus on functional and phylogenetic diversity in SAR/diversity–area relationship studies may help us better understand the processes that underpin SARs/diversity–area relationships and ultimately the mechanisms that underpin community assembly. In addition, it enables us to better predict the impact of habitat loss on these other facets of biodiversity, which may have some value for developing models of ecosystem function (see Chapter 5). However, as many FD and PD metrics generally co-vary strongly with taxonomic diversity (TD), discerning the additional information provided by these metrics requires assessment of whether and how the trends diverge from null expectations (e.g. Whittaker et al., 2014). Going forward, the evaluation of other diversity–area relationships alongside SARs represents a challenging but promising avenue for future research in the field.

1.4 The Present Volume

This book is loosely structured into five sections. The first section comprises the present chapter and a chapter on the history of the SAR (Chapter 2). The SAR has a long history; one that includes several early studies published by Scandinavian authors, which have been overlooked in the mainstream ecological literature until recently (Tjørve et al., 2018). Chapter 2 provides an overview of these early studies, the development of mathematical SAR theory and the subsequent application of the SAR in a wide range of fields, such as island biogeography and conservation biogeography.

More recently, the last ten to fifteen years have seen i) the publication of several meta-analyses that have aimed to draw out general SAR patterns (e.g. Triantis et al., 2012; Matthews et al., 2019) and ii) the expansion of SAR theory to other types of diversity (e.g. Mazel et al., 2015) as well as non-traditional species assemblages (e.g. alien species; Baiser & Li, 2018). The second section of the book covers these ideas, including, in regard to the former, an expansion of a recently published global model of ISARs that aims to explain variation in parameters of the power ISAR model between datasets (Chapter 3) and a more general discussion of the determinants of SAR shape, including the idea that SAR drivers can be organized as a hierarchy of different processes and factors (Chapter 4). In regard to the latter, reviews of functional diversity and phylogenetic diversity–area relationships (Chapter 5) and alien species–area relationships (Chapter 6) provide comprehensive accounts of these topics.

Due in part to increases in computer power, increased mathematical training in ecology and biology courses and the now wide range of available software packages for ecological data analysis, the last twenty years has seen a resurgence in theoretical SAR research: both in terms of pure SAR research questions and in the development of (macro) ecological theories in which the SAR is a fundamental component. The third section of the book, focused on theoretical advances in SAR research, includes chapters dedicated to both of these types of theoretical developments, including topics such as the search for mathematical SAR functions (Chapter 7), the scaling and geometric properties of SARs (Chapter 8), extreme value theory and the SAC (Chapter 9), trophic SARs (Chapter 12) and the role of SARs in two recently proposed unified macroecological and biogeographical theories: the maximum entropy theory of ecology (Chapter 10) and the unified neutral theory (Chapter 11).

The fields of conservation biology and conservation biogeography are concerned with impacts of environmental change on biodiversity and on generating useful applied information that can be used in conservation and sustainable biodiversity management. As a general rule, conservation biology is focused more on local scale questions, whilst conservation biogeography is the subset of that broader field that revolves around issues operating at coarser spatial (and often temporal) scales. However, the SAR is a fundamental tool in both disciplines and has been used, amongst other things, to estimate the number of species likely to become extinct as a result of habitat loss and climate change (e.g. Ladle, 2009; Triantis et al., 2010) and to identify potential biodiversity hotspots and protected areas in both the terrestrial and marine realms (e.g. Neigel, 2003; Guilhaumon et al., 2008). The fourth section of the book focuses on the applied uses of the SAR, including reviews of the aforementioned applications (hotspot identification – Chapter 13; extinction predictions – Chapter 14), as well as novel contributions such as the use of network theory in island biogeography research and the importance of intra-archipelago configuration (which is of both pure and applied interest) (Chapter 15), the importance of geometry in habitat loss research and extinction predictions (Chapter 16) and the introduction of the concept of relict species–area relationships to estimate the conservation value of reservoir islands in flooded forest landscapes (Chapter 17). The section concludes with a chapter on the application of SARs in the marine realm, with a particular focus on the issues involved in interpreting marine SARs at large spatial scales (Chapter 18).

The final section of the book, consisting of a single chapter (Chapter 19), is based around a metaphor M. L. Rosenzweig used in the preface of his 1995 book, comparing the study of species diversity patterns with a dinosaur that has come alive and is challenging us. This final chapter also brings together some of the general findings of the previous chapters and identifies outstanding research questions and areas where further work is needed; we hope providing a catalyst for future ISAR-based research.

Ultimately, our aim in this volume has been to bring together a diverse array of leading researchers on SAR issues in order to: i) generate a comprehensive treatment that is relevant to a wide range of fields, ii) provide a useful general resource for students and researchers interested in the SAR and iii) provide a substantial novel contribution to the literature. We hope that it may serve to increase interest in the SAR as a pattern (or set of patterns) and to stimulate further developments in the field.

Acknowledgements

We thank the many colleagues with whom we have worked on SAR related research over the last fifteen years and in particular acknowledge the contribution to our ideas of François Guilhaumon, François Rigal, Paulo Borges and Michael Borregaard. We are also grateful to the others with whom we have debated ideas about the form and meaning of the SAR, including Mark Lomolino, Bob Ricklefs, Mike Rosenzweig, José María Fernández-Palacios, Evan Economo, Robert Cameron, Ana Santos, Joaquin Hortal, Jürgen Dengler, Yasuhiro Kubota, Jon Sadler, Even Tjørve, Manuel Steinbauer, Spyros Sfenthourakis, Sakis Mylonas and Larry Heaney. Finally, much of our recent work on the SAR has been made possible by a Marie Curie Fellowship (Species–Area Relationship: New Challenges for an Old Pattern: https://cordis.europa.eu/project/rcn/83607/factsheet/en), awarded to one of us (KAT) in 2007, which funded the collection of a large number of island SAR datasets.

References

Baiser, B. & Li, D. (2018) Comparing species–area relationships of native and exotic species. *Biological Invasions*, **20**, 3647–3658.

Bolgovics, Á., Ács, É., Várbíró, G., Görgényi, J. & Borics, G. (2015) Species–area relationship (SAR) for benthic diatoms: A study on aquatic islands. *Hydrobiologia*, **764**, 91–102.

Connor, E. F. & McCoy, E. D. (1979) Statistics and biology of the species–area relationship. *The American Naturalist*, **113**, 791–833.

Dengler, J. (2009) Which function describes the species–area relationship best? A review and empirical evaluation. *Journal of Biogeography*, **36**, 728–744.

Dengler, J., Matthews, T. J., Steinbauer, M. J., Wolfrum, S., Boch, S., Chiarucci, A., Conradi, T., Dembicz, I., Marcenò, C., García-Mijangos, I., Nowak, A., Storch, D., Ulrich, W., Campos, J. A., Cancellieri, L., Carboni, M., Ciaschetti, G., De Frenne, P., Dolezal, J., Dolnik, C., Essl, F., Fantinato, E., Filibeck, G., Grytnes, J.-A., Guarino, R., Güler, B., Janišová, M., Klichowska, E., Kozub, L., Kuzemko, A., Manthey, M., Mimet, A., Naqinezhad, A., Pedersen, C., Peet, R. K., Pellissier, V., Pielech, R., Potenza, G., Rosati, L., Terzi, M., Valkó, O., Vynokurov, D., White, H., Winkler, M. & Biurrun, I. (2020) Species–area relationships in continuous vegetation: Evidence from Palaearctic grasslands. *Journal of Biogeography*, **47**, 72–86.

Drakare, S., Lennon, J. J. & Hillebrand, H. (2006) The imprint of the geographical, evolutionary and ecological context on species–area relationships. *Ecology Letters*, **9**, 215–227.

Gray, J. S., Ugland, K. I. & Lambshead, J. (2004) Species accumulation and species–area curves – A comment on Scheiner (2003). *Global Ecology & Biogeography*, **13**, 473–476.

Guilhaumon, F., Gimenez, O., Gaston, K. J. & Mouillot, D. (2008) Taxonomic and regional uncertainty in species–area relationships and the identification of richness hotspots. *Proceedings of the National Academy of Sciences USA*, **105**, 15458–15463.

Hubbell, S. P. (2001) *The unified neutral theory of biodiversity and biogeography*. Princeton, NJ: Princeton University Press.

Kraft, N. J. B., Valencia, R. & Ackerly, D. D. (2008) Functional traits and niche-based tree community assembly in an Amazonian forest. *Science*, **322**, 580–582.

Ladle, R. (2009) Forecasting extinctions: Uncertainties and limitations. *Diversity*, **1**, 133–150.

Lawton, J. H. (1999) Are there general laws in ecology? *Oikos*, **84**, 177–192.

Leveau, L. M., Ruggiero, A., Matthews, T. J. & Bellocq, M. I. (2019) A global consistent positive effect of urban green area size on bird richness. *Avian Research*, **10**, 30.

Lomolino, M. V. (2000) Ecology's most general, yet protean pattern: The species–area relationship. *Journal of Biogeography*, **27**, 17–26.

Lomolino, M. V., Riddle, B. R. & Whittaker, R. J. (2017) *Biogeography. Biological diversity across space and time*, 5th ed. Sunderland, MA: Sinauer Associates.

Losos, J. B. & Parent, C. E. (2010) The speciation–area relationship. *The theory of island biogeography revisited* (ed. by J. B. Losos and R. E. Ricklefs), pp. 415–438. Princeton, NJ: Princeton University Press.

MacArthur, R. H. & Wilson, E. O. (1967) *The theory of island biogeography*. Princeton, NJ: Princeton University Press.

Matthews, T. J., Cottee-Jones, H. E. & Whittaker, R. J. (2014) Habitat fragmentation and the species–area relationship: A focus on total species richness obscures the impact of habitat loss on habitat specialists. *Diversity and Distributions*, **20**, 1136–1146.

Matthews, T. J., Guilhaumon, F., Triantis, K. A., Borregaard, M. K. & Whittaker, R. J. (2016a) On the form of species–area relationships in habitat islands and true islands. *Global Ecology & Biogeography*, **25**, 847–858.

Matthews, T. J., Rigal, F., Triantis, K. A. & Whittaker, R. J. (2019) A global model of island species–area relationships. *Proceedings of the National Academy of Sciences USA*, **116**, 12337–12342.

Matthews, T. J., Triantis, K. A., Rigal, F., Borregaard, M. K., Guilhaumon, F. & Whittaker, R. J. (2016b) Island species–area relationships and species accumulation curves are not equivalent: An analysis of habitat island datasets. *Global Ecology & Biogeography*, **25**, 607–618.

Mazel, F., Renaud, J., Guilhaumon, F., Mouillot, D., Gravel, D. & Thuiller, W. (2015) Mammalian phylogenetic diversity–area relationships at a continental scale. *Ecology*, **96**, 2814–2822.

McGuinness, K. A. (1984) Equations and explanations in the study of species–curves. *Biological Reviews*, **59**, 423–440.

Neigel, J. E. (2003) Species–area relationships and marine conservation. *Ecological Applications*, **13**, 138–145.

Rosenzweig, M. L. (1995) *Species diversity in space and time*. Cambridge: Cambridge University Press.

Rosenzweig, M. L. (1998) Preston's ergodic conjecture: The accumulation of species in space and time. *Biodiversity dynamics: Turnover of populations, taxa, and communities* (ed. by M. L. McKinney and J. A. Drake), pp. 311–348. New York: Columbia University Press.

Rosenzweig, M. L. (2001) Loss of speciation rate will impoverish future diversity. *Proceedings of the National Academy of Sciences USA*, **98**, 5404–5410.

Rosenzweig, M. L. (2003) Reconciliation ecology and the future of species diversity. *Oryx*, **37**, 194–205.

Rosenzweig, M. L. (2004) Applying species–area relationships to the conservation of diversity. *Frontiers of biogeography: New directions in the geography of nature* (ed. by M. V. Lomolino and L. R. Heaney), pp. 325–343. Sunderland, MA: Sinauer Associates.

Scheiner, S. M. (2003) Six types of species–area curves. *Global Ecology & Biogeography*, **12**, 441–447.

Schoener, T. W. (1976) The species–area relations within archipelagoes: Models and evidence from island land birds. *Proceedings of the XVI International Ornithological Conference* (ed. by H. J. Firth and J. H. Calaby), pp. 629–642. Canberra: Australian Academy of Science.

Shmida, A. & Wilson, M. V. (1985) Biological determinants of species diversity. *Journal of Biogeography*, **12**, 1–20.

Si, X., Cadotte, M. W., Zhao, Y., Zhou, H., Zeng, D., Li, J., Jin, T., Ren, P., Wang, Y., Ding, P. & Tingley, M. W. (2018) The importance of accounting for imperfect detection when estimating functional and phylogenetic community structure. *Ecology*, **99**, 2103–2112.

Tjørve, E. & Tjørve, K. M. C. (2017) Species–area relationship. *eLS* (*Encyclopedia of Life Sciences Online*), pp. 1–9. Chichester: John Wiley & Sons.

Tjørve, E., Tjørve, K. M. C., Šizlingová, E. & Šizling, A. L. (2018) Great theories of species diversity in space and why they were forgotten: The beginnings of a spatial ecology and the Nordic early 20th-century botanists. *Journal of Biogeography*, **45**, 530–540.

Triantis, K. A., Borges, P. A. V., Ladle, R. J., Hortal, J., Cardoso, P., Gaspar, C., Dinis, F., Mendonça, E., Silveira, L. M. A., Gabriel, R., Melo, C., Santos, A. M. C., Amorim, I. R., Ribeiro, S. P., Serrano, A. R. M., Quartau, J. A. & Whittaker, R. J. (2010) Extinction debt on oceanic islands. *Ecography*, **33**, 285–294.

Triantis, K. A., Economo, E. P., Guilhaumon, F. & Ricklefs, R. E. (2015) Diversity regulation at macro-scales: Species richness on oceanic archipelagos. *Global Ecology & Biogeography*, **24**, 594–605.

Triantis, K. A., Guilhaumon, F. & Whittaker, R. J. (2012) The island species–area relationship: Biology and statistics. *Journal of Biogeography*, **39**, 215–231.

Triantis, K. A., Mylonas, M. & Whittaker, R. J. (2008) Evolutionary species–area curves as revealed by single-island endemics: Insights for the interprovincial species–area relationship. *Ecography*, **31**, 401–407.

Turner, W. R. & Tjørve, E. (2005) Scale-dependence in species–area relationships. *Ecography*, **28**, 721–730.

Wagner, C. E., Harmon, L. J. & Seehausen, O. (2014) Cichlid species–area relationships are shaped by adaptive radiations that scale with area. *Ecology Letters*, **17**, 583–592.

Wardle, D. A., Zackrisson, O., Hörnberg, G. & Gallet, C. (1997) The influence of island area on ecosystem properties. *Science*, **277**, 1296–1299.

Webb, C. O., Ackerly, D. D., McPeek, M. A. & Donoghue, M. J. (2002) Phylogenies and community ecology. *Annual Review of Ecology, Evolution, and Systematics*, **33**, 475–505.

Whittaker, R. J. & Fernández-Palacios, J. M. (2007) *Island biogeography: Ecology, evolution, and conservation*, 2nd ed. Oxford: Oxford University Press.

Whittaker, R. J. & Triantis, K. A. (2012) The species–area relationship: An exploration of that 'most general, yet protean pattern'. *Journal of Biogeography*, **39**, 623–626.

Whittaker, R. J., Araújo, M. B., Jepson, P., Ladle, R. J., Watson, J. E. M. & Willis, K. J. (2005) Conservation biogeography: Assessment and prospect. *Diversity and Distributions*, **11**, 3–23.

Whittaker, R. J., Fernández-Palacios, J. M., Matthews, T. J., Borregaard, M. K. & Triantis, K. A. (2017) Island biogeography: Taking the long view of nature's laboratories. *Science*, **357**, eaam8326.

Whittaker, R. J., Rigal, F., Borges, P. A. V., Cardoso, P., Terzopoulou, S., Casanoves, F., Pla, L., Guilhaumon, F., Ladle, R. J. & Triantis, K. (2014) Functional biogeography of oceanic islands and the scaling of functional diversity in the Azores. *Proceedings of the National Academy of Sciences USA*, **111**, 13709–13714.

Whittaker, R. J., Willis, K. J. & Field, R. (2001) Scale and species richness: Towards a general, hierarchical theory of species diversity. *Journal of Biogeography*, **28**, 453–470.

Wilson, E. O. (1961) The nature of the taxon cycle in the Melanesian ant fauna. *The American Naturalist*, **95**, 169–193.

2 · *The History of the Species–Area Relationship*

EVEN TJØRVE, THOMAS J. MATTHEWS
AND ROBERT J. WHITTAKER

2.1 Introduction

The discovery of the species–area relationship (SAR) cannot be attributed to a single person or time. Rather, and as true of the description and analysis of many patterns in nature, the story started with the realization of a phenomenon, which, over time and through many individual contributions, evolved into a developed theory. The history of the SAR thus concerns both the origins and the different forms and uses of SARs. We describe how the discovery of the phenomenon eventually led to the first proposed mathematical models of the relationship in the early twentieth century. This initiated the ongoing debate on the shape of species–area curves and the most appropriate model(s) for fitting. Alongside this debate, we review the history of the uses of the SAR and the varied purposes for fitting regression models to species–area data.

At the outset, it is necessary to recognize that there is not in fact a single type of relationship ('*the* SAR') to which species–area curves have been fitted; rather, there is an array of SAR types, each with different properties (see Chapter 1). These can be broadly grouped into those belonging to species accumulation curves (SACs based on nested and non-nested sample areas) and island species–area relationships (ISARs) (Chapter 1). SACs are often based on measures of sampling effort that are not area-based (e.g. trap hours) and so sample-areas based SACs (saSARs) can be distinguished as just a subset of SACs (see Chapter 4).

2.2 The Early History

2.2.1 The Forerunners

The realization of the SAR as a suite of natural phenomena can be traced back to the naturalists of the late eighteenth and early nineteenth centuries – notably including Johann Reinhold Forster and Georg Forster (father and son), who sailed as naturalists on James Cook's second Pacific voyage. In J. R. Forster's (1778) account, we find the observation that 'Islands only produce a greater or lesser number of species, as their circumference is more or less extensive', while G. Forster (1777) also made fleeting reference to both island size and isolation when he wrote: 'The small size of the island, together with its vast distance from either the eastern or western continent, did not admit of a great variety of animals.' Others were influenced by and built on these observations. The great German natural scientist Alexander von Humboldt (1807) was another who explicitly linked species diversity and area, as did the British botanist Hewett Cottrell Watson. With remarkable prescience he (Watson) wrote in 1835: 'On the average, a single county appears to contain nearly one half the whole number of species in Britain; and it would, perhaps, not be a very erroneous guess to say that a single mile may contain half the species of a county' (Watson, 1835, pp. 41–42). Work by Alphonse de Candolle (1855) was in turn referenced by Watson (1859), who wrote not only of the importance of controlling area in comparing floras, but who also provided more fully developed theoretical and empirical arguments for how richness scales with area (see Section 2.2.2).

Thus it was that, by early in the twentieth century, Alfred Russel Wallace (1914) was able to synthesize prior work, showing awareness not only that area matters in comparing the richness of different locales/regions, but also offering insights on the form of what we would now call the species accumulation curve within regions (and how it relates to the distributions of rare versus common species!), when he wrote: 'This characteristic of many small areas being often much richer in proportion to area than larger ones of which they form a part, is a necessary result of the great differences in the areas occupied by the several species *and the numbers of the individuals of each*' (p. 27, emphasis added). He goes on to comment on the causes of the patchiness of distributions and local diversity, arguing for the generality of these observations globally, before adding in connection with what he termed the 'peculiarities of vegetation', that '. . . this depends very much on diversities of climate and on the extent of land surface on which the entire flora has developed. The total number of species depends mainly on these two factors, and especially on the former'. In the space of one page of his book, he thus effectively bridges opposite ends of the range of scales at which we may look for area effects within and between different regions of the world, as well as identifying the key role of climate (controlling for area) in explaining large scale diversity differences!

Hence, we see that, between the Forsters' epic 1772–1775 voyage with Captain James Cook and the publication of Wallace's 1914 book, natural scientists had developed a basic understanding of many of the key properties of the relationship between species and area, including: the importance of area, isolation and other factors in driving island species diversity; key properties of species accumulation curves; how variation in

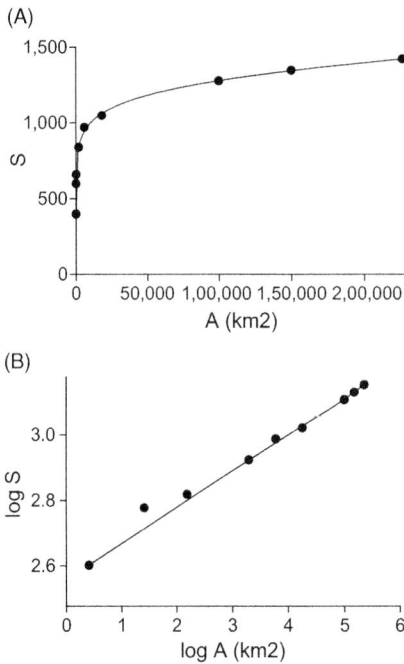

Figure 2.1 The graphs in panels (A) and (B) illustrate the SAR resulting from the dataset found in Watson (1859), where A = Area and S = Number of species. Watson never published a plot generated from the data. The power model was fitted with an untransformed dependent variable (S), which returned $S = 389.4\ A^{0.104}$, and $R^2 = 0.995$. Watson's dataset is plotted in (A) arithmetic space and (B) log–log space. For the finest scales in his dataset, one square mile and ten square miles, Watson did not use square (or quadrat) sample areas, but chose areas with irregular borders, as he argued the former would provide 'too unequal a comparison with ten miles of diversified surface'.

range size, abundance, equitability and niche requirements contribute to the latter; and the importance of area as a factor in analysing species richness differences at coarse scales of analysis.

2.2.2 A Quantitative Relationship

The first serious effort to quantify the form of a SAR is believed to be that of Watson (1859), who, in Volume IV of his *Cybele Britannica*, published in the same year as Darwin's *Origin*, provided a detailed analysis of the variation in richness as a function of area within and between land masses. Having first provided a general case for the non-linearity of saSARs, he stated: 'It is thus made very obvious that the smaller the area, the more numerous is the flora relative to the space' (p. 377). He illustrated this point by way of a large table providing comparisons of richness and density of the flora of different sized divisions, provinces and sub-provinces of Britain, concluding with a nested series from one mile of North Surrey up to the whole of Britain. When these data are plotted, we see that they are well described by a power model (Figure 2.1; the original published dataset is shown in Figure 2.2). Although he was evidently aware of the non-linearity of the relationship, he did not take the step of plotting the data graphically (first done by Rosenzweig, 1995) or describing it in formal

Series.				
1 All Britain	87,412	1425	61.34	0.16
2 England	57,812	1350	42.82	0.23
3 South Britain	38,474	1280	30.06	0.33
4 Province of Thames	7,007	1051	6.66	1.50
5 Sbp. of S. Thames	2,316	972	2.38	4.19
6 County of Surrey	760	840	0.90	11.0
7 Part of N. Surrey	60	660	.0900	110.
8 Ten miles of same	10	600	.0166	600.
9 One mile of same	1	400	.0025	....

Figure 2.2 Facsimile of Watson's (Watson, 1859) British plants SAR dataset. The first column (left) gives the names of the areas. The second column lists the sizes of the areas in square miles. The next column lists the number of species in each of these areas and the next column lists the species density in square miles to a species (area over number of species) with the last column listing species density as species per ten square miles (number of species over area). The figures of 600 and 400 species for the final two rows of the species richness column suggest to us that they may well be broad-brush estimates.

mathematical terms, and thus it fell to other, later workers to identify the relevance of the power model to SARs.

2.3 Towards a Mathematical Description

The next steps in the development of SAR theory largely emerged from the new field of ecology and, in particular, the early twentieth century developments in plant community ecology or phytosociology, in which the focus was on the description and classification of local vegetation associations, using quadrat-based sampling.

2.3.1 The New Botanists

Around the turn of the twentieth century, the focus of much botanical research switched from systematics and geographical distributions towards the emerging field of plant sociology (phytosociology), led by pioneering figures such as the Frenchman Charles Flahault (e.g. Flahaut & Schröter, 1910) and the American Henry Chandler Cowles (e.g. 1901). Around the same time, as the need for standardized, quantitative sampling methods became more widely recognized, a new breed of quantitative and theoretically inclined ecologists emerged, especially in the Nordic countries (Tjørve et al., 2018). The significance of their work has, until recently, been largely overlooked; in part because much of it was not published in English.

2.3.2 The Quadrat Method

The quadrat method was to become an important stepping-stone towards a mathematical description of saSARs, at least for plants. Just before the turn of the century, the American botanist and ecologist Frederic Edward Clements, together with Roscoe Pound, had refined the sampling design of small areas into the so-called quadrat method (Pound & Clements, 1898). Working on plants of the Minnesota prairie, they used five-by-five metre squares as sample areas to obtain the abundances and occupancies for the different species present. This method usually meant that quadrat sample areas were placed randomly within a plant community. The famous Danish botanist Christen Raun-kiær (1908, 1909, 1918) independently devised the same type of random sampling scheme, working with smaller one-by-one metre or fifty-by-fifty centimetre squares. In Danish he describes how a quadratic (metal) frame is literally thrown at random in the terrain: '... en kvaderatisk ramme der kastes på Maa og Faa'. The Swiss botanist Paul Jaccard (1908) instead placed his quadrats adjacent to each other, forming part of a cell grid. Another early phytobotanist and forest researcher, the Swedish botanist Torsten Lagerberg (1914), decided on a sampling design where the quadrats were placed with equal distances between them.

2.3.3 A Proposal for a Mathematical SAR

The step of applying a mathematical model to describe SARs resulting from quadrat-based sampling was first taken by the Swedish botanist Olof Arrhenius (1920a, b, c, 1921). In his first article, Arrhenius (1918) refers to Jaccard's (1908) and Raunkiær's (1909) use of sampling quadrats and observes that, based on such sampling schemes, one can calculate how much the number of species increases when the area is doubled or tripled, or alternatively what size of area is needed for a given number of species to occur. Subsequently, Arrhenius (1920c) proposed that SARs can be approximated by a power law, using the following formulation:

$$\frac{A}{A_1} = \left(\frac{S}{S_1}\right)^n, \tag{2.1}$$

where S is the number of species in area A and S_1 is the number of species in area A_1. If we change the notation and say that $n = 1/z$, A_1 is one unit of area (that is, $A_1 = 1$) and S_1 is the number of species in one unit of area (thus we set $S_1 = c$), then Equation (2.1) is written as:

$$A = \left(\frac{S}{c}\right)^n = \left(\frac{S}{c}\right)^{1/z}, \tag{2.2}$$

which means that the number of species as a function of area becomes

$$S = cA^{1/n} = cA^z, \tag{2.3}$$

which is the common form in which the power law SAR is expressed today, and where c and z (or $1/n$) are the parameters. It was the Swedish botanist Harald Kylin (1923, 1926) who first used this model form (Equation 2.3) in two articles written in Swedish and German, respectively, and published in Swedish journals. He also, correctly, described the c-value as the number of species in one unit of area, as well as providing the log-transformed form of the model:

$$\log S = \log c + z \log A, \qquad (2.4)$$

or

$$\log S = C + z \log A, \qquad (2.5)$$

where $C = \log c$. Arrhenius was not only the first to propose a mathematical SAR, but in his doctoral thesis (Arrhenius, 1920b) he also became the first to present graphs of species–area curves (Figure 2.3), which he plotted in log–log space, where the model conveniently becomes linear. In this plot, he compared saSARs for a grassy *Sesleria* meadow to one of Jaccard's (1908, p. 225) datasets, as well as to ISARs for two island groups in an archipelago outside Stockholm (Figure 2.3). He did not comment on the fact that he thus combined two saSARs with two ISAR datasets and, in his famous paper in the *Journal of Ecology* (Arrhenius, 1921) he only discussed SARs from sample areas, that is, saSARs.

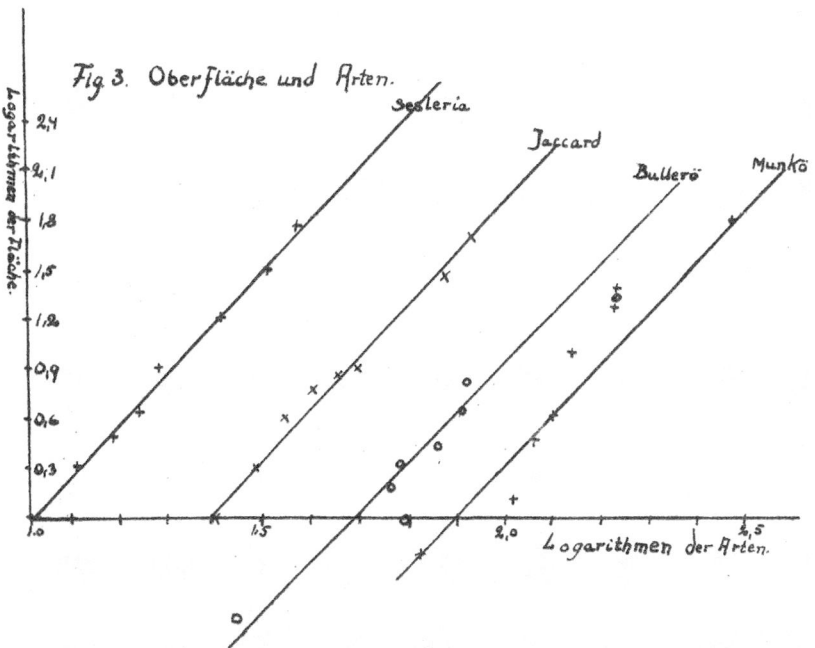

Figure 2.3 Arrhenius' (1920b) plant species–area curves (log–log space) from his doctoral thesis (written in German), where he also compares his own datasets to one of Jaccard's.

2.4 Discussions on the Shape of SARs

The debate about the shape and meaning of SARs, and what model best describes them, could be said to have begun in 1920, when Romell (1920) argued that the random distribution of individuals in space would generate a curved SAR and Arrhenius (1920c) proposed the power model. However, it was Arrhenius' (1921) subsequent paper, published in the *Journal of Ecology* and written in English, which generated wider attention, owing in part to a critical response from the American botanist and ecologist Henry Alan Gleason. The debate, which was originally focused on saSARs and embedded within arguments within phytosociology, continues to this day. In much of the literature the discussion has presumed a single model or shape might apply, despite the fact that there are different forms and manifestations of SARs and they have been estimated over massively different orders of scale. While it may turn out that a single model may be useful across a wide variety of SARs, the published literature to date shows there to be no single, universally applicable best model for this wide range of applications.

2.4.1 Arrhenius versus Gleason

As indicated above, following Arrhenius' (1921) publication of his case for the power model, Gleason (1922) wrote a reply in order to 'demonstrate the fallacy of Arrhenius' equation', arguing instead that an 'exponential ratio' is the most appropriate function, but without stating his model explicitly. Only in his response to Arrhenius' (1923) subsequent paper in the *Journal of Ecology* did Gleason (1925) provide an equation, the purpose of which was, however, to find the expected number of species (S_3) in an area (A_3) based on knowledge of the number in two different sized areas (S_2 and S_1 in A_2 and A_1, respectively):

$$\frac{\log A_2 - \log A_1}{\log A_3 - \log A_1} = \frac{S_2 - S_1}{S_3 - S_1}. \tag{2.6}$$

Using Equation (2.6), Gleason explained how we can calculate S_3 when we know all the other values. The first to present both the power model and the logarithmic SAR model in the form we know them today was the Swedish botanist Harold Kylin (1926). He expressed Gleason's logarithmic model as:

$$S = \log c + d \log A, \tag{2.7}$$

where c and d are parameters.

Gleason (1922) had argued that extrapolating Arrhenius' model will predict too many species and is therefore 'wholly erroneous'. Gleason provided justification for this line of argument by using a calculation corresponding to $z = 0.5$ ($n = 2$), and showing that, if there were thirty-three plant species per metre squared, Arrhenius' model predicts one kilometre squared should hold 33,000 species; as Gleason noted, this is obviously far too many. In his reply, Arrhenius (1923) countered by claiming that the power model also provided better fits to other datasets, such as those of Palmgren (1916) and Du Rietz (1921), as well as Gleason's (1922) data from contiguous sample areas (which Gleason used to advocate the logarithmic model). However, Du Rietz (1921) had also questioned the extrapolation capacity of the power model; arguing that, when fitted to smaller and medium spatial scale data, the model would not provide an accurate estimation of the number of species at large spatial scales.

Although both models are characterized by monotonically decelerating (convex-upward) non-asymptotic shapes in arithmetic space, the power curve is usually plotted in log–log space, whereas the logarithmic curve has usually been plotted in log–linear space, leading in both cases to straight lines. In Figure 2.4 we have fitted, in log–log space, the power model and the logarithmic model, respectively, to Gleason's (1922) two datasets from sample areas in aspen stands in Michigan. It can be seen that Gleason's logarithmic model provides the best fit to the scattered sample area data, while Arrhenius's power model provides a better fit to the continuous (nested) sample area data (Figure 2.4).

It may seem perplexing that Gleason, while arguing for the primacy of the logarithmic model, used a dataset for which a power model provided a better fit than the logarithmic. However, it was more difficult during that time to decide which model fitted better, as the method of ordinary-least squares for fitting and evaluating regression models was not widely used in the field. These findings agree with what we know today, namely that the power model is more likely to provide better fits to saSARs from nested (continuous) datasets compared to saSARs generated from randomly accumulated (scattered) sample areas. This is because two adjoining sample areas are likely to contain fewer species in total than two separated areas and, the further apart the separated samples are, the more difference in membership is to be expected (i.e. the distance decay in similarity). Therefore, if saSARs are accumulated from scattered sample areas, they will initially increase faster than saSARs from nested sample areas and, providing the two study systems have the same overall

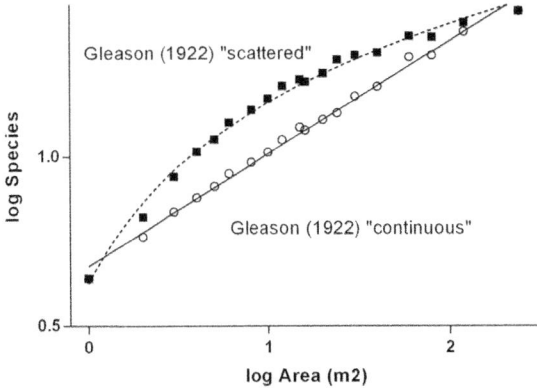

Figure 2.4 Sample–area curves of accumulated sample areas (saSARs) replotted from Gleason's (1922) datasets from scattered (discontinuous; black squares) sampling areas and continuous (nested; open circles) sample areas in aspen stands in Michigan. Gleason himself never provided any graphs of his data. The fit of the logarithmic model (dashed line) to the scattered dataset is shown, as is the fit of the power model (solid line) to the continuous dataset.

extent and are from the same region, the two will eventually approach each other, as the total accumulated area sampled becomes large (as can be seen in Figure 2.4).

Currently, Arrhenius seems to be 'winning' the argument, as the power model is generally preferred in contemporary SAR studies and is often the only model mentioned in ecology textbooks (e.g. Bush, 2003; Krebs, 2009). The power model has also been found, in various meta-analyses, to be the best performing model for both island (ISAR) and sample-area (saSAR) datasets (Dengler, 2009; Triantis et al., 2012; Matthews et al., 2016a, b; Dengler et al., 2020). Nonetheless, the logarithmic model sometimes produces a better fit (Triantis et al., 2012).

2.4.2 A Case for Randomness

The small group of Nordic botanists and ecologists previously mentioned not only introduced the power model and progressed the use of the logarithmic model, but were also the first to develop what we now call the random placement model as a basis for thinking about both saSARs and ISARs. Their insight grew from the observation that empirical SARs often resemble the curves that would result if individuals were distributed at random in space, generating species abundance distributions (SADs) from which, in turn, the SAR would emerge. For example, Romell (1920) argued that the distribution of individuals is typically more random than aggregated, while Arrhenius (1920c, 1921) actually proposed a model assuming random placement of individuals in space. This model was rediscovered (independently) by Coleman (1981) and can be stated as:

$$s_a = S - \sum_{i=1}^{S} \left(1 - \frac{a}{A}\right)^{n_i}, \tag{2.8}$$

where S_a is the expected number of species in the sample area a, S is the total number of species in the total study area (A), and n_i is the number of individuals of species i. Subsequent work has shown that random placement is rarely observed in nature, but the model remains, in theory, a useful null model for comparing the hypothetical case of random placement to empirical data. Unfortunately, the number of individuals of each species (the SAD) is usually unknown. This renders Arrhenius' and Coleman's random placement models of limited practical use, as they both require knowledge of species abundances. If we only know the occupancies of each species in a sample–area cell matrix, we may instead apply a model that assumes a random spatial distribution of occupied cells across the sample matrix: the hypergeometric distribution (see Tjørve et al., 2008; see also Chapter 4).

The Swedish botanist Lars Romell (1920) was the first to publish a SAR model based on the probability of occupancy for each species and fit it to data. He expressed his model as:

$$p(N) = 1 - (1 - p(1))^N, \tag{2.9}$$

where $p(N)$ is the probability of encountering a species in N (equally sized) sample units and $p(1)$ is the probability of encountering a species for each sample unit (calculated as the number of occupied sampling units over the total number of sampling units). However, this model does not necessarily reflect complete randomness in space; only that individuals are randomly distributed between units of area. Moreover, Romell's model is just an approximation, as it is based on sampling with replacement; arguably it should have been based on sampling without replacement (Tjørve et al., 2018).

Yet another Swede, Theodore (The) Svedberg (1922), proposed that random placement would cause single-species occupancy curves to follow the negative exponential model. Subsequently, Kylin (1923) suggested the negative exponential model for complete SARs (resulting from several single-species occupancy curves), expressed as:

$$S(A) = S_{tot}(1 - \exp(-bA)), \qquad (2.10)$$

where S_{tot} is the total number of species found in the sample areas and b is a fitted parameter. Kylin's first model (Equation 2.10) assumes equal abundances, although he also discussed how to model random placement SARs when abundances differ. Both Svedberg and Kylin understood that their random placement models only approach the negative exponential relationship at fine scales (see Tjørve et al., 2018). Thus, they precede both Coleman and colleagues (Coleman, 1981; Coleman et al., 1982), who today are commonly accredited as the originators of the random placement model, and Holdridge et al. (1971), who have been cited (e.g. Colwell & Coddington, 1994) as the first to fit the (asymptotic) negative exponential model to SAR data.

2.4.3 Preston's Legacy and the z-Value

The z-value, the exponent of the power model, is the most discussed parameter of the power law SAR (Equation 2.3). Indeed, it is the most discussed parameter of any SAR model. It is often referred to as the 'slope', although it only becomes the slope of the SAR when we linearize it by log-transforming both area and number of species. There is a large literature on z-values, of both ISARs and various types of saSARs, although once again many studies make insufficient effort to distinguish what sort of SAR is under consideration.

The seminal contribution to the discussion of z-values and their meaning started with English-American ecologist Frank W. Preston (1962). He was the first to put forward a 'canonical' z-value of 0.262, although many textbooks today indicate it as being usually about 0.3 (as suggested by Arrhenius, 1920a, b). MacArthur and Wilson (1967) based their theoretical z-value of 0.263 on the earlier work of Preston, simply changing the third digit after the decimal point. Reading his seminal work on SARs (Preston, 1960, 1962) today, one is struck with how Preston provided an answer to the majority of SAR questions that we currently consider important. The papers cover a staggering array of SAR topics, including the link between the SAR and the SAD (e.g. Martín & Goldenfeld, 2006), how multiple processes may affect the SAR (Whittaker & Fernández-Palacios, 2007), how the z-value is scale dependent (He & Legendre, 1996; Triantis et al., 2012) and, taken together, how these last two facts result in the triphasic SAR (Hubbell, 2001; Fridley et al., 2005). Preston (1962) was also the first to discuss explicitly the difference between SARs generated from what he termed isolates (which includes islands) and sample areas (or census areas, usually from mainlands), thus distinguishing between ISARs and saSARs.

While many studies have published z-values that are close to Preston's canonical value, there is a great deal of variation. Indeed, a large number of factors have been proposed/shown to explain the variation in z-values between studies, including isolation, the taxon studied, latitude, elevation, island type, matrix type, the range of island areas analysed, habitat diversity and spatial scale (e.g. Connor & McCoy, 1979; Rosenzweig, 1995; He & Legendre, 1996; Whittaker & Fernández-Palacios, 2007; Triantis et al., 2012; Matthews et al., 2016a, 2019a). The z-value has also been linked to self-similarity and scale invariance properties of the power SAR model (Harte et al., 1999, 2001; Tjørve & Tjørve, 2008).

A full evaluation of factors affecting the z-value is beyond the scope of this chapter (but see Chapters 3 and 4) and here we briefly discuss just three of the above factors. First, spatial scale. Williams (1943) may have been the first to explicitly note that the slope of the SAR changes with spatial scale. These ideas were properly codified in Rosenzweig's seminal book, where he proposed a scale-structured model of SARs based on four different biogeographical scales of relationship, from point to interprovincial. If we discount the point scale, Rosenzweig argued that interprovincial SARs (see also Triantis et al., 2008) have the highest z-values ($z = 0.9$), intraprovincial SARs the lowest ($z = 0.15$) and ISARs are characterized by intermediate values ($0.25 < z < 0.45$); ISAR z-values also increasing with the degree of isolation. In subsequent work, Rosenzweig (e.g. 2001) presented slightly different z-value ranges for the different categories. However, it is important to note that Rosenzweig's intraprovincial SARs are saSARs, whereas the other two classes are types of ISAR. Second, island type. Focusing exclusively on the ISAR, Matthews et al. (2016a), following Triantis et al. (2012), undertook an analysis of several hundred true island and habitat island ISAR datasets, reporting the median z-value for habitat islands to be 0.22 (first quartile (Q1) and third quartile (Q3) = 0.16 and 0.32; n = 135), while the median for oceanic islands was 0.35 (Q1 and Q3 = 0.24 and 0.49; n = 125). Finally, variation in z-values between taxa. In their analysis of true islands, Triantis et al. (2012) observed that mean z-values decreased from plants (0.36, $\pm$SE = 0.02) to invertebrates (0.32 $\pm$ 0.01) to vertebrates (0.29 $\pm$ 0.01), but only the difference between plants and vertebrates was significant. In contrast, Matthews et al. (2016a) found no significant (or even non-significant but observable) differences in average z-values between the same three taxa in their habitat island datasets. The coarse grouping of taxa into these three categories in both studies likely obscured finer taxonomic patterns. For example, the vertebrate category included both amphibians (relatively poor dispersers in general) and birds (relatively good dispersers in general). In terms of other taxa, several studies have shown that the z-values calculated from datasets of microbial taxa rarely exceed 0.1, likely due to this group's strong dispersal ability (e.g. Bolgovics et al., 2015).

In SAR research, most discussions have been focused on the z-value rather than c-value. The c-value is often referred to as the 'intercept', as it is the intercept when the power model is plotted in log–log space (where it becomes a straight line); in this case, the c-value represents the amount of species in one unit of area. The c-value has perhaps received less attention (Gould, 1979) because it is harder to interpret than the z-value, and has been argued to reflect a number of different properties, such as a measure of carrying capacity and the average size of the most common species (Gould, 1979; Fattorini, 2007; Whittaker & Fernández-Palacios, 2007; Triantis et al., 2012; Matthews et al., 2016a; Fattorini et al., 2017). In addition, when the c-value is also the sampling-unit size, for example in a survey cell grid, it can be seen as a measure of alpha diversity. As with the z-value, the c-value has been found to vary across datasets due to a variety of factors (Matthews et al., 2019a; Chapter 3). For example, in their analysis of multiple true and habitat island datasets, Matthews et al. (2016a; see also Triantis et al., 2012) found that the c-value (from the log–log power ISAR model) varied with taxa, island type and island size properties. The usefulness of the c-value is restricted by the fact that its value is dependent on the unit of measurement (e.g. hectares or square kilometres). As such, the unit of measurement has to be the same if we want to compare the c-values from different SAR curves and datasets (Chapter 3). A final point to add regarding the c-value is that, when calculated using the log–log form of the power model (i.e. the intercept), it is generally an extrapolated estimate (i.e. generally speaking an island of unit area is not within the dataset). If the smallest data point is considerably larger than a unit of area, then the estimate of the c-value should be treated with caution, particularly as the power model has been found to not always hold at very small spatial scales (Fattorini, 2007).

2.4.4 A Thicket of Models

Since the time of Arrhenius, Kylin and Gleason, a large number of alternative candidate SAR models have been proposed. Indeed, a survey of the literature reveals at least thirty non-linear models (see Chapter 7), not to mention the standard linear model (Triantis et al., 2012). Perhaps the most important grouping of these models is into 1) non-asymptotic and asymptotic SAR models (Williams, 1995; Flather, 1996; Lomolino, 2000; Williamson et al., 2001) and 2) convex upward versus sigmoidal models (Flather, 1996; Lomolino, 2000; Tjørve, 2003; Tjørve & Turner, 2009). Other than the power model and the logarithmic model, alternative SAR models of note include the negative exponential (Svedberg, 1922; Kylin, 1923; Holdridge et al., 1971), the Monod model (de Caprariis et al., 1976; Clench, 1979), the extreme-value function (Williams, 1995), the cumulative Weibull distribution (Flather, 1996), the Lomolino model (Lomolino, 2000; Tjørve, 2003), and the second persistence function (P2) (Ulrich & Buszko, 2003, 2004). The latter model is particularly interesting given that it is sigmoid but does not have an upper asymptote (Chapter 7). It is also interesting to note that both Lomolino (2000) and Ulrich and Buszko (2003, 2004) did not explicitly name their models, viewing them simply as extensions of other previously proposed models; only subsequently have these models been explicitly named (see Tjørve, 2003, 2009). All of the above models can be fitted to both ISAR and SAC-type datasets.

While a large variety of models have now been published, most models are rarely applied in practice. While there is no universal model that provides the best fit to all SAR datasets (Connor & McCoy, 1979; Dengler, 2009; Triantis et al., 2012; Matthews et al., 2016a; Dengler et al., 2020), the first two proposed models (the power and logarithmic functions) remain the most popular. This popularity stems from a combination of historical factors, simplicity, general performance and the fact that they are easily linearized. For further details of additional models see Tjørve (2003, 2009), Triantis et al. (2012) and Chapter 7, and for software to fit around 20 SAR models see Matthews et al. (2019b).

2.5 From Island Biogeography to Conservation

During the early twentieth century, the theoretical approach to SAR research was, as we have seen, headed by a group of spatial plant ecologists, based in particular in Sweden and the other Nordic countries. Zoologists only began to discuss SARs in the context of animal species and island biogeography from the middle of the twentieth century. Nowadays, much of the conversation on SARs (particularly in ecology textbooks) is through the lens of island biogeography and, in particular, how the SAR can help us better understand the increasing isolation of natural habitats. In this section, we briefly discuss how SAR research progressed from the aforementioned early ecological studies to form a key component in the development of the fields of island biogeography and conservation biogeography. It should be noted that the themes discussed in this section are covered in less detail than those in the preceding sections. This is because they are covered in many of the subsequent chapters in this volume (e.g. the drivers of island SARs – Chapters 3 and 4; hotspot detection – Chapter 13; habitat loss and extinctions – Chapters 14 and 16).

2.5.1 Theories of Island Biogeography

Modern island biogeography evolved as part of the new systems and theoretical ecology, dominated by American mathematically inclined zoologists. Many of the modern theories and models of (island) biogeography are linked to SARs. When Darlington (1957) presented his rule of thumb (for SARs) and MacArthur and Wilson (1963, 1967) their equilibrium theory of island biogeography, they presumed SARs – in this case mostly ISARs (number of species on islands of increasing sizes) – to be described by the power law.

The equilibrium theory of island biogeography proposes that an island's biota is the result of a dynamic balance between immigration, in situ diversification and extinction of species. It was proposed and developed by MacArthur and Wilson (1963, 1967) in part to explain regularities in the increase of species richness with island size. As an historical aside, the core of their model was independently derived in a 1948 doctoral thesis on Caribbean butterflies by the American zoologist Eugene Gordon Munroe, but he failed to publish or develop the idea into a fully-fledged theory of island biogeography (Brown & Lomolino, 1989; Lomolino & Brown, 2009).

Although Darlington and MacArthur and Wilson assumed ISARs to follow the power law, they were not the first to suggest this. As already discussed, Arrhenius (1920a) and the Finnish botanist Widar Brenner (1921) had earlier claimed the power model to be an accurate representation of ISARs, while studying island archipelagos in the Baltic Sea. Brenner (1921) was the first, to our knowledge, to plot a power law SAR in arithmetic space. As he used the version of the power model in Equation (2.1), he reported the n-value instead of a z-value, but his value corresponds to a z-value of 0.21. Using his own model (where n is calculated rather than z), Arrhenius (1920a, b) arrived at a 'general rule' corresponding to a z-value of about 0.3. Preston's (1962) canonical z-value of 0.26 (discussed above) therefore falls midway between Brenner's and Arrhenius' observations.

Following publication of MacArthur and Wilson's equilibrium theory, the SAR continued to play a central role in island biogeography. Indeed, many of the (albeit weak) tests of MacArthur and Wilson's model simply relied on the identification of a positive (generally power law) ISAR (see Whittaker & Fernández-Palacios, 2007). Those interested in the history of early island biogeography theoretical developments (including reference to the SAR) are also directed to Lomolino and Brown (2009).

More recently, the work of Mark Lomolino has been instrumental in expanding SAR theory in relation to islands (e.g. Lomolino, 2000; Lomolino & Weiser, 2001; Foreword). For example, Lomolino (2000) argued that the most frequently used ISAR models (power and logarithmic) were too simplistic for two reasons: 1) they do not have an asymptote and 2) they do not account for the small-island effect (i.e. the independent variation of species richness with island area on small islands). That is, both the far left-hand side and far-right hand side of the relationship are not well characterized by the non-asymptotic simple models; rather, a sigmoidal model is necessary to fully describe the relationship (e.g. see figure 1 in Lomolino, 2000). Both of these arguments (i.e. the presence of an asymptote and the small-island effect) have proved to be controversial (e.g. Williamson et al., 2001; Burns et al., 2009), and there has been only limited support for sigmoidal ISAR models in various analyses of multiple datasets (e.g. Triantis et al., 2012; Matthews et al., 2016a; but see Chapter 7). However, Tjørve and Turner (2008) reported that, for a number of datasets that included both the very finest scales (i.e. the smallest islands) and the islands that contain no species (i.e. zero richness values), a sigmoid regression model provided a better fit than the traditional models (power and logarithmic). More research on the prevalence of sigmoidal ISARs is clearly needed.

Lomolino (2000; see also the 'Conclusions' section in Lomolino and Weiser, 2001) also argued that, when very large (and isolated) islands are included, the slope of the ISAR should be 'deflected upward' due to the increasing importance of speciation processes on these islands. The logic here is similar to that discussed in Section 2.4.3 in regard to interprovincial SARs (e.g. Rosenzweig, 1995). These ideas remain the subject of discussion and debate and it appears that we are some way from a consensus on the patterns and causes of variation in the form of (I)SARs (see Chapter 4).

2.5.2 The SAR as a Conservation Tool

From the earliest days of conservation biogeography studies (long before the term conservation biogeography was coined; Whittaker et al., 2005) the SAR has been a key tool in the applied biogeographer's toolkit (e.g. Schoener, 1976; Rosenzweig, 1995, 2004; Lomolino, 2000). This can be traced back to the rise of the field of conservation biology during the 1960s and 1970s; the significance of this new discipline attracted theoretical ecologists such as Jared Diamond and Robert May (amongst many others), who put forward the SAR as a tool with a range of applied uses. In the following sections, we briefly discuss the use of the SAR in three applied situations: the design of nature reserves, extinction forecasting and habitat loss and fragmentation research. However, this is by no means an exhaustive list of applied SAR uses. The rise of conservation biogeography has also seen SARs being used to select biodiversity hotspots (Fattorini, 2007; Chapter 13), to construct baselines and targets for conservation strategies (Desmet & Cowling, 2004), to measure the effect of disturbance on ecological communities (Lawrey, 1991; Chapter 17) and to upscale diversity in areas too large to sample (Smith, 2010; Kunin et al., 2018; Matthews & Aspin, 2019). Rosenzweig (2004) provides a summary of various additional applied uses of the SAR, such as providing a cornerstone of his 'reconcilation ecology'.

2.5.3 Nature Reserve Design and the SLOSS Debate

Researchers began to apply island biogeography theory to the study of habitat islands shortly after the publication of MacArthur and Wilson's theory (Diamond, 1975, 1984; Wilson & Willis, 1975). A central component of this research agenda involved using the equilibrium theory as a basis for protected area design. Two early papers on the topic (Diamond, 1975; Wilson & Willis, 1975) presented a set of general principles for the optimal design of nature reserves, including recommendations about the size, shape and number of protected areas. These papers provided the catalyst for a set of long-lasting and often heated debates around protected area network design (e.g. Diamond,

1975; May, 1975; Williamson, 1975; Simberloff & Abele, 1976; Ter-borgh, 1976; Connor & McCoy, 1979; Boecklen & Gotelli, 1984). The SLOSS (single large or several small) protected area debate is perhaps the most notable example (Diamond, 1975; Simberloff, 1988). The debate arose from the argument that, due to the generality of positive ISARs in true and habitat islands, a single large nature reserve was to be preferred over a set of smaller reserves adding up to the same area (Diamond, 1975). A number of authors subsequently argued that the opposite was in fact more realistic. A review of the debate and the reasoning put forward by both sides is provided by Whittaker and Fernández-Palacios (2007). In essence, the answer to the SLOSS debate depends not only on the parameters of the ISAR but also on the degree of compositional overlap (nestedness/beta diversity) between the habitat islands being considered as potential reserves (Tjørve, 2010). Thus, determining the answer for a given area requires detailed information about the study area. Island theory, such as MacArthur and Wilson's (1967), turns out to be equivocal on the issue.

2.5.4 Extinction Forecasting

Diamond (1975) also provided an early example of using the SAR to predict the number of species that will go extinct following a reduction in habitat area (e.g. due to deforestation). This would become a primary use of the SAR in applied studies (He & Hubbell, 2011; Halley et al., 2014). A common procedure in such studies is to apply the power function backwards, that is, to calculate the number of species lost due to a reduction in habitat area (Brooks & Balmford, 1996; Brook et al., 2003; Smith, 2010; Whittaker & Matthews, 2014). Generally, a set z-value of between 0.10 and 0.30 is used. However, other SAR based methods have been proposed. Chapter 14 provides a detailed evaluation of the use of the SAR in extinction forecasting.

2.5.5 Habitat Loss and Fragmentation Research and the Rise of Countryside Biogeography

The destruction and fragmentation of natural habitat are the primary drivers of the current extinction crisis (Haddad et al., 2015). While they are distinct processes, habitat loss and fragmentation tend to occur together in real world systems. Together, they result in the loss of habitat area and the creation of areas/islands of natural habitat surrounded by anthropogenic land use. The reduction in habitat area is expected to increase extinction (and reduce colonization) rates (e.g. Fahrig, 2002) and the increased isolation of the remaining fragments is expected to reduce colonization (and increase extinction) rates (e.g. MacArthur & Wilson, 1967; Brown, 1971; Brown & Kodric-Brown, 1977; Hanski, 1999). Contrary to this theoretical expectation, Fahrig (e.g. 2013) has argued that fragmentation per se is relatively insignificant and, instead, losses can be predicted from the gross amount of habitat lost. Her habitat amount hypothesis has proved controversial and has, for example, been contested by Haddad et al. (2017). The debate on the relative importance of habitat loss per se and fragmentation (increased isolation of patches) continues. Notwithstanding such controversies, the SAR has been a central tool in assessing the impacts of habitat loss and fragmentation and there are hundreds of habitat fragmentation studies that apply the SAR in some form. Indeed, the SAR themes discussed in the two previous sub-sections (i.e. nature reserve design and extinction forecasting) are both strongly related to habitat loss and fragmentation and are treated as separate sub-sections here simply for presentational purposes.

The use of the SAR in fragmentation research is based on a long history of studies from various disciplines within biogeography. While not discussed in the context of anthropogenic habitat fragmentation, early biogeographers recognized the effect that the fragmentation of areas would have on species diversity. For example, very early on, the French-Swiss botanist Alphonse de Candolle (1855) wrote about the expected effect of fragmentation of large landmasses at the global scale. He proposed that: 'The break-up of a large landmass into smaller units would necessarily lead to the . . . extinction of (some) species and the preservation of others.' A few years later Alfred Russel Wallace (1869, 1876), in his books 'The Malay Archipelago' and 'The geographical distribution of animals', made several observations on area, species number and the effect of isolation of islands. As was stated above, the work of MacArthur and Wilson (1967) and other island biogeographers in the 1960s and 1970s (see Whittaker & Fernández-Palacios, 2007) then built on these early studies to derive, and test with empirical data, theoretical expectations of the effects of varying island area and isolation on species richness. The rise of metapopulation biology in the 1980s and 1990s (reviewed in Hanski, 1999) further explored the roles of area and isolation. Although here the focus was on the persistence of populations of individual species in habitat patches, rather than the richness of multiple species, it is germane to note that the total richness of an island (and by extension the ISAR of a set of islands) is simply the aggregation of the patterns of all individual species (Matter et al., 2002). Using these theoretical frameworks, conservation biogeographers and biologists have developed a wide range of techniques, many of which utilize the SAR, to study the impacts of habitat fragmentation on species richness (e.g. Ladle & Whittaker, 2011; Hanski et al., 2013). Here we have provided only a very cursory overview of these topics and readers are directed to a number of other sources (and the references within) for more in-depth information on the use of the SAR in habitat loss and fragmentation research (Watling & Donnelly, 2006; Whittaker & Fernández-Palacios, 2007; Ladle & Whittaker, 2011; Matthews, 2015; Chisholm et al., 2018).

The SAR has recently formed a cornerstone of the growing field of countryside biogeography: a research programme focused on anthropogenically fragmented landscapes (e.g. forest fragments embedded in a matrix of pasture land), within which the matrix habitat is regarded as of key importance. In particular, it is based on the idea that the matrix can support a number of species and provide resources that many species can utilize (Mendenhall et al., 2014). Thus, 'countryside' landscapes are argued to support a larger number of species and have lower community-level extinction rates than true island archipelagos (Daily et al., 2003). Tests of the countryside biogeography framework have frequently involved SAR-based evidence. For example, the countryside SAR model extends the power SAR model by accounting for i) the presence of multiple habitats in the surrounding matrix and ii) differential use of these habitats by species (Pereira & Daily, 2006). Tests of the countryside SAR have been reported to generate improved extinction predictions (Martins & Pereira, 2017).

A number of additional SAR extensions have been proposed to improve model fits for diversity patterns across multiple island or habitat patches and anthropogenic systems (e.g. fragmented landscapes). These include the Choros model (Triantis et al., 2003), the two-habitat and several-habitat models of Tjørve (2002, 2010) and of Bascompte et al. (2007), the matrix- and edge- calibrated SAR models of Koh et al. (2010), the species–fragmented area relationship of Hanski et al. (2013) and the lost-habitat SAR of De Camargo and Currie (2015). The testing of these models, in addition to the development of novel SAR adaptions, represents a promising trajectory for future SAR and conservation biogeography research.

2.6 Conclusions

In conclusion, the SAR was not discovered by one person alone. The realization of the various types of SAR relationship and interest in their shape developed slowly at first, via the first mathematical descriptions and towards a growing empirical and theoretical base of knowledge. Generally speaking, SAR research has advanced in waves. One such wave occurred in the early twentieth century with the theoretically inclined botanists, in particular those in the Nordic countries, such as Olof Arrhenius, Lars Romell, Widar Brenner, Harald Kylin and others. A second wave can be traced to the work of the American theoretical ecologists, zoologists and island biogeographers in the second half of the twentieth century, including Philip Jackson Darlington Jr., Frank Preston, Jared Diamond, Robert MacArthur, Edward O. Wilson and many more.

Although the shape of SARs has been extensively debated over the years, the best known and most commonly applied model remains the power model (Arrhenius, 1921; Preston, 1962), although the logarithmic curve (Gleason, 1922, 1925) is also frequently employed. This chapter has mostly focused on the early history of SARs; providing a background for the many topics that will be discussed in the subsequent chapters. The breadth of topics covered in these subsequent chapters illustrates how the SAR remains an important area of contemporary research in many fields of ecology, biogeography and macroecology. The fact that this is so, given that we have shown here that it is also one of the oldest studied ecological phenomena, highlights how central the SAR (in its many guises) has been to the development of these disciplines. It also suggests that SAR research will continue to be of interest for many years to come.

References

Arrhenius, O. (1918) En studie över yta och arter. *Svensk Botanisk Tidsskrift*, **12**, 180–188.

Arrhenius, O. (1920a) Distribution of the species over the area. *Meddelanden från Kungliga Vetenskapsakademiens Nobelinstitut*, **4**, 1–6.

Arrhenius, O. (1920b) Öcologisher Studien in den Stockholmer Scären. Dissertation, University of Stockholm, Stockholm (Svea).

Arrhenius, O. (1920c) Yta och arter. I. *Svensk Botanisk Tidsskrift*, **14**, 327–329.

Arrhenius, O. (1921) Species and area. *Journal of Ecology*, **9**, 95–99.

Arrhenius, O. (1923) On the relation between species and area – A reply. *Ecology*, **4**, 90–91.

Bascompte, J., Luque, B., Olarrea, J. & Lacasa, L. (2007) A probabilistic model of reserve design. *Journal of Theoretical Biology*, **247**, 205–211.

Boecklen, W. J. & Gotelli, N. J. (1984) Island biogeographic theory and conservation practice: Species–area or specious–area relationships? *Biological Conservation*, **29**, 63–80.

Bolgovics, Á., Ács, É., Várbíró, G., Görgényi, J. & Borics, G. (2015) Species area relationship (SAR) for benthic diatoms: A study on aquatic islands. *Hydrobiologia*, **764**, 91–102.

Brenner, W. (1921) Växtgeografiska studier i Barösunds skärgård. *Acta Sociatatis pro Fauna et Flora Fennica*, **49**, 1–151.

Brook, B. W., Sodhi, N. S. & Ng, P. K. L. (2003) Catastrophic extinctions follow deforestation in Singapore. *Nature*, **424**, 420–423.

Brooks, T. & Balmford, A. (1996) Atlantic forest extinctions. *Nature*, **380**, 115.

Brown, J. H. (1971) Mammals on mountaintops: Nonequilibrium insular biogeography. *The American Naturalist*, **105**, 467–478.

Brown, J. H. & Kodric-Brown, A. (1977) Turnover rates in insular biogeography: Effect of immigration on extinction. *Ecology*, **58**, 445–449.

Brown, J. H. & Lomolino, M. V. (1989) Independent discovery of the equilibrium theory of island biogeography. *Ecology*, **70**, 1954–1957.

Burns, K. C., Paul McHardy, R. & Pledger, S. (2009) The small-island effect: Fact or artefact? *Ecography*, **32**, 269–276.

Bush, M. B. (2003) *Ecology of a changing planet*, 3rd ed. Upper Saddle River, NJ: Prentice Hall.

Chisholm, R. A., Lim, F., Yeoh, Y. S., Seah, W. W., Condit, R., Rosindell, J. & He, F. (2018) Species–area relationships and biodiversity loss in fragmented landscapes. *Ecology Letters*, **21**, 804–813.

Clench, H. K. (1979) How to make regional lists of butterflies: Some thoughts. *Journal of the Lepidopterists' Society*, **33**, 216–231.

Coleman, B. (1981) On random placement and species–area relations. *Mathematical Biosciences*, **54**, 191–215.

Coleman, B. D., Mares, M. A., Willig, M. R. & Hsieh, Y.-H. (1982) Randomness, area and species richness. *Ecology*, **64**, 1121–1133.

Colwell, R. K. & Coddington, J. A. (1994) Estimating terrestrial biodiversity through extrapolation. *Philosophical Transactions of the Royal Society B: Biological Sciences*, **345**, 101–118.

Connor, E. F. & McCoy, E. D. (1979) The statistics and biology of the species–area relationship. *The American Naturalist*, **113**, 791–833.

Cowles, H. C. (1901) The plant societies of Chicago and vicinity. *The Geographical Society of Chicago*, Bulletin No. 2.

Daily, G. C., Ceballos, G., Pacheco, J., Suzán, G. & Sánchez-Azofeifa, A. (2003) Countryside biogeography of Neotropical mammals: Conservation opportunities in agricultural landscapes of Costa Rica. *Conservation Biology*, **17**, 1814–1826.

Darlington, P. J. (1957) *Zoogeography: The geographical distribution of animals*. New York: John Wiley.

De Camargo, R. X. & Currie, D. J. (2015) An empirical investigation of why species–area relationships overestimate species losses. *Ecology*, **96**, 1253–1263.

de Candolle, A. (1855) *Géographie botanique raisonnée: Ou l'exposition des faits principaux et des lois concernant la distribution géographique des plates de l'epoque actuelle*. Paris: Maisson.

de Caprariis, P., Lindemann, R. H. & Collins, C. M. (1976) A method for determining optimum sample size in species diversity studies. *Mathematical Geology*, **8**, 575–581.

Dengler, J. (2009) Which function describes the species–area relationship best? A review and empirical evaluation. *Journal of Biogeography*, **36**, 728–744.

Dengler, J., Matthews, T. J., Steinbauer, M. J., Wolfrum, S., Boch, S., Chiarucci, A., Conradi, T., Dembicz, I., Marcenò, C., García-Mijangos, I., Nowak, A., Storch, D., Ulrich, W., Campos, J. A., Cancellieri, L., Carboni, M., Ciaschetti, G., De Frenne, P., Dolezal, J., Dolnik, C., Essl, F., Fantinato, E., Filibeck, G., Grytnes, J.-A., Guarino, R., Güler, B., Janišová, M., Klichowska, E., Kozub, L., Kuzemko, A., Manthey, M., Mimet, A., Naqinezhad, A., Pedersen, C., Peet, R. K., Pellissier, V., Pielech, R., Potenza, G., Rosati, L., Terzi, M., Valkó, O., Vynokurov, D., White, H., Winkler, M. & Biurrun, I. (2020) Species–area relationships in continuous vegetation: Evidence from Palaearctic grasslands. *Journal of Biogeography*, **47**, 72–86.

Desmet, P. & Cowling, R. (2004) Using the species–area relationship to set baseline targets for conservation. *Ecology and Society*, **9**, article 11.

Diamond, J. M. (1975) The island dilemma: Lessons of modern biogeographic studies for the design of natural reserves. *Biological Conservation*, **7**, 129–146.

Diamond, J. M. (1984) 'Normal' extinctions of isolated populations. *Extinctions* (ed. by M. H. Nitecki), pp. 191–246. Chicago: Chicago Press.

Du Rietz, G. E. (1921) *Zur methodologischen Grundlage der modernen Pflanzensociologie, Akademishe Abhandlung.* Uppsala: Uppsala Universitet.

Fahrig, L. (2002) Effect of habitat fragmentation on the extinction threshold: A synthesis. *Ecological Applications*, **12**, 346–353.

Fahrig, L. (2013) Rethinking patch size and isolation effects: The habitat amount hypothesis. *Journal of Biogeography*, **40**, 1649–1663.

Fattorini, S. (2007) To fit or not to fit? A poorly fitting procedure produces inconsistent results when the species–area relationship is used to locate hotspots. *Biological Conservation*, **16**, 2531–2538.

Fattorini, S., Borges, P. A. V., Dapporto, L. & Strona, G. (2017) What can the parameters of the species–area relationship (SAR) tell us? Insights from Mediterranean islands. *Journal of Biogeography*, **44**, 1018–1028.

Flahaut, C. & Schröter, C. (1910) *Phytogeographischer Nomenklatur. Beriche Und Worschläge.* Zürich: Zürcher & Rurrer.

Flather, C. H. (1996) Fitting species-accumulation functions and assessing regional land use impacts on avian diversity. *Journal of Biogeography*, **23**, 155–168.

Forster, G. (1777) *A voyage around the world in his Majesty's sloop, Resolution, commanded by Captain James Cook, during the years 1772, 3, 4, and 5. B. Volume 1.* London: B. White, J. Robson, P. Elmsly, & G. Robinson.

Forster, J. R. (1778) *Observations made during a voyage round the world.* London: G. Robinson.

Fridley, J. D., Peet, R. K., Wentworth, T. R. & White, P. S. (2005) Connecting fine- and broad-scale species–area relationships of southeastern U.S. flora. *Ecology*, **86**, 1172–1177.

Gleason, H. A. (1922) On the relation between species and area. *Ecology*, **3**, 158–162.

Gleason, H. A. (1925) Species and area. *Ecology*, **6**, 66–74.

Gould, S. J. (1979) An allometric interpretation of species–area curves: The meaning of the coefficient. *The America Naturalist*, **114**, 335–343.

Haddad, N. M., Brudvig, L. A., Clobert, J., Davies, K. F., Gonzalez, A., Holt, R. D., Lovejoy, T. E., Sexton, J. O., Austin, M. P., Collins, C. D., Cook, W. M., Danschen, E. I., Ewers, R. M., Foster, B. L., Jenkins, C. N., King, A. J., Laurance, W. F., Levey, D. J., Margules, C. R., Melbourne, B. A., Nicholls, A. O., Orrock, J. L., Song, D.-X. & Townshend, J. R. (2015) Habitat fragmentation and its lasting impact on Earth's ecosystems. *Science Advances*, **1**, e1500052.

Haddad, N. M., Gonzalez, A., Brudvig, L. A., Burt, M. A., Levey, D. J. & Damschen, E. I. (2017) Experimental evidence does not support the Habitat Amount Hypothesis. *Ecography*, **40**, 48–55.

Halley, J. M., Sgardeli, V. & Triantis, K. A. (2014) Extinction debt and the species–area relationship: A neutral perspective. *Global Ecology & Biogeography*, **23**, 113–123.

Hanski, I. (1999) *Metapopulation ecology*. Oxford: Oxford University Press.

Hanski, I., Zurita, G. A., Bellocq, M. I. & Rybicki, J. (2013) Species–fragmented area relationship. *Proceedings of the National Academy of Sciences USA*, **110**, 12715–12720.

Harte, J., Blackburn, T. & Ostling, A. (2001) Self-Similarity and the relationship between abundance and range size. *The American Naturalist*, **157**, 374–386.

Harte, J., Kinzig, A. & Green, J. (1999) Self-similarity in the distribution and abundance of species. *Science*, **284**, 334–336.

He, F. & Hubbell, S. P. (2011) Species–area relationships always overestimate extinction rates from habitat loss. *Nature*, **473**, 368–371.

He, F. & Legendre, P. (1996) On species–area relations. *The American Naturalist*, **148**, 719–737.

Holdridge, L. R., Grenke, W. C., Hatheway, W. H., Liang, T. & Tosi, J. A. Jr. (1971) *Forest environments in tropical life zones. A pilot study*. Oxford: Pergamon Press.

Hubbell, S. P. (2001) *The unified neutral theory of biodiversity and biogeography*. Princeton, NJ: Princeton University Press.

Humboldt, A. von (1807) *Ideen zur einer Geographie der Pflanstzen nebst einem Naturgemälde der Tropenländer*. Tübringen: Cotta.

Jaccard, P. (1908) Nouvelles recherches sur la distribution florale. *Bulletin de la Societe Vaudoise des Sciences Naturelles*, **44**, 223–270.

Koh, L. P., Lee, T. M., Sodhi, N. S. & Ghazoul, J. (2010) An overhaul of the species–area approach for predicting biodiversity loss: Incorporating matrix and edge effects. *Journal of Applied Ecology*, **47**, 1063–1070.

Krebs, C. J. (2009) *Ecology, the experimental analysis of distribution and abundance*, 6th ed. San Francisco, CA: Benjamin Cummings.

Kunin, W. E., Harte, J., He, F., Hui, C., Jobe, R. T., Ostling, A., Polce, C., Šizling, A., Smith, A. B., Smith, K., Smart, S. M., Storch, D., Tjørve, E., Ugland, K.-I., Ulrich, W. & Varma, V. (2018) Upscaling biodiversity: Estimating the species–area relationship from small samples. *Ecological Monographs*, **88**, 170–187.

Kylin, H. (1923) Växtsociologiska randanmärkningar. *Botaniska Notiser*, **1923**, 161–234.

Kylin, H. (1926) Über begriffsbildung und statistik in der pflanzensoziologie. *Botaniska Notiser*, **1926**, 81–180.

Ladle, R. J. & Whittaker, R. J. (eds.) (2011) *Conservation biogeography*. Chichester: Wiley-Blackwell.

Lagerberg, T. (1914) Markflorans analys på objektiv grund. *Meddelanden från Statens Skogsförsöksanstalt*, **1914**, 129–200, XV–XXIV.

Lawrey, J. D. (1991) The species–area curve as an index of disturbance in saxicolous lichen communities. *The Bryologist*, **94**, 377–382.

Lomolino, M. V. (2000) Ecology's most general, yet protean pattern: The species–area relationship. *Journal of Biogeography*, **27**, 17–26.

Lomolino, M. V. & Brown, J. H. (2009) The reticulating phylogeny of island biogeography theory. *The Quarterly Review of Biology*, **84**, 357–390.

Lomolino, M. V. & Weiser, M. D. (2001) Towards a more general species–area relationship: Diversity on all islands, great and small. *Journal of Biogeography*, **28**, 431–445.

MacArthur, R. H. & Wilson, E. O. (1963) An equilibrium theory of insular zoogeography. *Evolution*, **17**, 373–387.

MacArthur, R. H. & Wilson, E. O. (1967) *The theory of island biogeography*. Princeton, NJ: Princeton University Press.

Martín, H. G. & Goldenfeld, N. (2006) On the origin and robustness of power-law species–area relationships in ecology. *Proceedings of the National Academy of Sciences USA*, **103**, 10310–10315.

Martins, I. S. & Pereira, H. M. (2017) Improving extinction projections across scales and habitats using the countryside species–area relationship. *Scientific Reports*, **7**, article 12899.

Matter, S. F., Hanski, I. & Gyllenberg, M. (2002) A test of the metapopulation model of the species–area relationship. *Journal of Biogeography*, **29**, 977–983.

Matthews, T. J. (2015) Analysing and modelling the impact of habitat fragmentation on species diversity: A macroecological perspective. *Frontiers of Biogeography*, **7**, 60–68.

Matthews, T. J. & Aspin, T. W. H. (2019) Model averaging fails to improve the extrapolation capability of the island species–area relationship. *Journal of Biogeography*, **46**, 1558–1568.

Matthews, T. J., Borregaard, M. K., Guilhaumon, F., Triantis, K. A. & Whittaker, R. J. (2016a) On the form of species–area relationships in habitat islands and true islands. *Global Ecology & Biogeography*, **25**, 847–858.

Matthews, T. J., Rigal, F., Triantis, K. A. & Whittaker, R. J. (2019a) A global model of island species–area relationships. *Proceedings of the National Academy of Sciences USA*, **116**, 12337–12342.

Matthews, T. J., Triantis, K. A., Rigal, F., Borregaard, M. K., Guilhaumon, F. & Whittaker, R. J. (2016b) Island species–area relationships and species accumulation curves are not equivalent: An analysis of habitat island datasets. *Global Ecology & Biogeography*, **25**, 607–618.

Matthews, T. J., Triantis, K., Whittaker, R. J. & Guilhaumon, F. (2019b) Sars: An R package for fitting, evaluating and comparing species–area relationship models. *Ecography*, **42**, 1446–1455.

May, R. M. (1975) Island biogeography and the design of wildlife preserves. *Nature*, **254**, 177–178.

Mendenhall, C. D., Karp, D. S., Meyer, C. F. J., Hadly, E. A. & Daily, G. C. (2014) Predicting biodiversity change and averting collapse in agricultural landscapes. *Nature*, **509**, 213–217.

Palmgren, A. (1916) Studier över lövengsområdena på Åland. *Acta Societatis pro Fauna et Flora Fennica*, **42**, 1–634.

Pereira, M. & Daily, G. C. (2006) Biodiversity dynamics in countryside landscapes. *Ecology*, **87**, 1877–1885.

Pound, R. & Clements, F. E. (1898) A method for determining the abundance of secondary species. *Minnesota Botanical Studies*, **2**, 19–24.

Preston, F. W. (1960) Time and space and the variation of species. *Ecology*, **41**, 611–627.

Preston, F. W. (1962) The canonical distribution of commonness and rarity: Part I & II. *Ecology*, **43**, 185–215, 410–432.

Raunkiær, C. (1908) Livsformenes Statistik som Grundlag for biologisk Plantegeografi. *Botanisk Tidsskrift*, **29**, 42–83.

Raunkiær, C. (1909) Formationsundersøgelse og formationsstatistik. *Botanisk Tidsskrift (København)*, **30**, 20–132.

Raunkiær, C. (1918) Recherches statistiques sur les formations vegetales. *Biologiske Meddelelser / Det Kongelige Danske Videnskabernes Selskab*, **1**, 1–80.

Romell, L. G. (1920) Sur la régle de distribution de fréquences. *Svensk Botanisk Tidsskrift*, **14**, 1–20.

Rosenzweig, M. L. (1995) *Species diversity in space and time*. Cambridge: Cambridge University Press.

Rosenzweig, M. L. (2001) Loss of speciation rate will impoverish future diversity. *Proceedings of the National Academy of Sciences USA*, **98**, 5404–5410.

Rosenzweig, M. L. (2004) Applying species–area relationships to the conservation of diversity. *Frontiers of biogeography: New directions in the geography of nature* (ed. by M. V. Lomolino and L. R. Heaney), pp. 325–343. Sunderland, MA: Sinauer Associates.

Schoener, T. W. (1976) The species–area relations within archipelagoes: Models and evidence from island land birds. *Proceedings of the XVI International Ornithological Conference* (ed. by H. J. Firth and J. H Calaby), pp. 629–642. Canberra: Australian Academy of Science.

Simberloff, D. (1988) The contribution of population and community biology to conservation science. *Annual Review of Ecology and Systematics*, **19**, 473–511.

Simberloff, D. S. & Abele, L. G. (1976) Island biogeography and conservation practice. *Science*, **191**, 285–286.

Smith, A. B. (2010) Caution with curves: Caveats for using the species–area relationship in conservation. *Biological Conservation*, **143**, 555–564.

Svedberg, T. (1922) Statistisk vegetationsanalys, några synspunkter. *Svensk Botanisk Tidsskrift*, **16**, 197–205.

Terborgh, J. (1976) Island biogeography and conservation: Strategy and limitations. *Science*, **193**, 1029–1030.

Tjørve, E. (2002) Habitat size and number in multi-habitat landscapes: A model approach based on species–area curves. *Ecography*, **25**, 17–24.

Tjørve, E. (2003) Shapes and functions of species–area curves: A review of possible models. *Journal of Biogeography*, **30**, 827–835.

Tjørve, E. (2009) Shapes and functions of species–area curves (II): A review of new models and parameterizations. *Journal of Biogeography*, **36**, 1435–1445.

Tjørve, E. (2010) How to resolve the SLOSS debate: Lessons from species-diversity models. *Journal of Theoretical Biology*, **264**, 604–612.

Tjørve, E. & Tjørve, K. M. C. (2008) The species–area relationship, self-similarity, and the true meaning of the z-value. *Ecology*, **89**, 3528–3533.

Tjørve, E. & Turner, W. R. (2009) The importance of samples and isolates for species–area relationships. *Ecography*, **32**, 391–400.

Tjørve, E., Kunin, W. E., Polce, C. & Tjørve, K. M. C. (2008) The species–area relationship: Separating the effects of species-abundance and spatial distribution. *Journal of Ecology*, **96**, 1141–1151.

Tjørve, E., Tjørve, K. C. M., Šizlingová, E. & Šizling, A. L. (2018) Great theories of species diversity in space and why they were forgotten: The beginnings of a spatial ecology and the Nordic early 20th-century botanists. *Journal of Biogeography*, **45**, 530–540.

Triantis, K. A., Guilhaumon, F. & Whittaker, R. J. (2012) The island species–area relationship: Biology and statistics. *Journal of Biogeography*, **39**, 215–231.

Triantis, K. A., Mylonas, M., Lika, K. & Vardinoyannis, K. (2003) A model for the species–area–habitat relationship. *Journal of Biogeography*, **30**, 19–27.

Triantis, K. A., Mylonas, M. & Whittaker, R. J. (2008) Evolutionary species–area curves as revealed by single-island endemics: Insights for the interprovincial species–area relationship. *Ecography*, **31**, 401–407.

Ulrich, W. & Buszko, J. (2003) Species–area relationships of butterflies in Europe and species richness forecasting. *Ecography*, **26**, 365–373.

Ulrich, W. & Buszko, J. (2004) Habitat reduction and patterns of species loss. *Basic and Applied Ecology*, **5**, 231–240.

Wallace, A. R. (1869) *The Malay Archipelago: The land of the orang-utan, and the bird of paradise. A narrative of travel, with studies of man and nature.* London: Macmillan and Co.

Wallace, A. R. (1876) *The geographical distribution of animals: With a study of the relations of living and extinct Faunas Volume 1.* Cambridge: Cambridge University Press.

Wallace, A. R. (1914) *The world of life: A manifestation of creative power, directive mind and ultimate purpose.* London: Chapman and Hall.

Watling, J. I. & Donnelly, M. A. (2006) Fragments as islands: A synthesis of faunal responses to habitat patchiness. *Conservation Biology*, **20**, 1016–1025.

Watson, H. C. (1835) *Remarks on the geographical distribution of British plants, chiefly in connection with latitude, elevation, and climate.* London: Longman.

Watson, H. C. (1859) *Cybele Britannica, or British plants and their geographical relations.* London: Longman and Company.

Whittaker, R. J. & Fernández-Palacios, J. M. (2007) *Island biogeography: Ecology, evolution, and conservation*, 2nd ed. Oxford: Oxford University Press.

Whittaker, R. J. & Matthews, T. J. (2014) The varied form of species–area relationships. *Journal of Biogeography*, **41**, 209–210.

Whittaker, R. J., Araújo, M. B., Jepson, P., Ladle, R. J., Watson, J. E. M. & Willis, K. J. (2005) Conservation biogeography: Assessment and prospect. *Diversity and Distributions*, **11**, 3–23.

Williams, C. B. (1943) Area and the number of species. *Nature*, **152**, 262–265.

Williams, M. R. (1995) An extreme-value function model of the species incidence and species–area relations. *Ecology*, **76**, 2607–2616.

Williamson, M. (1975) The design of wildlife preserves. *Nature*, **256**, 519.

Williamson, M., Gaston, K. J. & Lonsdale, W. M. (2001) The species–area relationship does not have an asymptote! *Journal of Biogeography*, **28**, 827–830.

Wilson, E. O. & Willis, E. O. (1975) Applied biogeography. *Ecology and evolution of communities* (ed. by M. L. Cody and J. M. Diamond), pp. 522–534. Cambridge: Belknap Press.

Part II

Diversity–Area Relationships: The Different Types and Underlying Factors

3 · Explaining Variation in Island Species–Area Relationship (ISAR) Model Parameters between Different Archipelago Types: Expanding a Global Model of ISARs

THOMAS J. MATTHEWS, FRANÇOIS RIGAL, KONSTANTINOS PROIOS, KOSTAS A. TRIANTIS AND ROBERT J. WHITTAKER

3.1 Introduction

"The literature of species–area curves is replete with discussion, debate, and a good deal of puzzlement about the meaning of parameters c and z" (Gould, 1979, p. 335). This statement was written in 1979 but could just as easily have been written today (e.g. Triantis et al., 2012; Fattorini et al., 2017; Tjørve & Tjørve, 2017; Whittaker et al., 2017; see Chapter 4) – while much progress has been made, we still lack a comprehensive understanding of what the parameters of the power island species–area relationship (ISAR) model mean. Indeed, the central question remains can the power model parameters be interpreted biologically? At this stage, it is necessary to outline what we mean by power model parameters. The power model, which is given by the equation $S = cA^z$ (where S is number of species, A is area and c and z are fitted parameters), is the most widely used ISAR model and in previous meta-analyses has been found to provide the best fit to the most datasets (75 per cent of 612) and to be the best general model from twenty SAR models tested (Triantis et al., 2012; Matthews et al., 2016). Often the logarithmic form of the power model, which is given by $\log S = \mathrm{Log}C + z\log A$, is used as it can be fitted using simple linear regression (Rosenzweig, 1995). In this model, $\mathrm{Log}C$ is the intercept (in log–log space) and z is the slope of the relationship.

Following this usage, many studies have attempted to explain variation in ISAR slope across island datasets (e.g. Connor & McCoy, 1979; Rosenzweig, 1995; Watling & Donnelly, 2006; Triantis et al., 2012; Sólymos & Lele, 2012), with rather fewer studies focusing attention on variation in the intercept (e.g. Gould, 1979; Triantis et al., 2012; Matthews et al., 2016; Fattorini et al., 2017).

It should be noted that in the non-log or arithmetic version of the power model ($S = cA^z$), z is the rate at which the slope of the curve decelerates with increasing area. A steeper slope in the log–log power relation (higher z) typically corresponds to depressed SAR curvature in arithmetic space, although this is also dependent on the c-value. That is, in the arithmetic version, the 'slope' of the curve is dependent on both c and z, while in log–log space the slope is characterized by z alone (see Chapter 4). The remainder of this chapter is focused on the log–log form of the power model and, thus, when we refer to $LogC$ and z we are referring to the intercept and slope of the relationship in log–log space. It is perhaps as important to note that we are solely concerned with ISARs as opposed to nested forms of species–area relationship, for which different considerations apply (see Chapter 1).

3.1.1 Variation in LogC and z

Part of the confusion that is alluded to in the quote at the beginning of this chapter has arisen, we would argue, from the almost universal tendency to study the $LogC$ and z parameters independently, as we ourselves have done in several previous papers, including two substantial meta-analyses (Triantis et al., 2012; Matthews et al., 2016). In these analyses we documented variation in each parameter between archipelagos of differing origin (habitat, inland, continental shelf, oceanic) and also noted, for example, variation in z in relation to the range of island area encompassed within each archipelago. While significant trends were found, a great deal of variation remained unexplained and, when we subsequently came to scrutinize a simple plot of $LogC$ versus z-values for 596 island datasets sourced from these meta-analyses in relation to archipelago type (oceanic, continental-shelf, inland and habitat), the most striking impression was of the extent to which the parameter space occupied overlaps, notwithstanding that the central tendencies of the three subsets differ slightly (Figure 3.1). This troublingly indistinct pattern was something that we repeatedly returned to thinking about.

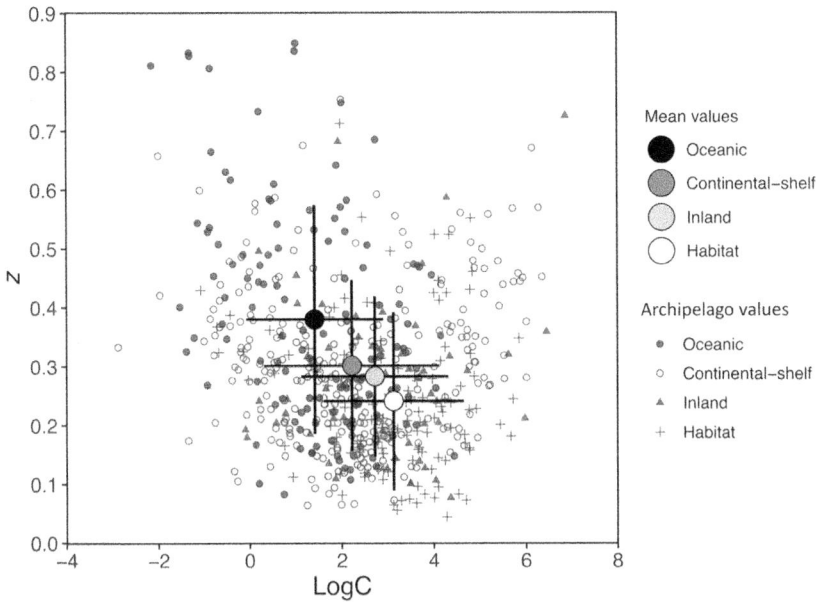

Figure 3.1 Relationship between the parameters of the power ISAR model (z and LogC) calculated for 596 ISARs. Only parameter values from datasets with significant z-values (P < 0.05) were included: 136 habitat island, 125 oceanic island, 58 inland water-body island ('Inland') and 277 continental shelf island datasets. For each group of islands, horizontal and vertical error bars around the mean are the standard deviation for LogC and z, respectively. Eight data points lying outside the main range of z and LogC were removed for graphical convenience.

Eventually, analyses of these data we undertook for a presentation at the January 2017 International Biogeography Society conference prompted us to dig a little deeper into how LogC and z might vary together in relation to other properties of each archipelago, starting with the realization that there appeared to be a form of trade-off in the values for the two parameters in relation to archipelago richness (Gamma; Figure 3.2A). Specifically, it appeared that, for given values of Gamma, there may be a negative relationship between LogC and z (Figure 3.2A). Identification of this pattern provided the catalyst for us to undertake an extensive evaluation of the variables underpinning variation in LogC and z, but with the logic that any such analysis needed to study the two parameters in tandem (i.e. in the same model with explicit consideration of their covariation). To do so we used structural equation models (SEMs), which provide a multivariate statistical approach that enables the user to infer causality from observational data (Shipley, 2016). SEMs also enable the total effect (i.e. the direct effect and the indirect effects) of a predictor variable on a response variable to be calculated (Grace & Bollen, 2005; Shipley, 2016). A SEM typically comprises a mix of

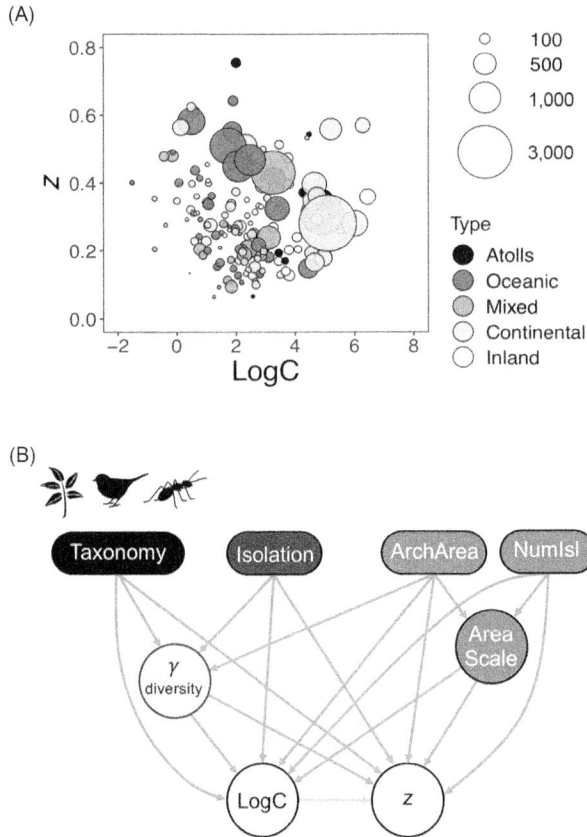

Figure 3.2 (A) Archipelago richness (Gamma) in relation to the parameters (z and Log$C$) of each ISAR, from the analysis of Matthews et al. (2019). The five archipelago types are distinguished for the all-ISARs dataset (n = 151). (B) The theoretical causal model structure from Matthews et al. (2019). Three categories of taxa were considered: Plants (i.e. vascular plants), Vertebrates (e.g. birds, mammals) and Invertebrates (e.g. land snails, beetles). Archipelago configuration was represented by: NumIsl = Number of islands, AreaScale = the ratio between the largest and the smallest islands within each archipelago and ArchArea = the total land area of the archipelago. Isolation = archipelago distance from the mainland. Diversity properties are represented by: Gamma diversity = the total species richness of an archipelago; and Log$C$ and z (the parameters of the power ISAR model). Both figures are from Matthews et al. (2019)

endogenous (a variable with parents and children, i.e. arrows going in and coming out from the variable in a classical path model diagram) and exogenous variables (a variable with just children, i.e. only arrows coming out). The resulting analyses provided some novel insights into the interrelationships between the parameters of the ISAR power model, which we presented in an earlier paper (Matthews et al., 2019), on which we build in the present chapter.

3.1.2 A Global Model of ISARs

Within the Matthews et al. (2019) study, we selected a subset of 151 of our datasets, each for archipelagos of true islands (i.e. islands surrounded by water), specifically: i) those for which we could retrieve Gamma values, ii) those for which the power model provided a significant fit and iii) removing any essentially duplicative datasets. Based on theoretical arguments (e.g. MacArthur & Wilson, 1967; Schoener, 1976; Holt et al., 1999; Gascuel et al., 2016; see also Chapters 2, 4, 12 and 15) and previously documented empirical patterns (e.g. Connor & McCoy, 1979; Triantis et al., 2012, 2015), we developed a theoretical causal SEM to explain variation in $LogC$ and z across datasets (Figure 3.2B). This causal model contained three broad sets of explanatory variables: archipelago isolation, taxonomy and archipelago configuration (Figure 3.2B), with four variables identified as endogenous (Gamma, the ratio of the largest island area to the smallest island area [AreaScale], $LogC$ and z) and all others exogenous (e.g. number of islands [NumIsl], archipelago area [ArchArea]). Notably, the model contained a link from $LogC$ to z but not *vice versa* (this directionality was confirmed analytically). We then used piecewise structural equation modelling (an extension of standard SEM that allows for the inclusion of random effects; Lefcheck, 2016) in combination with a backward stepwise selection procedure and the 151 datasets to test and simplify our theoretical causal model. Archipelago identity (i.e. the name of the archipelago) was included as a random effect.

The model selection exercise was undertaken three times, using: i) all datasets; ii) just oceanic island datasets and iii) just continental island datasets. The results were broadly similar across the three sets and provided substantial support for our general model. The models explained a remarkable amount of variation in $LogC$ and z (e.g. marginal R^2 values of 0.82 and 0.55 for z and $LogC$, respectively, for the oceanic island analysis). We also observed a strong negative effect of $LogC$ on z as a function of Gamma in all cases.

The global model (Matthews et al., 2019) is revealing in many ways and provides a solid foundation for understanding ISAR variation across archipelagos. First, there were significant effects of the higher taxon (plants, vertebrates or invertebrates) a particular island dataset represents in the different best models. This is unsurprising as different taxa have very different densities and thus richness per island and archipelago, and also differing propensities to disperse over long distances. Were we able to calibrate models for particular sub-taxa of vertebrates and invertebrates we would expect further improvements in model fit. Second, the expected increases in slope and decreases in intercept with geographical distance from source area failed to feature within the models. We speculated that this may have been due in part to the simple isolation metric we used (i.e. distance from the archipelago to the nearest mainland). Other, more complex metrics, such as the amount of land within a buffer of a given size around an island, may be more informative (Weigelt & Kreft, 2013). It is also likely that variation in archipelago configuration (the amount and disposition of area across islands within an archipelago) confounds the pattern expected as a function of distance from a single mainland source.

Hence, one way of interpreting the models reported in Matthews et al. (2019) is to hypothesize that archipelago area and location largely set the richness for the archipelago, and that the trade-off between slope and intercept as a function of Gamma then reflects how the archipelago area is subdivided. Of course, the control is not simply top-down, but is also and simultaneously bottom-up, as ecological and evolutionary processes of turnover and of speciation on each island result in modification to Gamma. Further complexity is likely added by meta-archipelago effects (cf. Whittaker et al., 2018), that is, the exchange of propagules between islands belonging to different archipelagos – exchanges which occur in addition to within-archipelago and island–mainland exchanges.

3.1.3 Expanding the Global Model

The Matthews et al. (2019) global ISAR model is of course incomplete and here we pick on two issues for further scrutiny: i) improved quantification of island isolation and configuration and ii) addition of habitat islands. i) Archipelagos are subdivided into multiple islands and these are, to extents depending on their configuration and the ecology of the taxon, linked by dispersal and metapopulation type dynamics (Hanski, 1998). The way in which islands within an archipelago are configured geographically, and, in turn, the degree to which they are isolated from one another, are thus expected to be important drivers of island richness and endemism (Gascuel et al., 2016; Chapter 15). However, intra-archipelago isolation has only recently started to be included extensively in empirical island biogeography studies (e.g. Weigelt & Kreft, 2013; Cabral et al., 2014; Gascuel et al., 2016). ii) The global model was developed and tested using purely true islands and thus its applicability to habitat islands is unknown. Habitat islands (patches of habitat surrounded by a non-aquatic matrix) differ from true islands in many respects, such as, for example, inter-island dispersal being a function of matrix permeability (Laurance, 2008; Matthews, 2015). Thus, if island theory and our global ISAR model, in particular, is to be of use in the study of habitat islands and fragmented landscapes, it is necessary to further adapt and test the model with an explicit consideration of habitat island datasets. Such an advancement is important as ISARs are a key tool in the conservation biogeographer's toolkit (Rosenzweig, 2004; Whittaker et al., 2005; Chapters 13, 14, 16 and 17) and are often used in conservation applications (e.g. predicting the number of species extinctions due to habitat loss; Halley et al., 2013; Chapter 14).

Here, we expand on our previous work (Matthews et al., 2019) to address these two points. First, for 144 true island datasets we calculated the mean distance between pairs of islands, a measure of intra-archipelagic isolation. We also calculated alternative measures of archipelagic isolation: the proportion of landmass in buffers around the archipelago (buffer distances of 1,000 and 2,500 km^2). We extended our original theoretical causal model to include these variables and tested the model using the 144 true island datasets. Second, we simplified the global model so that it was applicable to habitat islands. We then tested this habitat island model using 65 habitat island datasets, as well as a full analysis of the 144 true island datasets in combination with the 65 habitat island datasets.

3.2 Methods

3.2.1 Data Collection and Compilation

For true island datasets, we took 144 of the 151 datasets in Matthews et al. (2019) for which we were able to acquire spatial information (i.e. latitude and longitude coordinates) of all islands in the dataset. These datasets each pertained to archipelagos or geographically coherent groups of archipelagos of true geographical islands, that is, areas of land surrounded by water. For each of these datasets, we also took the exogenous (i.e. Taxon, ArchArea, NumIsl and distance from an archipelago to the mainland [Isolation1]) and endogenous (i.e. Gamma, AreaScale, LogC and z) variables used in Matthews et al. (2019).

For habitat island datasets, we took 65 datasets from Matthews et al. (2016) for which we were able to obtain estimates of Gamma (i.e. the total diversity across the islands in a dataset) and that had at least six islands. We excluded datasets that were focused at very small spatial scales (e.g. insects on individual plants). The sourced datasets each pertained to a set of discrete habitat patches surrounded by contrasting matrix habitat. The island area values in all true and habitat island datasets were converted to km^2.

Our final dataset comprised several different types of archipelagos: oceanic (39 ISARs), continental (64 ISARs), atoll (8 ISARs), inland (15 ISARs), mixed (archipelagos that included both oceanic and either continental or atoll islands; 18 ISARs) and habitat islands (64 ISARs). One of the original 65 datasets (a dataset of butterflies from small experimental grassland fragments) was removed as an outlier, with a LogC of 11.

3.2.2 Archipelagic and Intra-archipelago Isolation Metrics

Geospatial data for all true islands in the 144 datasets were obtained from the Global Island Database, provided by the United Nations Environment Program's World Conservation Monitoring Centre (UNEP-WCMC, 2013). Mainland coastline data were taken from the Global Self-consistent, Hierarchical, High-resolution Geography Database (GSHHG) version 2.3.7 (Wessel & Smith, 1996).

For the true island datasets, the distance from each island to the mainland (Isolation1) was calculated (see Figure 3.3 for details). We then separately calculated the isolation for each dataset using the proportion of a buffer around the archipelago that contained land (Weigelt & Kreft, 2013). Two buffer distances (i.e. the radius length) were used: 1,000 km^2 (Isolation2) and 2,500 km^2 (Isolation3) (see Figure 3.3). These two distances were chosen i) as they were considered to create buffers within which the majority of dispersal (for the various taxa in our datasets) would occur and ii) for practical reasons; much smaller than this and often oceanic archipelagos would have no surrounding land in a buffer, much larger and the buffer would become very large, encompassing large proportions of nearby continents. These calculations were undertaken using functions in the 'sf' R package (Pebesma, 2018).

Isolation1: Distance to mainland

The geographical isolation of the archipelago (Isolation) in metres from the nearest mainland (or lake edge for islands within lakes). The distance between each island (using the island centroid) in the archipelago and the mainland was calculated, and the shortest distance taken as the Isolation metric. As 'mainland', we included the world's continents, Madagascar (the world's largest continental fragment island at 587,040 km^2) and the largest (>130,000 km^2) of the world's land-bridge islands that were relevant to our study systems (New Guinea, Java, Sumatra, Borneo, Great Britain). See Matthews et al. (2019).

Isolation2 and Isolation3: Surrounding landmass

The amount (or proportion) of land within a circular buffer, of a given radius (arrow in the image to the left), around an island. The area of the island itself was not included. To obtain an archipelago value, the mean of all the individual island values was taken. This metric was found to be one of the superior archipelago isolation metrics assessed by Weigelt & Kreft (2013). Two buffer distances (i.e. the radius length) were used: 1,000km^2 (Isolation2) and 2,500km^2 (Isolation3).

MeanDist: Intra-archipelagic mean distance

All shortest pairwise distances between islands (mean nearest neighbour) were calculated and the mean distance then taken. The shortest distance between island edges was used. The metric is similar to the mean nearest taxon distance (MNTD) metric and gives an idea of how islands are packed within the archipelago.

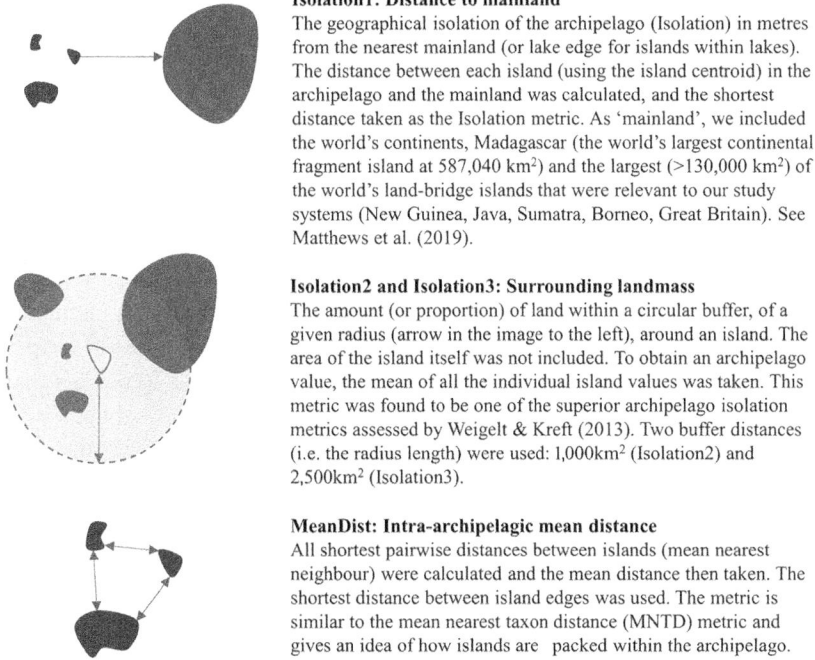

Figure 3.3 The four isolation metrics used in the analyses. Isolation1, Isolation2 and Isolation3 are measures of archipelago isolation, while MeanDist is a measure of intra-archipelago isolation.

While the island–mainland (or simply land–sea) dichotomy makes sense in the context of true islands, such a clear distinction does not exist for habitat islands (Matthews, 2015). For habitat islands, the matrix type (and permeability) is also an important consideration (Daily et al., 2003) and to further confuse matters there will likely be an interaction effect between matrix type and the focal taxon. For these reasons we did not attempt to calculate any of the archipelagic isolation metrics for the habitat island datasets.

As a measure of intra-archipelago isolation, the mean distance between all pairs of islands in an archipelago (MeanDist) was calculated for the 144 true island datasets (Figure 3.3). This involved calculating the distances among all pairs of islands within an archipelago using the 'Near' tool in ArcGIS 10 (ESRI, 2012), and then taking the mean of these values as MeanDist. This metric is based on the logic that the more spaced out the islands are within an archipelago (i.e. with increasing MeanDist), the greater the difference in species composition between islands (i.e. reflecting the distance decay in similarity). This in turn results in a larger Gamma (Cabral et al., 2014). As we did not have location information for individual habitat islands, it was not possible to calculate the intra-archipelago isolation metric for the habitat island datasets.

3.2.3 Theoretical Causal Structural Equation Models for LogC and z

Here, we took the primary theoretical causal model of Matthews et al. (2019; see Figure 3.2B) as the start-point of our analyses. As already stated, we then expanded on this model by including i) extra isolation (archipelago [Isolation2 and Isolation3] and intra-archipelago [MeanDist]) metrics and ii) habitat island datasets (see Figure 3.4). However, due to the fact that we could not calculate any of the isolation metrics for the habitat island datasets, we had to construct slightly different theoretical causal SEMs for the different subsets of datasets. In total, we constructed five theoretical causal SEMs comprising two different model architectures:

1) All-ISARs model (i.e. true island and habitat islands) (n = 208); not including any of the archipelago or intra-archipelago isolation metrics.
2) Habitat-ISARs model (n = 64); not including any of the archipelago or intra-archipelago isolation metrics. Same model architecture as (1).
3) True-ISARs model (n = 144), with MeanDist and either one of Isolation1, Isolation2 or Isolation3.
4) Oceanic-ISARs model (n = 39), with MeanDist and either one of Isolation1, Isolation2 or Isolation3. Same model architecture as (3).
5) Continental-ISARs model (n = 64), with MeanDist and either one of Isolation1, Isolation2 or Isolation3. Same model architecture as (3).

The first architecture, used in Models 1−2 (in the above list), is different from that used in Matthews et al. (2019), in that there is no Isolation variable. The second architecture, used in Models 3−5, differs for three reasons. First, MeanDist is included as an exogenous variable within the 'Archipelago configuration' group of variables. We have included links from MeanDist to Gamma (based on the aforementioned logic) and also to LogC and z. Second, since Gamma was directly linked to MeanDist, it was logical for us to also include the direct links with the other remaining Archipelago configuration metrics, namely NumIsl and AreaScale. Third, each model selection analysis (discussed below) was run three times, using one of Isolation1, Isolation2 or Isolation3 as the Isolation variable (see Figure 3.4).

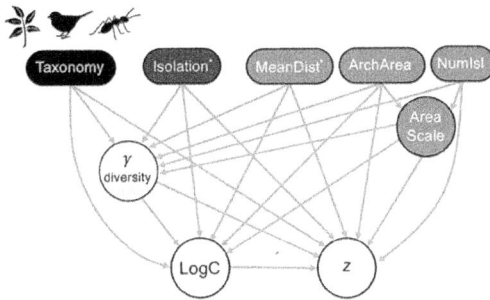

Figure 3.4 The theoretical causal model(s) for the main analyses. Variable names are defined in the legend of Figure 3.2 and in Figure 3.3. Archipelago identity (not shown) is included as a random factor as some archipelagos are represented by multiple separate datasets (for different taxa), with the exception of the habitat-ISARs model. ⋆ Indicates the exogenous variables (Isolation and MeanDist) that were not included in the all-ISARs and habitat-ISARs models.

3.2.4 Model Fitting and Validation

Before fitting the models, the variable Taxon was turned into dummy binary variables: two dummy variables were created that classified a dataset as an Invertebrate or Plant dataset (Vertebrate was the base category). All variables, with the exception of LogC and the two dummy variables, were log-transformed (natural logarithms) to approximate normality. We also standardized all variables to a mean of zero and standard deviation of one. We tested for multicollinearity between predictors using Pearson's correlation, with a threshold $|r < 0.7|$, following Dormann et al. (2013).

Models 2−5 were fitted using piecewise structural equation modelling (piecewiseSEM R package; Lefcheck, 2016). Archipelago identity (i.e. the archipelago name) was used as a random effect within linear mixed models (LMMs) fitted using restricted maximum likelihood (Lefcheck, 2016; Matthews et al., 2019). For Model 2 (habitat-ISARs model), it was not necessary to include a random effect as almost all datasets related to distinct habitat island systems. Thus, simple linear models were fitted instead, using the piecewiseSEM R package.

A given model fit was assessed using direct separation tests (*d*-sep) based on Fisher's C statistic, with the model being accepted where the associated $P > 0.05$ (Shipley, 2009). For each of the five theoretical causal SEMs, we simplified the model using a backward stepwise selection procedure (Grace, 2006; Ando et al., 2017; Matthews et al., 2019). Briefly, the stepwise procedure worked by, at each iteration, identifying and dropping the path with the highest P-value. This process carried on until all paths in the model were significant. Each model along this process was evaluated using Fisher's C statistic. All satisfactory models (i.e. those with non-significant Fisher's C) in the set were then compared using AIC_c. The model with the lowest AIC_c was taken to be the best model. In all five best models, the effect of each predictor on each of the endogenous variables was evaluated using their standardized path coefficients. Both the direct and indirect effects of the predictors on z and $LogC$ were reported using the standardized path coefficients (Grace & Bollen, 2005). For a given predictor A, the strength of its indirect effect on C through B is obtained by multiplying the direct standardized path coefficients of A on B and B on C. The total effect of A on C is then calculated by summing the standardized path coefficient of its direct effect and the sum of its indirect effects. For the best models with a random effect included, we calculated both the conditional R^2c (all factors including the random effect) and marginal R^2m (fixed factors only) for all endogenous variables (z, $LogC$, Gamma and AreaScale), following Nakagawa and Schielzeth (2013). For the model without a random effect (Model 2), the standard R^2 was calculated.

Four of the five causal models (1, 2, 4 and 5 in the previous section), as well as their respective best models, had satisfactory fits (i.e. a P-value of Fisher's C > 0.05). However, for the true-ISARs model (Model 3), neither the causal model nor its respective best model was satisfactory (i.e. in both cases the P-value of Fisher's C < 0.05). To deal with this, we used the piecewiseSEM R package to identify the most important missing link. This was found to be AreaScale←MeanDist. Clearly, this link does not make ecological or geological sense: how spread out the islands in an archipelago are in space is not a driver of the sizes of those islands. However, when this link was included both the full and best model were satisfactory. In addition, with the exception of the added link, the paths (and signs of the coefficients) in the best model including and excluding this link are identical. Thus, including the link improves the statistical properties of the best model without changing the story. As such, we decided to retain this link in the full causal true-ISARs model, and only the results with this link included are discussed below (for similar reasoning, see Blackburn et al., 2016). It is likely that this link is due to the correlation between MeanDist and a missing covariate, but further research is needed to explore this possibility.

It is worth noting that we were not able to estimate the AIC_c of the full oceanic-ISARs model. This is due to the fact that our sample size was relatively low (n = 39) in comparison to the number of predictor variables. The AIC_c formula used in piecewiseSEM takes as a denominator n-K-1, with K being the number of parameters estimated in the full SEM model (see Shipley, 2013, 2016). For the oceanic island datasets, K was 38, leading therefore to a denominator of 0. However, we were able to properly calculate AIC_c for all of the remaining steps of the backward selection procedure, and thus to successfully complete the model selection procedure. Overall, each of the island type subset analyses (i.e. oceanic, continental and habitat) should be viewed with slight caution due to the smaller number of datasets involved, relative to Models 1 and 3.

As a final step, for each of the five best models we assessed the model's predictive power using k-fold cross-validation, where $k = 10$. For each run, Pearson's correlation was used to test the association between the predicted (using the model fit) and observed values of the four endogenous variables and the correlation coefficient averaged across the 10-folds. This 10-fold cross-validation process was then repeated 100 times and the mean correlation value (along with its associated 95 per cent confidence interval) taken (for more details see Matthews et al., 2019). All analyses were undertaken using R (version 3.6; R Core Team, 2019).

3.3 Results

Pairwise Pearson's correlation tests between explanatory variables did not detect any problematic multicollinearity $|r < 0.7|$, except between archipelago isolation metrics (Isolation1 − Isolation2, r = 0.78; Isolation2 − Isolation3, r = 0.79). Therefore, as outlined in Section 3.2, we performed each analysis three times, each time with a different isolation metric. Since results were identical (isolation was never retained in our best model regardless of the selected metric), we only present results obtained with Isolation1, that is, distance to the mainland (Figure 3.3). Table 3.1 provides the summary statistics of the full and best models resulting from backward stepwise selection for each of the five models listed in Section 3.2.3. The standardized path coefficients for each model are provided in Table 3.2 and the direct and indirect effects of the predictors on the endogenous variables z and $LogC$ in Table 3.3. The best SEMs for each of the five models as well as the overall net effect of each predictor on z and $LogC$ (sum of the direct and indirect effects) are illustrated in Figure 3.5. The plots showing the relationship between

$LogC$ and z as a function of Gamma for Models 1, 2, 4 and 5 are provided in Figure 3.6. Definitions of terms (e.g. ArchArea) are provided in the legend of Table 3.2.

The best all-ISARs (Model 1; Figure 3.5A and B), true-ISARs (Model 3; Figure 3.5E and F) and continental-ISARs (Model 5; Figure 3.5I and J) models are very similar to the best models in Matthews et al. (2019; see their figure 2). Overall, Gamma increased

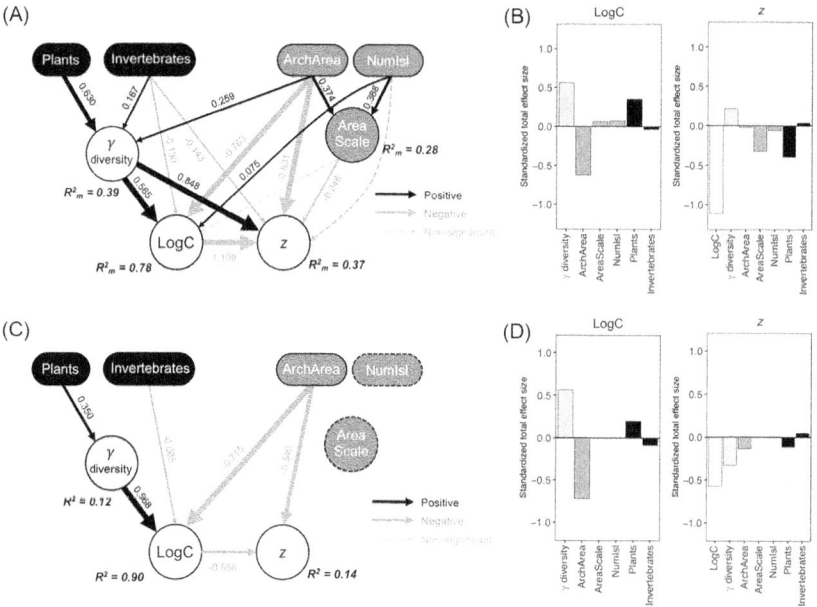

Figure 3.5 Best all-ISARs (n = 208) (A, B), habitat-ISARs (n = 64) (C, D), true-ISARs (n = 144) (E, F), oceanic-ISARs (n = 39) (G, H), and continental-ISARs (n = 64) (I, J) path models. The best path models were obtained using a backward stepwise selection procedure and AIC_c. Pathways show how taxon (with vertebrates as the base level), isolation, archipelago configuration (ArchArea, NumIsl, AreaScale, and MeanDist) and Gamma influence $LogC$ and, together, z. Note that the exogenous variables Isolation and MeanDist were not included in the all-ISARs and habitat-ISARs models. Arrow widths are proportional to the standardized path coefficients (values are also given) and R^2 values are given for each endogenous variable, which for the linear mixed models is the marginal R^2 (R^2_m, fixed factors only). Right panels (B, D, F, H, J) show the standardized total effect size of variables on $LogC$ and z (see Section 3.2.4 and Table 3.3). Note that the full true-ISARs model (E) was re-specified by adding the missing link AreaScale←MeanDist to improve the fit (see Section 3.2.4).

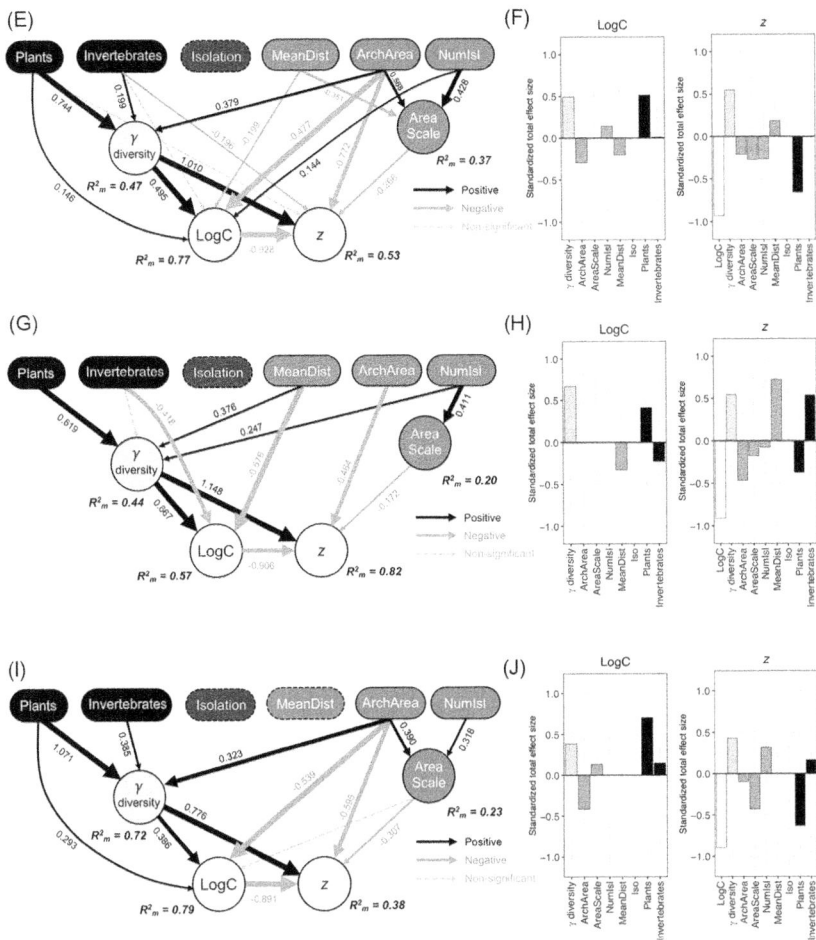

Figure 3.5 (cont.)

from Vertebrates to Invertebrates to Plants (the largest effect), and was also a positive function of ArchArea. We also found a strong negative relationship between $LogC$ and z, driven by increasing values of Gamma. $LogC$ was a positive function of Gamma and a negative function of ArchArea. ISAR slope (z) values decreased in response to $LogC$, ArchArea, AreaScale, and increased in response to Gamma. Finally, AreaScale increased with NumIsl and with ArchArea. Mean-Dist had a weak but significant negative effect on $LogC$ in the best true-ISARs model (Figure 3.5E and F).

The best habitat-ISARs model (Model 2) was the least robust of the five best models and had a substantially different structure from the others, characterized by very few links (Figure 3.5C and D); Gamma being only a function of Plants, z being only a function of LogC and ArchArea, and LogC being explained by Gamma and ArchArea.

The best oceanic-ISARs model (Model 4; Figure 3.5G and H) differed slightly from the equivalent model reported in Matthews et al. (2019), which was due to the inclusion of MeanDist here. Mean-Dist was found to have a significant and strong negative impact on LogC, and a weaker but still significant positive impact on Gamma. As a consequence of the inclusion of MeanDist, the negative effect of ArchArea on LogC previously reported in Matthews et al. (2019) was no longer present.

By combining both direct and indirect effects, it is apparent that, for the best all-ISARs, true-ISARs, oceanic-ISARs and continental-ISARs models, taxon, archipelago configuration and Gamma are all important in explaining variation in LogC. For z, the interplay between LogC and Gamma is of substantial importance, as are the effects of archipelago configuration and taxon (Figure 3.5B, F, H and J). For the best oceanic-ISARs model, MeanDist had a larger total effect (i.e. direct and indirect) on both z and LogC (Figure 3.5H). For the best habitat-ISARs model, only Gamma and archipelago configuration are important in explaining variation in LogC, while, for z, only LogC had a relatively large total effect, with Gamma and ArchArea having smaller total effects (Figure 3.5D).

The percentage of explained variation in the three endogenous variables of interest (Gamma, LogC and z) varied considerably across the five best models (Table 3.1). For Gamma, R^2_m/R^2 ranged from 12 per cent in the best habitat-ISARs model to 72 per cent in the best continental-ISARs model; for LogC, from 57 per cent in the best oceanic-ISARs model to 91 per cent in the best habitat-ISARs model; and for z, from 14 per cent in the best habitat-ISARs model to 82 per cent in the best oceanic-ISARs model.

In the repeated k-fold cross-validation analyses, the mean coefficient of the Pearson's correlation between the observed and the predicted endogenous variable was >0.5 for the three primary endogenous variables (i.e. z, LogC and Gamma) for all five best models, with the exception of z (mean r $=$ 0.29) for the best habitat-ISARs model (Table 3.4).

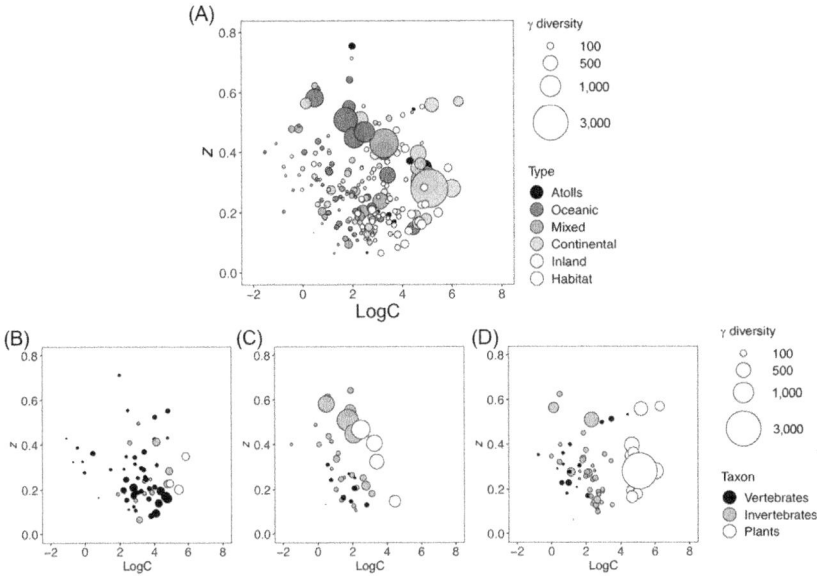

Figure 3.6 Archipelago richness (Gamma) in relation to the power model parameters of each ISAR. The six archipelago types are distinguished for the all-ISARs dataset (n = 208) (A) and vascular plants, vertebrates and invertebrates (note that animal datasets represent sub-sets, such as birds or spiders) for the habitat-ISARs (n = 64) (B), oceanic-ISARs (n = 39) (C) and continental-ISARs datasets (n = 64) (D).

3.4 Discussion

Despite the addition of new variables and datasets representing different types of island, our best models are similar to those presented in Matthews et al. (2019): a result that points to the robustness of the global model of ISARs. We consider it noteworthy how much of the variation in LogC and z can be explained, in most cases, by simple models that include variables representing archipelago configuration, taxon effects and Gamma. In the present work, we aimed to expand on our previous study by including i) habitat island datasets and ii) alternative measures of archipelago and intra-archipelago isolation. We now discuss the results of each of these variants in turn.

3.4.1 Expanding the Global Model to Include Habitat Island Datasets

The SAR is widely used in both the theoretical and applied study of fragmented landscapes (e.g. He & Hubbell, 2011; Hanski et al., 2013;

Table 3.1 A brief summary of the backward stepwise selection procedure starting from the full theoretical causal models (see Figure 3.4), for the all-ISARs model (n = 208), the habitat-ISARs model (n = 64), the true-ISARs model (n = 64), the oceanic-ISARs model (n = 144) and the continental-ISARs model (n = 39). For all-ISARs and habitat-ISARs, the exogenous variables Isolation and MeanDist were not included in the analysis. See Section 3.4.2 for a full description of the method. Only the parameters of the hypothesized causal model (row full model) and the model with the lowest AIC_c (row best model) are given. The values of Fisher's C statistic (C), the associated degrees of freedom (df) and P-values (P), values of R^2 for the endogenous variables (z, LogC, Gamma, AreaScale) and the AIC_c values are reported. For the linear mixed models, reported R^2 values are the marginal R^2 values. Note that for the true-ISARs, the full model was re-specified by adding the missing link AreaScale—MeanDist to improve the fit (see Section 3.2.4). We were not able to calculate AIC_c for the full oceanic-ISARs model as there were too many parameters (see Section 3.2.4).

	C	df	P	$R^2 z$	R^2LogC	R^2Gamma	R^2AreaScale	AIC_c
All-ISARs								
Full model	6.122	4	0.190	0.367	0.781	0.384	0.278	83.345
Best model	10.857	12	0.541	0.369	0.781	0.385	0.278	77.689
Habitat-ISARs								
Full model	6.395	8	0.603	0.240	0.913	0.139	0.067	101.008
Best model	10.235	12	0.595	0.142	0.910	0.123		42.962
True-ISARs								
Full model	8.640	6	0.195	0.531	0.771	0.485	0.365	119.963
Best model	15.378	16	0.497	0.525	0.770	0.467	0.365	96.057
Oceanic-ISARs								
Full model	9.366	8	0.312	0.819	0.564	0.502	0.235	–
Best model	27.714	30	0.586	0.820	0.570	0.442	0.196	191.656
Continental-ISARs								
Full model	10.793	8	0.214	0.436	0.799	0.726	0.231	222.190
Best model	16.060	18	0.588	0.383	0.791	0.722	0.231	111.259

Table 3.2 Standardized path coefficients from the best all-ISARs (n = 208), habitat-ISARs (n = 64), true-ISARs (n = 64), oceanic-ISARs (n = 144), oceanic-ISARs (n = 39) and continental-ISARs (n = 64) models. For each path, the arrow indicates the direction of the relationship and the estimated standardized path coefficients (Est.), the associated standard errors (SE) and P-values (P) are reported. NumIsl = Number of islands; Inverts = invertebrates; AreaScale is the ratio between the largest and the smallest islands within each archipelago; ArchArea is the total area of the archipelago and MeanDist is the mean distance between all pairs of islands. NI indicates that the link was not included in the model (see main text). Note that for the true-ISARs model, the full model was re-specified by adding the missing link AreaScale←MeanDist to improve the fit (last row; see Section 3.2.4)

Paths	All-ISARs			Habitat-ISARs			True-ISARs			Oceanic-ISARs			Continental-ISARs		
	Est.	SE	P	Est.	SE	P	Est.	SE	P	Est.	SE	P	Est.	SE	P
z←LogC	-1.109	0.116	<0.001	-0.568	0.186	0.003	-0.928	0.113	<0.001	-0.906	0.084	<0.001	-0.891	0.216	0.002
z←Gamma	0.848	0.087	<0.001	–	–	–	1.010	0.098	<0.001	1.148	0.094	<0.001	0.776	0.163	0.001
z←ArchArea	-0.831	0.110	<0.001	-0.540	0.186	0.005	-0.772	0.100	<0.001	-0.464	0.089	<0.001	-0.595	0.178	0.008
z←NumIsl	0.121	0.064	0.063	–	–	–	–			–			–		
z←AreaScale	-0.246	0.070	0.001	–			-0.266	0.062	<0.001	-0.172	0.082	0.049	-0.307	0.113	0.022
z←Inverts	-0.143	0.061	0.023	–			-0.196	0.069	0.006	–			–		
z←Plants	–		–	–			-0.179	0.094	0.063	–			–		
LogC←Gamma	0.565	0.034	<0.001	0.568	0.039	<0.001	0.495	0.058	<0.001	0.667	0.121	<0.001	0.386	0.094	0.002
LogC←ArchArea	-0.763	0.037	<0.001	-0.715	0.041	<0.001	-0.477	0.064	<0.001	–			-0.539	0.072	0.000
LogC←NoIsl	0.075	0.036	0.041	–			0.144	0.045	0.002	–			–		
LogC←AreaScale	0.066	0.039	0.096	–			–		–	–			0.134	0.062	0.055
LogC←MeanDist	NI	NI	NI	NI	NI	NI	-0.199	0.062	0.002	-0.576	0.124	<0.001	–		–
LogC←Inverts	-0.130	0.034	<0.001	-0.085	0.040	0.039	-0.086	0.049	0.085	-0.418	0.110	0.001	0.293	0.100	0.015
LogC←Plants	–		–	–		–	0.146	0.064	0.028	–		–	0.323	0.075	0.001
Gamma←ArchArea	0.259	0.058	<0.001	0.350	0.119	0.005	0.379	0.068	<0.001	–			–		–
Gamma←MeanDist	NI	NI	NI	NI	NI	NI	–		–	0.376	0.135	0.011	–		–
Gamma←Inverts	0.167	0.060	0.007	–			0.199	0.071	0.007	0.288	0.145	0.059	0.385	0.081	0.001
Gamma←Plants	0.630	0.057	<0.001	–			0.744	0.071	<0.001	0.619	0.140	<0.001	1.071	0.085	0.000
AreaScale←ArchArea	0.374	0.067	<0.001	–			0.568	0.098	<0.001	–		–	0.390	0.118	0.006
AreaScale←NumIsl	0.388	0.060	<0.001	–			0.428	0.074	<0.001	0.411	0.122	0.003	0.318	0.117	0.018
AreaScale←MeanDist	NI	NI	NI	NI	NI	NI	-0.351	0.098	0.001	NI	NI	NI	NI	NI	NI

Table 3.3 *Estimates of the direct and indirect effects of the predictors on the endogenous variables z and LogC. Estimates are given for the best all-ISARs (n = 208), habitat-ISARs (n = 64), true-ISARs (n = 144), oceanic-ISARs (n = 39) and continental-ISARs (n = 64) models. The direct effects are standardized path coefficients, while the indirect effects are calculated by multiplying the direct path coefficients along the path mediated by associated variables. The total effect is calculated by summing the direct and indirect effects where both routes of influence apply. Gamma and AreaScale are not included because of the absence of indirect paths to the exogenous variables. Therefore, effects of the predictors on Gamma and AreaScale are only direct effects and correspond to the standardized path coefficients reported in Table 3.2. Variable names as in Table 3.2.*

Endogenous	Exogenous	All-ISARs			Habitat-ISARs			True-ISARs			Oceanic-ISARs			Continental-ISARs		
		Direct	Indirect	Total	Direct	Indirect	Total	Direct	Indirect	Total	Direct	Indirect	Total	Direct	Indirect	Total
z	LogC	-1.109	—	—	-0.568	—	—	-0.928	—	—	—	—	—	-0.891	—	—
	Gamma	0.848	-0.627	0.221	—	-0.323	—	1.010	-0.459	0.551	1.148	-0.604	0.544	0.776	-0.344	0.432
	ArchArea	-0.831	0.812	-0.019	-0.540	0.406	-0.134	-0.772	0.563	-0.209	-0.464	—	—	-0.595	0.620	0.025
	AreaScale	-0.246	-0.073	-0.319	—	—	—	-0.266	—	—	-0.172	—	—	-0.307	—	—
	NumIsl	0.121	-0.179	-0.058	NI	NI	NI	—	-0.262	—	—	-0.071	—	—	-0.098	—
	MeanDist	NI	NI	NI	NI	NI	NI	—	0.185	—	—	0.726	—	—	—	—
	Isolation	NI	NI	NI	NI	NI	NI	—	—	—	—	—	—	—	—	—
	Plants	—	-0.395	—	—	-0.113	—	-0.179	-0.477	-0.656	—	-0.374	—	—	0.202	—
	Inverts	-0.143	0.181	0.038	—	0.048	—	-0.196	0.189	-0.007	—	0.535	—	—	0.166	—
LogC	Gamma	0.565	—	—	0.568	—	—	0.495	—	—	0.667	—	—	0.386	—	—
	ArchArea	-0.763	0.146	-0.617	-0.715	—	—	-0.477	0.188	-0.289	—	—	—	-0.539	0.125	-0.414
	AreaScale	0.066	—	—	—	—	—	—	—	—	—	—	—	0.134	—	—
	NumIsl	0.075	—	—	NI	NI	NI	0.144	—	—	—	—	—	—	—	—
	MeanDist	NI	NI	NI	NI	NI	NI	-0.199	—	—	-0.576	0.251	-0.325	—	—	—
	Isolation	NI	NI	NI	NI	NI	NI	—	—	—	—	—	—	—	—	—
	Plants	—	0.356	—	—	0.199	—	0.146	0.368	0.514	—	0.413	—	0.293	0.413	0.706
	Inverts	-0.130	0.094	-0.036	-0.085	—	—	-0.086	0.099	0.013	-0.418	0.192	-0.226	—	0.149	—

Table 3.4 *Results of the repeated 10-fold cross validation sensitivity analyses using the best all-ISARs (n = 208), habitat-ISARs (n = 64), true-ISARs (n = 144), oceanic-ISARs (n = 39) and continental-ISARs (n = 64) models. The mean Pearson's r between the predicted and observed values and the associated 95 per cent confidence interval values are given for the four endogenous variables: z, LogC, Gamma and AreaScale. Note that, for the habitat-ISARs model, 10-fold cross validation was not performed for AreaScale, as no predictor significantly explained AreaScale*

	Endogenous variables			
	z	LogC	Gamma	AreaScale
All-ISARs	0.590	0.875	0.597	0.411
	[0.550;0.617]	[0.862;0.887]	[0.556;0.637]	[0.379;0.446]
Habitat-ISARs	0.293	0.933	0.617	–
	[0.124;0.420]	[0.874;0.957]	[0.482;0.815]	–
True-ISARs	0.694	0.854	0.649	0.413
	[0.655;0.732]	[0.823;0.871]	[0.602;0.688]	[0.358;0.462]
Oceanic-ISARs	0.869	0.655	0.545	0.369
	[0.748;0.946]	[0.489;0.822]	[0.318;0.711]	[0.125;0.539]
Continental-ISARs	0.560	0.857	0.800	0.255
	[0.396;0.689]	[0.751;0.915]	[0.681;0.870]	[0.118;0.390]

Chisholm et al., 2018; Chapters 11, 14, 16 and 17) and it (the power model in particular) has been frequently adopted as a tool to predict the number of extinctions resulting from habitat loss (Halley et al., 2013; Chapter 14). As such, it would be useful if the global model of ISARs of Matthews et al. (2019) could be expanded to predict the parameters of the power model in habitat island systems. Unfortunately, while the best habitat-ISARs model explained most of the variation in LogC ($R^2 = 0.91$), the explanatory power with regard to z was very low ($R^2 = 0.14$), suggesting that we are some way from developing a predictive model for use in conservation applications (cf. Chapter 14). As with the study of many other macroecological patterns in habitat islands (Matthews, 2015), it is likely that the 'noisy' nature of habitat island datasets, due for example to the prevalence of anthropogenic disturbance within patches (e.g. hunting; Benchimol & Peres, 2013) and factors such as edge and matrix effects (Ewers & Didham, 2006; Laurance, 2008), does not allow for effective predictive diversity modelling in such systems, at least with simple predictor variables like those used here. The high explanatory power with regard to Log*C* indicates that this parameter can almost be fully explained as an increasing function of Gamma and a negative function of ArchArea (Figure 3.5C). These paths (and coefficient signs) are a feature in all of our best models (and all of the models in Matthews et al., 2019), with the exception of the oceanic-ISARs model here.

3.4.2 Incorporating Additional Metrics of Archipelago and Intra-archipelago Isolation

Based on previous work (e.g. Weigelt & Kreft, 2013), which has shown that alternative archipelago isolation metrics to the commonly used 'distance to mainland' are better predictors of island diversity patterns, we hypothesized that the inclusion of alternative archipelago isolation metrics might lead to us finding a signal for isolation in our best models. However, this did not occur. This is not to say that no isolation signal can be detected, but rather that when placed into a model with alternative explanatory variables, an isolation effect does not emerge as a significant component of the best models and that this holds regardless of how we have assessed the isolation of the archipelago. In addition, we have not used an exhaustive set of archipelago isolation metrics here and perhaps other metrics may be more revealing (e.g. see Chapter 15). It is likely that island/archipelago isolation is much harder to capture in a single metric than island/archipelago area.

In contrast, intra-archipelago isolation (i.e. MeanDist) had a much more pronounced effect, at least in the oceanic-ISARs model. This indicates that (oceanic) archipelago configuration and resultant intra-archipelago dynamics (e.g. metapopulation style rescue effects, inter-island dispersal, gene flow between islands; Gascuel et al., 2016; Chapter 15) are of considerable importance in determining the parameters of the ISAR. Islands that are closer together (i.e. lower MeanDist) will have, all else being equal, more inter-island dispersal and thus greater gene flow between islands, resulting in lower beta diversity, lower rates of speciation and lower rates of extinction (due to greater rescue effects) (e.g. Brown & Kodric-Brown, 1977; Whittaker & Fernández-Palacios, 2007; Cabral et al., 2014; Gascuel et al., 2016; Chapter 15). The effects of these processes are then apparent in Gamma, z and Log*C*: larger MeanDist equates to a higher z and a lower Log*C* (based on the total effects). The fact that MeanDist was not included in the best continental-ISARs model and only had a small effect in the true-ISARs model, suggests that these intra-island processes are proportionally less important among less isolated archipelagos. Interestingly, for the best oceanic-ISARs model, the total effect of MeanDist was stronger for z than for $LogC$, although MeanDist did not have any direct effect on z (Figure 3.5H). This finding illustrates the benefits of using SEMs rather than standard multiple regression approaches in island biogeographical analyses of this nature.

3.4.3 Conclusions and Scope for Further Expansion of the Global Model

Analysis of the global model thus far has been informative and the fact that we can already explain over 80 per cent of the variation in z between oceanic archipelagos illustrates its explanatory capacity. However, we suggest that there is still much we can learn from this general approach and we now discuss some interesting potential avenues to explore in future work.

As always, testing the model on additional datasets would be revealing, particularly as our oceanic best model was fitted to only thirty-nine datasets. Additional datasets could be used to cross-validate the model: using the model as parameterized here to predict the endogenous variables in the new datasets. While we have used k-fold cross-validation here (and the results are promising), cross-validation using fully independent datasets is arguably a much stronger test of model performance (Shipley, 2016).

Another interesting approach, although one that is clearly limited by the availability of suitable data, would be to test the global model using multiple datasets for one particular, narrower taxonomic group (e.g. snakes, snails, passerines). As there would not be any variation attributable to taxon effects, we would expect to see more constrained and consistent behaviour of ISAR parameters in such cases and a clearer relationship between z and $LogC$ as a function of Gamma. Further tests of the model using different groups of species and types of diversity would also likely prove rewarding. Examples include archipelago endemics and single island endemics (see Chapter 19), alien species (Chapter 6) and model parameters derived from functional and phylogenetic diversity–area relationships (Chapter 5).

In addition, this approach appears ripe for exploration via simulation modelling. The notion that the significance of intra-archipelago configurations effects only manifests clearly when archipelagos are truly isolated (and have always been isolated) provides a problem: there aren't enough such archipelagos encompassing a good range of island areas with well sampled faunas and floras. Simulation approaches, as exemplified by Cabral et al. (2019a, b; see also Chapter 15) provide an obvious route to testing the ideas developed herein.

With regard to habitat islands, it appears more challenging to develop a general model of ISAR parameters of the form more successfully developed here for true islands. This is disappointing given the importance of understanding the implications of habitat fragmentation and the unresolved debate on the significance of habitat fragmentation as opposed to simply reductions in the overall amount of habitat (Fahrig, 2017; Horváth et al., 2019). We nonetheless emphasize that our results support the importance of considering both parameters of the ISAR in the search for predictive models of habitat island diversity. For habitat islands, additional data are required that are specific to habitat island systems, including inter alia matrix type, some measure of matrix permeability (a function of matrix type but also the focal taxon) and disturbance dynamics within the patches (see also Chapter 14).

The quote at the beginning of this chapter describes the 'puzzlement' regarding the meaning of $\mathrm{Log}C$ and z that has characterized study of the ISAR for at least the last fifty years (see Chapter 2). In part this reflects the fact that the ISAR can be represented by several alternative models of different form and to varying degrees of success (Chapter 7). The work we have undertaken over the last decade and herein suggests to us the following key generalizations. First, the power model provides the best general model for ISAR datasets and, given that it provides a linear fit when applied in its log–log implementation, holds potential as a diagnostic and potentially also a predictive tool. Second, it is important to consider both parameters of the model simultaneously in order to unlock the biological meaning of the model. Third, in general, for a given archipelago richness (Gamma), there is a clear tendency for a negative relationship between slope and intercept, conditioned by the biology of the taxon of interest and the configuration of the archipelago. Fourth, these trends are most clearly seen on Darwinian islands: those which have emerged from the oceans and have never been connected to mainland areas. It is on these islands that we see perhaps most clearly the outcome of the theoretically expected interplay of the three key biogeographical processes of migration, speciation and extinction, through their impact on ISAR parameters. These findings suggest to us that there is further as yet untapped potential in the analysis of island species–area relationships.

References

Ando, Y., Utsumi, S. & Ohgushi, T. (2017) Aphid as a network creator for the plant associated arthropod community and its consequence for plant reproductive success. *Functional Ecology*, **31**, 632–641.

Benchimol, M. & Peres, C. A. (2013) Anthropogenic modulators of species–area relationships in Neotropical primates: A continental-scale analysis of fragmented forest landscapes. *Diversity and Distributions*, **19**, 1339–1352.

Blackburn, T. M., Delean, S., Pyšek, P. & Cassey, P. (2016) On the island biogeography of aliens: A global analysis of the richness of plant and bird species on oceanic islands. *Global Ecology & Biogeography*, **25**, 859–868.

Brown, J. H. & Kodric-Brown, A. (1977) Turnover rates in insular biogeography: Effect of immigration on extinction. *Ecology*, **58**, 445–449.

Cabral, J. S., Weigelt, P., Kissling, W. D. & Kreft, H. (2014) Biogeographic, climatic and spatial drivers differentially affect α-, β- and γ-diversities on oceanic archipelagos. *Proceedings of the Royal Society B: Biological Sciences*, **281**, 20133246.

Cabral, J. S., Whittaker, R. J., Wiegand, K. & Kreft, H. (2019a) Assessing predicted isolation effects from the general dynamic model of island biogeography with an eco-evolutionary model for plants. *Journal of Biogeography*, **46**, 1569–1581.

Cabral, J. S., Wiegand, K. & Kreft, H. (2019b) Interactions between eco-logical, evolutionary, and environmental processes unveil complex dynamics of insular plant diversity. *Journal of Biogeography*, **46**, 1582–1597.

Chisholm, R. A., Lim, F., Yeoh, Y. S., Seah, W. W., Condit, R. & Rosindell, J. (2018) Species–area relationships and biodiversity loss in fragmented landscapes. *Ecology Letters*, **21**, 804–813.

Connor, E. F. & McCoy, E. D. (1979) Statistics and biology of the species–area relationship. *The American Naturalist*, **113**, 791–833.

Daily, G. C., Ceballos, G., Pacheco, J., Suzán, G. & Sánchez-Azofeifa, A. (2003) Countryside biogeography of neotropical mammals: Conservation opportunities in agricultural landscapes of Costa Rica. *Conservation Biology*, **17**, 1814–1826.

Dormann, C. F., Elith, J., Bacher, S., Buchmann, C., Carl, G., Carré, G., Marquéz, J. R. G., Gruber, B., Lafourcade, B., Leitão, P. J., Münkemüller, T., McClean, C., Osborne, P. E., Reineking, B., Schröder, B., Skidmore, A. K., Zurell, D. & Lautenbach, S. (2013) Collinearity: A review of methods to deal with it and a simulation study evaluating their performance. *Ecography*, **36**, 27–46.

ESRI (2012) *ArcGIS Desktop*, version 10. Redlands, CA: ESRI.

Ewers, R. M. & Didham, R. K. (2006) Confounding factors in the detection of species responses to habitat fragmentation. *Biological Reviews*, **81**, 117–142.

Fahrig, L. (2017) Ecological responses to habitat fragmentation per se. *Annual Review of Ecology, Evolution, and Systematics*, **48**, 1–23.

Fattorini, S., Borges, P. A. V., Dapporto, L. & Strona, G. (2017) What can the parameters of the species–area relationship (SAR) tell us? Insights from Mediterranean islands. *Journal of Biogeography*, **44**, 1018–1028.

Gascuel, F., Laroche, F., Bonnet-Lebrun, A.-S. & Rodrigues, A. S. L. (2016) The effects of archipelago spatial structure on island diversity and endemism: Predictions from a spatially-structured neutral model. *Evolution*, **70**, 2657–2666.

Gould, S. J. (1979) An allometric interpretation of species–area curves: The meaning of the coefficient. *The American Naturalist*, **114**, 335–343.

Grace, J. B. (2006) *Structural equation modeling and natural systems*. Cambridge: Cambridge University Press.

Grace, J. B. & Bollen, K. A. (2005) Interpreting the results from multiple regression and structural equation models. *The Bulletin of the Ecological Society of America*, **86**, 283–295.

Halley, J. M., Sgardeli, V. & Monokrousos, N. (2013) Species–area relationships and extinction forecasts. *Annals of the New York Academy of Sciences*, **1286**, 50–61.

Hanski, I. (1998) Metapopulation dynamics. *Nature*, **396**, 41–49.

Hanski, I., Zurita, G. A., Bellocq, M. I. & Rybicki, J. (2013) Species–fragmented area relationship. *Proceedings of the National Academy of Sciences USA*, **110**, 12715–12720.

He, F. & Hubbell, S. P. (2011) Species–area relationships always overestimate extinction rates from habitat loss. *Nature*, **473**, 368–371.

Holt, R. D., Lawton, J. H., Polis, G. A. & Martinez, N. D. (1999) Trophic rank and the species–area relationship. *Ecology*, **80**, 1495–1504.

Horváth, Z., Ptacnik, R., Vad, C. F. & Chase, J. M. (2019) Habitat loss over six decades accelerates regional and local biodiversity loss via changing landscape connectance. *Ecology Letters*, **22**, 1019–1027.

Laurance, W. F. (2008) Theory meets reality: How habitat fragmentation research has transcended island biogeographic theory. *Biological Conservation*, **141**, 1731–1744.

Lefcheck, J. S. (2016) piecewiseSEM: Piecewise structural equation modelling in R for ecology, evolution, and systematics. *Methods in Ecology and Evolution*, **7**, 573–579.

MacArthur, R. H. & Wilson, E. O. (1967) *The theory of island biogeography*. Princeton, NJ: Princeton University Press.

Matthews, T. J. (2015) Analysing and modelling the impact of habitat fragmentation on species diversity: A macroecological perspective. *Frontiers of Biogeography*, **7**, 60–68.

Matthews, T. J., Guilhaumon, F., Triantis, K. A., Borregaard, M. K. & Whittaker, R. J. (2016) On the form of species–area relationships in habitat islands and true islands. *Global Ecology & Biogeography*, **25**, 847–858.

Matthews, T. J., Rigal, F., Triantis, K. A. & Whittaker, R. J. (2019) A global model of island species–area relationships. *Proceedings of the National Academy of Sciences USA*, **116**, 12337–12342.

Nakagawa, S. & Schielzeth, H. (2013) A general and simple method for obtaining R^2 from generalized linear mixed-effects models. *Methods in Ecology and Evolution*, **4**, 133–142.

Pebesma, E. (2018) Simple features for R: Standardized support for spatial vector data. *The R Journal*, **10**, 439–446.

R Core Team (2019) *R: A language and environment for statistical computing*. Vienna, Austria: R Foundation for Statistical Computing.

Rosenzweig, M. L. (1995) *Species diversity in space and time*. Cambridge: Cambridge University Press.

Rosenzweig, M. L. (2004) *Applying species–area relationships to the conservation of diversity. Frontiers of biogeography: New directions in the geography of nature* (ed. by M. V. Lomolino and L. R. Heaney), pp. 325–343. Sunderland, MA: Sinauer Associates.

Schoener, T. W. (1976) The species–area relations within archipelagoes: Models and evidence from island land birds. *Proceedings of the XVI International Ornithological Conference* (ed. by H. J. Firth and J. H. Calaby), pp. 629–642. Canberra: Australian Academy of Science.

Shipley, B. (2009) Confirmatory path analysis in a generalized multilevel context. *Ecology*, **90**, 363–368.

Shipley, B. (2013) The AIC model selection method applied to path analytic models compared using a d-separation test. *Ecology*, **94**, 560–564.

Shipley, B. (2016) *Cause and correlation in biology: A user's guide to path analysis, structural equations and causal inference with R*. Cambridge: Cambridge University Press.

Sólymos, P. & Lele, S. R. (2012) Global pattern and local variation in species–area relationships. *Global Ecology & Biogeography*, **21**, 109–120.

Tjørve, E. & Tjørve, K. M. C. (2017) Species–area relationship. *eLS* (*Encyclopedia of Life Sciences Online*), pp. 1–9. Chichester: John Wiley & Sons.

Triantis, K. A., Economo, E. P., Guilhaumon, F. & Ricklefs, R. E. (2015) Diversity regulation at macro-scales: Species richness on oceanic archipelagos. *Global Ecology & Biogeography*, **24**, 594–605.

Triantis, K. A., Guilhaumon, F. & Whittaker, R. J. (2012) The island species–area relationship: Biology and statistics. *Journal of Biogeography*, **39**, 215–231.

UNEP-WCMC (2013) *Global distribution of islands*. Global Island Database (version 2). Based on Open Street Map data (© OpenStreetMap contributors). www .unep-wcmc.org.

Watling, J. I. & Donnelly, M. A. (2006) Fragments as islands: A synthesis of faunal responses to habitat patchiness. *Conservation Biology*, **20**, 1016–1025.

Weigelt, P. & Kreft, H. (2013) Quantifying island isolation – insights from global patterns of insular plant species richness. *Ecography*, **36**, 417–429.

Wessel, P. & Smith, W. H. F. (1996) A global, self-consistent, hierarchical, high-resolution shoreline database. *Journal of Geophysical Research*, **101**, 8741–8743.

Whittaker, R. J. & Fernández-Palacios, J. M. (2007) *Island biogeography: Ecology, evolution, and conservation*, 2nd ed. Oxford: Oxford University Press.

Whittaker, R. J., Araújo, M. B., Jepson, P., Ladle, R. J., Watson, J. E. M. & Willis, K. J. (2005) Conservation biogeography: Assessment and prospect. *Diversity and Distributions*, **11**, 3–23.

Whittaker, R. J., Fernández-Palacios, J. M., Matthews, T. J., Borregaard, M. K. & Triantis, K. A. (2017) Island biogeography: Taking the long view of nature's laboratories. *Science*, **357**, eaam8326.

Whittaker, R. J., Fernández-Palacios, J. M., Matthews, T. J., Rigal, F. & Triantis, K. A. (2018) Archipelagos and meta-archipelagos. *Frontiers of Biogeography*, **10**, e41470.

4 · *Determinants of the Shape of Species–Area Curves*

EVEN TJØRVE, KATHLEEN M. C. TJØRVE,
EVA ŠIZLINGOVÁ AND ARNOŠT L. ŠIZLING

In order to predict and explain the curve shape of the species–area relationship (SAR) it is necessary to understand the patterns and processes that drive it. However, after almost a century, the shapes of species–area curves and the factors and mechanisms that underpin them are still hotly debated (see e.g. Tjørve & Tjørve, 2017, for review) and there remains little consensus on anything other than the fact that larger areas contain, on average, more species than smaller ones.

Most introductions to work on SARs rarely discuss more than a few determinants of SAR shape. Authors usually neither reflect on the incompleteness of their lists of likely patterns and processes nor attempt to systemize them. There are, however, numerous hypotheses or theories that singly attempt to account for the shape of SARs (e.g. McGuinness, 1984; Hill et al., 1994; Rosenzweig, 1995; Connor & McCoy, 2001; Turner & Tjørve, 2005). Moreover, we should assume that more than one underlying process combines to form the observed SAR, though it may not be possible to quantify the contribution of these processes to the observed data. Still, we may group them in several ways, to include such processes as sampling design, extinction, immigration, evolution, population dynamics, random placement and the aggregation and segregation of individuals. In addition, each process is parametrized by several factors, such as the distance between samples, species spatial density, habitat diversity and immigration and extinction rates. This parametrization affects the importance of the processes to the observation. However, because many processes and patterns are determined or affected by one or several others, it is almost impossible to derive completely unequivocal systematics for the various determinants of SAR shape. In reality, the SAR is the result of a complex network of interacting and overlapping processes and it is only by understanding how these processes interact that we will develop a mechanistic understanding of SARs.

There is a growing literature aiming to predict and explain SARs mechanistically from patterns and mechanisms. Such mechanistic approaches rely on underlying universal principles and the processes reflect fundamental laws drawn from the fields of kinetics and dynamics (namely energetic), as well as the basic biochemistry of aerobic metabolism (Brown et al., 2003). However, very few rules have been proposed for underlying systems of processes causing species diversity and environmental patterns, such as topography, soil, climate and land cover.

In an attempt to present a general overview, recognizing a hierarchy of determinants (processes and factors), we have grouped these determinants into three categories. Sampling and data handling design, which also includes statistical effects such as the central limit theorem, will be discussed first. Second, we will discuss constraints imposed by geometry, energy flow and evolution. Finally, we will discuss the role of species spatial patterns in driving SARs and the processes causing these relationships.

It is necessary to briefly summarize the different types of SAR focused on in this chapter (see also Chapter 1). SARs from islands and other types of isolates (ISARs) are generated by comparing species numbers in independent areas of different sizes, whereas SARs from sample areas are usually generated as the accumulated number of species from a sample area that increases in size (a nested design) or by adding sampling units. We can define such sample-area SARs (saSARs) as a particular type of species accumulation curve (SAC; see Chapter 1).

4.1 A Systematic Approach to SAR Determinants

Explanations for SAR shape are often regarded as mutually exclusive, even when affecting the SAR at different levels. However, the different driving mechanisms rarely operate individually (Kohn & Walsh, 1994; Rosenzweig, 1995; Ricklefs & Lovette, 1999; Brown et al., 2003; Triantis et al., 2003). For example, one may ask whether hypotheses of evolutionary history and ecosystem productivity represent alternative explanations (Qian et al., 2005; Tjørve et al., 2008). The debate of the effect of 'habitat diversity versus area per se' represents another example (see e.g. Williams, 1964; Connor & McCoy, 1979; Newmark, 1986; Nilsson et al., 1988). Many have discussed these two hypotheses individually, but very few have discussed how habitat diversity and area might interact to affect species richness (but see Kohn & Walsh, 1994; Ricklefs & Lovette, 1999; Tjørve, 2002; Triantis et al., 2006). In reality, these two hypotheses do not constitute mutually exclusive explanations for curve shape but complementary ones (see also Williamson, 1988; Ricklefs & Lovette, 1999).

4.1.1 Interpretation of Curve Shape

Fallacies regarding how different patterns and processes can explain the shape of the SAR curve are exacerbated by widespread misinterpretations of the actual observed shapes of fitted SAR models. Such incorrect interpretations often result from the visual interpretation of SAR curves that have undergone an axial transformation, either by log-transforming area (log–linear plot) or by log-transforming both area and species richness (log–log plot) (Tjørve, 2003). In the literature, differences between SAR curves are usually discussed by comparing 'slopes'. This so-called slope refers to the z-parameter of the power-law SAR, $S = cA^z$ (where S is number of species, A is area and c and z are fitted parameters), which becomes a straight line in log–log (double logarithmic) space, $\log S = \log c + z \log A$. Here, the z-parameter becomes the slope of the line. Therefore, the z-value is usually said to represent the 'slope of the SAR'. For example, oceanic archipelagos are expected to have 'steeper SARs' (here, ISARs) than mainland areas (saSARs), meaning higher z-values in the former. However, what is then referred to as a 'steeper slope' (for a given c-value) in reality corresponds to a lower, depressed or more linear (less curvilinear) SAR curvature in arithmetic (untransformed) space. This is because the z-parameter controls the rate at which the increase in diversity with area decelerates (along A). Still, to find the actual rate or 'speed' at which the curve decelerates, we need to include the c-parameter. Mathematically, this is evident through the inspection of the first derivative of the SAR as A increases. Thus, for a power-law SAR the 'speed' is its first derivative, expressed as:

$$\frac{d}{dA} cA^z = czA^{z-1}. \qquad (4.1)$$

We see from Equation (4.1) that a higher z-value means that the accumulation of species with A slows down more quickly (at a given c-value). This is because $z-1 < 0$, which causes the second derivative of the SAR to become negative. However, we also notice that what we call the 'speed' not only depends on z but also on the c-value, meaning that both parameters affect the steepness of the curve and how it decelerates with increasing A. This causes the power-law SAR in arithmetic space to be monotonically increasing but decelerating, meaning an upward convex curve.

If it is believed that what is often referred to as a steeper 'slope' (meaning larger z-value) indicates greater species loss per unit of area lost than a shallower 'slope' (smaller z-value), confusion arises when we compare two power-law SARs. Then the above is true only at larger scales. If we plot the two curves in arithmetic space (Figure 4.1) we see

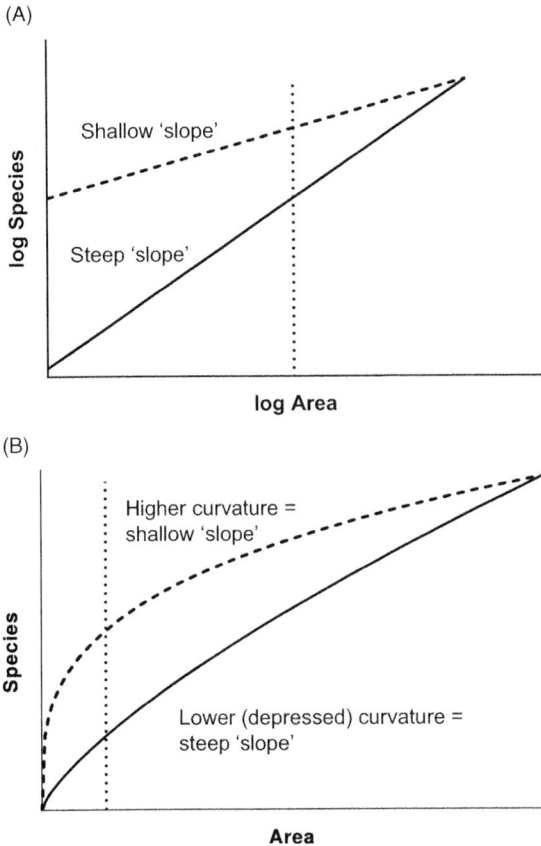

Figure 4.1 The two panels show the same two SARs, the first in log–log space and the second in arithmetic space. The dashed line illustrates a SAR with a shallow 'slope' and the unbroken line a SAR with a steep 'slope', meaning low and high z-values, respectively. For areas larger than the area marked by the vertical dotted line, species accumulate at a faster rate (S/A) along the steep 'slope' (high z-value) compared to the shallow 'slope' (low z-value). For areas smaller than the area marked by the dotted line, species accumulate faster along the shallow 'slope' compared to the steep 'slope' (for information on how this transition point is calculated; see the solution in Equation 4.2).

that when a certain (and equal) fraction of each area is lost, then further shrinking will cause more species to be lost per unit of area in the SAR with a so-called shallow slope compared to a 'steep' slope. Figure 4.1 shows the scale (vertical dotted line) where species loss per unit of area is the same in both curves, which corresponds to the solution of:

$$\frac{d}{dA} c_1 A^{z_1} = \frac{d}{dA} c_2 A^{z_2}, \text{which is } A = \sqrt[z_2-z_1]{(c_1 z_1)/(c_2 z_2)}. \qquad (4.2)$$

This and similar observations may go unnoticed if we only discuss SAR curves in log–log space. The same holds for log–linear space, where only area (A) is log-transformed. Ultimately, this means that it becomes difficult to interpret or forecast the effects of various factors if we rely solely on interpreting SAR curves in transformed space.

An additional issue with log-transforming species numbers (S) is that zeros have to be excluded, unless they are given some value, for example $S = 0.1$ (which gives $\log_{10}(S) = -1$); both approaches may introduce bias. Log-transforming species richness before fitting a regression model to the data may also cause statistical problems, such as the spurious detection of small-island effects when fitting breakpoint regressions to ISAR data (Burns et al., 2009).

The scale window studied, meaning the range from the finest to the coarsest scales surveyed, may also affect the resulting SAR shape. If an ISAR dataset from a narrow scale window displays a convex upward shaped curve (in arithmetic space), an ISAR from a wider scale window may instead display a sigmoid curve (Tjørve & Turner, 2009; but see also Chapter 7). However, sample-area SARs (saSARs) should, as a rule, emerge as convex curves in arithmetic space (Chapter 7).

4.2 Two Types of SARs

There are two basic or generic types of SARs, namely (as mentioned previously) the saSAR and the ISAR. Preston (1962) made this fundamental distinction over half a century ago, distinguishing between sample areas and isolates. The latter includes oceanic islands, lakes, mountain tops ('sky islands'; Brown, 1971) and other types of isolates resulting from habitat loss and fragmentation. Any discussion of the drivers of SARs needs to consider how these drivers may vary across these different types of SAR.

4.2.1 Convex and Sigmoid Shapes

In the discussion about SAR shape, an important question has been whether ISARs are inherently sigmoid rather than strictly decelerating (convex upward; the second derivative is negative for all values of A) shapes (see e.g. Williams, 1995, 1996; He & Legendre, 1996; He et al., 1996; Scheiner et al., 2000; Lomolino, 2001; Tjørve, 2003; Chapter 7).

Species are not expected to be able to sustain a viable population as isolated areas become smaller because of minimum-area requirements and/ or resource restrictions. Such restraining factors set a minimum isolate area in which a species can sustain itself. This minimum-area effect (MAE) is expected to lower the curvature of an isolate curve (ISAR) so that the shape becomes sigmoid (Turner & Tjørve, 2005; Tjørve & Turner, 2009; Chapter 7). Moreover, if the ISAR curve captures the expected number of species, this must result from the arithmetic summation of probabilities of occurrence for each species across all islands studied. Diamond (1975) refers to such single-species probability–area relationships as incidence functions and holds that they are sigmoid, determined by colonization and extinction, reflecting the limit of island size where the species can survive. Though the shape of the incidence curve varies between species, their summation logically results in a j-shaped part at the lower end (finest scales) of the ISAR (see Chapter 7 for a more detailed discussion about sigmoid SAR models).

The realization that there are different types of SARs, broadly divided into ISARs and saSARs (Figure 4.2), is key to understanding why SARs can have different shapes (Lomolino, 2001; Tjørve & Turner, 2009). The saSAR is expected to be inherently convex upward in arithmetic space, whereas the ISAR is expected to be inherently sigmoid. The fact that most ISAR meta-analyses have found the power model to provide the best fit to the most datasets (e.g. Triantis et al., 2012) is due, we would argue, to the fact that most ISAR datasets do not include a wide enough range of island areas, particularly smaller islands.

4.2.2 Factors Shaping saSARs and ISARs

The distinction between saSARs and ISARs is also important when we discuss the factors underpinning SARs, as some patterns and/or processes are only relevant for a particular SAR type. For example, equilibrium island biogeography theory and metapopulation dynamics are only relevant with respect to isolate systems (MacArthur &

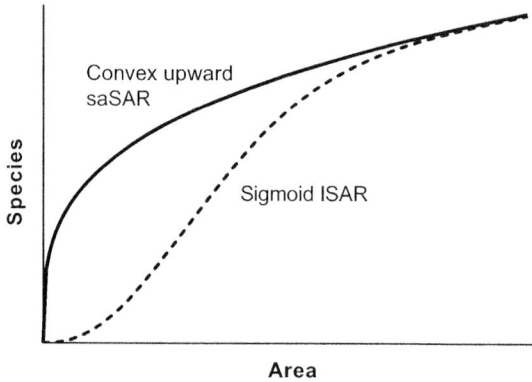

Figure 4.2 Illustration of the expected difference between the curve shapes of sample-area species–area relationships (saSARs) and isolate (or island) species–area relationships (ISARs). Oceanic islands and other types of isolates typically have fewer species than sample areas of the same size (because of minimum-area effects), especially at fine scales. This should cause the expected shape of the ISAR to become sigmoid in arithmetic space, at least when very small islands (or isolates) are included. The saSAR, however, should remain an increasing but decelerating (convex upward) relationship at all scales.

Wilson, 1963, 1967; Hanski & Gyllenberg, 1997; Matter et al., 2002; Ovaskainen & Hanski, 2003). Numerous other biological mechanisms and theories have been proposed to explain features of ISARs that are not as relevant when discussing saSARs (see Shmida & Wilson, 1985; Rosenzweig, 1995; Turner & Tjørve, 2005; Whittaker & Fernandéz-Palacios, 2007, pp. 87–88), such as species minimum-area requirements.

The completeness of sampling affects the shape of the SAR, but the way in which this occurs is also something that differs between saSARs and ISARs (Cam et al., 2002; Turner & Tjørve, 2005; Kolasa et al., 2012). In particular, to understand saSARs we need to consider the sampling design and how the curve is generated from the samples. For example, the saSAR can result from randomly placed sample areas, which are then accumulated in random order or from a sample (cell) matrix of bordering sampling squares, which are in turn accumulated in a nested manner. The curve from these two procedures will differ given that the similarity between sample areas decays with the distance between them. If the saSAR is generated by accumulating species from sample areas added in a random order, the curve will have a higher curvature than if the sample areas are added in a nested

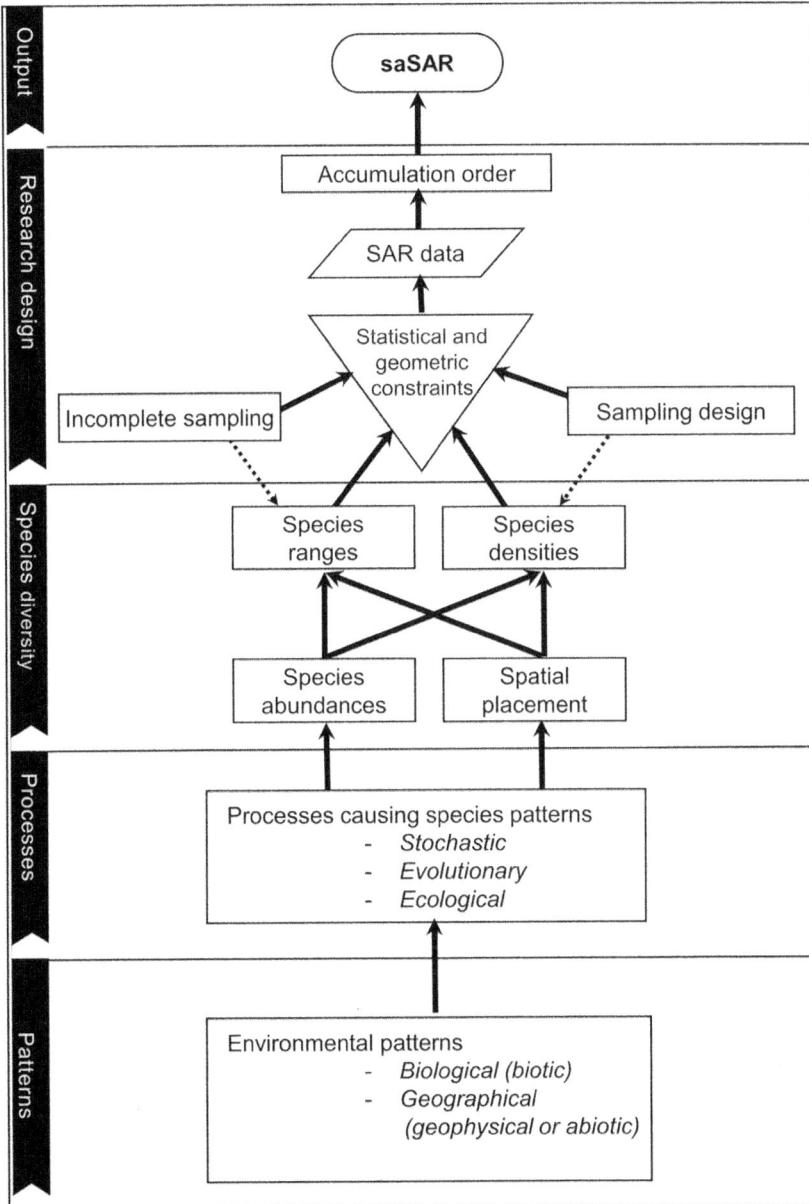

Figure 4.3 Hierarchy of factors (patterns and processes – 'rectangles') and some relationships (connections – 'arrows') affecting the shape of the saSAR curve. Not only does the sampling design affect the resulting data through statistical and/or

order (Palmer et al., 2002; Barnett & Stohlgren, 2003). The plot shape will also affect the curve shape. For example, rectangular sample areas (plots) typically result in a higher curvature than quadrats (Keeley & Fotheringham, 2005).

Based on the above discussion, we may arrange SAR determinants into a hierarchy of patterns and processes (mechanisms), as indicated in Figure 4.3. At the top of the hierarchy is 1) research design; below it 2) species diversity spatial patterns and 3) processes causing species diversity patterns and at the bottom 4) underlying environmental patterns. Each of these is now discussed in turn.

4.3 Research Design

We are not going to go into detail regarding the effect of research design on SAR shape, as this is treated in more detail elsewhere, including in Section 7.1. However, we note that research design includes: 1) the sampling design, including what are considered spatially explicit sample areas (i.e. the sampling unit) and 2) the accumulation order (e.g. random), which is the order in which we examine the sample areas (or sampling units) for new species (not previously encountered) and add them to the saSAR. The research design is usually of less interest in ISARs, as these are generated simply by comparing the number of species on islands of different sizes. In saSARs, both sampling design and the order in which the diversity of each sample area is added together affects the curve shape, though present SAR typologies only consider the sampling scheme (see Scheiner, 2003; Gray et al., 2004; Dengler, 2008, 2009; Scheiner et al., 2011).

The size of the sampling unit (or smallest sample area) is also part of the sampling design and this size affects, for example, the measured species density (Figure 4.3). Moreover, the sampling unit size also combines with

---

Figure 4.3 (cont.) geometric constraints, but sampling unit size (as part of the sampling design) also directly affects the measured value of species density (the larger the sampling unit, the fewer recorded species per unit of area (S/A)). Species patterns (patterns of species diversity) are here represented by species ranges, species densities, species abundances and the spatial distribution of individuals. Environmental patterns include biological patterns (biotic factors), such as the presence of plant cover, predators and prey and parasites and zoonoses, as well as geographical patterns (abiotic factors), such as weather, soils and landforms.

geometric constraints to influence the saSAR curve (see e.g. Colwell & Lees, 2000; Šizling et al., 2017). The effects of geometric constraints on the saSAR (see Chapter 8) differ between scales.

4.4 Species Spatial Patterns

The saSAR can be explained and, thus, derived mathematically from only two patterns: 1) the density or abundance of individuals of each species, that is, the species abundance distribution (SAD) and 2) how the individuals of each species are distributed in space (He & Legendre, 1996, 2002). The latter can be calculated using the occupancies in a sampling-cell matrix (Tjørve et al., 2008). Together these patterns can also provide mechanistic explanations for the shape of saSARs.

Some claim we (in addition to the SAD and the spatial distribution of individuals) also need to know the total number of species and/or species densities at different scales to construct the saSAR (e.g. He & Legendre, 2002; Tjørve et al., 2008; McGill, 2010, 2011; Powell, 2013). However, if we know the SAD of the total expanse studied rather than just a sample, we of course have the total number of species. Equally, if we know the spatial distribution of individuals, we can deduce the species density at any scale. However, if we do not know the size of the area sampled, then we need to know the species density.

There are several ways to show that species densities directly affect the saSAR (e.g. Hurlbert, 2004; Harte et al., 2009; Šizling et al., 2011) and that the spatial distribution of individuals contributes to the shape of the curve. In most cases the individuals of species are aggregated, although they can sometimes be evenly distributed, especially towards very fine scales, where ultimately the size of individuals will determine the spatial pattern (e.g. Hopkins, 1955; Pielou, 1977; Condit et al., 1996; Plotkin et al., 2000; Tjørve et al., 2008). In reality, there are very few datasets with information about the placement of single individuals or with complete SADs or SADs for the same system at different scales. Thus, we are usually restricted to work with information about species occupancies in a cell grid (sample-area matrix) or in randomly distributed sample areas. Still, the occupancy distribution and the spatial aggregation of occupancies in a sample-area matrix may be used as substitutes to provide proximate explanations for the saSAR curve.

4.4.1 Patterns of Random Placement

The random placement hypothesis (Arrhenius, 1920, 1921; Romell, 1920, 1930; Coleman, 1981; see Chapter 2) assumes a random distribution of individuals in space rather than aggregated patterns. If individuals were randomly distributed, we would only have to consider the SAD in the discussion of how saSAR curve shape differs between datasets. Svedberg (1922) and Kylin (1923) (see also Chapter 2) were the first to make the observation that the random-placement model returns a SAR that approaches the negative-exponential model (Tjørve, 2003) regardless of the observed SAD (Tjørve et al., 2018).

Several studies have predicted the random distribution of individuals in space (i.e. providing support for the random placement hypothesis; e.g. Arrhenius, 1921; Preston, 1948, 1960; Coleman, 1981; Coleman et al., 1982). However, Romell (1930) realized early on that individuals of a species may be either over-dispersed (more evenly dispersed than random) or under-dispersed (aggregated) at different scales and that the effect of dispersal on saSAR shape would therefore 'vary with the size of the sample areas'. As mentioned above, numerous studies have since shown that individuals of a species tend to be aggregated. As such, the saSAR shape resulting from random placement may simply be seen as a null model, which deviates more or less from our observed saSAR. Using this approach, He and Legendre (2002) derived theoretically how spatial distributions of individuals deviating from random placement (i.e. aggregation) affect the shape of the saSAR curve; Tjørve et al. (2008) subsequently demonstrated this effect of aggregation on saSAR curves using empirical saSARs.

4.4.2 Linking SADs to SARs

There is a long-standing tradition of discussing the relationship between SADs and SARs (e.g. May, 1975; Pielou, 1975; Gray, 1986; Magurran, 1988; Tokeshi, 1993). Fisher et al. (1943) and Williams (1943) are often cited as the first to try and connect SADs and SARs, by linking logarithmic SARs (sometimes called exponential SARs) to log-series SADs. Subsequently, May (1975) and later Engen (1977), as well as other authors, discussed how logarithmic SARs result from Fisher's log-series SAD model (see also Hubbell, 2001; He & Legendre, 2002; McGill & Collins, 2003). May (1975) also provided the mathematical form of the SAR resulting from MacArthur's (1957) broken-stick SAD model. Preston (1960, 1962) proposed that a power-law

SAR would result from a lognormal SAD and, over the years, most attempts have linked SADs to power-law SARs (Harte et al., 1999; Martín & Goldenfeld, 2006; Pueyo, 2006; Irie & Tokita, 2012). However, almost all of these attempts were undertaken under the assumption of random placement. Recently, it has been acknowledged that the SAD is not the sole master pattern underpinning saSARs and that the power-law saSAR cannot be predicted with any confidence without considering patterns regarding the spatial distribution of individuals (see Solow & Smith, 1991; Plotkin et al., 2000; He & Legendre, 2002; Green & Ostling, 2003; Olszewski, 2004; Picard et al., 2004; Martín & Goldenfeld, 2006; Green & Plotkin, 2007).

A mechanistic explanation for the shape of the SAR is offered by the maximum entropy theory of ecology (METE; Harte, 2011; Chapter 10), where a link between SADs and SARs is used to derive SARs from a first principle predicting energy flow through an ecosystem. Šizling et al. (2011) discussed the limitations of this method and demonstrated that similar, but less restricted, results can be derived from considering geometric constraints or autocorrelation. Importantly, realistic SADs, including their spatial scaling properties, can also be derived from the local properties of spatial autocorrelation (Šizling et al., 2009a, b; Kůrka et al., 2010). No other theory has been able to explain the spatial scaling of the SAD, which suggests that the assumed link between the SAR and SAD is an artificial mathematical construct that does not capture the real mechanisms acting in ecosystems. It also suggests that both of these large-scale patterns (SARs and SADs) are generated simultaneously by local-scale processes.

4.4.3 Partitioning the Effects of SADs and the Spatial Distribution of Individuals on the saSAR

By comparing the observed saSAR curve to that which would have resulted from even abundances and random placement of individuals in space as a null model, we can assess the extent to which uneven abundances and the spatial aggregation of individuals contribute to curve shape. Several authors have demonstrated the effect of aggregation relative to random placement and how this effect varies in relation to common and rare species (He & Legendre, 2002; Tjørve et al., 2008; He & Hubbell, 2013). For example, Tjørve et al. (2008), using observed occupancies in a cell matrix, illustrated how the curve shape of empirical SARs reflects the distribution of species occupancies and spatial

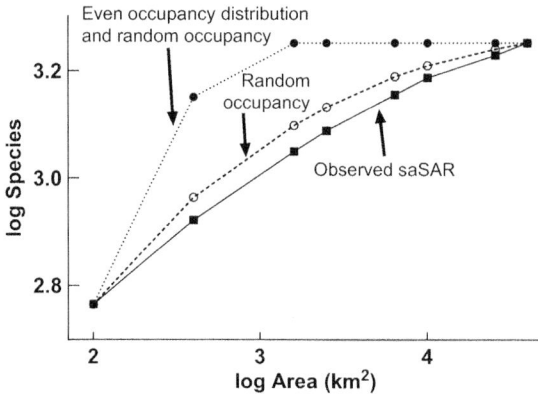

Figure 4.4 saSAR generated from the number of plant species in 10×10 km quadrat cells (in a 200×200 km matrix covering central England and touching Wales), replotted from Tjørve et al. (2008). Data from the *New atlas of the British & Irish flora* (Preston et al., 2002). Number of species is calculated for eight scales ($A = 1$, 4, 16, 25, 64, 100, 256 and 400 quadrats) of nested quadrats. The observed number of species (saSAR) is shown as the unbroken line and filled squares. The other two relationships represent models where each occupancy of a species is placed randomly (random occupancy) in quadrats within the grid but with equal (dotted line, filled circles) and observed (dashed line, open circles) numbers of occupied quadrats for each of the species.

patterning of individuals. In particular, they showed why aggregation patterns depress SARs relatively more at finer scales and when abundances are uneven, and that this effect is strongest for rare species (see Figure 4.4). He and Legendre (2002) and He and Hubbell (2013) published similar findings using simulations.

Picard et al. (2004) modelled the SAR based on the positions of tree and shrub individuals in a 50×100 m plot of tree savannah in Mali. Comparing the impact of uneven occupancy (instead of abundance) distributions and aggregation patterns, they reported that the spatial distributions of individuals modified the shape of the species–area curve as much as the SAD. However, Tjørve et al. (2008) and Qiao et al. (2012), who also partitioned the effects of aggregation and commonness, found that the effect of uneven occurrence distributions and SADs was greater than that of the spatial aggregation of individuals. Figure 4.4 compares the nested saSAR curve generated from a cell matrix with that of the curve where the sample units (cells) are aggregated randomly (equalling random placement of occupancies), to separate out the effect of non-random spatial distributions, at the scale of the sampling units. The random occupancy saSAR can be expressed using a hypergeometric distribution:

$$S_A = S_k - \sum_{i=1}^{S_k} \frac{\binom{k-A}{a_i}}{\binom{k}{a_i}}, \tag{4.3}$$

where S_k is the number of species in the total area studied, k, S_A the number of species in A units of area and a_i is the number of units where the i-th species is observed ($i = 1, \ldots, S_k$). The observed saSAR curve and the curve representing each occupancy of a species being placed randomly (random occupancy) in quadrats within the grid (Equation 4.3) are then compared to a curve resulting from random occupancy, but with an even frequency distribution of occupancies (i.e. every species is present in the same number of sample cells). This curve, where all species occupy the same amount of area (all occupancies a_i are equal to a common occupancy a), is a special case of Equation (4.3) and can be expressed as:

$$S_A = S_k \left[1 - \frac{\binom{k-A}{a}}{\binom{k}{a}} \right]. \tag{4.4}$$

Such assessments of the effect of both uneven abundances and aggregation on saSAR shape provide a useful first step when choosing a regression model(s) for analysis of the saSAR. As discussed in more detail in Chapter 7, it is essential that one has an understanding of what kind of curve shape to expect before selecting a model to fit to the data. This is particularly important given that the power model, the most widely used model in SAR studies, is unlikely to be the 'best' or most useful model in all cases.

4.5 Processes and Environmental Patterns Affecting SARs

A list of the processes that are expected to influence SARs is provided in Table 4.1. We have systemized the processes driving SARs into stochastic, evolutionary, ecological and geographical, where the latter two make up environmental processes (Table 4.1). Certain processes may only be relevant to saSARs rather than ISARs and vice versa (see Table 4.1).

Table 4.1 *List of processes and environmental patterns that are expected to affect the SAR curve. Information on whether a pattern/process affects saSARs or ISARs (or both) is provided. The patterns and processes listed here often overlap to some extent and likely act in combination*

Category	Specific pattern or process	Sample areas or isolates
Stochastic processes	Random placement	Both
	Random dispersal (immigration)	Isolates
	Random population fluctuations	Isolates
	Random disturbance event	Isolates
	Random extinction	Isolates
Evolutionary processes (causing speciation)	Evolutionary independence	Both
	Geographic isolation	Both (isolates more)
	Niche vacancy (from immigration and extinction)	Both (isolates more)
Ecological factors	Species interactions	Both
	Dispersal	Both
	Habitat diversity (heterogeneity)	Both
	Species traits (life histories)	Both
	Niche relations	Both
	Intraspecific interactions (*predation, competition, mutualism, commensalism, parasitism*)	Both
		Both
		Isolates
	Interspecific interactions (*competition, cooperation, dispersal, population dynamics and more*)	Isolates
	Immigration and extinction (*as in island biogeography theory*)	
	Minimum-area effects	
Biological patterns	Productivity	Both
	Vegetation and land cover	Both
	Available energy in ecosystem	Both
Geographical (geophysical) patterns	Topography	Both
	Soil	Both
	Climate	Both
	Ambient energy	Both
	Water availability	Both
	Natural catastrophes	Both
Other	Matrix quality (both ecological and geographical factors)	Isolates (specifically habitat islands)

Most SAR theories and SAR models are phenomenological, though mechanistic ones do exist, such as MacArthur and Wilson's (1963, 1967) equilibrium theory of island biogeography and Levins' (1969, 1970) ideas on metapopulation dynamics (Hanski, 1994; Hastings & Harrison, 1994; see Hanski, 1999, for a broad review). These theories all relate to ISARs. Mechanistic modelling of saSARs includes random-placement approaches (Arrhenius, 1920, 1921; Romell, 1920; Coleman, 1981; Storch et al., 2003; Chapter 2), Hubbell's (2001; see also Chapter 11) neutral theory of biodiversity, fractal based models (Harte et al., 1999; Šizling & Storch, 2004; but see also appendix S1 in Kunin et al., 2018), a model based on energetic constraints (METE) (Harte et al., 2009; Chapter 10) and approaches based on geometric principles (Šizling & Storch, 2004; Šizling et al., 2011; Chapter 8).

Key processes driving ISAR shape include immigration, colonization, population variation, disturbance, extinction and speciation (Table 4.1), and these processes are related to environmental patterns such as isolation (geographic distance and other barriers including matrix quality) and target (island) size; both of which have been argued to affect immigration and extinction rates (see e.g. MacArthur & Wilson, 1963, 1967; Gilpin & Diamond, 1976; Ricklefs & Lovette, 1999). Moreover, ISARs generated from habitat patches such as forest fragments are affected by the surrounding matrix, whether natural or impacted by anthropogenic activities (see Halley et al., 2014; Freeman et al., 2018). A more inhospitable matrix is expected to result in a depressed SAR curve compared to an ISAR of habitat fragments imbedded within a natural matrix or to a saSAR curve constructed from plots in a continuous habitat (see e.g. Halley et al., 2014).

Species loss from habitat loss and fragmentation is expected to be greatest in the smallest fragments or isolates (e.g. Pardini et al., 2010; Öckinger et al., 2012; Haddad et al., 2015). One may even envision a situation where, following habitat loss and/or fragmentation, an ecologically supportive/permeable matrix (e.g. plantation forest surrounding native forest) increases the number of species in the remaining habitat, causing the SAR curvature to rise (Figure 4.5). However, these effects will differ between taxa. For example, Freeman et al. (2018) studied birds in forest fragments. They reported ISARs that are more depressed for forest specialists than for generalists, in forest fragments surrounded by an anthropogenic matrix.

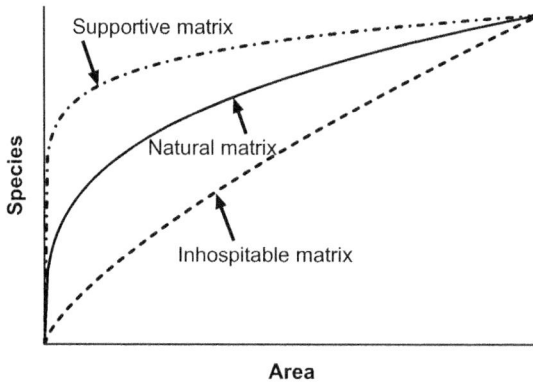

Figure 4.5 The expected effect of matrix quality on the SAR. The solid line represents a SAR constructed from patches within a natural matrix, whereas the dashed line represents a SAR from patches within an inhospitable anthropogenic matrix, such as urban land use. The dash-dotted line represents the SAR from patches within a supportive/more permeable anthropogenic matrix, such as plantation forest. The curves can also be seen as an illustration of the continuum between ISARs and saSARs (where a natural matrix represents a sample-area situation, whereas an inhospitable matrix represents an isolate situation).

4.5.1 Stochastic Processes

Stochastic processes comprise random dispersal (and immigration), random placement, population fluctuations, disturbance events and stochastic extinction, as well as processes where population fluctuations occur together with stochastic extinction and ecological drift. Hubbell (2001) argued, as part of his neutral theory, that random immigration and ecological drift are important processes contributing to differences in species composition between islands (or isolates). If this argument holds, one should, therefore, expect insular or isolate communities, and thereby ISARs, to depend more on stochastic processes than mainland saSARs, where extinction events are rarer and ecological drift weaker. Even if all of the assumptions of neutral theory do not hold (e.g. Ricklefs, 2006), it is still likely that ISARs are more affected by stochastic processes. One benefit of the debate surrounding neutral theory has been the greater focus placed on stochastic factors in ecology and biogeography (Rosindell et al., 2012; Chapter 11).

Stochastic extinction is a key determinant of species diversity on true islands and other types of isolates. Stochastic extinction occurs because there is always a probability a species (within an isolated area) can go extinct during a certain time frame, either due to chance events and/or by random (or cyclic) fluctuations in population size. Chance events include pandemics and natural catastrophes such as tropical storms, floods, tsunamis, draughts, fires and volcanic eruptions. Extinction on islands (and other isolates) is a function of population size and therefore has a greater probability of occurring on smaller islands. However, it is not only MacArthur and Wilson's (1963, 1967) equilibrium theory of island biogeography that is based on stochastic immigration and extinction; for example, metacommunity models are also based on similar types of dynamics.

4.5.2 Evolutionary Processes

Speciation is the main evolutionary process affecting SARs (Rosenzweig, 1995; Pigolotti & Cencini, 2009). It is governed by evolutionary independence (see Preston, 1960; Rosenzweig, 1995), which is facilitated by restricted gene flow, and is enhanced by genetic drift, mutation and natural selection. Speciation is expected to become increasingly important to the ISAR with increasing island (or isolate) area and, above a threshold size, speciation should become more important than immigration as a source of new species (Losos & Schluter, 2000). Speciation may also act to obscure the ISAR (or at least create noise) of an archipelago, particularly if it occurs in an area-independent manner (Losos & Parent, 2009); for example, the small Pacific island of Rapa supports sixty-seven species of *Miocalles* weevils, presumably descended from a single colonist (see Losos & Parent, 2009). We note that extinction rates and extinction events together with ecological drift, isolation and niche vacancies on islands (e.g. caused by island emergence, isolation and extinctions) affect the speciation process. While SARs are also controlled by species life histories and traits (see e.g. Franzén et al., 2012), these biological characteristics are ultimately the result of evolutionary processes.

4.5.3 Ecological Processes

In addition to stochastic and evolutionary processes, we expect species diversity patterns to be driven by fundamental ecological processes such as competition, predation, herbivory, symbiosis, mutualism, commensalism and succession (Begon et al., 1990). However, these processes will affect the saSAR only if they alter the probabilities of occupancy across scales, which means the total area of each species' spatial range has to change. If the ecological processes only reshape species' ranges, while the sizes remain the same, the saSAR will not change. Ecological processes may be grouped into categories such as environmental filtering, source–sink dynamics, landscape complementation/supplementation and the neighbourhood effect (e.g. Pulliam, 1988; Dunning et al., 1992; Kraft et al., 2014). Compared to the literature on stochastic processes, studies outlining or discussing the effect of ecological (and indeed evolutionary) processes on SAR shape are scarce.

4.5.4 Environmental Patterns

If species spatial patterns, meaning both species abundances and how individuals are distributed in space, can be argued to shape saSARs through stochastic, evolutionary and ecological processes, then we should also discuss how these processes are affected by underlying environmental patterns. Such patterns may be divided into biological (biotic) and geographical (geophysical or abiotic). Biotic patterns include productivity, vegetation (or land) cover and available energy in the ecosystem. Key geographical patterns are topography, soil and climate, which control, for example, ambient energy and water availability (see Hawkins et al., 2003) and provide space and shelter. Together, geographical and biological environmental patterns determine productivity, including energy availability within the ecosystem, vegetation characteristics and ultimately patterns of habitat diversity. The latter is typically recognized as resulting from the total environmental heterogeneity (see e.g. Stein et al., 2014). The apparent effect of habitat heterogeneity is that it reduces species occupancies, meaning that we get more rare species. Consequently, as both the ISAR and saSAR originate from the summing of occupancies, habitat heterogeneity lowers the curvature of the SAR.

4.6 Linking Proximate and Ultimate Factors

It is interesting to discuss how ultimate (stochastic, evolutionary, ecological and geographical) factors affect the proximate factors such as the spatial aggregation of individuals and SADs, as well as how the latter (aggregation and SADs) determine species occupancies at different scales, which directly affects SARs. In particular, linking the literature on SADs (and the ultimate processes affecting SADs) to that of SARs may greatly improve our understanding of how various factors drive SAR shape. For example, it has been shown that ecological interaction factors such as interspecific competition, as well as disturbance, can cause more even SADs (Connell, 1978), which in turn results in a higher curvature in the saSAR. This has also been shown by He and Legendre (2002), who illustrated how *interspecific* competition can cause higher saSAR curvature caused by even SADs and also how, in contrast, *intraspecific* competition can result in lower saSAR curvature through more uneven SADs.

Moreover, higher saSAR curvature (i.e. more convex upward in arithmetic space) may result not only from more even SADs, but also from a more even distribution of individuals in space. We may regard this as the effect of the SAD on the saSAR being 'modified' by spatial aggregation. Even a more uneven SAD may result in a higher saSAR curvature, if some rare species are overdispersed and/or some common species are very aggregated, producing a more even frequency distribution of occupancies.

We may also consider geographical patterns, such as latitude and altitude, as ultimate (underlying) factors that contribute to explaining saSAR shape through their effect on proximate factors, such as SADs and species spatial distribution (aggregation) patterns. For example, it is typically reported that SADs become more uneven with higher latitude and altitude (Whittaker, 1965; Tolimieri, 2007; Qiao et al., 2012; Ulrich et al., 2016), which will cause a lower saSAR curvature.

Unfortunately, unravelling the effects of underlying (ultimate) factors on SARs may not be as straightforward as simply discussing proximate factors, however. To take one example of why this is so, in colder regions, communities and ecosystems have a tendency to be dominated by just a handful of species (Whittaker, 1965; Currie & Fritz, 1993), which explains the higher saSAR curvature often observed. However, Qiao et al. (2012) report (for tree species) not only more uneven SADs and occupancy distributions, but also higher spatial aggregation of individuals in colder regions (both these patterns are in contrast to what other studies have found). The latter (aggregation) will drive the saSAR towards a lower curvature, whereas the former (uneven SAD) will drive it towards a higher one. Separating the effects of the two in non-experimental data is a complicated task.

Another problem with determining which factors underpin the shape of the SAR is that few, if any, patterns or processes are ruled by a single factor alone. It is likely that the SAR is controlled by several ultimate factors (evolutionary, ecological and geographical). For example, immigration is affected not only by species traits (ecological and evolutionary), but also by isolation (geographical). The quality and type of matrix that is to be crossed is also important. In addition, not only do the processes and patterns causing SARs act in combination, but the relative importance of each factor, proximate or ultimate, will differ between scales and between species. Below a minimum island (or isolate) size almost all species will face resource restrictions, meaning reduced access to, for example, fresh water, food, shelter and breeding habitat. However, each species has a different specific minimum–area requirement, that is, a different minimum area required in order for a viable population to persist. This in turn affects the shape of the ISAR at different scales, depending on what taxon or species guild is the focus of study.

4.7 Area as a Proxy

Many attempts have been made to disentangle the effect of area on richness from various other ecological, evolutionary and environmental factors using techniques such as multiple regression (e.g. Sfenthourakis, 1996; Welter-Schultes & Williams, 1999; Fattorini, 2002, 2006; Panitsa et al., 2006, 2010; Triantis et al., 2008; Weigelt & Kreft, 2013; Economo et al., 2017). Such approaches are problematic as many factors (e.g. habitat diversity and maximum island altitude) co-vary strongly with area. Area is, in reality, acting as a proxy for a large number of other factors affecting the number of species. Therefore, in many cases it is difficult to separate these indirect effects from the effect of area (McGuinness, 1984; Hart & Horwitz, 1991; Kohn & Walsh, 1994; Rosenzweig, 1995), without the use of experimental approaches. A recent study by Chase et al. (2019) provides a possible framework for disentangling the mechanisms underpinning the ISAR based on the use of rarefaction curves and analysis of beta diversity patterns. Further research is needed to assess the utility of this framework in a variety of systems.

4.8 Conclusions

In this chapter, we have discussed how various factors can affect SARs. The SAR can be seen as an algorithmic expression arising from a combination of factors. To present a systematic account of all driving factors (or determinants) that underpin SARs is close to an impossible task, as such factors may be grouped in different ways into types or classes. Some factors may represent proximate explanations, whereas others represent ultimate ones and many likely without constituting competing explanations for SAR shape. At the same time, each factor, whether process or pattern, will be correlated with other factors. Often a single factor can even be divided into several, or may overlap with, other factors. The matter is complicated further by the fact that the effect of each factor will vary between taxa and systems and on whether saSARs or ISARs are being studied; most factors are also scale dependent and the pattern of scaling will likely differ between ISARs and saSARs. Looking to the future, we hope we will move towards understanding how the combined effect of several patterns and processes cause shapes of SAR curves to differ, rather than putting all our effort into the search for canonical curve shapes or trying to compare (what are perceived as) mutually exclusive theories of SAR shape (e.g. the 'habitat diversity versus area per se' debate). Only then will we be able to develop the SAR into a fully effective tool; not only for descriptive, but also for explicative and predictive purposes.

Acknowledgements

The work was supported by the Czech Science Foundation GA ČR EXPRO 20-29554X board No. EX5.

References

Arrhenius, O. (1920) Yta och arter. I. *Svensk Botanisk Tidsskrift*, **14**, 327–329.

Arrhenius, O. (1921) Species and area. *Journal of Ecology*, **9**, 95–99.

Barnett, D. T. & Stohlgren, T. J. (2003) A nested-intensity design for surveying plant data. *Biodiversity and Conservation*, **12**, 255–278.

Begon, M., Harper, J. L. & Townsend, C. R. (1990) *Ecology: Individuals, populations and communities*, 2nd ed. Boston, MD: Blackwell Scientific Publications.

Brown, J. H. (1971) Mammals on mountaintops: Nonequilibrium insular biogeography. *The American Naturalist*, **105**, 467–478.

Brown, J. H., Gillooly, J. F., West, G. B. & Savage, V. M. (2003) The next step in macroecology: From general empirical patterns to universal ecological laws. *Macroecology: Concepts and consequences* (ed. by T. M. Blackburn and K. J. Gaston), pp. 408–423. Malden, MA: Blackwell.

Burns, K. C., McHardy, P. & Pledger, S. (2009) The small-island effect: Fact or artefact? *Ecography*, **32**, 269–276.

Cam, E., Nichols, J. D., Hines, J. E., Sauer, J. R., Alpizar-Jara, R. & Flather, C. H. (2002) Disentangling sampling and ecological explanations underlying species–area relationships. *Ecology*, **83**, 1118–1130.

Chase, J. M., Gooriah, L., May, F., Ryberg, W. A., Schuler, M. S., Craven, D. & Knight, T. M. (2019) A framework for disentangling ecological mechanisms underlying the island species–area relationship. *Frontiers of Biogeography*, **11**, e40844.

Coleman, B. (1981) On random placement and species–area relations. *Mathematical Biosciences*, **54**, 191–215.

Coleman, B. D., Mares, M. A., Willig, M. R. & Hsieh, Y.-H. (1982) Randomness, area and species richness. *Ecology*, **64**, 1121–1133.

Colwell, R. K. & Lees, D. C. (2000) The mid-domain effect: Geometric constraints on the geography of species richness. *Trends in Ecology & Evolution*, **15**, 70–76.

Condit, R., Hubbell, S. P., LaFrankie, J. V., Sukumar, R., Monokaran, N., Foster, R. B. & Ashton, P. S. (1996) Species–area and species–individual relationships for tropical trees: A comparison of three 50-ha plots. *Journal of Ecology*, **84**, 549–562.

Connell, J. H. (1978) Diversity in tropical rainforests and coral reefs. *Science*, **199**, 1302–1310.

Connor, E. F. & McCoy, E. D. (1979) The statistics and biology of the species–area relationship. *The American Naturalist*, **113**, 791–833.

Connor, E. F. & McCoy, E. D. (2001) Species–area relationships. *Encyclopedia of biodiversity*, vol. 5 (ed. by S. A. Levin), pp. 397–411. San Diego, CA: Academic Press.

Currie, D. J. & Fritz, J. T. (1993) Global patterns of animal abundance and species energy use. *Oikos*, **67**, 56–68.

Dengler, J. (2008) Sampling-design effects on properties of species–area relation-ships – A case study from Estonian dry grassland communities. *Folia Geobotanica*, **43**, 289–304.

Dengler, J. (2009) Which function describes the species–area relationship best? A review and empirical evaluation. *Journal of Biogeography*, **36**, 728–744.

Diamond, J. M. (1975) The island dilemma: Lessons of modern biogeographic studies for the design of natural reserves. *Biological Conservation*, **7**, 129–146.

Dunning, J. B., Danielson, B. J. & Pulliam, H. R. (1992) Ecological processes that affect populations in complex landscapes. *Oikos*, **65**, 169–175.

Economo, E. P., Janda, M., Guénard, B. & Sarnat, E. (2017) Assembling a species–area curve through colonization, speciation and human-mediated introduction. *Journal of Biogeography*, **44**, 1088–1097.

Engen, S. (1977) Exponential and logarithmic species–area curves. *The America Naturalist*, **111**, 591–594.

Fattorini, S. (2002) Biogeography of the tenebrionid beetles (Coleoptera, Tenebrionidae) on the Aegean Islands (Greece). *Journal of Biogeography*, **29**, 49–67.

Fattorini, S. (2006) Spatial patterns of diversity in the tenebrionid beetles (Coleoptera, Tenebrionidae) of the Aegean Islands (Greece). *Evolutionary Ecology Research*, **8**, 237–263.

Fisher, R. A., Corbet, A. S. & Williams, C. B. (1943) The relation between the number of species and the number of individuals in a random sample of an animal population. *Journal of Animal Ecology*, **12**, 42–58.

Franzén, M., Schweiger, O. & Betzholtz, P.-E. (2012) Species–area relationships are controlled by species traits. *PLoS One*, **7**, e37359.

Freeman, M. T., Oliver, P. I. & van Aarde, R. J. (2018) Matrix transformation alters species–area relationships in fragmented coastal forests. *Landscape Ecology*, **33**, 307–322.

Gilpin, M. E. & Diamond, J. M. (1976) Calculation of immigration and extinction curves from the species–area–distance relation. *Proceedings of the National Academy of Sciences USA*, **73**, 4130–4134.

Gray, J. S. (1986) Species-abundance patterns. *Organization of communities past and present, the 27th symposium of the British Ecological Society, Aberystwyth* (ed. by J. H. R. Gee and P. S. Giller), pp. 53–67. Oxford: Blackwell Science.

Gray, J. S., Ugland, K. I. & Lambshead, J. (2004) On species accumulation and species–area curves. *Global Ecology & Biogeography*, **13**, 567–568.

Green, J. L. & Ostling, A. (2003) Endemics–area relationships: The influence of species dominance and spatial aggregation. *Ecology*, **84**, 3090–3097.

Green, J. L. & Plotkin, J. B. (2007) A statistical theory for sampling species abundances. *Ecology Letters*, **10**, 1037–1045.

Haddad, N. M., Bruvig, L. A., Clobert, J., Davies, K. F., Gonzales, A., Holt, R. D., Lovejoy, T. E., Sexton, J. O., Austin, M. P., Collins, C. D., Cook, W. M., Damschen, E. I., Ewers, R. M., Foster, B. L., Jenkins, C. N., King, A. J., Laurance, W. F., Levey, D. J., Margules, C. R., Melbourne, B. A., Nicholls, A. O., Orrock, J. L., Song, D.-X. & Townshend, J. R. (2015) Habitat fragmentation and its lasting impact on Earth's ecosystems. *Science Advances*, **1**, e1500052.

Halley, J. M., Sgardeli, V. & Triantis, K. A. (2014) Extinction debt and the species–area relationship: A neutral perspective. *Global Ecology & Biogeography*, **23**, 113–123.

Hanski, I. (1994) A practical model of metapopulation dynamics. *Journal of Animal Ecology*, **63**, 151–162.

Hanski, I. (1999) *Metapopulation ecology*. Oxford: Oxford University Press.

Hanski, I. & Gyllenberg, M. (1997) Uniting two general patterns in the distribution of species. *Science*, **275**, 397–400.

Hart, D. D. & Horwitz, R. J. (1991) Habitat diversity and the species–area relationship: Alternative models and tests. *Habitat structure, population and community biology series*, vol. **8** (ed. by S. S. Bell, E. D. McCoy and H. R. Mushinsky), pp. 47–68. Dordrecht: Springer.

Harte, J. (2011) *Maximum entropy and ecology: A theory of abundance, distribution, and energetics*. Oxford: Oxford University Press.

Harte, J., Kinzig, A. & Green, J. (1999) Self-similarity in the distribution and abundance of species. *Science*, **284**, 334–336.

Harte, J., Smith, A. B. & Storch, D. (2009) Biodiversity scales from plots to biomes with a universal species–area curve. *Ecology Letters*, **12**, 789–797.

Hastings, A. & Harrison, S. (1994) Metapopulation dynamics and genetics. *Annual Review of Ecology and Systematics*, **25**, 167–188.

Hawkins, B. A., Field, R., Cornell, H. V., Currie, D. J., Guégan, J.-F., Kaufman, D. M., Kerr, J. T., Mittelbach, G. G., Oberdorff, T., O'Brien, E. M., Porter, E. E. & Turner, J. R. G. (2003) Energy, water, and broad-scale geographic patterns of species richness. *Ecology*, **84**, 3105–3117.

He, F. & Hubbell, S. (2013) Estimating extinction from species–area relationships: Why the numbers do not add up. *Ecology*, **94**, 1905–1912.

He, F. & Legendre, P. (1996) On species–area relations. *The American Naturalist*, **148**, 719–737.

He, F. & Legendre, P. (2002) Species diversity patterns derived from species–area models. *Ecology*, **83**, 1185–1198.

He, F., Legendre, P. & LaFrankie, V. (1996) Spatial patterns of diversity in a tropical rain forest of Malaysia. *Journal of Biogeography*, **23**, 57–74.

Hill, J. L., Curran, P. J. & Fookolady, G. M. (1994) The effect of sampling on the species–area curve. *Global Ecology & Biogeography Letters*, **4**, 97–106.

Hopkins, B. (1955) The species–area relations of plant communities. *Journal of Ecology*, **43**, 409–426.

Hubbell, S. P. (2001) *The unified neutral theory of biodiversity and biogeography*. Princeton, NJ: Princeton University Press.

Hurlbert, A. H. (2004) Species–energy relationships and habitat complexity in bird communities. *Ecology Letters*, **7**, 714–720.

Irie, H. & Tokita, K. (2012) Species–area relationship for power-law species abundance distribution. *International Journal of Biomathematics*, **5**, 1260014.

Keeley, J. E. & Fotheringham, C. J. (2005) Plot shape effects on plant species diversity measurements. *Journal of Vegetation Science*, **16**, 249–256.

Kohn, D. D. & Walsh, D. M. (1994) Plant species richness – the effect of island size and habitat diversity. *Journal of Ecology*, **82**, 367–377.

Kolasa, J., Manne, L. L. & Pandit, S. N. (2012) Species–area relationships arise from interaction of habitat heterogeneity and species pool. *Hydrobiologia*, **685**, 135–144.

Kraft, N. J. B., Adler, P. B., Godoy, O., James, E. C., Fuller, S. & Levine, J. M. (2014) Community assembly, coexistence and the environmental filtering metaphor. *Functional Ecology*, **29**, 592–599.

Kunin, W. E., Harte, J., He, F, Hui, C., Jobe, R., Ostling, A., Polce, C., Šizling, A. L., Smith, A. B., Smith, K., Smart, S. M., Storch., D, Tjørve, E., Ugland, K.-I., Ulrich, W. & Varma, V. (2018) Upscaling biodiversity: Estimating the species–area relationship from small samples. *Ecological Monographs*, **88**, 170–187.

Kůrka, P., Šizling, A. L & Rosindell, J. (2010) Analytical evidence for scale-invariance in the shape of species abundance distributions. *Mathematical Biosciences*, **223**, 151–159.

Kylin, H. (1923) Växtsociologiska randanmärkningar. *Botaniska Notiser*, **1923**, 161–234.

Levins, R. (1969) Some genetic and demographic consequences of environmental heterogeneity for biological control. *Bulletin of the Entomological Society of America*, **15**, 237–240.

Levins, R. (1970) Extinction. *Some mathematical questions in biology*, vol. 2 (ed. by M. Gerstenhaber). Providence, RI: American Mathematical Society.

Lomolino, M. V. (2001) The species–area relationship: New challenges for an old pattern. *Progress in Physical Geography*, **25**, 1–21.

Losos, J. B. & Parent, C. E. (2009) The speciation–area relationship. *The theory of island biogeography revisited* (ed. by J. B. Losos and R. E. Ricklefs), pp. 415–438. Princeton, NJ: Princeton University Press.

Losos, J. B. & Schluter, D. (2000) Analysis of an evolutionary species–area relationship. *Nature*, **408**, 847–850.

MacArthur, R. H. (1957) On the relative abundance of bird species. *Proceedings of the National Academy of Sciences USA*, **43**, 293–295.

MacArthur, R. H. & Wilson, E. O. (1963) An equilibrium theory of insular zoogeography. *Evolution*, **17**, 373–387.

MacArthur, R. H. & Wilson, E. O. (1967) *The theory of island biogeography*. Princeton, NJ: Princeton University Press.

Magurran, A. E. (1988) *Ecological diversity and its measurement*. London: Croom Helm.

Martín, H. G. & Goldenfeld, N. (2006) On the origin and robustness of power-law species–area relationships in ecology. *Proceedings of the National Academy of Sciences USA*, **103**, 10310–10315.

Matter, S. F., Hanski, I. & Gyllenberg, M. (2002) A test of a metapopulation model of the species–area relationship. *Journal of Biogeography*, **29**, 977–983.

May, R. M. (1975) Patterns of species abundance and diversity. *Ecology and evolution of communities* (ed. by M. L. Cody and J. M. Diamond), pp. 81–120. Cambridge, MA: Harvard University Press.

McGill, B. J. (2010) Towards a unification of unified theories. *Ecology Letters*, **13**, 627–642.

McGill, B. J. (2011) Species abundance distributions. *Biological diversity: Frontiers in measurement and assessment* (ed. by A. E. Magurran and B. J. McGill), pp. 105–122. Oxford: Oxford University Press.

McGill, B. & Collins, C. (2003) A unified theory for macroecology based on spatial patterns of abundance. *Evolutionary Ecology Research*, **5**, 469–492.

McGuinness, K. A. (1984) Species–area relationships of communities on intertidal boulders: Testing the null hypothesis. *Journal of Biogeography*, **11**, 439–456.

Newmark, W. D. (1986) Species–area relationship and its determinants for mammals in western North American national parks. *Biological Journal of the Linnean Society*, **28**, 83–98.

Nilsson, S. G., Bengtson, J. & Ås, S. (1988) Habitat diversity or area per se? Species richness of woody plants, carabid beetles and land snails on islands. *Journal of Animal Ecology*, **57**, 685–704.

Öckinger, E., Lindborg, R., Sjödin, N. E. & Bommarco, R. (2012) Landscape matrix modifies richness of plants and insects in grassland fragments. *Ecography*, **35**, 259–267.

Olszewski, T. D. (2004) A unified mathematical framework for the measurement of richness and evenness within and among multiple communities. *Oikos*, **104**, 377–387.

Ovaskainen, O. & Hanski, I. (2003) The species–area relationship derived from species-specific incidence functions. *Ecology Letters*, **6**, 903–909.

Palmer, M. W., Earls, P. G., Hoagland, B. W., White, P.S. & Wohlgemuth, T. (2002) Quantitative tools for perfecting species lists. *Environmetrics*, **13**, 121–137.

Panitsa, M., Trigas, P., Iatrou, G. & Sfenthourakis, S. (2010) Factors affecting species richness and endemism on land-bridge islands – an example from the East Aegean archipelago. *Acta Oecologica*, **36**, 431–437.

Panitsa, M., Tzanoudakis, D., Triantis, K. A. & Sfenthourakis, S. (2006) Patterns of species richness on very small islands: The plants of the Aegean archipelago. *Journal of Biogeography*, **33**, 1223–1234.

Pardini, R., Bueno, A. D. A., Gardner, T. A., Prado, P. I. & Metzger, J. P. (2010) Beyond the fragmentation threshold hypothesis: Regime shifts in biodiversity across fragmented landscapes. *PLoS One*, **5**, e13666.

Picard, N., Karambé, M. & Birnbaum, P. (2004) Species–area curve and spatial pattern. *Écoscience*, **11**, 45–54.

Pielou, E. C. (1975) *Ecological diversity*. New York: Wiley-Interscience.

Pielou, E. C. (1977) *Mathematical ecology*. New York: Wiley.

Pigolotti, S. & Cencini, M. (2009) Speciation-rate dependence in species–area relationships. *Journal of Theoretical Biology*, **260**, 83–89.

Plotkin, J. B., Potts, M. D., Leslie, N., Manokaran, N., LaFrankie, J. & Ashton, P. S. (2000) Species–area curves, spatial aggregation, and habitat specialization in tropical forests. *Journal of Theoretical Biology*, **207**, 81–99.

Powell, K. I. (2013) Invasive plants have scale-dependent effects on diversity by altering species–area relationships. *Science*, **339**, 316–318.

Preston, C. D., Pearman, D. A. & Dines, T. D. (2002) *New atlas of the British & Irish flora*. Oxford: Oxford University Press.

Preston, F. W. (1948) The commonness, and rarity, of species. *Ecology*, **29**, 254–283.

Preston, F. W. (1960) Time and space and the variation of species. *Ecology*, **41**, 611–627.

Preston, F. W. (1962) The canonical distribution of commonness and rarity: Part I & II. *Ecology*, **43**, 185–215, 410–432.

Pueyo, S. (2006) Self-similarity in species–area relationship and in species abundance distribution. *Oikos*, **112**, 156–162.

Pulliam, H. R. (1988) Sources, sinks, and population regulation. *The American Naturalist*, **132**, 652–661.

Qian, H., Ricklefs, R. E. & White, P. S. (2005) Beta diversity of angiosperms in temperate floras of eastern Asia and eastern North America. *Ecology Letters*, **8**, 15–22.

Qiao, X., Tang, Z., Wang, S., Liu, Y. & Fang, J. (2012) Effects of community structure on the species–area relationship in China's forests. *Ecography*, **35**, 1117–1123.

Ricklefs, R. E. (2006) The unified neutral theory of biodiversity: Do the numbers add up? *Ecology*, **87**, 1424–1431.

Ricklefs, R. E. & Lovette, I. J. (1999) The roles of island area *per se* and habitat diversity in the species–area relationships of four Lesser Antillean faunal groups. *Journal of Animal Ecology*, **68**, 1142–1160.

Romell, L. G. (1920) Sur la régle de distribution de fréquences. *Svensk Botanisk Tidsskrift*, **14**, 1–20.

Romell, L. G. (1930) Comments on Raunkiær's and similar methods of vegetation analysis and the 'law of frequency'. *Ecology*, **11**, 598–596.

Rosenzweig, M. L. (1995) *Species diversity in space and time*. Cambridge: Cambridge University Press.

Rosindell, J., Hubbell, S. P., He, F., Harmon, L. J. & Etienne, R. S. (2012) The case for ecological neutral theory. *Trends in Ecology & Evolution*, **27**, 203–208.

Scheiner, S. M. (2003) Six types of species–area curves. *Global Ecology & Biogeography*, **12**, 441–447.

Scheiner, S. M., Chiarucci, A., Fox, G. A., Helmus, M. R., McGlinn, D. J. & Willig, M. R. (2011) The underpinnings of the relationship of species richness with space and time. *Ecological Monographs*, **81**, 195–213.

Scheiner, S. M., Cox, S. B., Willig, M., Mittelbach, G. G., Osenberg, C. & Kaspari, M. (2000) Species richness, species–area curves and Simpson's paradox. *Evolutionary Ecology Research*, **2**, 791–802.

Sfenthourakis, S. (1996) The species–area relationship of terrestrial isopods (Isopoda; Oniscidea) from the Aegean archipelago (Greece): A comparative study. *Global Ecology & Biogeography Letters*, **5**, 149–157.

Shmida, A. & Wilson, M. V. (1985) Biological determinants of species diversity. *Journal of Biogeography*, **12**, 1–20.

Šizling, A. L. & Storch, D. (2004) Power-law species–area relationships and self-similar species distributions within finite areas. *Ecology Letters*, **7**, 60–68.

Šizling, A. L., Kunin, W. E., Šizlingová, E., Reif, J. & Storch, D. (2011) Between geometry and biology: The problem of universality of the species–area relationship. *The American Naturalist*, **178**, 602–611.

Šizling, A. L., Šizlingová, E., Tjørve, E., Tjørve, K. M. C. & Kunin, W. E. (2017) How to allow SAR collapse across local and continental scales: A resolution of the controversy between Storch et al. (2012) and Lazarina et al. (2013). *Ecography*, **40**, 971–981.

Šizling, A. L., Storch, D., Reif, J. & Gaston, K. J. (2009b) Invariance in species-abundance distributions. *Theoretical Ecology*, **2**, 89–103.

Šizling, A. L., Storch, D., Šizlingová, E. D., Reif, J. & Gaston, K. J. (2009a) Species abundance distribution results from a spatial analogy of central limit theorem. *Proceedings of the National Academy of Sciences USA*, **106**, 6691–6695.

Solow, A. R. & Smith, W. (1991) Detecting cluster in a heterogeneous community sampled by quadrats. *Biometrics*, **47**, 311–217.

Stein, A., Gerstner, K. & Kreft, H. (2014) Environmental heterogeneity as a universal driver of species richness across taxa, biomes and spatial scales. *Ecology Letters*, **17**, 866–880.

Storch, D., Šizling, A. L. & Gaston, K. J. (2003) Geometry of the species–area relationship in central European birds: Testing the mechanism. *Journal of Animal Ecology*, **72**, 509–519.

Svedberg, T. (1922) Statistisk vegetationsanalys, några synspunkter. *Svensk Botanisk Tidsskrift*, **16**, 197–205.

Tjørve, E. (2002) Habitat size and number in multi-habitat landscapes: A model approach based on species–area curves. *Ecography*, **25**, 17–24.

Tjørve, E. (2003) Shapes and functions of species–area curves: A review of possible models. *Journal of Biogeography*, **30**, 827–835.

Tjørve, E. & Tjørve, K. (2017) Species–area relationship. *eLS (Encyclopedia of Life Sciences Online)*, pp. 1–9. Chichester: John Wiley & Sons.

Tjørve, E. & Turner, W.R. (2009) The importance of samples and isolates for species–area relationships. *Ecography*, **32**, 391–400.

Tjørve, E., Kunin, W. E., Polce, C. & Tjørve, K. M. C. (2008) The species–area relationship: Separating the effects of species-abundance and spatial distribution. *Journal of Ecology*, **96**, 1141–1151.

Tjørve, E., Tjørve, K. M. C., Šizlingová, E. & Šizling, A. L. (2018) Great theories of species diversity in space and why they were forgotten: The beginnings of a spatial ecology and the Nordic early 20th-century botanists. *Journal of Biogeography*, **45**, 530–540.

Tokeshi, M. (1993). Species abundance patterns and community structure. *Advances in Ecological Research*, **24**, 111–186.

Tolimieri, N. (2007) Patterns in species richness, species density, and evenness in groundfish assemblages on the continental slope of the U.S Pacific coast. *Environmental Biology of Fishes*, **78**, 241–256.

Triantis, K. A., Guilhaumon, F. & Whittaker, R. J. (2012) The island species–area relationship: Biology and statistics. *Journal of Biogeography*, **39**, 215–231.

Triantis, K. A., Mylonas, M., Lika, K. & Vardinoyannis, K. (2003) A model for the species–area–habitat relationship. *Journal of Biogeography*, **30**, 19–27.

Triantis, K. A., Sfenthourakis, S. & Mylonas, M. (2008) Biodiversity patterns of terrestrial isopods from two island groups in the Aegean Sea (Greece): Species–area relationship, small island effect and nestedness. *Écoscience*, **15**, 169–181.

Triantis, K. A., Vardinoyannis, K., Tsolaki, E. P., Botsaris, I., Lika, K. & Mylonas, M. (2006) Re-approaching the small island effect. *Journal of Biogeography*, **33**, 914–923.

Turner, W. R. & Tjørve, E. (2005) Scale-dependence in species–area relationships. *Ecography*, **28**, 721–730.

Ulrich, W., Kusumoto, B., Shiono, T. & Kubota, Y. (2016) Climatic and geographic correlates of global forest tree species-abundance distributions and community evenness. *Journal of Vegetation Science*, **27**, 295–305.

Weigelt, P. & Kreft, H. (2013) Quantifying island isolation – insights from global patterns of insular plant species richness. *Ecography*, **36**, 417–429.

Welter-Schultes, F. W. & Williams, M. R. (1999) History, island area and habitat availability determine land snail species richness of Aegean islands. *Journal of Biogeography*, **26**, 239–249.

Whittaker, R. H. (1965) Dominance and diversity in land plant communities. *Science*, **147**, 250–260.

Whittaker, R. J. & Fernandéz-Palacios, J. M. (2007) *Island biogeography: Ecology, evolution, and conservation*, 2nd ed. Oxford: Oxford University Press.

Williams, C. B. (1943) Area and number of species. *Nature*, **152**, 264–267.

Williams, C. B. (1964) *Patterns in the balance of nature and related problems in quantitative ecology*. London: Academic Press.

Williams, M. R. (1995) An extreme-value function model of the species incidence and species–area relations. *Ecology*, **76**, 2607–2616.

Williams, M. R. (1996) Species–area curves: The need to include zeroes. *Global Ecology & Biogeography Letters*, **5**, 91–93.

Williamson, M. (1988) Relationship of species number to area, distance and other variables. *Analytical biogeography* (ed. by A. A. Myers and P. S. Giller), pp. 91–115. London: Chapman and Hall.

5 · *Functional and Phylogenetic Diversity–Area Relationships*

FLORENT MAZEL AND WILFRIED
THUILLER

5.1 Introduction

The species–area relationship (SAR) describes the tendency of species richness (SR) to increase with the sampled area (Rosenzweig, 1995; Chapter 1). SARs have been documented for a wide range of organisms, ecosystems and spatial scales and can therefore be considered as one of the few universal laws of ecology (Lawton, 1999). They have been used to predict extinction from habitat loss (He & Hubbell, 2011; Chapter 14) and to understand the processes underlying community assembly (MacArthur & Wilson, 1967; Rosenzweig, 1995; Hubbell, 2001; Kadmon & Allouche, 2007; Rosindell & Cornell, 2007; Wagner et al., 2014; e.g. Chapters 10, 11 and 15). To measure the relative importance of different assembly processes (e.g. dispersal or environmental filtering) (Vellend, 2010; Leibold & Chase, 2018), quantitative predictions of the SAR are derived from these processes and compared with empirical data (Hubbell, 2001; Rosindell & Cornell, 2007; Shen et al., 2009). The processes yielding the prediction that best matches empirical data are then considered to be the main drivers of community assembly. However, different processes may predict similar SAR patterns (Chase et al., 2005), so that SARs are not always truly informative.

One potential reason for the limited ability of SR patterns to detect community assembly processes is that they rely on species identity only and ignore species' functional traits. Functional traits mediate the relationship between organisms and their biotic and abiotic environment and the study of their geographical or temporal distribution (e.g. functional diversity (FD) patterns) can help unravel community assembly processes (Wright et al., 2005; Violle et al., 2007; Kraft et al., 2008). For example, theory predicts that, under strong competition for multiple resources, even if SR is constant, trait values within a community should be more evenly spaced than a random set of traits from the species pool

(MacArthur & Levins, 1967). This is expected to occur because species with similar trait values are unlikely to coexist since they compete more strongly with each other than with species with dissimilar traits. The additional insights gained from empirical FD patterns compared to SR patterns come from a variety of biological systems, from microbes in the global ocean (Louca et al., 2016), to plants at local scales (de Bello et al., 2013), to vertebrates at large scales (Mazel et al., 2017a).

It is often hypothesized that the importance of different assembly processes varies with spatial scale (Cavender-Bares et al., 2006; Swenson et al., 2006). For example, competition is sometimes thought to dominate at fine spatial scales, where individuals interact, while environmental filtering (the process by which only organisms adapted to a given environment colonize this environment) might be more important at larger spatial scales (e.g. see Leibold & Chase, 2018). Patterns of FD might help to reveal such processes; for example, describing how FD scales with area (i.e. constructing a FD–area relationship, FDAR hereafter) might reveal the spatial scale dependencies of assembly processes (Weiher & Keddy, 1995; Leibold & Chase, 2018). From a conservation perspective, functional traits are related to ecosystem services (Loreau, 2010; Cadotte et al., 2011; Lavorel et al., 2011) and might represent interesting targets for conservation prioritization. As a consequence, developing tools that predict FD loss from habitat loss may be useful. In a manner similar to the use of the SAR to predict species loss from habitat loss (Chapter 14), it has been suggested that the FDAR can be used to predict FD loss from habitat loss (Keil et al., 2015).

In practice, it is challenging to measure FD: we have imperfect knowledge about which, and how many, traits and functions are important in a given context (Carscadden et al., 2017). Many researchers have therefore advocated the use of molecular phylogenies (Figure 5.1) as a surrogate for unmeasured and unmeasurable traits (Webb et al., 2002). Analysing patterns of phylogenetic diversity (PD hereafter) has rapidly generated interest among ecologists as a means to understand community assembly processes (notably after the seminal paper of Webb et al., 2002), but also among conservationists to prioritize species and areas to protect (Faith, 1992). As for FD, studying how PD scales with area, that is, constructing PD–area relationships (PDAR hereafter), might allow us to identify scale-dependent ecological processes and to predict the loss of PD with habitat loss. However, it seems that the initial appeal of using PD in ecology and conservation is losing strength. Indeed, an increasing number of researchers are actually questioning its value and caution its

(A)

(B)

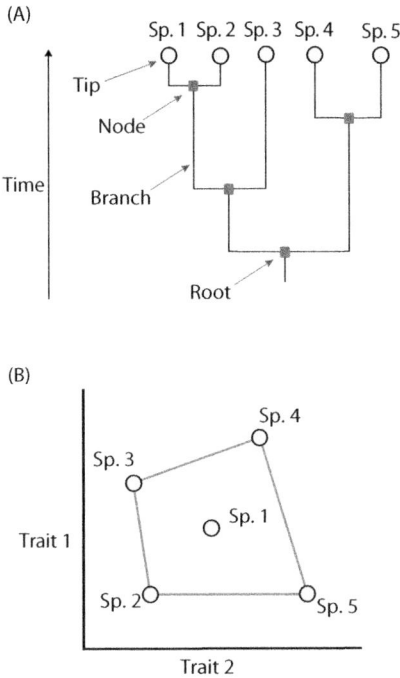

Figure 5.1 Phylogenies and traits. (A) A phylogenetic tree and its constituents: a root, a set of branches, a set of nodes and a set of tips. (B) Species in the functional space defined by two functional traits. The segments that connect species two-to-five define the convex hull volume (see Table 5.1 for details) of the set of five species.

use, whether in conservation (Kelly et al., 2014; Mazel et al., 2017b, 2018, 2019) or in community ecology (Narwani et al., 2013; Gerhold et al., 2015; Venail et al., 2015). This is because the ability of PD to be a good surrogate for the trait of interest (either FD in community ecology or other types of traits of interest in conservation) is – in most cases – not tested or, when it is tested, it sometimes proves to be a poor surrogate for the traits of interest (e.g. Mazel et al., 2018).

In summary, describing and understanding how FD (and perhaps PD) scales with area represents a promising research programme to unravel assembly mechanisms across spatial scales and to predict the loss of FD (and PD) from habitat loss. In this review, our aim is to provide 1) practical recommendations to measure FD and PD and to construct FDARs and PDARs, 2) a theoretical background on how to interpret FDAR and PDAR patterns and 3) an overview of the empirical FDARs and PDARs published to date. We conclude by proposing a series of recommendations for future work.

5.2 Functional and Phylogenetic Diversity: An Overview

Biodiversity is not only the mere sum of species but can also be viewed as an ensemble of organisms with different phenotypic traits and evolutionary history. For example, let us assume two islands both with two species. On island A, they comprise two rodents, while island B holds one rodent and one carnivore species. Intuitively, while they have an equal species richness, island B is more biologically diverse since it is both more functionally and phylogenetically diverse than island A. Here, we briefly describe phylogenetic and functional traits and then provide an overview of the different approaches to measure PD and FD.

5.2.1 Phylogenies and Functional Traits

A rooted phylogenetic tree represents the historical branching structure (evolutionary relationships) among species or other units (Figure 5.1A) and is constructed based on a set of characters and an evolutionary model. Usually, the characters used are molecular markers such as protein, RNA or DNA sequences alignments and the evolutionary model is constructed using maximum likelihood or Bayesian inference (Felsenstein, 2004). The rooted tree contains a root node, a set of branches, internal nodes and tips (here, species). A node represents the most common ancestor of its descending tips and the root node is the common ancestor of all tips in the tree and represents the starting point in time from which the tree flows towards the tips. Branch lengths depict inferred evolutionary changes of the characters used to construct the tree. They are sometimes rescaled to time (e.g. using fossils), so that the tree becomes ultrametric (as in Figure 5.1A; all distances from root to tips are equal).

Traits are defined as any 'morphological, physiological or phenological feature measurable at the individual level, from the cell to the whole-organism level, without reference to the environment or any other level of organization' (Violle et al., 2007, p. 884). Functional traits are those traits that are assumed or known to 'impact fitness indirectly via their effects on growth, reproduction and survival, the three components of individual performance' (Violle et al., 2007, p. 882). These functional traits define a multi-dimensional space (Figure 5.1B) within which each species can be located. Examples of functional traits include body mass, diet, beak size or leaf surface. In this chapter, we will focus on functional traits to provide a better understanding of the processes behind community assembly and to link to ecosystem functions and services, but the approaches presented in Section 5.3 can theoretically be applied to other traits.

5.2.2 PD and FD Metrics

PD and FD capture the phylogenetic and functional structure of a set of species (e.g. local communities or entire biota) and are quantified by different metrics (sometimes also referred to as indices). Here, we limit ourselves to the description of some commonly used phylogenetic and functional alpha diversity metrics, that is, metrics characterizing the diversity of a set of species (referred to as PD and FD hereafter) as opposed to metrics quantifying beta diversity, the dissimilarity between pairs of species sets. Rather than listing the plethora of PD and FD metrics available, we take advantage of several synthetic frameworks (Chao et al., 2010; Pavoine & Bonsall, 2011; Tucker et al., 2017) to summarize the three main factors that are important to consider when choosing a PD or FD metric. Depending on the metric chosen, one can: 1) quantify different 'aspects' of the structure of the functional or phylogenetic space filled by a set of species, 2) take species abundances into account or not and 3) focus on different phylogenetic and functional resolutions. We describe each of these points and we provide examples of commonly used metrics, although, for a more comprehensive overview of the diversity of PD and FD metrics available, readers should refer to specialized reviews (e.g. Mouchet et al., 2010; Pavoine & Bonsall, 2011; Tucker et al., 2017).

FD and PD metrics can be classified into three broad categories: richness, regularity and divergence (Pavoine & Bonsall, 2011). Following Tucker et al. (2017), richness can be defined as 'the sum of accumulated phylogenetic/functional differences among taxa'. FD examples include the convex hull (Cornwell et al., 2006), that is, the surface of the polygon represented in Figure 5.1B, and, for PD, the sum of the branch lengths connecting all species or Faith's PD (Faith, 1992; Table 5.1). Richness metrics are thought to be sensitive to environmental filtering (Cornwell et al., 2006) and other metrics have thus been recommended. Divergence can be defined as 'the mean phylogenetic/functional relatedness among taxa' (e.g. the mean distance between species in an assemblage, such as the Mean Phylogenetic

Distance (MPD) and the Mean Functional Distance (MFD) metrics; Table 5.1). Regularity can be defined as 'the variance in differences among taxa, representing how regular the phylogenetic/functional differences between taxa in an assemblage are' (e.g. the FEve metric; Villéger et al., 2008). Regularity metrics are often used to quantify trait spacing within a community to detect the effect of competition (Kraft et al., 2008; see also Section 5.3.3).

Within each of these categories, metrics can incorporate or ignore species abundances (i.e. only use presence/absence). Species abundances are often an important source of information, either for conservation (Vane-Wright et al., 1991) or for fundamental aspects of ecology, as they may reflect species performance in a given environment. Species abundances are often estimated as the ratio between species absolute abundance (e.g. cover, number of individuals or biomass) and the total absolute abundance of the community (e.g. total cover, total number of individuals or total biomass). We have listed only presence/absence metrics (Table 5.1) but some are easily modified to take into account abundances (Mouchet et al., 2010; Pavoine & Bonsall, 2011; Tucker et al., 2017). The impact of species abundances on a given metric can be fine-tuned along a continuum of importance given to abundant versus rare species using the 'Hill numbers' framework (Hill, 1973; Chao et al., 2010).

Finally, metrics are also sensitive to the phylogenetic and functional resolution considered (Webb et al., 2002; Mazel et al., 2016). For example, the MNTD metric quantifies the mean phylogenetic distance between nearest related species within an assemblage, while the MPD metric quantifies the mean phylogenetic distance between all possible species within an assemblage. MNTD is sensitive to phylogenetic structure toward the tips of the phylogenetic tree, that is, at fine phylogenetic resolution, while MPD is more influenced by patterns at the root of the phylogenetic tree. The MNTD/MPD example is easily expandable to FD, using functional distances instead of phylogenetic distances.

Many PD and FD metrics are based on phylogenetic or functional distances between species, so we provide a brief overview on how to compute them. Phylogenetic distance between species is straightforward to calculate as it simply represents (for an ultrametric tree) twice the age of the most common ancestor of these two species (see Figure 5.1A). The functional distances between species can be quantified by many different metrics depending on the category of traits considered (e.g. continuous, categorical, ordered) and whether different categories are mixed together or not (Pavoine et al., 2009). Sometimes, functional distances are used to construct a functional dendrogram (Petchey & Gaston, 2007). This dendrogram is similar to a phylogenetic tree and PD metrics can be applied to measure FD based on the functional dendrogram. For example, the Petchey and Gaston (2007) dendrogram-based FD metric is conceptually similar to Faith's PD (Table 5.1). As a cautionary note, we mention that the use of functional dendrograms has generated much debate in the literature (e.g. Villéger et al., 2017) and there is currently no clear consensus on how to construct and use them. Therefore, we recommend that these metrics are always used in conjunction with non-dendrogram-based metrics.

Table 5.1 *The formulas of some commonly used FD and PD metrics. Bt is the set of branches in the phylogenetic tree,* L_b *is the length of branch b, S is the assemblage species number or species richness,* d_{ij} *is the phylogenetic (or functional) distance between two species i and j and* $d_{i\ min}$ *is the phylogenetic (or functional) distance of a given species i to its closest relative in the assemblage*

Category	Type	Metric	Formula	Reference
Richness	PD	PD_{Faith}	$\sum_{b \in Bt} L_b$	Faith (1992)
Richness	FD	Convex hull volume	See reference	Cornwell et al. (2006)
Divergence	PD	MPD (Mean Phylogenetic Distance)	$\dfrac{\sum_{ij} d_{ij}}{S(S-1)}$	Webb et al. (2002)
Divergence	PD	MNTD (Mean Nearest Taxon Distance)★	$\dfrac{1}{S}\sum_{i} d_{i\,min}$	Webb et al. (2002)
Divergence	FD	MFD (Mean Functional Distance)	$\dfrac{\sum_{ij} d_{ij}}{S(S-1)}$	Webb et al. (2002); Cianciaruso et al. (2012)
Regularity	PD	VNPD (Variance of Nearest Taxon Distance)★	$\dfrac{1}{S}\sum_{i=1}^{S}[(d_{i\,min}-MNTD)^2\,]$	Tucker et al. (2017)
Regularity	FD	FEve (Functional Evenness)	See reference	Villéger et al. (2008)

★ These metrics can be used with functional distances instead of phylogenetic distances.

5.3 PDARs and FDARs: Methods and Theoretical Expectations

5.3.1 The Various SAR Sampling Designs Can Be Expanded to the PDAR and the FDAR

Multiple types of SAR can be constructed depending on the type of data collected. They can be summarized into four broad types (Scheiner, 2003) that can all be applied to construct PDARs and FDARs. Type I curves correspond to nested designs where 'each data point is based on a single measurement for a given size' (Scheiner, 2003). An example of a PDAR Type I curve is provided by Morlon et al.'s (2011) local plot-based sampling of Mediterranean plants. Type II and III curves are estimated by the mean diversity for a given area. An example of a PDAR type II curve would be that calculated by Mazel et al. (2015) for mammals globally based on grid-cell sampling and an example of a PDAR type III curve is provided by Helmus and Ives' (2012) non-continuous grid-based sampling of temperate forests trees. Finally, type IV curves correspond to classic island species–area relationships and are built from single data points (as for type I curves) that are each from a unique area and which differ from the other curves in the diversity metric being calculated separately for each area. Examples of FDAR type IV curves include those calculated by Whittaker et al. (2014) for spiders and for beetles in the Azores.

5.3.2 Relationships between the SAR, PDAR and FDAR

The SAR and FDAR (or PDAR) often have similar shapes because the underlying FD (or PD) metrics are mathematically constrained by SR. Specifically, a positive correlation between PD (or FD) and area can emerge if 1) there is a positive correlation between SR and area and 2) the PD (or FD) metric chosen is mathematically constrained to be positively correlated with SR. This is expected to be the case, especially with metrics belonging to the richness aspect and, to some extent, the regularity aspect (Table 5.1). For example, the PD_{Faith} metric (Faith, 1992; Table 5.1) sums up the branch length of the phylogenetic tree linking all species within a community and will inevitably be positively related to SR. Adding a species to a given community will add a branch to the phylogenetic tree of the community and PD will increase. One simple approach to avoid this bias is to choose a FD (or PD) metric that is known to be theoretically unconstrained by SR, such as MPD (see Table 5.1). This approach has the advantage of being simple but the disadvantage of restricting the set of potential PD and FD metrics available. It has been used, for example, to describe how PD scales with island area in the Caribbean (Helmus & Ives, 2012).

When using a FD (or PD) metric that is known to be mathematically constrained by SR, such as PD_{Faith}, standardization is required to meaningfully interpret FDARs (or PDARs) and compare it to SARs. Plotting the SAR and FDAR (or PDAR) on the same graph is misleading as diversity is measured in different units (Figure 5.2A). For example, SAR measures diversity in terms of the species count, while a PDAR constructed with PD_{Faith} measures diversity in millions of years. One potential approach to make the PDAR and SAR comparable is to standardize them by the maximum diversity value (Mazel et al., 2015) (Figure 5.2B). To produce FDAR (or PDAR) expectations based on SR, we suggest using a null model approach (Gotelli, 2001). Null models randomize some aspect of the data while keeping another aspect constant (here, SR) and computing associated PD (or FD). For example, a simple null model shuffles the tip labels of the

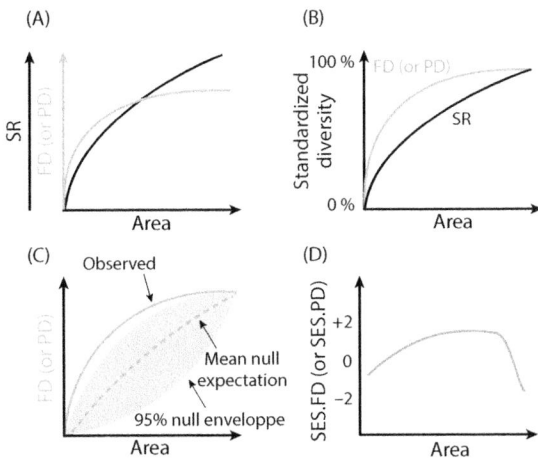

Figure 5.2 Comparing the SAR, PDAR and FDAR. (A) Hypothetical SAR (in black) and PDAR or FDAR (in grey) in their original units (e.g. species richness for the SAR). (B) Hypothetical SAR (in black) and PDAR or FDAR (in grey) in rescaled units (values rescaled by their maximum). (C) Hypothetical observed PDAR or FDAR (plain grey curve) along with a null mean expectation (dotted grey curve) and a 95% confidence null envelope (shaded grey zone). (D) Hypothetical spatial scaling of SES values of PD or FD (see Equation 5.1).

phylogenetic tree or trait dendrogram while keeping the site–species matrix constant. The randomizations can be repeated a large number of times (e.g. 1,000) to produce a distribution of expected PD (or FD) for a given species richness. The null PDAR (or FDAR) envelope can be plotted alongside the observed PDAR (see Figure 5.2C for an example).

The null FDAR (or PDAR) expectation is influenced by the structure of the underlying phylogenetic or functional data (Mazel et al., 2015). For example, the shape of the null PDAR is influenced by the shape of the phylogenetic tree: a 'tippy' tree (i.e. with long terminal branches) will produce a null PDAR close to the SAR shape, while a 'stemmy' tree (i.e. with long internal branches) will produce a null PDAR that is different to the SAR (Figure 5.3A). The shape of the null FDAR is influenced by the degree of functional redundancy among species: low redundancy will produce a null FDAR close to the SAR shape, while high redundancy will produce a null FDAR that is different to the SAR (Figure 5.3B). Observed PD (or FD) values can then be compared to this null distribution and standardized using, for example, standardized effect sizes (SES):

$$SES.FD = \frac{FD_{observed} - mean(FD_{null})}{SD(FD_{null})}. \qquad (5.1)$$

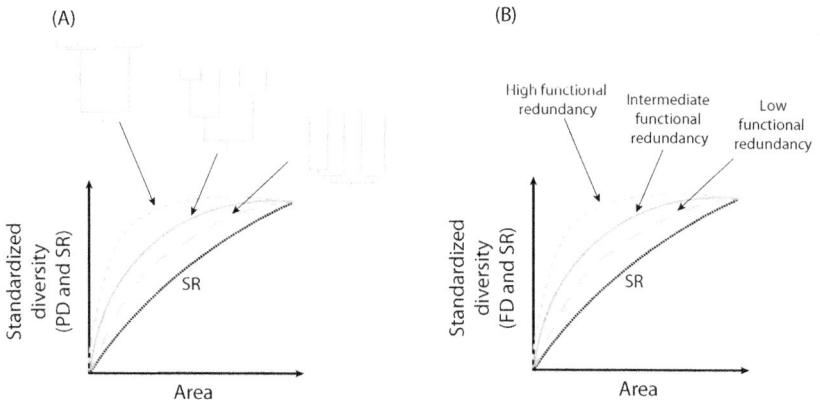

Figure 5.3 The effect of tree shape and degree of functional redundancy on the PDAR and FDAR. (A) Hypothetical SAR (dotted black curve) and three hypothetical PDARs (in grey) corresponding to three different tree shapes. (B) Hypothetical SAR (dotted black curve) and three hypothetical FDARs (in grey) corresponding to three different degrees of functional redundancy among species. The hypothetical SARs, PDARs and FDARs presented in this figure are standardized by their maximum values.

These SES values can then be plotted against area (Figure 5.2D). Section 5.3.3 provides some theory to interpret deviations from the null expectation. It is important to note that carefully designing the null model and the species pool from which species are sampled is of prime importance in the detection of assembly processes (Lessard et al., 2012a; Münkemüller et al., 2014).

5.3.3 Theoretical Expectations for PDARs and FDARs

Multiple ecological and evolutionary processes can cause PDARs and FDARs to depart from random expectations, from classical community assembly processes (filtering, competition, facilitation, dispersal) to bio-geographical and evolutionary processes (e.g. allopatric speciation, vicariance events, adaptive radiations) (Morlon et al., 2011; Helmus & Ives, 2012; Smith et al., 2013; Mazel et al., 2015).

In community ecology, the filtering framework posits that abiotic factors (e.g. temperature) and biotic factors (e.g. competition for light) determine species' ability to colonize and persist in a given community (e.g. see Leibold & Chase, 2018). The abiotic factors filter species from a regional species pool into the local species pool and biotic interactions determine which species from the local pool can coexist in the community (Weiher & Keddy, 1999; Vellend, 2010; HilleRisLambers et al., 2012; Leibold & Chase, 2018). This approach assumes that different ecological processes result in distinct patterns in diversity so that we can deduce processes from patterns. Specifically, coexistence theory predicts that biotic and abiotic filtering lead to different patterns of FD (or PD). In a given location, abiotic factors (e.g. temperature) are thought to filter a given species with a specific set of trait values from the species pool. This restricted set of traits values are linked to the fundamental niche of the species, that is, its ability to tolerate certain environmental conditions, independently of biotic interactions. Because only a limited range of trait values are represented in this community, the resulting FD (e.g. the convex hull volume) is expected to be lower than the FD of a random sample of trait values from the pool. This is sometimes referred to as a 'clustered' pattern. The same result is expected for PD if phylogenetic distances between species are related to relevant niche differences between species. In contrast, biotic interactions and, in particular, competition are expected to lead to a different pattern of FD (or PD). Differences in trait values are expected to foster coexistence of species because of resource partitioning: species with similar trait values will be unlikely to coexist because they compete and exclude each other. In this case, trait values within a community are expected to be more evenly spaced than a random set of traits from the species pool. This is sometimes referred as to an 'overdispersed' pattern. FD metrics that best capture trait spacing are regularity metrics (see Table 5.1). The same reasoning applies for PD if relevant niche differences are related to phylogenetic distances.

It is often hypothesized that the importance of alternative ecological and evolutionary processes can vary with spatial scale (Cavender-Bares et al., 2006; Swenson et al., 2006). As mentioned in Section 5.1, biotic interactions (e.g. competition, predation, mutualism) are thought to dominate at fine spatial scales, where individuals interact, while environment filtering (the process by which only organisms adapted to a given environment colonize this environment) might be more important at larger spatial scales. As patterns of FD might help to reveal such processes (providing they are not a trivial outcome of the dependency of the FD metric on species richness), describing how FD scales with area (i.e. constructing FDARs) might reveal the spatial scale dependencies of ecological and evolutionary processes (Weiher & Keddy, 1995; Smith et al., 2013; Leibold & Chase, 2018).

We should caution that the conceptual framework described above, where patterns of over-dispersed traits or phylogenies are interpreted as the result of competitive exclusion, while patterns of clustered traits or phylogenies are interpreted as the result of environmental filtering, is simplistic and ignores some recent advances in modern coexistence theory. In particular, Mayfield and Levine (2010) used the Chesson framework (Chesson, 2000) to point out that competition can lead to both overdispersion and clustering if the trait considered is related to either stabilizing niche differences or equalizing fitness differences (Kraft et al., 2015a; Cadotte & Tucker, 2017). Using annual plants as a study system and based on competition experiments, Kraft et al. (2015b) showed that multiple individual traits (e.g. maximum height, leaf area) are related to equalizing fitness differences, while no individual traits correlated with niche differences (but models including multiple traits do correlate with niche differences). Overdispersion is expected for traits related to stabilizing niche differences, whereas clustering is expected for traits related to equalizing fitness differences. Mixing both type of traits thus limits the interpretability of FD patterns. However, the filtering framework represents a valuable tool to interpret FDARs and PDARs, if we keep in mind that clustered FD patterns are not necessarily the product of filtering but could also be caused by competition (see Figure 5.4 for a summary of the potential processes).

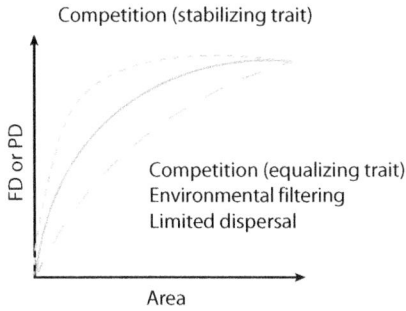

Figure 5.4 Ecological and evolutionary mechanisms impacting PDARs and FDARs. The figure depicts a hypothetical PDAR or FDAR (solid grey curve) and the mechanisms that would cause the FDAR to depart from this hypothetical example (dotted grey curves).

Other sets of processes can theoretically influence the shape of FDARs and PDARs. Specifically, biogeographical processes occurring at large temporal and spatial scales can influence the shape of FDARs and PDARs. For example, Mazel et al. (2015) constructed PDARs at the continental scale for mammals (starting at a resolution of 200 × 200 km) and found that, for Eurasia, observed PD values were much lower than expected from random for all the spatial scales considered. This pattern could in theory result from environmental filtering acting at all spatial scales, but an alternative explanation based on the presence of multiple zoogeographic regions in Eurasia is also plausible (Mazel et al., 2015). Zoogeographic regions represent relatively compositionally homogenous biotas separated by sharp zoogeographic boundaries (Wallace, 1876; Ficetola et al., 2017). In Eurasia, several biogeographic regions and boundaries exist. Since different biogeographic regions harbour different clades of mammals, observed local PD is expected to be low compared to a random set of species sampled from the entire continental pool (i.e. from all the regions). As zoogeographic regions and boundaries typically correspond with sharp environmental transitions or with geographical barriers that prevent successful dispersal (Ficetola et al., 2017), the non-random PDAR reported by Mazel et al. (2015) is likely to be the result of multiple processes, including limited dispersal over geological time scales (and not only filtering). The importance of such processes is likely to become more important as the grain of the study becomes larger and encompasses regional species pools. Disentangling the effect of ecological and biogeographical processes will remain difficult as they can both produce the same PD patterns (Warren et al., 2014). Again, there is a clear limitation in pinpointing particular processes from PD and FD patterns and, thus, we suggest that future studies should carefully discuss the potential effect of biogeographical versus ecological processes when interpreting their PDAR and FDAR results.

5.3.4 Fitting PDARs and FDARs

Section 5.3.3 provided some theory to interpret deviations of PDARs and FDARs from random expectations across spatial scales. However, the spatial dependency of these deviations (e.g. overdispersion at fine scales and clustering at large scales) were only qualitatively explored (e.g. Figure 5.2D). A better way to rigorously quantify these trends is to fit mathematical models to the observed and expected FDARs (or PDARs) and compare the coefficient estimates of the expected and observed fits. For example, the power model (Equation 5.2) (Triantis et al., 2012; Chapter 3) has been shown to adequately fit many empirical SARs:

$$SR = cA^z, \tag{5.2}$$

where A represents the sampled area and c and z are fitted constants.

Comparing observed and expected model coefficients can quantify the scale dependencies of PD and FD deviations from null expectations. For example, Mazel et al. (2015) fitted a log–log power model to observed and expected PDARs for mammals at a continental scale (same analysis and data as reported in Section 5.3.3). For Eurasia, they found that the observed z-value was undistinguishable from random, meaning that deviation of PD from random does not depend on area. Indeed, they found that the observed c-value was always lower than expected from random, meaning that observed PD is lower than expected from random at all spatial scales (see Section 5.3.3 for a detailed interpretation of this pattern).

The power function is not the only model that has been proposed to describe the SAR and other models should be systematically tested. There are few theoretical works that provide expectations for the fit of different mathematical models to observed PDARs and FDARs (but see Morlon et al., 2011) and there is no reason to believe that the power model should be a priori favoured, especially for those PD and FD metrics that are not mathematically constrained by SR. Empirical work suggests that different models fit the data for mammals equally well at the biome scale (Mazel et al., 2014) and that the power model does not always provide the best fit to the data (White et al., 2018). As a consequence, we suggest to always test multiple models when fitting PDARs and FDARs (e.g. using the 'sars' R package; Matthews et al., 2019). Multiple methods are available to construct SARs, such as island-type species–area curves (Type IV in Scheiner's 2003 framework) and species accumulation curves (e.g. Type IIIa in the Scheiner framework). As it has been shown that different methods can yield very different SAR parameter estimates, even for the same dataset (Matthews et al., 2016), we suggest caution is employed when comparing FDAR and PDAR parameter estimates across studies that use different methods to construct the curve.

5.4 Empirical PDAR and FDAR Patterns

So far, we have only provided methods to construct PDARs and FDARs and a set of expectations to interpret them. But how do PD and FD scale with area in the real world? A good review on this subject is provided by Leibold and Chase (2018), who set out to summarize the empirical evidence supporting the Weiher and Keddy (1995) hypothesis: communities are over-dispersed at local scales due to competition, but clustered at larger scales due to filtering. Leibold and Chase (2018) report nineteen studies from a wide range of biological systems (tropical and temperate trees, fishes, mammals, birds and frogs) and continents that compare FD and/or PD at local versus regional scales. Fifteen studies (out of nineteen) report a non-random scale dependency of PD and/or FD with area, but the way in which the data departs from random is not consistent across studies. Only four studies clearly confirm the Weiher and Keddy (1995) hypothesis, while five report results that clearly differ from it. Leibold and Chase (2018) proposed several non-mutually exclusive reasons for why those studies do not report consistent results. Specifically, they suggest that the choice of regional versus local scale is somewhat arbitrary, so that it is actually difficult to compare studies (i.e. the regional scale of one study might correspond to the local scale of another study). One way to overcome this is to control for scale difference. Here, we present a collection of seventeen analyses from fourteen studies (some studies conducted multiple analyses) that explicitly constructed FDARs and/or PDARs (Table 5.2). These studies encompass many biological systems (e.g. alpine meadow and tree communities, mammal and bird assemblages), multiple sampling designs (nested or non-nested plots, grid cells, different PD or FD metrics; fifteen in total) and focus on a wide variety of spatial scales (from less than a metre squared to entire continents). Two main objectives are pursued in these studies: the majority assessed the degree of clustering and overdispersion of FD and/or PD across spatial scales and a minority compared the fit of alternative mathematical models to observed PDARs and FDARs. The studies pursuing the latter objective generally reported that the power model is not necessarily the best fitting model and that others need to be considered (Table 5.2). An important result is that the relative fit of different models heavily depends on the PD and FD metric considered. Also, the type of curve constructed (e.g. type I or III *sensu* Scheiner, 2003) is known to impact the resulting SAR shape (Matthews et al., 2016) and it is likely that this result extends to PDARs and FDARs as well. More theoretical work is needed to better understand the influence of the type of PDAR/FDAR sampling design on the relative fit of different mathematical models. The studies focusing on measuring the degree of FD and/or PD clustering across spatial scales generally detected some degree of scale dependency in PD and FD. In other words, PD and FD generally depart from random expectation at some scales but not all. For example, Smith et al. (2013) found that the trait range of serpentine soil plants was generally smaller than expected (i.e. clustered) at small scales but random at larger scales. However, the direction of the deviation of PDARs and FDARs from random expectation does not generally validate the Weiher and Keddy (1995) hypothesis. In particular, no studies found over-dispersed patterns at small scales and clustering at large scales. Rather, PD and FD patterns were clustered at small scales and random at large scales (e.g. Helmus & Ives, 2012; Smith et al., 2013), random at small scales and clustered at large scales (Smith et al., 2013) or random at all scales (Morlon et al., 2011; Mazel et al., 2015). These mixed results are in line with the aforementioned results of Leibold and Chase (2018) for PD and FD patterns reported at two spatial scales only (local and regional).

There are multiple explanations for the discrepancy between the empirical results and the theoretical predictions of Weiher and Keddy (1995). First, methodological differences between studies could potentially explain the mixed results we report. In particular, the choice of FD and PD metrics can have a dramatic impact on the observed patterns (Miller et al., 2016). While beyond the scope of this chapter, an interesting future research avenue would be to assemble multiple datasets and compare PDARs and FDARs with a consistent methodology. In particular, overdispersion at small scales is, in theory, expected to be detected with regularity metrics (see Table 5.1), but those metrics are

Table 5.2 *Overview of published examples of PDARs and FDARs. PSV = Phylogenetic Species Variability (Helmus & Ives, 2012); FRich = Functional Richness (Villéger et al., 2008); FDis = Functional Dispersion (Laliberté & Legendre, 2010); FDiv = Functional Divergence (Villéger et al., 2008); RaoQE = Rao Quadratic Entropy (Rao, 1982); FD$_{Petchey}$ = Petchey and Gaston's dendrogram-based FD (Petchey & Gaston, 2007). For additional PD and FD metrics, see Table 5.1*

Study	Biological system	Sampling type*	Area covered (in m²)**	PD and/or FD	Clustering pattern	Model fit
White et al. (2018)	Grassland plants (Scotland)	Nested plots (I)	0.0016–25	FD (FRic and FDis)	NA	Semi-logarithmic favoured over the power model
Smith et al. (2013)	Subalpine meadow plants (Colorado, USA)	Non-continuous plots (IIIa)	0.25–16	FD (trait range and a measure of regularity)	Mixed results depending on disturbance level	NA
Morlon et al. (2011)	Plants (Mediterranean ecosystems)	Nested plots (I)	6–20	PD (PD$_{Faith}$)	Random PD structure at all scales	Power model
Smith et al. (2013)	Serpentine soil plants (California, USA)	Nested plots (IIa)	0.25–64	FD (trait range and a measure of regularity)	Trait range: random (large scale) and clustered (small scale); Regularity metric: random at all scales	NA
Smith et al. (2013)	Coastal plants (California, USA)	Non-continuous plots (IIIa)	0.016–4	FD (Convex Hull)	Random (small scale) and clustered (large scale)	NA

(cont.)

Table 5.2 (*cont.*)

Study	Biological system	Sampling type[*]	Area covered (in m²)[**]	PD and/or FD	Clustering pattern	Model fit
Karadimou et al. (2016)	Plants (Greece)	Nested Plots (I)	1–128	FD (FRic, FEve, FDiv, FDis and RaoQ)	NA	Different indices and traits provide different fits (7 models)
Helmus and Ives (2012)	Temperate forest trees (Oregon, USA)	Plots (IIIa,b)	500–10^5	PD (PSV)	Negative relationship	NA
Wang et al. (2013)	Temperate forest trees (China; USA)	Random quadrats of varying sizes (?)	1–10^5	PD (PD_{Faith}) and FD ($FD_{Petchey}$)	NA	Comparisons with four mechanistic models
Zhang et al. (2018)	Plants (China)	Plots (?)	400–10^4	FD (Rao QE)	Random (large scale) and clustered (small scale)	NA
Ding et al. (2013)	Birds (China)	Islands (IV)	10^6–10^7	FD (FRic, FDiv and Feve)	Generally clustered; FRic more clustered on large islands	Logarithmic model
Carvajal-Endara et al. (2017)	Plants (Galápagos)	Islands (IV)	10^6–10^9	PD (MPD)	Clustering on small islands; overdispersion on large islands	Logarithmic model
Smith et al. (2013)	Mammals (North America)	Grid cells (IIa)	10^{10}–10^{16}	FD (trait range and a measure of regularity)	Both metrics: random (large scale) and clustered (small scale)	NA

Helmus and Ives (2012)	Anolis lizards (Caribbean)	Islands (IV)	10^4–10^5	PD (PSV)	No relationships	NA
Mazel et al. (2015)	Mammals (Global)	Grid cells (IIa)	10^10–10^12	PD (PD_Faith)	Random for all continents except Eurasia (clustered at all scales)	NA
Mazel et al. (2014)	Mammals (Global)	Ecoregions (IV)	10^8–10^12	PD and FD (Hill numbers)	NA	No best model among 19 models
Whittaker et al. (2014)	Beetles and spiders	Islands (IV)	17^6–757^6	FD (FRich)	Depends on species pool. Habitat species pool yielded clustered patterns for beetles	Power model only
Li et al. (2018)	Vascular plants and birds (USA)	US National Parks (IV)	10^5–10^10	PD (PSV)	No relationship for plants, positive relationship for birds	Power model

* With Scheiner (2003) classification given in parentheses. ** For large areas, only orders of magnitude are given.

not always used. Second, the null model and the species pool used to randomize observed communities are not equivalent in all the studies we reviewed but are of prime importance to the detection of clustering and overdispersion (Lessard et al., 2012b). Third, the spatial scales (both grain and extent) encompassed by the studies we report vary tremendously. For example, White et al. (2018) focused on grains ranging from 1 to 25 m^2, while Mazel et al. (2015) focused on grains ranging from 10^8 to 10^{12} m^2. This scale discrepancy between studies might also explain why studies report different patterns. Fourth, and as noted by Leibold and Chase (2018), the Weiher and Keddy (1995) theoretical predictions are rather simplistic and make a number of assumptions that are not necessarily met in the real world (Chesson, 2000; Mayfield & Levine, 2010). Fifth, different traits can produce different patterns (e.g. stabilizing versus equalizing traits) so that mixing traits together when measuring FD can obscure significant patterns at the individual trait level.

5.5 PD/FD Loss from Habitat Loss and PDARs/FDARs

Historically, SARs have been a key tool to estimate species loss from habitat loss (Chapter 14). Indeed, by reversing the SAR and extrapolating backward, one can compute expected species loss from habitat loss (assuming that the SAR function is unchanged by the habitat loss and isolation). This method has been extended to PD and FD (Keil et al., 2015; Mazel et al., 2015). Both studies suggest that PD and FD might be more robust to habitat loss than SR, mostly because there is some degree of functional and phylogenetic redundancy between species (Nee & May, 1997). In addition, both studies show that the degree of robustness of PD and FD to habitat loss is related to the degree of redundancy. However, there has been much controversy about the use of the SAR (or PDAR and FDAR) to predict diversity loss from habitat loss, both because most predictions do not empirically match the data and because the approach appears to be theoretically flawed (He & Hubbell, 2011; see Chapter 14). As this debate is not yet settled, we suggest using the SAR, PDAR and FDAR with caution when inferring diversity loss from habitat loss. In particular, we note that, when estimating species FD and PD loss, the geometry of habitat loss appears to be important (simulations by Keil et al., 2015; Chapters 14 and 16). In order to estimate PD and FD loss from habitat loss, we thus recommend simulating various scenarios of habitat loss and recording the corresponding loss of PD and FD (as in Keil et al., 2015).

5.6 Conclusions, Perspectives and Future Directions

This chapter provides a methodological, theoretical and empirical overview of PDARs and FDARs. PDARs and FDARs can be constructed in many ways depending on the sampling used, the FD (or PD) metric used and the null model used. The choice of the methods used critically impacts the reported pattern, so users should carefully design their study and ideally provide results for multiple metrics (e.g. a richness metric, a divergence metric and a regularity metric) and appropriate null models, carefully defining the species pool. A review of the literature reveals some degree of scale dependency in PD and FD, that is, PD and FD generally depart from random expectation at some scales but not all. However, the direction of the deviation of PDARs and FDARs from random expectation does not generally validate the Weiher and Keddy (1995) hypothesis and often differs from one study to another. The discrepancy in the methods used in the literature is one potential explanation for the inconsistent results we found. Along with the description of new PDAR and FDAR patterns, we believe that one potential way to reach a synthesis is to conduct large meta-analyses based on standardized sampling and methods.

Acknowledgements

We thank K. Davis, R. J. Whittaker and an anonymous reviewer for comments on an earlier version of this manuscript. F. M. is supported by a postdoctoral Banting fellowship.

References

Cadotte, M. W. & Tucker, C. M. (2017) Should environmental filtering be abandoned? *Trends in Ecology & Evolution*, **32**, 429–437.

Cadotte, M. W., Carscadden, K. & Mirotchnick, N. (2011) Beyond species: Functional diversity and the maintenance of ecological processes and services. *Journal of Applied Ecology*, **48**, 1079–1087.

Carscadden, K. A., Cadotte, M. W. & Gilbert, B. (2017) Trait dimensionality and population choice alter estimates of phenotypic dissimilarity. *Ecology and Evolution*, **7**, 2273–2285.

Carvajal-Endara, S., Hendry, A. P., Emery, N. C. & Davies, T. J. (2017) Habitat filtering not dispersal limitation shapes oceanic island floras: Species assembly of the Galápagos archipelago. *Ecology Letters*, **20**, 495–504.

Cavender-Bares, J., Keen, A. & Miles, B. (2006) Phylogenetic structure of floridian plant communities depends on taxonomic and spatial scale. *Ecology*, **87**, S109–S122.

Chao, A., Chiu, C.-H. & Jost, L. (2010) Phylogenetic diversity measures based on Hill numbers. *Philosophical Transactions of the Royal Society B: Biological Sciences*, **365**, 3599–3609.

Chase, J. M., Amarasekare, P., Cottenie, K., Gonzalez, A., Holt, R. D., Holyoak, M., Hoopes, M. F., Leibold, M. A., Loreau, M., Mouquet, N., Shurin, J. B. & Tilman, D. (2005) Competing theories for competitive metacommunities. *Metacommunities: Spatial dynamics and ecological communities* (ed. by M. Holyoak, M. A. Leibold and R. D. Holt), pp. 335–354. Chicago, IL: University of Chicago Press.

Chesson, P. (2000) Mechanisms of maintenance of species diversity. *Annual Review of Ecology and Systematics*, **31**, 343–366.

Cianciaruso, M. V., Silva, I. A., Batalha, M. A., Gaston, K. J. & Petchey, O. L. (2012) The influence of fire on phylogenetic and functional structure of woody savannas: Moving from species to individuals. *Perspectives in Plant Ecology, Evolution and Systematics*, **14**, 205–216.

Cornwell, W. K., Schwilk, D. W., Ackerly, D. D. & Schwilk, L. (2006) A trait-based test for habitat filtering: Convex hull volume. *Ecology*, **87**, 1465–1471.

de Bello, F., Lavorel, S., Lavergne, S., Albert, C. H., Boulangeat, I., Mazel, F. & Thuiller, W. (2013) Hierarchical effects of environmental filters on the functional structure of plant communities: A case study in the French Alps. *Ecography*, **36**, 393–402.

Ding, Z., Feeley, K. J., Wang, Y., Pakeman, R. J. & Ding, P. (2013) Patterns of bird functional diversity on land-bridge island fragments. *Journal of Animal Ecology*, **82**, 781–790.

Faith, D. P. (1992) Conservation evaluation and phylogenetic diversity. *Biological Conservation*, **61**, 1–10.

Felsenstein, J. (2004) *Inferring phylogenies*. Sunderland, MA: Sinauer.

Ficetola, G. F., Mazel, F. & Thuiller, W. (2017) Global determinants of zoogeographical boundaries. *Nature Ecology & Evolution*, **1**, 0089.

Gerhold, P., Cahill, J. F., Winter, M., Bartish, I. V. & Prinzing, A. (2015) Phylogenetic patterns are not proxies of community assembly mechanisms (they are far better). *Functional Ecology*, **29**, 600–614.

Gotelli, N. J. (2001) Research frontiers in null model analysis. *Global Ecology & Biogeography*, **10**, 337–343.

He, F. & Hubbell, S. P. (2011) Species–area relationships always overestimate extinction rates from habitat loss. *Nature*, **473**, 368–371.

Helmus, M. R. & Ives, A. R. (2012) Phylogenetic diversity–area curves. *Ecology*, **91**, 31–43.

Hill, M. O. (1973) Diversity and evenness: A unifying notation and its consequences. *Ecology*, **54**, 427–432.

HilleRisLambers, J., Adler, P. B., Harpole, W. S., Levine, J. M. & Mayfield, M. M. (2012) Rethinking community assembly through the lens of coexistence theory. *Annual Review of Ecology, Evolution, and Systematics*, **43**, 227–248.

Hubbell, S. P. (2001) *The unified neutral theory of biodiversity and biogeography*. Princeton, NJ: Princeton University Press.

Kadmon, R. & Allouche, O. (2007) Integrating the effects of area, isolation, and habitat heterogeneity on species diversity: A unification of island biogeography and niche theory. *The American Naturalist*, **170**, 443–454.

Karadimou, E. K., Kallimanis, A. S., Tsiripidis, I. & Dimopoulos, P. (2016) Functional diversity exhibits a diverse relationship with area, even a decreasing one. *Scientific Reports*, **6**, 35420.

Keil, P., Storch, D. & Jetz, W. (2015) On the decline of biodiversity due to area loss. *Nature Communications*, **6**, 8837.

Kelly, S., Grenyer, R. & Scotland, R. W. (2014) Phylogenetic trees do not reliably predict feature diversity. *Diversity and Distributions*, **20**, 600–612.

Kraft, N. J. B., Adler, P. B., Godoy, O., James, E. C., Fuller, S. & Levine, J. M. (2015a) Community assembly, coexistence and the environmental filtering metaphor. *Functional Ecology*, **29**, 592–599.

Kraft, N. J. B., Godoy, O. & Levine, J. M. (2015b) Plant functional traits and the multidimensional nature of species coexistence. *Proceedings of the National Academy of Sciences USA*, **112**, 797–802.

Kraft, N. J. B., Valencia, R. & Ackerly, D. D. (2008) Functional traits and niche-based tree community assembly in an Amazonian forest. *Science*, **322**, 580–582.

Laliberté, E. & Legendre, P. (2010) A distance-based framework for measuring functional diversity from multiple traits. *Ecology*, **91**, 299–305.

Lavorel, S., Grigulis, K., Lamarque, P., Colace, M.-P., Garden, D., Girel, J., Pellet, G. & Douzet, R. (2011) Using plant functional traits to understand the landscape distribution of multiple ecosystem services. *Journal of Ecology*, **99**, 135–147.

Lawton, J. H. (1999) Are there general laws in ecology? *Oikos*, **84**, 177–192.

Leibold, M. A. & Chase, J. M. (2018) *Metacommunity ecology*. Princeton, NJ: Princeton University Press.

Lessard, J.-P., Belmaker, J., Myers, J. A., Chase, J. M. & Rahbek, C. (2012a) Inferring local ecological processes amid species pool influences. *Trends in Ecology & Evolution*, **27**, 600–607.

Lessard, J.-P., Borregaard, M. K., Fordyce, J. A., Rahbek, C. & Sanders, N. J. (2012b) Strong influence of regional species pools on continent-wide structuring of local communities. *Proceedings of the Royal Society B: Biological Sciences*, **279**, 266–274.

Li, D., Monahan, W. B. & Baiser, B. (2018) Species richness and phylogenetic diversity of native and non-native species respond differently to area and environmental factors. *Diversity and Distributions*, **24**, 853–864.

Loreau, M. (2010) Linking biodiversity and ecosystems: Towards a unifying ecological theory. *Philosophical Transactions of the Royal Society B: Biological Sciences*, **365**, 49–60.

Louca, S., Parfrey, L. W. & Doebeli, M. (2016) Decoupling function and taxonomy in the global ocean microbiome. *Science*, **353**, 1272–1277.

MacArthur, R. H. & Levins, R. (1967) The limiting similarity, convergence, and divergence of coexisting species. *The American Naturalist*, **101**, 377–385.

MacArthur, R. H. & Wilson, E. O. (1967) *The theory of island biogeography*. Princeton, NJ: Princeton University Press.

Matthews, T. J., Triantis, K. A., Rigal, F., Borregaard, M. K., Guilhaumon, F. & Whittaker, R. J. (2016) Island species–area relationships and species accumulation curves are not equivalent: An analysis of habitat island datasets. *Global Ecology & Biogeography*, **25**, 607–618.

Matthews, T. J., Triantis, K. A., Whittaker, R. J. & Guilhaumon, F. (2019) sars: An R package for fitting, evaluating and comparing species–area relationship models. *Ecography*, **42**, 1446–1455.

Mayfield, M. & Levine, J. (2010) Opposing effects of competitive exclusion on the phylogenetic structure of communities. *Ecology Letters*, **13**, 1085–1093.

Mazel, F., Davies, T. J., Gallien, L., Renaud, J., Groussin, M., Münkemüller, T. & Thuiller, W. (2016) Influence of tree shape and evolutionary time-scale on phylogenetic diversity metrics. *Ecography*, **39**, 913–920.

Mazel, F., Guilhaumon, F., Mouquet, N., Devictor, V., Gravel, D., Renaud, J., Cianciaruso, M. V., Loyola, R., Diniz-Filho, J. A. F., Mouillot, D. & Thuiller, W. (2014) Multifaceted diversity–area relationships reveal global hotspots of mammalian species, trait and lineage diversity. *Global Ecology & Biogeography*, **23**, 836–847.

Mazel, F., Mooers, A. O., Riva, G. V. D. & Pennell, M. W. (2017b) Conserving phylogenetic diversity can be a poor strategy for conserving functional diversity. *Systematic Biology*, **66**, 1019–1027.

Mazel, F., Pennell, M. W., Cadotte, M. W., Diaz, S., Dalla Riva, G. V., Grenyer, R., Leprieur, F., Mooers, A. O., Mouillot, D., Tucker, C. M. & Pearse, W. D. (2018) Prioritizing phylogenetic diversity captures functional diversity unreliably. *Nature Communications*, **9**, 2888.

Mazel, F., Pennell, M. W., Cadotte, M. W., Diaz, S., Dalla Riva, G. V., Grenyer, R., Leprieur, F., Mooers, A. O., Mouillot, D., Tucker, C. M. & Pearse, W. D. (2019) Reply to 'Global conservation of phylogenetic diversity captures more than just functional diversity'. *Nature Communications*, **10**, 859.

Mazel, F., Renaud, J., Guilhaumon, F., Mouillot, D., Gravel, D. & Thuiller, W. (2015) Mammalian phylogenetic diversity–area relationships at a continental scale. *Ecology*, **96**, 2814–2822.

Mazel, F., Wüest, R. O., Gueguen, M., Renaud, J., Ficetola, G. F., Lavergne, S. & Thuiller, W. (2017a) The geography of ecological niche evolution in mammals. *Current Biology*, **27**, 1369–1374.

Miller, E. T., Farine, D. R. & Trisos, C. H. (2016) Phylogenetic community structure metrics and null models: A review with new methods and software. *Ecography*, **40**, 461–477.

Morlon, H., Schwilk, D. W., Bryant, J. A., Marquet, P. A., Rebelo, A. G., Tauss, C., Bohannan, B. J. M. & Green, J. L. (2011) Spatial patterns of phylogenetic diversity. *Ecology Letters*, **14**, 141–149.

Mouchet, M. A., Villéger, S., Mason, N. W. H. & Mouillot, D. (2010) Functional diversity measures: An overview of their redundancy and their ability to discriminate community assembly rules. *Functional Ecology*, **24**, 867–876.

Münkemüller, T., Gallien, L., Lavergne, S., Renaud, J., Roquet, C., Abdulhak, S., Dullinger, S., Garraud, L., Guisan, A., Lenoir, J., Svenning, J.-C., Van Es, J., Vittoz, P., Willner, W., Wohlgemuth, T., Zimmermann, N. E. & Thuiller, W. (2014) Scale decisions can reverse conclusions on community assembly processes. *Global Ecology & Biogeography*, **23**, 620–632.

Narwani, A., Alexandrou, M. A., Oakley, T. H., Carroll, I. T. & Cardinale, B. J. (2013) Experimental evidence that evolutionary relatedness does not affect the ecological mechanisms of coexistence in freshwater green algae. *Ecology Letters*, **16**, 1373–1381.

Nee, S. & May, R. M. (1997) Extinction and the loss of evolutionary history. *Science*, **288**, 328–330.

Pavoine, S. & Bonsall, M. B. (2011) Measuring biodiversity to explain community assembly: A unified approach. *Biological Reviews*, **86**, 792–812.

Pavoine, S., Vallet, J., Dufour, A.-B., Gachet, S. & Daniel, H. (2009) On the challenge of treating various types of variables: Application for improving the measurement of functional diversity. *Oikos*, **118**, 391–402.

Petchey, O. L. & Gaston, K. J. (2007) Dendrograms and measuring functional diversity. *Oikos*, **116**, 1422–1426.

Rao, R. C. (1982) Diversity and dissimilarity coefficients: A unified approach. *Theoretical Population Biology*, **21**, 24–43.

Rosenzweig, M. L. (1995) *Species diversity in space and time*. Cambridge: Cambridge University Press.

Rosindell, J. & Cornell, S. J. (2007) Species–area relationships from a spatially explicit neutral model in an infinite landscape. *Ecology Letters*, **10**, 586–595.

Scheiner, S. M. (2003) Six types of species–area curves. *Global Ecology & Biogeography*, **12**, 441–447.

Shen, G., Yu, M., Hu, X.-S., Mi, X., Ren, H., Sun, I.-F. & Ma, K. (2009) Species–area relationships explained by the joint effects of dispersal limitation and habitat heterogeneity. *Ecology*, **90**, 3033–3041.

Smith, A. B., Sandel, B., Kraft, N. J. B. & Carey, S. (2013) Characterizing scale-dependent community assembly using the functional-diversity–area relationship. *Ecology*, **94**, 2392–2402.

Swenson, N. G., Enquist, B. J., Pither, J., Thompson, J. & Zimmerman, J. K. (2006) The problem and promise of scale dependency in community phylogenetics. *Ecology*, **87**, 2418–2424.

Triantis, K. A., Guilhaumon, F. & Whittaker, R. J. (2012) The island species–area relationship: Biology and statistics. *Journal of Biogeography*, **39**, 215–231.

Tucker, C., Cadotte, M. W., Carvalho, S. B., Davies, J. T., Ferrier, S., Fritz, S., Grenyer, R., Helmus, M. R., Jin, L., Mooers, A. O., Pavoine, S., Purschke, O., Redding, D. W., Rosauer, D., Winter, M. & Mazel, F. (2017) A guide to phylogenetic metrics for conservation, community ecology and macroecology. *Biological Reviews*, **92**, 698–715.

Vane-Wright, R. I., Humphries, C. J. & Williams, P. H. (1991) What to protect? – Systematics and the agony of choice. *Biological Conservation*, **55**, 235–254.

Vellend, M. (2010) Conceptual synthesis in community ecology. *The Quarterly Review of Biology*, **85**, 183–206.

Venail, P., Gross, K., Oakley, T. H., Narwani, A., Allan, E., Flombaum, P., Isbell, F., Joshi, J., Reich, P. B., Tilman, D., van Ruijven, J. & Cardinale, B. J. (2015) Species richness, but not phylogenetic diversity, influences community biomass production and temporal stability in a re-examination of 16 grassland biodiversity studies. *Functional Ecology*, **29**, 615–626.

Villéger, S., Maire, E. & Leprieur, F. (2017) On the risks of using dendrograms to measure functional diversity and multidimensional spaces to measure phylogenetic diversity: A comment on Sobral *et al.* (2016). *Ecology Letters*, **20**, 554–557.

Villéger, S., Mason, N. & Mouillot, D. (2008) New multidimensional functional diversity indices for a multifaceted framwork in functional ecology. *Ecology*, **89**, 2290–2301.

Violle, C., Navas, M.-L., Vile, D., Kazakou, E., Fortunel, C., Hummel, I. & Garnier, E. (2007) Let the concept of trait be functional! *Oikos*, **116**, 882–892.

Wagner, C. E., Harmon, L. J. & Seehausen, O. (2014) Cichlid species–area relationships are shaped by adaptive radiations that scale with area. *Ecology Letters*, **17**, 583–592.

Wallace, A. (1876) *The geographical distribution of animals*. Cambridge: Cambridge University Press.

Wang, X., Swenson, N. G., Wiegand, T., Wolf, A., Howe, R., Lin, F., Ye, J., Yuan, Z., Shi, S., Bai, X., Xing, D. & Hao, Z. (2013) Phylogenetic and functional diversity area relationships in two temperate forests. *Ecography*, **36**, 883–893.

Warren, D. L., Cardillo, M., Rosauer, D. F. & Bolnick, D. I. (2014) Mistaking geography for biology: Inferring processes from species distributions. *Trends in Ecology & Evolution*, **29**, 572–580.

Webb, C. O., Ackerly, D. D., McPeek, M. A. & Donoghue, M. J. (2002) Phylogenies and community ecology. *Annual Review of Ecology, Evolution, and Systematics*, **33**, 475–505.

Weiher, E. & Keddy, P. A. (1995) Assembly rules, null models, and trait dispersion: New questions from old patterns. *Oikos*, **74**, 159–164.

Weiher, E. & Keddy, P. A. (eds.) (1999) *Ecological assembly rules: Perspectives, advances, retreats*. Cambridge: Cambridge University Press.

White, H. J., Montgomery, W. I., Pakeman, R. J. & Lennon, J. J. (2018) Spatio-temporal scaling of plant species richness and functional diversity in a temperate semi-natural grassland. *Ecography*, **41**, 845–856.

Whittaker, R. J., Rigal, F., Borges, P. A. V., Cardoso, P., Terzopoulou, S., Casanoves, F., Pla, L., Guilhaumon, F., Ladle, R. J. & Triantis, K. A. (2014) Functional biogeography of oceanic islands and the scaling of functional diversity in the Azores. *Proceedings of the National Academy of Sciences USA*, **111**, 13709–13714.

Wright, I. J., Reich, P. B., Cornelissen, J. H. C., Falster, D. S., Groom, P. K., Hikosaka, K., Lee, W., Lusk, C. H., Niinemets, Ü., Oleksyn, J., Osada, N., Poorter, H., Warton, D. I. & Westoby, M. (2005) Modulation of leaf economic traits and trait relationships by climate. *Global Ecology & Biogeography*, **14**, 411–421.

Zhang, H., Chen, H. Y. H., Lian, J., John, R., Ronghua, L., Liu, H., Ye, W., Berninger, F. & Ye, Q. (2018) Using functional trait diversity patterns to disentangle the scale-dependent ecological processes in a subtropical forest. *Functional Ecology*, **32**, 1379–1389.

6 · *Species–Area Relationships in Alien Species: Pattern and Process*

TIM M. BLACKBURN, PHILLIP CASSEY
AND PETR PYŠEK

6.1 Introduction

Earth is home to a remarkable diversity of life. We share our planet with trillions of individual animals and plants and unimaginable numbers of microbes. These organisms belong to millions of species, most of which remain undescribed (Mora et al., 2011), distributed over land and through the aquatic realms in a staggering array of forms. It is the job of biologists to understand the rules that underpin this complexity, yet it has not proved an easy task. Gradually, however, we have started to identify regularities in the diversity and distribution of life. We understand the broad process by which diversification occurs (Darwin, 1859), even if we are still haggling over the details (Sherratt & Wilkinson, 2009). We have also managed to identify regularities in how the resulting diversity is distributed across the planet (Brown, 1995; Rosenzweig, 1995; Whittaker & Fernández-Palacios, 2007). Species richness is generally greater at low compared with high latitudes, for example, and in larger relative to smaller geographic areas. This latter pattern is the species–area relationship (SAR), of course, and is our focus here.

The fact that species richness tends to increase with area has been known for more than a century (Chapter 2) and there is an enormous scientific literature documenting the form(s) of the relationship and proposing and testing hypotheses for the underlying mechanism(s) (e.g. Chapters 3 and 4). For most of the history of life, patterns in the distribution of species have been the result of natural processes of speciation, extinction and migration and the wide range of evolutionary, ecological, environmental, life history and behavioural factors that determine variation in those processes. In the last few centuries, however, these natural processes have increasingly been perturbed and supplanted by anthropogenic influences on the diversity and distribution of life. Rates of speciation, extinction and migration have all been affected by human activities, including habitat destruction and fragmentation, over-harvesting, agriculture and trade. Rates of extinction and migration in particular having increased worldwide in the last 500 years or so, by an estimated two-to-three orders of magnitude above standard background levels (Lawton & May, 1995; Gaston et al., 2003). Patterns of variation in these rates have also been affected, with rates especially elevated on islands (Moser et al., 2018).

The impacts of humanity on rates of migration have led to a new line of research into SARs. Humans have primarily affected migration by deliberately or accidentally translocating species beyond the natural biogeographic limits of their distributions, to areas where they do not naturally occur. These 'alien' species may establish viable populations in their new recipient areas. Currently, emergent populations of alien species are being recorded worldwide at an average rate of around one a day, with no sign that the rate of accumulation is slowing down (Seebens et al., 2017). Alien species that spread widely from the original location of establishment and which have negative impacts on the environment or socio-economic activities in their new range are termed invasive (CBD, 2002). The negative impacts of some alien species, coupled with the rate at which new aliens are accumulating, have created a strong impetus to understand the process by which some species invade and has led to the burgeoning research field of invasion biology (Richardson, 2011). One interesting question here is the extent to which alien species follow the same rules as native species when it comes to patterns of diversity. In this chapter, we review the increasing body of research exploring alien species richness in the context of geographic area.

First, we assess the extent to which the richness of alien and native species respond in similar ways to geographic area. This job has been facilitated by the recent publication of a review and analysis of studies that have assessed alien and native SARs for the same taxon in the same set of areas, based on twenty-three studies reporting thirty-six native-exotic pairs for both plant (twenty-two) and animal (fourteen) assemblages (Baiser & Li, 2018). Second, we assess how the addition of alien species to areas affects the overall patterns of SARs. Third, we assess what the relationship between area and richness for alien species tells us about the mechanisms controlling the species–area relationship.

6.2 Alien SARs

While studies in the ecological literature are near-unanimous in finding that the number of species is a positive function of area, there is a range of forms that the positive function can take (Chapters 4 and 7). Studies that report SARs for alien species, however, as far as we are aware, are unanimous in reporting species richness as a power function of area (although that is not to say another SAR function may provide a better fit in certain cases), such that:

$$S = cA^z, \tag{6.1}$$

or to linearize the relationship for convenience:

$$\log S = \log c + z \, \log A, \tag{6.2}$$

where S is species richness, A is area, z is the exponent (slope) of the relationship and c is the intercept.

While it is possible to plot the relationship using nested areas (type I, *sensu* Scheiner, 2003), such that smaller areas are subsets of larger ones, only two studies of alien SARs that we are aware of have adopted this approach (Hulme, 2008; Tarasi & Peet, 2017). This is too small a sample to draw conclusions about the effect of nesting areas on the form of alien versus native SARs (although typically type I SARs as a group have shallower slopes than other forms; Rosenzweig, 1995). All other studies plot variation in species richness across discrete areas of different sizes, with a slight preponderance of studies analysing areas that are true islands (i.e. areas of land separated by water) versus different sized areas of a contiguous landmass (e.g. different ecoregions within a continent; Baiser & Li, 2018). Traditionally, island SARs are thought to have steeper slopes than relationships from different sized areas of mainland (Rosenzweig, 1995).

Power function SARs can vary in terms of slope, intercept or both, but most attention is typically focused on the slope. Baiser and Li (2018) found that the slopes for alien SARs did not differ, on average, from the slopes for native species plotted from the same taxon in the same area (across studies, mean $z = 0.233$ versus 0.248, respectively). Thus, alien species richness increases with area at the same rate as does native species richness: in both cases, a ten-fold increase in area leads roughly to a 1.7-fold increase in the number of species present. Further analyses incorporating information on taxon and location revealed SAR slopes to be steeper for studies of plants than for animals and across islands versus different-sized patches of mainlands (mainly island SARs but two nested

SAR datasets were included), but still not to differ between alien and native relationships in these taxa and locations (Baiser & Li, 2018). Thus, plant richness accumulates faster with area than does animal richness and faster with area on islands than on mainlands, but in each case still at the same rate for alien and native species.

Comparison of SAR intercepts, on the other hand, reveals consistent differences between alien and native species: in all cases, intercepts were on average higher for native species than for aliens (Baiser & Li, 2018). Intercepts were also higher for plants than for animals. Thus, a given area in general has more species of plants than of animals and more native species than alien species. However, it is worth noting that all bar three of the fourteen studies for animals relate to vertebrates; we would expect areas of a given size to have fewer species of plants than of some invertebrate taxa, given the relative richness of different groups (Mora et al., 2011).

The resulting average SARs reported for native and alien animals and plants by Baiser and Li (2018) are presented in Figure 6.1, in the range 10 to 10,000 km^2. Slopes for island SARs would be slightly steeper than these depictions and slopes for mainland SARs slightly less steep. Overall, the relationships suggest a general trend for there to be more alien species

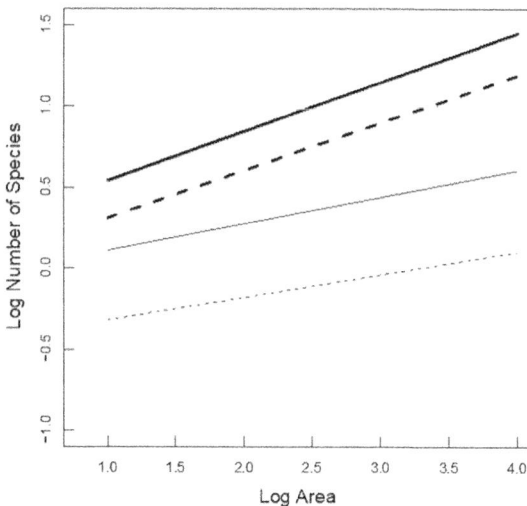

Figure 6.1 Depiction of the average species–area relationships identified by Baiser and Li (2018) in their review of the literature, calculated from the average c- and z-values for native and alien plants and animals, between areas of 10 and 10,000 km^2. Thick lines = plants, thin lines = animals, solid lines = natives, dashed lines = aliens.

of plant in a given area than native species of animals in the studied groups (though see the caveat about the lack of invertebrate studies) and for the difference to be larger in larger areas. A total of 13,168 plant species are known to have been naturalized somewhere in the world, corresponding to 3.9% of the extant global vascular flora (van Kleunen et al., 2015; Pyšek et al., 2017), versus just over 400 bird species (Dyer et al., 2017a) and around 150 mammal species (Capellini et al., 2015; Blackburn et al., 2017). In fact, there are more species of naturalized plant than there are native species of bird (around 11,000; HBW & BirdLife International, 2017) or mammal (c. 5,400; Wilson & Reeder, 2005). Given that the species richness of all these groups increases with area, we would expect alien plants to outnumber the native species of birds and mammals and many other taxa.

Additional studies document SARs for alien species, but without corresponding comparative data for native species. For example, global data on the combined richness of more than 15,000 alien species from eight taxonomic groups (vascular plants, ants, spiders, fishes, amphibians, reptiles, birds and mammals) across 446 regions of the world also showed a general positive effect of area ($z = 0.35$), with a steeper slope for island (0.53) versus mainland (0.25) regions (Dawson et al., 2017). The lack of equivalent slopes for native species in this and other studies makes them of lesser interest for a comparative assessment of SARs in aliens versus natives and so we do not consider them further here.

6.3 Aliens in SARs

The general similarity in the slopes of alien and native SARs for a given taxon and the larger intercepts of the latter imply that SAR analyses that do not account for the origins of the species included will produce results that differ little from analyses that exclude aliens. The average slope and intercept values for alien and native plant and animal SARs shown in Figure 6.1 translate into five times as many native as alien plant species in a given area, on average, and 3.5–4.0 times as many native as alien animal species. Thus, SARs that do not distinguish native and alien species will, in general, tend to have slightly higher intercept values, and minimally different slope values. That said, there is a reasonable amount of variation in the slopes and intercepts of alien and native SARs across studies, so that in some cases the presence of alien species can make a substantial difference to the observed relationship.

For example, Whittaker et al. (2014) plotted SARs for spiders and beetles of different provenance across nine islands in the Azores. The overall SAR for all spider species ($c = 2.915$, $z = 0.25$) was heavily driven by the presence of alien species ($c = 2.38$, $z = 0.268$), which constituted two thirds of the species in the analysis and similar proportions of the richness of individual islands. The SAR for indigenous spiders has both a lower intercept (2.034) and slope (0.220). Thus, the presence of alien spiders substantially alters the intercept of the SAR for Azorean spiders and to some degree also the slope. Aliens similarly dominate the beetle fauna of these islands, constituting more than 60 per cent of species. They subsequently elevate the Azorean beetle SAR substantially and again affect the slope to some degree (Whittaker et al., 2014).

Sax and Gaines (2005) found that SARs for native and alien plants on oceanic islands showed common slopes (0.31) and intercepts (1.46), such that these islands house equivalent richness of both groups. The addition of alien species effectively doubles the plant richness of these islands. Sax and Gaines (2005) plot SARs for plants in five sets of areas before and after human intervention (i.e. before versus after human-mediated extinction and naturalization). They found that in all cases, the slopes of the relationships remained effectively constant, but the intercepts increased. However, the magnitude of the increases relates to the richness and isolation of the areas. Isolated, species-poor oceanic islands increase the most (100 per cent increase), followed by California Channel Islands (44 per cent), with well-connected, species-rich Californian mainland counties increasing in plant richness the least (17 per cent). In this example, aliens are thus causing the richness of species-poor and species-rich areas to converge.

6.4 Mechanisms Underlying Alien SARs

The burgeoning number of alien species establishing populations worldwide provides strong motivation on its own to study patterns of variation in their richness. Alien species are a classic example of a natural experiment (Diamond, 1986), in which human actions intentionally or accidentally bring about changes in natural systems that can be considered analogous to experimental manipulations. As such, they represent an opportunity to explore the mechanisms that underpin variation in diversity: i) through the extent to which they show similar or divergent patterns to native species and ii) given similarities and differences in the way different hypothesized mechanisms may act on alien compared to native species (Sax et al., 2005; Cadotte et al., 2006). What then do the regularities in alien versus native SARs tell us about the determinants of richness in both groups?

6.4.1 Sampling Effects

Arguably, the simplest answer to this question is 'nothing'. This position arises from the observation that the number of alien species in an area is typically a strong and positive function of the number of alien species that have been introduced to that area (termed colonization pressure; Lockwood et al., 2009). For example, Blackburn et al. (2008) showed that the SAR for alien birds established on a sample of thirty-five islands worldwide had a more or less identical slope (but lower intercept) to the SAR for alien birds introduced to those islands (0.18 versus 0.20). This suggests that alien bird richness was effectively a constant proportion of the number of species humans had translocated to an island, with variation in colonization pressure therefore being the primary driver of the alien SAR slope. Dyer et al. (2017b) showed that colonization pressure was by far the strongest predictor of alien bird species richness worldwide, albeit with other anthropogenic (time since first introduction, distance to a historic port) and environmental (native species richness) factors explaining additional variation in richness. A positive relationship between alien species richness and colonization pressure is expected because, at least in closed systems such as many oceanic islands or archipelagos, the latter sets a ceiling on the former (Lockwood et al., 2009): thus, the relationship is between Y and X + Y, where Y is alien species richness (the number of introduced populations that establish in an area) and X is the number of introduced populations that fail. The null expectation for relationships of this form, which are termed 'spurious' (Prairie & Bird, 1989; Brett, 2004), is positive rather than zero. While alien species can in theory disperse between geographic regions, in practice few aliens spread widely enough that the richness gains from spread outweigh losses due to the failure of introduced populations to establish (most populations fail; Williamson, 1996). Therefore, any SAR we observe for established alien species may actually be a SAR for introduced alien species, with no additional processes required to generate the established alien SAR beyond random extinction.

Unfortunately, it is currently difficult to establish the generality of the relationship between alien species richness and colonization pressure, because, aside from birds, data on colonization pressure are few and far between. The fact that the relationship is spurious suggests that alien SARs will necessarily be due at least in part to colonization pressure, regardless of taxon. How large that part is depends on how many introduced populations subsequently fail to establish, because, in spurious relationships of this type, the correlation is small when Y is much smaller than X. For example, Brett (2004) used simulations to show that X + Y explains around 50 per cent of the variation in Y (r ≈ 0.7) when X and Y are equal, falling to around 5 per cent (r ≈ 0.22) when Y/X = 0.2, with only modest effects of sample size (the absolute magnitudes of X and Y). Thus, alien SARs will be less likely to be a function of colonization pressure when most introductions fail (Y≪X). In the island bird data analysed by Blackburn et al. (2008), the numbers of successes and failures were about equal and so we would expect the alien SAR to be strongly determined by colonization pressure. It might not be unreasonable to assume that establishment success rates are relatively high for large and adaptable homeothermic taxa like birds, which therefore may not be representative of the extent to which colonization pressure influences alien SARs in other taxa. Yet again, the lack of information on colonization pressure hampers a robust assessment of the situation, although such data as do exist show that establishment success rates can be highly variable within taxa (Jeschke & Strayer, 2005).

If alien taxa show SARs largely because colonization pressure is a positive function of area, this begs the question of why the number of species introduced should increase with area. The most likely answer is that human population size (Pyšek, 1998; Blackburn et al., 2008) and the associated volume of traded goods imported (Moser et al., 2018) both increase with area. These relationships mean that the number of alien species introduced accidentally through trade would be expected to increase with area and so too would opportunities for deliberate introductions. Indeed, multivariate analysis for alien bird introductions to islands showed that area did not explain variation in colonization pressure if human population size was also included in the model (Blackburn et al., 2008). What is interesting, however, is that, for island data, both human population size and trade increase with area with exponents close to 1 (Blackburn et al., 2008; Moser et al., 2018), while the exponent for colonization pressure is close to the typical z-value for island SARs ($z = 0.2$; Blackburn et al., 2008). In fact, we might reasonably expect such a relationship from sampling effects alone.

In a classic paper, Preston (1962) derived an expected z-value for island SARs of around 0.27, assuming a specific ('canonical') lognormal form for the species abundance distribution, a given minimum number of individuals necessary for a species to persist and that the number of individuals on an island is proportional to its area. Larger areas house more individuals and, as a result, more species on the basis of the underlying species abundance distribution. The canonical lognormal form of this distribution means that, while the number of individuals scales proportionally to area, the number of species scales with z close to 0.25 (different species abundance distributions give different predicted z-values; May, 1975). Given that the introduction of alien individuals, especially those introduced by chance, may be thought of as a random sampling process from an underlying pool of species with a given species abundance distribution, it is interesting that this process produces a slope for the introduced alien SAR that is close to the theoretical expectation. This implies that islands are sampling introduced individuals in proportion to their area. The alien SAR could then arise because a proportion of those introduced species goes extinct that is unrelated to area. However, two unanswered questions are: i) does random sampling from an underlying species abundance distribution give a slope for the introduced alien SAR that is close to that observed given that, at this stage, we have no constraint imposed for a minimum number of individuals and ii) would we expect extinction rate to be unrelated to area if the number of individuals sampled by a location is proportional to its area?

6.4.2 Simulating the Establishment Process

To answer these questions and assess whether a sampling model can produce patterns consistent with those we observe in alien SARs, we conducted a bespoke simulation experiment (Figure 6.2) in the R software environment (Version 3.5.1) for statistical and graphical computing (R Core Team, 2018). A species abundance distribution was constructed in the package *mobsim* (May et al., 2018) from a lognormal distribution with K species and $I = 1 \times 10^7$ individuals; where for the examples shown $K = 100$ (Figure 6.3A–C), 1,000 (Figure 6.3D–F) or 10,000 (Figure 6.3G–I).

For 100 iterations we randomly introduced individuals from our species abundance distribution to ten islands in varying total numbers (100, 200, 500, 1,000, 2,000, 5,000, 10,000, 12,500, 15,000, 20,000 individuals introduced in total across all species), where we assume that the larger numbers of individuals are introduced to larger islands. For each simulation, we extracted the total number of species that were

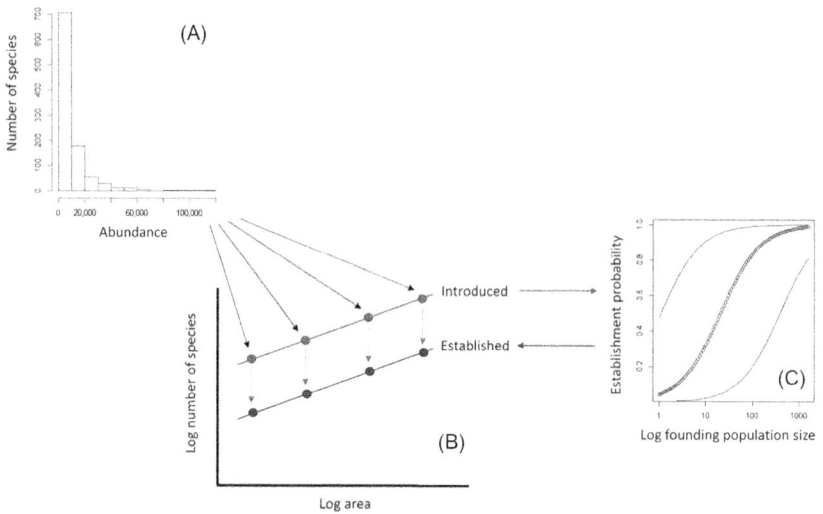

Figure 6.2 An illustration of the simulation model. For each of a set of islands (four in this example), a number of individuals is sampled at random from an underlying lognormal species abundance distribution (A), with the number of individuals proportional to island area. The number of species in those random samples gives the introduced species–area relationship for the islands (upper points and line in B). The probability that each species establishes a viable population is a stochastic function of its founding population size (C), such that some introduced species fail to establish (predominantly species with low founding population sizes). Those failures transform the introduced alien species–area relationship into the established alien species–area relationship (lower points and line in B). (A) depicts the actual species abundance distribution with 10,000,000 individuals and $K = 1,000$ and (C) depicts the actual functions for establishment as a function of propagule pressure, with the central line (comprised of open circles) corresponding to Equation (6.3) (Cassey et al., 2018) and the lines either side to Equation (6.3), but adding 3 or subtracting 3 from the intercept.

introduced to a given island across those introduced individuals, to calculate an introduced alien SAR (upper points; Figures 6.2 and 6.3), assuming that island area was proportional to the number of introduced individuals.

Each introduced species has a founding population size (the total number of individuals of that species introduced in that iteration of the simulation, i.e. propagule pressure *sensu* Cassey et al., 2018). We then evaluated the probability that a species introduced with a propagule pressure of q individuals would be likely to establish successfully on an island by comparing the mean probability of establishment for a given

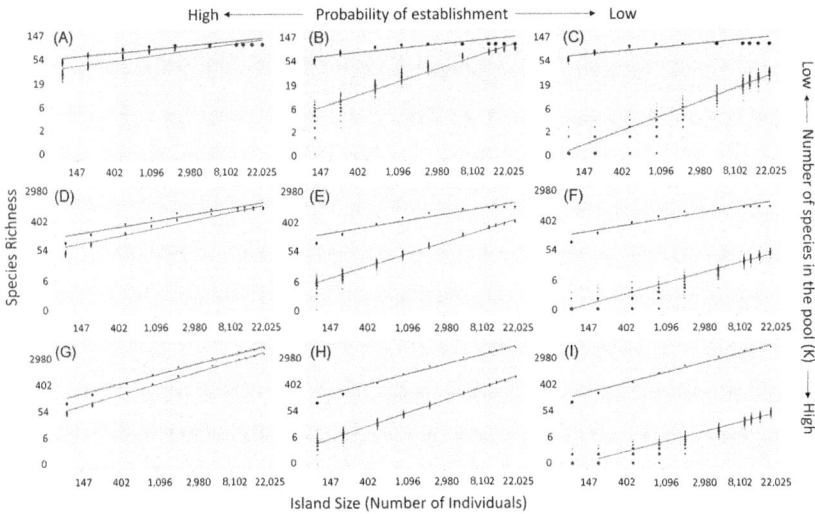

Figure 6.3 The outcome of simulations to model introduced (upper lines and points in each plot) and established (lower lines and points in each plot) alien species–area relationships (SARs). Introduced alien SARs were produced by random samples from an underlying species abundance distribution (Figure 6.2A) with 1×10^7 individuals and $K = 100$ (A, B, C; Low), 1,000 (D, E, F) or 10,000 (G, H, I; High) species. Established alien SARs were produced by the failure of some introduced species to establish, where establishment probability can be high (A, D, G), medium (B, E, H) or low (C, F, I). The equations for the regression lines are given in Table 6.1.

propagule pressure q (taken from the logistic function provided in figure 3 in Cassey et al. (2018) and reproduced here as Figure 6.2C) as follows:

$$\text{logit(Establishment Success)} = -3.10 + 2.35 * \log_{10}(q), \qquad (6.3)$$

with a random uniform integer between zero and one. If the probability of establishment (for a given propagule pressure) was greater than the random integer then the introduced species was considered to be successfully established on the island. Otherwise the species failed to establish. For each simulation we extracted the total number of species that were successfully established on a given island to calculate the established alien SAR (lower points; Figures 6.2 and 6.3). We examined variation in the form of the logistic relationship between establishment success and propagule pressure (Figure 6.3B, E and H) by adding 3 to (Figure 6.3A, D and G) or subtracting 3 from, the intercept (Figure 6.3C, F and I) in Equation (6.3).

Real data for island bird introductions (Blackburn et al., 2008) show more or less parallel slopes for introduced and established alien SARs. Our simulations show that the precise form of the SARs for introduced and established species depends upon both the number of species in the pool, K, and how establishment success relates to propagule pressure (Table 6.1; Figure 6.3). Fewer introduced species establish when establishment probability for a given propagule pressure is lower (e.g. cf. Figure 6.3A–C), as expected. However, how this affects the slope of the established alien SAR also depends on K (cf. Figure 6.3C, F and I).

The difference between the slope values of introduced and established alien SARs increases as establishment probability decreases, and decreases as K increases (Table 6.1). The first of these two effects is because decreasing establishment success has its largest impacts on alien species richness on small islands (cf. Figure 6.3A and C), where propagule pressures will be lower, dragging down the slope of the established alien SAR relative to that for introduced species. This leads to greater differences in the intercepts for introduced and established alien SARs. Increasing K increases the number of species that get introduced, but decreases the average propagule pressure for each of these species, thus decreasing the likelihood that they will establish. The effect of higher K in lowering establishment success is greatest at larger island sizes, because the much higher introduced alien richness on these islands is largely accompanied by smaller propagule pressures. This leads to relatively higher failure rates on large islands as K increases (cf. Figure 6.3C and I). In combination, these two effects lead to similar slope values for introduced and established alien SARs when establishment probability is high and K is low (Figure 6.3A) – most species on most islands succeed – and when establishment probability is low and K is high (Figure 6.3I) – most species on most islands fail. In other words, parallel SARs for introduced and established species are more likely when establishment success is equalized across islands.

Simulated slope values (Table 6.1) are generally steep relative to those observed for introduced (Blackburn et al., 2008) and established (Baiser & Li, 2018) alien SARs. The introduced SAR slope depends on K alone and, hence, the ratio of individuals to species in the simulations. When the underlying species pool consisted of $I = 1 \times 10^7$ individuals and $K = 100$ species, this slope approximated that observed for island birds (c. 0.20), but steepened as the number of species increased. This suggests that a low ratio of K/I might approximate the species pool from which birds introduced to islands were drawn. What that ratio might be

Table 6.1 *The intercept and slope values for the regression lines shown in Figure 6.3. The probability of establishment (P_{est}) does not affect the form of the SAR for introduced aliens, which is therefore the same (given stochastic variation) for all P_{est} for a given K. Medium P_{est} derives from Equation (6.3), with high and low P_{est} based on this equation but adding or subtracting 3 from the intercept (see Section 6.4.2). All slopes and intercepts are significantly different from zero at $P < 0.001$.*

K	Invasion stage	High P_{est}		Medium P_{est}		Low P_{est}	
		Intercept	Slope	Intercept	Slope	Intercept	Slope
100	Introduction	3.33 ± 0.08	0.16 ± 0.01	3.33 ± 0.08	0.16 ± 0.01	3.29 ± 0.07	0.16 ± 0.01
	Establishment	2.73 ± 0.07	0.21 ± 0.01	−0.47 ± 0.12	0.51 ± 0.01	−2.64 ± 0.27	0.59 ± 0.03
1,000	Introduction	3.16 ± 0.06	0.41 ± 0.01	3.13 ± 0.07	0.41 ± 0.01	3.08 ± 0.07	0.42 ± 0.01
	Establishment	1.84 ± 0.06	0.52 ± 0.01	−2.05 ± 0.08	0.82 ± 0.01	−3.39 ± 0.24	0.71 ± 0.01
10,000	Introduction	1.69 ± 0.04	0.73 ± 0.01	1.64 ± 0.04	0.74 ± 0.01	1.63 ± 0.04	0.74 ± 0.01
	Establishment	0.21 ± 0.03	0.84 ± 0.01	−2.88 ± 0.06	0.96 ± 0.01	−3.62 ± 0.25	0.74 ± 0.01

in reality is unknown, although for the British breeding avifauna $K/I \sim$ $225/1.6 \times 10^8$ or around one tenth that in Figure 6.3A (Musgrove et al., 2013; Blackburn & Gaston, 2018). In general, establishment failure steepened these SAR slopes, so that z-values for established species were generally in the range of 0.5 and above (Table 6.1). This is because, in most cases, failure was proportionally higher on smaller islands. Again, however, slopes were closer to those typically observed in alien SARs when K was low. Introduced alien SARs in general appear more curvilinear on log–log axes than established alien SARs (Figure 6.3), but we did not formally explore this element further.

Our simulations are simple and only scratch the surface of possible parameter space in terms of K, I, the form of the species abundance distribution, and how individuals might be sampled from that distribution. They also assume that alien individuals reach islands in proportion to island area, which is speculative if not unreasonable, and do not address alien population dynamics beyond establishment. That SARs can be derived by sampling from underlying species abundance distributions is well known since the classic work of Preston (1962; see review in McGill et al., 2007), but to our knowledge no one has previously explored how establishment probability might transform introduced SARs into established SARs. It is clear that some simple assumptions allow realistic introduced SARs to be modelled as sampling effects and realistic alien SARs to then arise from introduced population failure. Sampling effects can therefore potentially explain both native and alien SARs and, if native and alien colonists are drawn from underlying species abundance distributions of similar form, why they have very similar slopes (Baiser & Li, 2018).

6.4.3 Other Mechanisms

If the similarity between native and alien SARs is due to a common mechanism, some of the other explanations proposed to underpin native SARs seem likely to be ruled out. For example, MacArthur and Wilson (1963, 1967) proposed that SARs are the result of size-dependent variation in the rates of colonization and extinction across islands. Larger islands intercept more colonists and the species that colonize have lower extinction rates than on small islands. However, for aliens, while the number of colonist species does increase with area, extinction does not seem (on the limited evidence available) to be a negative function of area as MacArthur and Wilson posit – a more or less constant proportion (and hence a larger number) of introduced aliens goes extinct from larger islands (Blackburn et al., 2008). That element of the Equilibrium Theory of Island Biogeography is not supported, at least by island alien birds. Metapopulation-based explanations also seem unlikely, inasmuch as extinction probability in such models is also a negative function of patch area (Hanski & Gyllenberg, 1997).

One oft-cited mechanism for SARs in native species is that areas with a wider range of habitats can support more species, while habitat diversity is a positive function of area (Williams, 1964; Pyšek et al., 2002). A related idea is that smaller areas may possess different habitats to larger ones as a consequence of their smallness (Whittaker & Fernández-Palacios, 2007). Leaving aside the question of how habitats are defined (which in practice is often in terms of their biological communities, implying a degree of circularity), native and alien SARs could show similar slopes if native and alien species respond to habitat diversity in the same way. Once again, it seems surprising under this mechanism that more of the alien populations introduced to larger areas should go extinct, as one might expect introduced species to be more likely to find suitable habitats in larger areas. However, it is not beyond the bounds of possibility that a fairly constant proportion of species would fail to be translocated to suitable habitats in areas of different sizes, given that more species are translocated to larger areas (i.e. the introduced alien SAR).

In a related vein, Baiser and Li (2018) suggest that, if alien richness is determined by the likelihood of disturbance, aliens and natives may show similar SAR exponents if disturbance scales with area in the same way as habitat diversity. This seems an unlikely coincidence across the range of studies they reviewed. However, disturbance has itself been proposed as a general driver of SARs, either if higher disturbance in small areas makes such areas unsuitable for some species in the source pool (McGuinness, 1984) or, conversely, if greater disturbance in some areas creates opportunities for species that otherwise would not be present (Whittaker & Fernández-Palacios, 2007). Traditionally, species richness is argued to be highest at intermediate levels of disturbance, as this allows the coexistence of species that are good dispersers but poor competitors (and hence that thrive in disturbed areas) with species that are good competitors but poor dispersers, which would come to dominate communities in the absence of disturbance (Connell, 1978). In effect, this is a restatement of the habitat diversity hypothesis, but where diversity is maintained by some areas (but not all) being subject to disturbance. Nevertheless, if some disturbance leads to increased native species richness and allows alien species to establish, this could in theory lead to similar scaling of native and alien richness with area. We know of no tests of this idea for alien species, although the hypothesis of a relationship between species richness and disturbance is arguably unsupported on both empirical and theoretical grounds (Fox, 2013).

6.5 Conclusions

The species–area relationship is one of the most general patterns in ecology (Chapters 1–3) and it is little surprise to see that alien species also conform to this general pattern (Baiser & Li, 2018). With all else being equal, we would expect to find more species in larger geographic areas even if life were distributed randomly across the surface of the planet. Yet, native and alien species are not just similar in showing positive SARs – the slope values of their respective relationships would seem to be too close to arise by coincidence. Alien species also exhibit other macroecological patterns, including Bergmann's Rule (Blackburn et al., 2019) and Rapoport's Rule (Sax, 2001; Dyer et al., 2020), suggesting that drivers of the distribution of native species across the planet also force alien species distributions to adhere to their rules. Given that alien species represent a massive natural experiment, encompassing more or less every habitat on every significant continent and island, this appears to open up an unprecedented opportunity to study the mechanisms underpinning the distribution of biodiversity worldwide.

And yet, alien SARs are the result of a sequential process involving the transportation of individuals beyond their natural geographic range limits, the introduction of some proportion of those individuals to a place where they do not naturally occur and the establishment of a viable population (only) from those introduced individuals (Blackburn et al., 2011). The outcomes of later stages in this process depend on inputs from the earlier stages. In particular, the alien species richness of an area is typically some proportion (<1) of the number of alien species that were introduced to the area; that is, colonization pressure. In the one data set we have, island birds, this proportion is essentially constant across islands of different areas, suggesting that the slope of the SAR for established aliens is a consequence of the slope of the SAR for introduced aliens, with the intercept modified by random extinction. Colonization pressure similarly influences Bergmann's and Rapoport's Rules in alien birds (Blackburn et al., 2019; Dyer et al., 2020) and may also explain when alien species do not match the macroecological patterns of native species, for example in terms of spatial gradients of species richness (Dawson et al., 2017; Dyer et al., 2017b).

That said, there are still many interesting and unresolved questions about the similarities between native and alien diversity patterns. In terms of SARs, for example, we would expect positive relationships if life were distributed randomly across the surface of the planet, which is a state that may at least be approximated by some alien species groups. The non-deliberate transport and introduction of many alien species may well approximate a random sampling process, leading to widespread and abundant native species being more likely to be introduced (Colautti et al., 2006; Leung & Roura-Pascual, 2012); this is also true for deliberate introductions (Blackburn & Duncan, 2001). Yet, this process generally leads to SARs with near-identical slopes to those for the native species in the same taxon and location, once we factor in a biologically realistic establishment process (Figure 6.2C). Can we identify drivers of establishment failure related to any underlying factor? Could native SARs ultimately be a consequence of random sampling of colonists from a species pool, with the same establishment process (driven by founding population size or propagule pressure) whittling the colonists down to the SAR? What would a 'colonist SAR' look like for native species? Given that we only really have one dataset underpinning our knowledge of how introduced and established alien SARs might be related (island birds; Blackburn et al., 2008), how representative is the introduced alien SAR in those data? Parallel introduced and established alien SAR slopes seems a less likely outcome of random sampling processes than a steeper alien SAR slope (Figure 6.3), as we would generally expect higher extinction rates on smaller islands. However, our simulations barely begin to explore the possible parameter space of relevant processes.

The similar slopes of alien and native SARs, for the same taxa over the same areas, are also interesting in terms of their implications and consequences for the overall biodiversity of the areas concerned. While areas are losing species to extinction at an elevated rate, as well as gaining species through alien establishment, aliens in general add to the species richness of areas, with gains more or less in proportion to native species richness. Thus, areas with more species tend to gain more species (Stohlgren et al., 2006). What does this mean for the drivers of species richness? Does it mean that areas generally are not saturated in terms of their species richness and can readily gain new species without consequences? Does it mean that human activities in those areas have facilitated higher levels of richness, perhaps through the effects of disturbance

(Kolar & Lodge, 2001)? Or is biodiversity now out of equilibrium in many areas and, in the long term, we can expect more extinctions, perhaps in both the native and alien biotas? Certainly, many species are teetering on the brink (IUCN, 2018). Time will tell. In the meantime, it will be helpful to try to gather more information on the process of invasion by alien species, especially how introduced richness transforms into established richness, both for what it tells us about how colonization processes might operate in practice to generate richness and for mitigating that transformation in the future.

Acknowledgements

We thank Ben Baiser for letting us see a copy of his key paper before publication, and Tom Matthews for inviting us to contribute to this volume.

References

Baiser, B. & Li, D. (2018) Comparing species–area relationships of native and exotic species. *Biological Invasions*, **20**, 3647–3658.

Blackburn, T. M. & Duncan, R. P. (2001) Establishment patterns of exotic birds are constrained by non-random patterns in introduction. *Journal of Biogeography*, **28**, 927–939.

Blackburn, T. M. & Gaston, K. J. (2018) Abundance, biomass and energy use of native and alien breeding birds in Britain. *Biological Invasions*, **20**, 3563–3573.

Blackburn, T. M., Lockwood, J. L. & Cassey, P. (2008) The island biogeography of exotic bird species. *Global Ecology & Biogeography*, **17**, 246–251.

Blackburn, T. M., Pyšek, P., Bacher, S., Carlton, J. T., Duncan, R. P., Jarošík, V., Wilson, J. R. U. & Richardson, D. M. (2011) A proposed unified framework for biological invasions. *Trends in Ecology & Evolution*, **26**, 333–339.

Blackburn, T. M., Redding, D. W. & Dyer, E. E. (2019) Bergmann's Rule in alien birds. *Ecography*, **42**, 102–110.

Blackburn, T. M., Scrivens, S. L., Heinrich, S. & Cassey, P. (2017) Patterns of selectivity in introductions of mammal species worldwide. *NeoBiota*, **33**, 33–51.

Brett, M. T. (2004) When is correlation between non-independent variables 'spurious'? *Oikos*, **105**, 647–656.

Brown, J. H. (1995) *Macroecology*. Chicago, IL: University of Chicago Press.

Cadotte, M. W., McMahon, S. M. & Fukami, T. (2006) *Conceptual ecology and invasion biology: Reciprocal approaches to nature*. Dordrecht, the Netherlands: Springer.

Capellini, I., Baker, J., Allen, W. L., Street, S. E. & Venditti, C. (2015) The role of life history traits in mammalian invasion success. *Ecology Letters*, **18**, 1099–1107.

Cassey, P., Delean, S., Lockwood, J. L., Sadowski, J. S. & Blackburn, T. M. (2018) Dissecting the null model for biological invasions: A meta-analysis of the propagule pressure effect. *PLoS Biology*, **16**, e2005987.

CBD (2002) *COP 6 Decision VI/23. Alien species that threaten ecosystems, habitats or species.* www.cbd.int/decision/cop/default.shtml?id=7197.

Colautti, R. I., Grigorovich, I. A. & MacIsaac, H. J. (2006) Propagule pressure: A null model for biological invasions. *Biological Invasions*, **8**, 1023–1037.

Connell, J. H. (1978) Diversity in tropical rain forests and coral reefs. *Science*, **199**, 1302–1310.

Darwin, C. (1859) *On the origin of species.* London: Murray.

Dawson, W., Moser, D., Van Kleunen, M., Kreft, H., Pergl, J., Pyšek, P., Weigelt, P., Winter, M., Lenzner, B., Blackburn, T. M., Dyer, E. E., Cassey, P., Scrivens, S. L., Economo, E. P., Guénard, B., Capinha, C., Seebens, H., García-Díaz, P., Nentwig, W., García-Berthou, E., Casal, C., Mandrak, N. E., Fuller, P., Meyer, C. & Essl, F. (2017) Global hotspots and correlates of alien species richness across taxonomic groups. *Nature Ecology & Evolution*, **1**, 0186.

Diamond, J. (1986) Overview: Laboratory experiments, field experiments, and natural experiments. *Community ecology* (ed. by J. Diamond and T. J. Case), pp. 3–22. New York: Harper Row.

Dyer, E. E., Cassey, P., Redding, D. W., Collen, B., Franks, V., Gaston, K. J., Jones, K. E., Kark, S., Orme, C. D. L. & Blackburn, T. M. (2017b) The global distribution and drivers of alien bird species richness. *PLoS Biology*, **15**, e2000942.

Dyer, E. E., Redding, D. W., Cassey, P., Collen, B. & Blackburn, T. M. (2020) Evidence for Rapoport's rule and latitudinal patterns in the distribution of alien bird species. *Journal of Biogeography*, **47**, 1362–1372.

Dyer, E. E., Redding, D. W. & Blackburn, T. M. (2017a) The global avian invasions atlas, a database of alien bird distributions worldwide. *Scientific Data*, **4**, 170041.

Fox, J. W. (2013) The intermediate disturbance hypothesis should be abandoned. *Trends in Ecology & Evolution*, **28**, 86–92.

Gaston, K. J., Jones, A. G., Hanel, C. & Chown, S. L. (2003) Rates of species introduction to a remote oceanic island. *Proceedings of the Royal Society B: Biological Sciences*, **270**, 1091–1098.

Hanski, I. & Gyllenberg, M. (1997) Uniting two general patterns in the distribution of species. *Science*, **275**, 397–400.

HBW & BirdLife International (2017) *Handbook of the birds of the World and BirdLife International digital checklist of the birds of the world. Version 2.* datazone.birdlife .org/userfiles/file/Species/Taxonomy/HBW-BirdLife_Checklist_v2% 20Dec17.zip.

Hulme, P. E. (2008) Contrasting alien and plant species–area relationships: The importance of spatial grain and extent. *Global Ecology & Biogeography*, **17**, 641–647.

IUCN (2018) *The IUCN red list of threatened species. Version 2018-2.* www.iucnredlist .org/.

Jeschke, J. M. & Strayer, D. L. (2005) Invasion success of vertebrates in Europe and North America. *Proceedings of the National Academy of Sciences USA*, **102**, 7198–7202.

van Kleunen, M., Dawson, W., Essl, F., Pergl, J., Winter, M., Weber, E., Kreft, H., Weigelt, P., Kartesz, J., Nishino, M., Antonova, L. A., Barcelona, J. F.,

Cabezas, F. J., Cárdenas, D., Cárdenas-Toro, J., Castaño, N., Chacón, E., Chatelain, C., Ebel, A. L., Figueiredo, E., Fuentes, N., Groom, Q. J., Henderson, L., Inderjit, Kupriyanov, A., Masciadri, S., Meerman, J., Morozova, O., Moser, D., Nickrent, D. L., Patzelt, A., Pelser, P. B., Baptiste, M. P., Poopath, M., Schulze, M., Seebens, H., Shu, W., Thomas, J., Velayos, M., Wieringa, J. J. & Pyšek, P. (2015) Global exchange and accumulation of non-native plants. *Nature*, **525**, 100–103.

Kolar, C. S. & Lodge, D. M. (2001) Progress in invasion biology: Predicting invaders. *Trends in Ecology & Evolution*, **16**, 199–204.

Lawton, J. H. & May, R. M. (eds.) (1995) *Extinction rates*. Oxford: Oxford University Press.

Leung, B. & Roura-Pascual, N. (2012) TEASIng apart alien species risk assessments: A framework for best practices. *Ecology*, **15**, 1475–1493.

Lockwood, J. L., Cassey, P. & Blackburn, T. M. (2009) The more you introduce the more you get: The role of colonization pressure and propagule pressure in invasion ecology. *Diversity and Distributions*, **15**, 904–910.

MacArthur, R. H. & Wilson, E. O. (1963) An equilibrium theory of insular zoogeography. *Evolution*, **17**, 373–387.

MacArthur, R. H. & Wilson, E. O. (1967) *The theory of island biogeography*. Princeton, NJ: Princeton University Press.

May, F., Gerstner, K., McGlinn, D. J., Xiao, X. & Chase, J. M. (2018) mobsim: An R package for the simulation and measurement of biodiversity across spatial scales. *Methods in Ecology and Evolution*, **9**, 1401–1408.

May, R. M. (1975) Patterns of species abundance and diversity. *Ecology and evolution of communities* (ed. by M. L. Cody and J. M. Diamond), pp. 81–120. Cambridge, MA: Harvard University Press.

McGill, B. J., Etienne, R. S., Gray, J. S., Alonso, D., Anderson, M. J., Benecha, H. K., Dornelas, M., Enquist, B. J., Green, J. L., He, F., Hurlbert, A. H., Magurran, A. E., Marquet, P. A., Maurer, B. A., Ostling, A., Soykan, C. U., Ugland, K. I. & White, E. P. (2007) Species abundance distributions: Moving beyond single prediction theories to integration within an ecological framework. *Ecology Letters*, **10**, 995–1015.

McGuinness, K. A. (1984) Species–area curves. *Biological Reviews*, **59**, 423–440.

Mora, C., Tittensor, D. P., Adl, S., Simpson, A. G. B. & Worm, B. (2011) How many species are there on Earth and in the ocean? *PLOS Biology*, **9**, e1001127.

Moser, D., Lenzner, B., Weigelt, P., Dawson, W., Kreft, H., Pergl, J., Pyšek, P., van Kleunen, M., Winter, M., Capinha, C., Cassey, P., Dullinger, S., Economo, E. P., García-Díaz, P., Guénard, B., Hofhansl, F., Mang, T., Seebens, H. & Essl, F. (2018) Remoteness promotes biological invasions on islands worldwide. *Proceedings of the National Academy of Sciences USA*, **115**, 9270–9275.

Musgrove, A., Aebischer, N., Eaton, M., Hearn, R., Newson, S., Noble, D., Parsons, M., Risely, K. & Stroud, D. (2013) Population estimates of birds in Great Britain and the United Kingdom. *British Birds*, **106**, 64–100.

Prairie, Y. T. & Bird, D. F. (1989) Some misconceptions about the spurious correlation problem in the ecological literature. *Oecologia*, **81**, 285–288.

Preston, F. W. (1962) The canonical distribution of commonness and rarity: Part I and II. *Ecology*, **43**, 185–215, 410–432.

Pyšek, P. (1998) Alien and native species in Central European urban floras: A quantitative comparison. *Journal of Biogeography*, **25**, 155–163.

Pyšek, P., Kučera, T. & Jarošík, V. (2002) Plant species richness of nature reserves: The interplay of area, climate and habitat in a central European landscape. *Global Ecology & Biogeography*, **11**, 279–289.

Pyšek, P., Pergl, J., Essl, F., Lenzner, B., Dawson, W., Kreft, H., Weigelt, P., Winter, M., Kartesz, J., Nishino, M., Antonova, L. A., Barcelona, J. F., Cabezas, F. J., Cárdenas, D., Cárdenas-Toro, J., Castaño, N., Chacón, E., Chatelain, C., Dullinger, S., Ebel, A. L., Figueiredo, E., Fuentes, N., Genovesi, P., Groom, Q. J., Henderson, L., Inderjit, Kupriyanov, A., Masciadri, S., Maurel, N., Meerman, J., Morozova, O., Moser, D., Nickrent, D., Nowak, P. M., Pagad, S., Patzelt, A., Pelser, P. B., Seebens, H., Shu, W., Thomas, J., Velayos, M., Weber, E., Wieringa, J. J., Baptiste, M. P. & van Kleunen, M. (2017) Naturalized alien flora of the world: Species diversity, taxonomic and phylogenetic patterns, geographic distribution and global hotspots of plant invasion. *Preslia*, **89**, 203–274.

R Core Team (2018) *R: A language and environment for statistical computing*. Vienna, Austria: R Foundation for Statistical Computing.

Richardson, D. M. (ed.) (2011) *Fifty years of invasion ecology: The legacy of Charles Elton*. Oxford: Wiley-Blackwell.

Rosenzweig, M. L. (1995) *Species diversity in space and time*. Cambridge: Cambridge University Press.

Sax, D. F. (2001) Latitudinal gradients and geographic ranges of exotic species: Implications for biogeography. *Journal of Biogeography*, **28**, 139–150.

Sax, D. F. & Gaines, S. D. (2005) The biogeography of naturalised species and the species–area relationship: Reciprocal insights to biogeography and invasion biology. *Conceptual ecology and invasions biology: Reciprocal approaches to nature* (ed. by M. W. Cadotte, S. M. McMahon and T. Fukami), pp. 449–479. Dordrecht, Netherlands: Kluwer.

Sax, D. F., Stachowicz, J. J. & Gaines, S. D. (eds.) (2005) *Exotic species: A source of insight into ecology, evolution, and biogeography*. Sunderland, MA: Sinauer Associates.

Scheiner, S. M. (2003) Six types of species–area curves. *Global Ecology & Biogeography*, **12**, 441–447.

Seebens, H., Blackburn, T. M., Dyer, E. E., Genovesi, P., Hulme, P. E., Jeschke, J. M., Pagad, S., Pyšek, P., Winter, M., Arianoutsou, M., Bacher, S., Blasius, B., Brundu, G., Capinha, C., Celesti-Grapow, L., Dawson, W., Dullinger, S., Fuentes, N., Jäger, H., Kartesz, J., Kenis, M., Kreft, H., Kühn, I., Lenzner, B., Leibhold, A., Mosena, A., Moser, D., Nishino, M., Pearman, D., Pergl, J., Rabitsch, W., Rojas-Sandoval, J., Roques, A., Rorke, S., Rossinelli, S., Roy, H. E., Scalera, R., Schindler, S., Štajerová, K., Tokarska-Guzik, B., van Kleunen, M., Walker, K., Weigelt, P., Yamanaka, T. & Essl, F. (2017) No saturation in the accumulation of alien species worldwide. *Nature Communications*, **8**, 14435.

Sherratt, T. N. & Wilkinson, D. M. (2009) *Big questions in ecology and evolution*. Oxford: Oxford University Press.

Stohlgren, T. J., Barnett, D., Flather, C., Fuller, P., Peterjohn, B., Kartesz, J. & Master, L. L. (2006) Species richness and patterns of invasion in plants, birds, and fishes in the United States. *Biological Invasions*, **8**, 427–447.

Tarasi, D. D. & Peet, R. K. (2017) The native-exotic species richness relationship varies with spatial grain of measurement and environmental conditions. *Ecology*, **98**, 3086–3095.

Whittaker, R. J. & Fernández-Palacios, J. M. (2007) *Island biogeography: Ecology, evolution, and conservation*, 2nd ed. Oxford: Oxford University Press.

Whittaker, R. J., Rigal, F., Borges, P. A. V., Cardoso, P., Terzopoulou, S., Casanoves, F., Pla, L., Guilhaumon, F., Ladle, R. J. & Triantis, K. A. (2014) Functional biogeography of oceanic islands and the scaling of functional diversity in the Azores. *Proceedings of the National Academy of Sciences USA*, **111**, 13709–13714.

Williams, C. B. (1964) *Patterns in the balance of nature*. London: Academic Press.

Williamson, M. (1996) *Biological invasions*. London: Chapman and Hall.

Wilson, D. E. & Reeder, D. M. (eds.) (2005) *Mammal species of the world. A taxonomic and geographic reference*, 3rd ed. Baltimore, MD: Johns Hopkins University Press.

Part III
Theoretical Advances in Species–Area Relationship Research

7 · *Mathematical Expressions for the Species–Area Relationship and the Assumptions behind the Models*

EVEN TJØRVE AND KATHLEEN M. C. TJØRVE

7.1 Introduction

Ever since Gleason (1922, 1925) criticized Arrhenius' (1920a, b, 1921, 1923) proposal for a mathematical expression for the species–area relationship (SAR), ecologists have discussed the shape of SARs and what function should be chosen to model them. These questions are complicated by the fact that we have different types of SARs, which might exhibit different types of curvatures. Preston (1962) was the first to distinguish between curves compiled from surveys of isolates and curves compiled from surveys of sample areas (and sampling units). Isolates can be, for example, oceanic islands, lakes, forest fragments or mountain-tops ('sky islands'; Brown, 1971). SARs from islands and other types of isolates (ISARs) are generated by comparing species numbers in independent areas of different sizes, whereas SARs from sample areas are usually generated as the accumulated number of species from a sample area that increases in size (a nested design) or by adding sampling units (Tjørve & Turner, 2009). We can define such sample-area SARs (saSARs) as a particular type of species accumulation curve (SAC). SACs, also called collector's curves, describe the increase in species number (as the response variable) resulting from increasing sampling effort, such as number of traps (or trap hours), sampling of habitat (units of area) or simply number of individuals (as the predictor variable). Thus, the saSAR becomes a special case (subset) of the SAR and the SAC based on sampling area (saSAC) becomes a special case of the SAC, where the subsets form the intersection of two terms, the SAR and the SAC (see also Chapter 1).

A key theme of discussion has been whether SARs, both ISARs and saSARs, are power laws or whether they are better fit by the logarithmic or some other model (see e.g. Triantis et al., 2012; Tjørve et al., 2018; Chapter 4). There has been less attention paid to whether there are inherent differences between species–area curves generated from sampling units placed within some expanse (according to a sample design) and curves generated from oceanic islands or other types of isolates.

Much attention has been paid to explaining variation in the exponent (z-value) of the power model, which was first stated in the form we are used to seeing it, $S = cA^z$ (where S is number of species, A is area, and c and z are parameters), by Kylin (1923; see Chapter 2). Prior to this, Brenner (1921) had suggested that the power model would also fit ISARs. However, it was applied mostly to saSAR data until the middle of the twentieth century, when Darlington (1957) and Wilson (1961) claimed that the power model also holds for describing the ISAR. MacArthur and Wilson (1963, 1967) then presented their equilibrium theory of island biogeography based on the power model applied to ISARs. It thus became commonplace to treat isolates in the same way as sample areas. This is unfortunate because these two types of SAR have inherently different shapes.

Connor and McCoy (1979), in their discussion of 'the statistics and biology' of both ISARs and saSARs, argued that the interpretation of z-values (from the power model) is difficult, perhaps even futile. Several authors have since noted that z-values estimated over small and large scales often differ (Rosenzweig, 1995; Harte et al., 1999; Plotkin et al., 2000; Crawley & Harral, 2001; Hubbell, 2001; Storch et al., 2003; Ulrich & Buszko, 2003; Fridley et al., 2005; Triantis et al., 2012). Connor and McCoy (1979) also concluded that there is no universal model that will fit all species–area datasets; noting that 'Preston's work has subsequently led to the near-uniform acceptance of the power function as the model best describing the species–area relationship' (p. 794).

Here, we review the large number of mathematical models proposed to describe the shapes of SARs. We build on the reviews of models and discussions of the expected shapes of SARs found in Tjørve (2003, 2009), Tjørve and Turner (2009), Dengler (2009), Williams et al. (2009) and Triantis et al. (2012). We also discuss the purpose of describing the SAR mathematically, as well as examining the general assumptions behind the shape of the SAR curve, including a discussion on the prevalence of convex versus sigmoid curves and the presence or not of an asymptote in the SAR.

7.2 Fundamentals of Looking for a 'Best' SAR Model

Not only do the shapes of ISARs and saSARs differ, but so do the respective steps of data collection and curve generation. Both these steps affect the pattern (curve) of the plotted data and will therefore also influence which model returns the best fit. The botanists who first produced species–area plots typically plotted the accumulated area of their square or rectangular sampling units against the number of species observed (e.g Arrhenius, 1921; Gleason, 1922, 1925; Braun-Blanquet & Jenny, 1926; Cain, 1934, 1938; Hopkins, 1955; Chapter 2). Gleason (1922, 1925) argued ardently that the exponential (logarithmic) curve gives a better fit than the power curve proposed by Arrhenius (1921). However, it seems that Gleason reached this conclusion because he applied a dispersed (discontinuous) sampling scheme, with 240 squares of 1×1 m, distributed within a 25 km^2 tract (or focal area) (Tjørve & Tjørve, 2017; Chapter 2). Herein lies an explanation for why he found the logarithmic curve to be more realistic for his data: namely, that discontinuous sampling units cause a higher SAR curvature in arithmetic space than nested sampling (i.e. for continuous sample areas; see Chapter 4). The logarithmic SAR has a higher curvature in arithmetic space than the power law SAR (Tjørve et al., 2008) and should therefore be expected to fit SARs generated from discontinuous sample areas better than the power model.

7.2.1 Sampling Designs and Accumulation Order

One of the main problems when fitting regression models or comparing species–area data is that both sampling design and how the data are plotted (e.g. by plotting area cumulatively or each area separately) affect curve shape. For ISARs, the shape of the resulting curve is affected by what scales – from the finest to the coarsest – are surveyed. For saSARs, not only the extent of the total area (arena) studied, but also the sizes, shapes and placement of the sampling units may affect the shape of the curve. However, for a given scale window, sampling-unit sizes will only affect saSARs that are generated from discontinuous sampling units. This is because the similarity between two sampling units typically decreases with the distance between them. Moreover, logically the use of larger sampling units (in a discontinuous design) results in more of the sampled area lying closer together and thereby altering the curvature relative to smaller sampling units (set out in the same type of pattern). For SARs generated from continuous sampling units, although curve shape will not be affected by sampling-unit size, it will still be affected by the geometric shape of the sampling unit or total sampled area. Because all these factors affect curve shape, it becomes more difficult to compare SARs or regression curves between datasets, unless all the above conditions are equal.

Sample areas (or sampling units) may be discontinuous or continuous, and discontinuous sampling units may be scattered randomly within the surveyed area or located according to any other type of pattern. Sometimes researchers compile SARs from sampling designs that even change from one scale to another. This will also change the shape of the curve. In addition to variations in sampling schemes, SARs may (as already mentioned) differ in accumulation (or assembly) order, meaning how the sampling units are ordered when counting occurrences of new species. Again, all of this affects the shape of the SAR curve, as does incomplete surveying (Fisher et al., 1943; Turner & Tjørve, 2005; Dengler, 2009; Tjørve & Turner, 2009). Even though the combined effect of both sampling design and accumulation order should be taken into consideration when generating and studying saSARs, this is not approached systematically in the suggested SAR typologies found in the existing literature (see e.g. Scheiner, 2003, 2004; Dengler, 2009).

For saSARs, sampling designs with quadrat sampling units are most common, although both rectangular (e.g. Harte et al., 1999) and circular units (He & Hubbell, 2011) have been discussed. The sampling design can affect the shape of the saSAR curve through both the size of the sample areas (units) and their shape (e.g. Keeley & Fotheringham, 2005; Turner & Tjørve, 2005), as well as by their placement in space: either contiguous or non-contiguous. Non-contiguous sampling units are placed randomly or in any other pattern, within a larger area studied, sometimes called a focal area or an arena. If a given number of non-contiguous sampling units are spaced out to cover the arena (or study area) as well as possible, then the size of the arena will affect the curve shape. Many sampling designs other than random have been suggested and used; for example, evenly distributed sample areas (or sampling units) and Whittaker plots (e.g. Scheiner, 2003, 2004; Keeley & Fotheringham, 2005; Dengler, 2008). However, most of these sampling designs are better suited when the purpose is simply to find as many species as possible in the study arena, but not when the aim is to study SARs.

For practical purposes, we would argue that saSARs should be generated from nested (or cell matrix) sample areas or from random samples with or without replacement. Even the two forms of random-sample-area designs are problematic choices, in that the resulting curve is affected not only by the shape of the sample areas, their placement pattern and accumulation order, but also by their sizes. Therefore, with randomly placed sample areas, it is almost impossible to compare curves between datasets or studies unless one has agreed on the same sample-area size, equally sized arenas and the same accumulation (or assembly) order.

7.2.2 Expected Curve Shape and Scale Window Surveyed

As a result of ecological patterns and sampling design, nested saSAR curves are expected to be constantly increasing but at a decelerating rate (i.e. convex upward without an upper asymptote) in arithmetic space. These curves contrast with ISARs, which are generated by comparing species numbers found within each of a series of areas (isolates) of different sizes, whereby a larger area may sometimes have fewer species than a smaller one. Whereas both saSAR and ISAR data are typically well fitted by convex upwards models, such as the power model (e.g. Drakare et al., 2006; Dengler, 2009; Triantis et al., 2012; Matthews et al., 2016a, b), it has been argued that ISARs are inherently sigmoid (Lomolino, 2000, 2001; Turner & Tjørve, 2005; Tjørve & Turner, 2009), although the notion of an upper asymptote has been challenged (Williamson et al., 2001; Tjørve & Turner, 2009). A major difference between sample areas and islands (or other types of isolates) is that islands have to have a minimum size to be able to sustain, with any probability, a viable population, whereas a rare species may be present within a sample area without having a viable population contained within it. Small populations are also prone to extinction events. Thus, the probability that we will find a given species on an island increases with island size from a lower base and at a faster rate than for comparable non-isolated sample areas.

Given the probability that an island holds a given species increases with area, it can be argued that the probability density distribution for the 'first-time occurrence' of the species with increasing island size (meaning the smallest island on which we find the species) should be bell-shaped (although it may be left- or right-skewed). Such bell-shaped probability density distributions are common in nature, where environmental clines are characterized by gradual probability distributions. The resulting cumulative probability distribution for any given species (single-species relationship), meaning the proportion of islands in each size category on which a given species occurs, is often termed an incidence curve

(Diamond, 1975). If the probability density distribution is bell-shaped, the incidence function must be sigmoidal (in arithmetic space) and reach an asymptote as area becomes large (e.g. Cody et al., 2002). Consequently, the ISAR curve (as an expression of the probability of encountering a number of species) becomes the additive result of several such (sigmoid) curves. By this logic, the expected curve shape of ISARs is sigmoidal. The fact that only few ISAR datasets are well characterized by sigmoid curves is, by the same logic, a consequence of these data rarely including the finest scales (the smallest islands and islands with zero species), which therefore results in ISARs with monotonically decelerating growth (Tjørve & Turner, 2009). Consequently, these ISARs are better fitted with convex regression models, such as the power model.

Several authors have recently argued that the shape of saSARs, and nested ones in particular, should be triphasic in log–log space, meaning that the curve consists of three segments (Williams, 1964; Hubbell, 2001; Fridley et al., 2005; Storch, 2016), where the first (at finer scales) is decelerating (convex upwards and often reported to follow the logarithmic model). The second segment (at intermediate scales) follows a power law and the third (at coarser scales) also follows a power law but with a higher z-value (i.e. steeper slope in log–log space).

The concepts of sigmoidal ISARs and triphasic saSARs outlined earlier in this section are based on the inclusion in a dataset of samples/islands that cover a wide range in area, from the finest to the coarsest scales. With areas from finer and/or coarser scales missing, we may not correctly model shapes or shape changes that occur outside the studied scales. If the observed scale window studied is too narrow, even changes in curve shape that go on inside the scale window may go undetected. Consequently, which regression model returns a better fit may depend on the range in area included in the dataset. This should encourage us to consider models with expected curve shapes, rather than simply the ones that return a better fit.

7.2.3 Available Models

A large number of functions and mathematical models have been suggested to fit one or several types of SARs, and the statistical fit to observations is usually used to identify what is then considered to be the best model. Fitting regression models to SAR data relies mostly on the unmentioned underlying assumption that the best fitting model is the correct one. Connor and McCoy (1979) proposed that the best fitting model should be determined empirically, according to statistical criteria. This is traditionally done by comparing R^2 or AIC values between models or by looking at bias by inspecting regression residuals. Almost all studies aimed at finding the best model agree that it should be selected according to some best fit criterion (see He & Legendre, 1996, for discussion of how to select a model). However, we hold that the choice of mathematical model must go beyond just identifying the best fit model, as, for example, fitted using ordinary least squares regression. One may instead take into account the expected shape of the SAR when a wide range of areas are included or discuss the choice of model relative to the purpose of fitting the model to the data. The purpose may be 'descriptive' (describing the SAR mathematically), 'explicative' (explained from patterns in nature) and/or 'predictive' (to predict diversity or diversity changes).

Each mathematical (regression) model also comes with a set of assumptions about how individuals are distributed in space and the form of the species abundance distribution. These underlying assumptions (or constraints) will not be the same for saSARs as for ISARs, as the former are usually accumulative while the latter are comparative by nature. As such, the underlying assumptions should be explicitly considered when choosing a SAR model.

7.3 Proposed SAR Models

The many mathematical models that have been proposed to describe SAR shape can be broadly divided into those that are convex upward in arithmetic (i.e. untransformed) space and those that are sigmoid in arithmetic space. The models may also be divided into those with an upper asymptote and those that are monotonically increasing (at least within the scales of interest). In addition, some models do not (or only sometimes do) intersect the axes at the origin, whereas others always intersect the origin.

Models proposed to fit different types of SAR are reviewed by Tjørve (2003, 2009), Dengler (2009), Williams et al. (2009) and Triantis et al. (2012). Table 7.1 lists thirty models. Dengler's (2009) 'quadratic' and 'general' models are included, as well as Tjørve's hybrid model (Tjørve, 2012), despite the latter being applicable only to nested saSARs (where values are typically averaged for each scale). However, we have not included Hanski et al.'s (2013) model for SARs generated from fragmented habitats. Neither have we included polynomial models based on single-species incidence curves, nor models that build on additional

Table 7.1 *Mathematical models suggested to fit species–area datasets. The list was compiled from the reviews of Tjørve (2003, 2009), Fattorini (2007b), Dengler (2009), Williams et al. (2009) and Triantis et al. (2012), in addition to Malyshev (1991) and Fattorini (2007a). The shape is described for arithmetic space. In addition, the number of fitted parameters, whether there is an upper asymptote (plus possible maxima) and if the model curve intersects the axes at the origin, are also listed. Shapes are only discussed for positive values of the fitted parameters. Traits in brackets are not consistent for all parameter values. In the model column, A = area, and the other letters refer to fitted parameters. Capital B denotes upper asymptotes.*

	Model name	Model	Parameters	Shape	Upper asymptote	Intersects origin
1	Linear	$b + cA$	2	Linear	No	No unless $b = 0$
2	Power model	cA^z	2	Convex	No	Yes
3	Power Rosenzweig	$b + cA^z$	3	Convex	No	No
4	Power-plus-linear model	$b + cA^z + dA$	3	Convex	No but max. when $d<0$	No
5	Power model quadratic	$10^{b + c\,\log A + d(\log_2 A)^2}$	3	(Several)	(Varies)	(Varies)
6	EPM 1 model (extended power model 1)	$cA^{b \cdot A^{-d}}$	3	Sigmoid/Convex	No / may have max.	Yes
7	EPM 2 model (extended power model 2)	$cA^{b-(d/A)}$	3	Sigmoid	No / have min.	No
8	P1 model (first persistence model)	$cA^b \exp(-dA)$	3	Convex	No but have max.	Yes
9	P2 model (second persistence model)	$cA^b \exp(-d/A)$	3	Sigmoid	No	Yes
10	Logarithmic model	$c + b \log A$	2	Convex	No	No

11	Logarithmic model – simplest	$b \log A$	Convex	1	No	No
11	Logarithmic model – quadratic	$(c + b \log A)^2$	Sigmoid with min.	2	No	No
12	Logarithmic model – general power	$(c + b \log A)^d$	Sigmoid with min.	3	No	(No)
13	Connor-McCoy exponential	$10^{(b + cA)}$	Partly convex downward with min.	2	No	(No)
14	Log-series logarithmic	$b \log (1 + A/c)$	Convex	2	No	Yes
15	Negative exponential	$B(1-\exp(-cA))$	Convex	2	Yes	Yes
16	Common logistic	$B/(1 + \exp(-cA + d))$	Sigmoid	3	Yes	No
17	Archibald logistic (He-Legendre function)$\star\star$	$B/(c + A^{-d})$	Sigmoid	3	Yes	Yes
18	Uranov function	$bA/(c + A^d)$	Convex for $0<d<1$	3	No	Yes
19	Gompertz	$B \exp(-\exp(-cA + d))$	Sigmoid	3	Yes	No
20	Extreme-value function (EVF)	$B (1-\exp(-\exp(cA + d)))$	Sigmoid	3	Yes	No
21	Monod (Michaelis-Menten) function	$B(A/(c + A))$	Convex	2	Yes	Yes
22	Asymptotic regression	$B - cd^{-A}$	Convex	3	Yes	Possible$\star\star\star$
23	Rational function	$(b + cA)/(1 + dA)$	Convex	3	Yes	Possible$\star\star\star$
24	Chapman–Richards	$B(1-\exp(-cA))^d$	Sigmoid	3	Yes	Yes
25	Cumulative Weibull distribution – 3 param.	$B(1-\exp(-cA^d))$	Sigmoid	3	Yes	Yes

(cont.)

Table 7.1 (*cont.*)

Model name	Model	Shape	Parameters	Upper asymptote	Intersects origin
26	Cumulative Weibull distribution – 4 param.	$B(1-\exp(-cA^d))^m$	Sigmoid	3	Yes	Yes
27	Morgan–Mercer–Flodin (Hill-function)	$BA^d/(c + A^d)$	Sigmoid	3	Yes	Yes
28	Lomolino function	$B/(1 + (c^{\log(d/A)}))$	Sigmoid	3	Yes	Yes
29	Cumulative beta-P distribution	$B\left(1 - (1 + (A/c)^d)^{-m}\right)$	Sigmoid	4	Yes	Yes
30a	Tjorve hybrid 1	$(c_1 + b \log A)^{dz/(A + m)} \cdot (c_2 A^z)^{1-(dA/(A + m))}$	Convex★	6 (2)^	No	No
30b	Tjorve hybrid 2	$(c_1 + b \log A)^{1-(dA/(A + m))} \cdot (c_2 A^z)^{dA/(A + m)}$	Convex★	6 (2)^	No	No

★ Convex (for practical purposes), as long as the component models (logarithmic and power) stay within the ranges the model is intended for.

★★ Note that the 'Archibald logistic' shown in Tjørve (2009) is incorrect, as the base and the exponent are mixed up.

★★★ These models will only intersect the origin of axes for certain parameter values.

^ See Section 7.5.1 for an explanation of how the number of parameters is reduced from six to two.

variables, such as information about habitats, as exemplified by the Choros model (Triantis et al., 2003) and the Countryside model (Pereira & Daily, 2006). The Choros model attempts to correct the independent variable (A) for the number of habitats (H) (by substituting A for AH). The Countryside model attempts to correct not only for H, but also for the proportion of each habitat utilized by different groups of species. Neither have we included piecewise (also called breakpoint or segmental) regression models that have been proposed to identify datasets with a small island effect (SIE) (Lomolino & Weiser, 2001; Barrett et al., 2003; Gentile & Argano, 2005; Matthews et al., 2014). We note that SIEs and their detection are controversial (Tjørve & Tjørve, 2011).

Recently, a number of SAR studies have assessed the utility of models other than the power and logarithmic models, many of which are listed in Table 7.1 (e.g. Williams, 1995; Flather, 1996; Lomolino, 2000; Veech, 2000; Triantis et al., 2003; Fattorini & Fowles, 2005; Gentile & Argano, 2005; Fattorini, 2006, 2007a, b). However, recent meta-analyses all report the power model to be the 'best' (Drakare et al., 2006; Dengler, 2009; Triantis et al., 2012; Matthews et al., 2016a, b).

7.4 Why Fit Curves to SARs?

Before we can choose a regression model to fit to empirical species–area datasets, the purpose of the fitting exercise must be established, as this will affect the choice of model. The best fit criterion is usually used when adopting a mathematical model. However, the purpose of fitting a model to data may go beyond just obtaining the best fitting curve. Another important, but often overlooked, consideration is choosing a model based on the expected curve shape rather than simply the goodness of fit.

The study of SARs has always mostly been *descriptive*, focusing on identifying curve shape, the best fitting model and model parameters for different types of areas. Only more recently has the interest shifted towards *explicative* purposes, focusing on the link between the observed SAR, the model curve and patterns in nature (e.g. He & Legendre, 2002; Tjørve, 2003; Šizling & Storch, 2004, 2007; Turner & Tjørve, 2005; Tjørve et al., 2008; Storch, 2016; Šizling et al., 2017), as well as the interpretation of parameters and parameter values (e.g. Harte et al., 1999, 2001; Tjørve & Tjørve, 2008). The recent focus on extinction caused by habitat loss and the desire to predict the richness of large unsampled areas has increased the drive to apply the species–area curve for *predictive* purposes (Kilburn, 1966; Dony, 1977; Kangas, 1987; Gitay et al., 1991;

Simberloff, 1992; Pimm & Raven, 2000; Chapter 14). In the remainder of this section we discuss the purposes of fitting a curve, and in Section 7.5 we deliberate on how this guides us to choose a model.

7.4.1 Descriptive Purposes

Descriptive study of the SAR generally involves comparisons between areas or studies. The z-value of the power model (generally the slope of the linear regression line in log–log space is used; see Chapter 4) has more or less become a standard parameter for comparing SARs of different types of areas (e.g. Chapter 3). For example, Rosenzweig (1995) described three different types of species–area curves (interprovincial, intraprovincial and island) defined by this parameter (see Chapter 19). Sometimes the second parameter in the power model, the so-called intercept or c-value, is also used in such comparisons (Chapter 3).

7.4.2 Explicative Purposes

Explicative study of the species–area curve covers the understanding of how patterns in nature affect the observed curve. SAR shape, as stated, reflects such factors as species abundances, the distribution of individuals in space, minimum-area requirements, habitat diversity and more (see Turner & Tjørve, 2005, for review). Several researchers have recently paid much attention to the self-similarity properties of the power model based on the argument that patterns of self-similarity in nature justify its use. However, this line of enquiry has thus far proved inconclusive (Harte et al., 1999, 2001; Lennon et al., 2002, 2007; Ulrich & Buszko, 2003; Krishnamari et al., 2004; Maddux, 2004; Šizling & Storch, 2004).

There have been extensive discussions on the nature of the relationship between species abundance distributions (SADs) and SARs. Usually, this is discussed in the context of saSARs, as SAD patterns are expected to differ between islands (or between isolates) of different sizes (which greatly complicates the relationship). Both the occurrence of power law SARs and logarithmic SARs have been attributed to specific SADs. It has been claimed that lognormal SADs result in power SARs (Preston, 1960, 1962; Williams, 1964), and that logarithmic (log series) SADs result in logarithmic SARs (Fisher et al., 1943; Kobayashi, 1975). The discussions of such relationships involve comparisons of empirical species–area curves with those that should theoretically result from given SADs. However, such mathematical translations of SADs to saSARs (see also

Chapter 4) have typically assumed random placement of individuals, despite the fact that individuals are not usually randomly distributed but tend to be aggregated in space (e.g. Goodall, 1952; Hopkins & Skellam, 1954; Taylor et al., 1978; Veech et al., 2003) or, occasionally at fine scales, tend towards regular or even spacing (e.g. Hopkins, 1955; Condit et al., 1996; Plotkin et al., 2000).

If individuals are randomly distributed, SAD models can be translated directly to saSARs, as the number of individuals can be substituted with area multiplied by some constant. One such model is Fisher et al.'s (1943) log-series SAD model, $S = a \log (1 + N/a)$ (where S is number of species, N is total number of individuals and a is a derived constant), which was first evaluated in this context by Williams (1950) and was later explored by Hopkins (1955) and Kobayashi (1975), who showed that it was straightforward to reformulate the model as $S = b \log(1 + A/c)$. The latter is a logarithmic model (where b and c are fitted parameters) that approaches Gleason's logarithmic SAR model (model 10 in Table 7.1) when A/c becomes large.

Several researchers have attempted to derive the power model directly from SADs (e.g. Irie & Tokita, 2012). The problems with this type of approach are that it assumes patterns that are known not to be universal (i.e. log-series and lognormal SADs), and that, as mentioned above, individuals are randomly distributed in space. More complex approaches taking both abundance and the spatial distribution of individuals into account typically do not yield such simple expressions that are as useful as curve fitting models (see modelling in He & Legendre, 2002; Green & Ostling, 2003; Olszewski, 2004; Picard et al., 2004; Green & Plotkin, 2007). Yet, Martín and Goldenfeld (2006) report that the power law saSAR is a robust consequence of lognormal SADs with higher rarity, together with aggregated spatial patterns (clustering). Because the shape of the SAR curve is affected by a very large number of factors, to derive a mathematical species–area model directly from these factors seems an insurmountably complex task. However, it would be a useful exercise to discuss and link various patterns and processes in nature to candidate models by reason and observation.

7.4.3 Predictive Purposes

Predictive uses of species–area curves include extrapolations to predict total biodiversity (e.g. Palmer, 1990; Colwell & Coddington, 1994; Colwell et al., 2004; Kunin et al., 2018), extinction forecasting (e.g. Pimm & Askins, 1995; Pimm & Raven, 2000; He & Hubbell, 2011; Keil et al., 2015, 2018; Chapter 14) and hotspot detection (e.g. Veech, 2000; Ulrich & Buszko, 2005; Fattorini, 2006; Guilhaumon et al., 2008; Chapter 13). The usefulness of such approaches relies on the quality of the fitted curve and is, therefore, dependent on fitting the 'right' model to the data. In addition, one can use species–area curves to predict total species number in several areas combined or in landscape mosaics comprising several types of habitats (Tjørve, 2002). Moreover, predictions from such species diversity models hold the answer to the SLOSS question, that is, understanding why a single nature reserve will sometimes hold more species than two or more smaller nature reserves (making up the same total area) or vice versa (Tjørve, 2010; Chapter 2).

7.5 A Shortlist of Models

To facilitate the effective uses of SARs listed above, a shortlist of useful models should be identified. Table 7.1 provides a good starting point in this regard. The models in Table 7.1 can be divided into convex upward and sigmoid, as well as into asymptotic and non-asymptotic, producing four classes. Which regression models are useful depends, of course, on the purpose of fitting a model. If the goal is to find the model with the best fit to whatever dataset is at hand, then all models should be considered, irrespective of the shape outside the surveyed scale window. If the purpose is to compare parameters between datasets from different sites or different studies, a single model, the one that provides the best fit to the most datasets for example, may be chosen. This seems to be the power model for both ISARs (Drakare et al., 2006; Dengler, 2009; Triantis et al., 2012; Matthews et al., 2016a, b) and for saSARs from nested sample areas (Drakare et al., 2006; Dengler, 2009), although we argue that sigmoid models should provide better fits to ISARs if certain conditions are met. We note, however, that the logarithmic model is often reported to fit saSARs better than the power model at finer scales (e.g. Hopkins, 1955; Williams, 1964). The explanation generally provided to explain this observation is that individuals have a tendency to be less aggregated at finer scales. Several authors have shown mathematically how less aggregation at finer scales causes a more upward convex curve (in log–log space) (He & Legendre, 2002; Tjørve et al., 2008; He & Hubbell, 2013) that is better approximated by the logarithmic model. Finally, it has also been shown that the logarithmic model often performs better than the power model when fitted to (whole) saSARs from random sample areas (and accumulated in random order) (e.g. Tjørve & Tjørve, 2017).

7.5.1 Models without an Upper Asymptote

Where the purpose of fitting a regression model to SAR data is to discuss the resulting curves across multiple scales or to explain inherent curve shapes, then expectations about the shape characteristics outside the sampled scale window should be considered. One important issue is whether the SAR at hand is expected to have an upper asymptote (see Figure 7.1), that is, whether it levels out towards a finite number of species as the area becomes relatively large (e.g. Miller & Wiegert, 1989; Colwell & Coddington, 1994; Lomolino, 2000, 2002; Williamson et al., 2001, 2002). Some authors have advocated fitting convex upward models that have an asymptote to SAR data (Clench, 1979; Miller & Wiegert, 1989). For example, Lomolino (2000, 2002) argued that SARs can be asymptotic because of finite species pools. This view assumes that species pools never increase, with the logical consequence that the pools do not accumulate more species even if we extrapolate the SAR to much larger total areas than the ones surveyed or to continents instead of islands. However, as Williamson et al. (2001, 2002) point out, not only is the number of species finite, but so is the area on a finite planet. Thus,

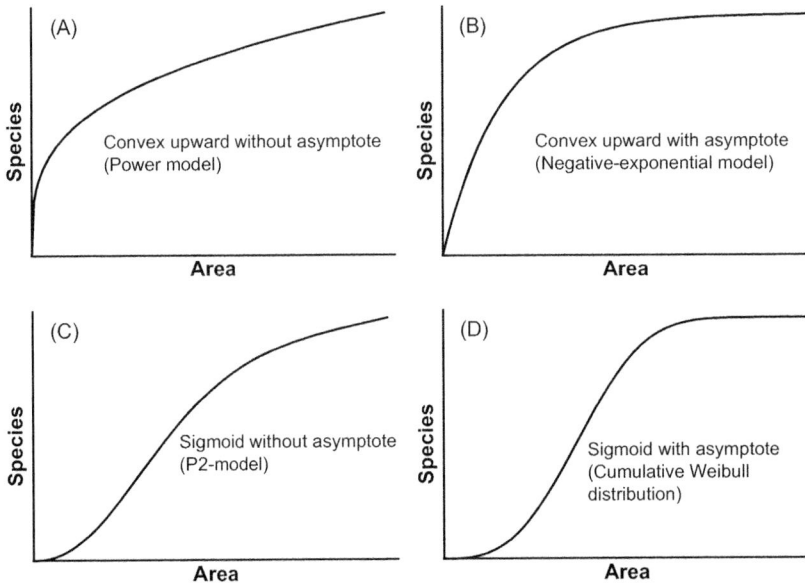

Figure 7.1 Examples of four different types of SAR regression models, including convex (A, B) and sigmoid models (C, D), and models with (B, D) and without (A, C) an asymptote.

Williamson et al. argue, ordinary ISARs do not have asymptotes, as larger islands have more species than smaller, continents have more species than islands and larger continents have more than smaller ones.

If we discard the SAR models (either saSARs or ISARs) based on species pools that do not increase with area, then saSARs may have asymptotes only under one of two special assumptions. The first is that subsamples are taken randomly and with replacement from some expanse of area (arena) and the second is that sample areas are taken from a completely homogenous expanse or homogenous areas containing the same habitat or community type. The latter assumption seems unlikely to hold in most empirical systems. Thus, the only practical sampling design that will cause an asymptotic SAR is randomly placed sample areas where the same area (or part of the area) can be surveyed again. For instance, Kobayashi (1974) and Engen (1978) independently showed how the combining of sample areas (e.g. quadrats of the same size) with replacement results in an asymptotic species–area curve.

As highlighted at the outset, saSARs can be seen as a special type of SAC. Traditionally, SACs – in which some type of sampling effort (e.g. total number of individuals, trap-time or trap occasions) is plotted against the number of species – are usually considered to be asymptotic (see e.g. Gotelli & Colwell, 2001, for review). These SACs approach the total number of species in a given community, habitat or nature type as the number of species added by continued sampling effort approaches zero. If we adopt the aforementioned extreme view, then neither of these should be truly asymptotic, since the sampling effort is typically exhaustive (i.e. without replacement) and areas are not truly homogenous. Nevertheless, with relatively large, homogenous areas, it should be possible for practical purposes to consider SACs as asymptotic. Note that, although 'area' could be defined as a measure of sampling effort, thus making the SAR a special case of the SAC, because the conditions behind these sampling approaches differ we should study them separately.

In conclusion, neither ISARs nor the two types of saSARs (from nested sampling and from discontinuous sampling units) discussed in this chapter are expected to have asymptotic curves. Therefore, if we want to discuss these SARs from finer to coarser scales, we may want to consider only mathematical models that are monotonically increasing. In Table 7.1 we find six ordinary models that are always monotonically increasing (for positive values of their parameters) and thus lack asymptotes or maxima.

In addition, there is Rosenzweig's (1995) version of the power model, which in reality is just a power model plus a constant, and Tjørve's (2012) hybrid models. There are two complementary forms of the latter, bridging the gap between the logarithmic and the power model. The two forms are given as:

$$S_A = \left(c_1 + b\log A\right)^{\frac{dA}{A+n}} \cdot \left(c_2 A^z\right)^{1-\frac{dA}{A+n}} \tag{7.1}$$

and

$$S_A = \left(c_1 + b\log A\right)^{1-\frac{dA}{A+n}} \cdot \left(c_2 A^z\right)^{\frac{dA}{A+n}}, \tag{7.2}$$

where S_A is the number of species in an area, A, d and n are fitted parameters, and z, b, c_1 and c_2 are parameters found by calculation before the model is fitted to data. These four calculated parameters are found by accepting two predetermined data points (A_0, S_0) and (A_k, S_k) as part of the model curve (here '0' represents a single sampling unit and k represents the total number of sampling units). The first of these (A_0, S_0) represents the area of one sampling unit and the average number of species encountered in one such unit, whereas the second (A_k, S_k) represents the area of all sampling units surveyed and the total number of species encountered. The calculated parameter values are then entered into the model. Therefore, the number of parameters is reduced from six to two, fixing the model at either end of the scale window of the dataset. This is justified when the model produces a very good fit to the data.

The two model versions have been found to provide good fits to nested saSAR data and averaged continuous areas from sample matrices (Tjørve, 2012). They successfully describe datasets that fall between the power model and the logarithmic model and should be expected to fit particularly well to what are termed triphasic SARs (datasets that are sigmoid in log–log space). Here, we have illustrated this by fitting the hybrid model to a dataset generated from surveying tree species in fifty sampling units of 100x100 m, laid out in the Pasoh Forest Reserve in Malaysia, and taken from He and Legendre (1996). The hybrid curve in Figure 7.2 illustrates visually how the Tjørve hybrid (in this case Equation 7.1) is able to trace the data points between the finest and the coarsest scales much better than the logarithmic model and the power model. Because this model is able to return a better fit, it is potentially better suited for extrapolations of saSAR curves than both the logarithmic and power models.

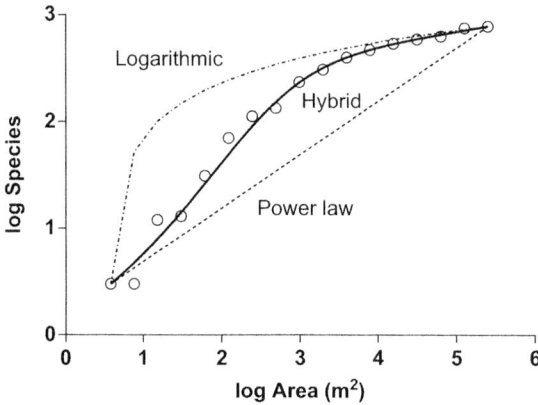

Figure 7.2 Tjørve's (2012) hybrid model (solid line) compared to a logarithmic curve (dash-dotted line) and a power law curve (dashed line), both constructed between the data points for the finest scale (grain) and the coarsest scale. The hybrid model, which in this case is the first model form (Equation 7.1), was fitted with an untransformed dependent variable (number of species). The data were sampled from a lowland tropical rain forest tree community in the Pasoh Forest Reserve, Malaysia, and were extracted from He and Legendre (1996).

7.5.2 Convex and Sigmoid Models

According to MacArthur and Wilson's (1967) equilibrium theory, owing to enhanced isolation (distance from the mainland) and extinction risk (due to small populations) we should expect islands to have fewer species than equally-sized sample areas on the mainland. This effect should be relatively greater for smaller islands, because few species can maintain viable populations in small areas and because the frequency of arrival of populations of other species is so much lower than for patches within contiguous habitats. The resulting minimum-area effect will consequently depress the SAR curve more at smaller scales, resulting in a sigmoid shape with an inflection point (see Figure 7.1; see also Tjørve & Turner, 2009; Tjørve, 2010; Tjørve & Tjørve, 2011). At the lower j-shaped part, the curve is expected to taper off toward the smallest scales, where the chances of any species occurring are very low. There are not that many mathematical models that are sigmoid but do not have an upper asymptote. Of all the models in Table 7.1, only the P2 model, the EPM2 model and the quadratic logarithmic model consistently satisfy both these criteria. Of these three, only the P2 model goes through the origin of axis. In contrast, the quadratic logarithmic model and the EPM2 model have minima which, for the quadratic logarithmic, falls on the x-axis. This minimum may serve as a starting point for the SAR, meaning that one may wish to restrict the curve to values above the minimum, that is, the smallest scale at which any species are found.

The arguments presented above, that the ISAR should be inherently sigmoid (but without an upper asymptote), are not supported by much empirical evidence. In most studies comparing models it is reported that convex models perform better, both for saSARs and ISARs. Recent meta-analyses (see above) have concluded that saSARs and ISARs are both best represented by the power model (Drakare et al., 2006; Dengler, 2009; Triantis et al., 2012; Matthews et al., 2016a, b). However, Tjørve and Turner (2009) found that a sigmoid model (the P2 model) fitted better than any convex upward model to three datasets that also included the very smallest scales, not only when comparing R^2 values, but also according to AIC_C, which penalizes for the extra parameter(s) usually found in sigmoid compared to convex models. One of the datasets analysed, a study of plant diversity in a Canadian archipelago (Deshaye & Morisset, 1988), covers an unrivalled scale window width, with a ratio of 18,428 between the areas of the largest (92.14 ha) and smallest (0.005 ha) islands surveyed (Figure 7.3).

A lower j-shaped part of the ISAR seems to be found only when the finest scales are included and noticeably when isolates are sufficiently small to contain no species. Thus, we believe that the scarcity of evidence for sigmoid ISARs is the result of surveying mostly narrow scale windows, where the finest scales in particular are not included (Tjørve, 2009, 2010; Tjørve & Turner, 2009). Not only are such empirical examples of sigmoid SARs rare because the scale window surveyed is narrow, with the inflection point falling outside the surveyed scales, but an additional difficulty arises from the fact that the inflection point may (as in Deshaye & Morisett's dataset) fall so far toward the finer scales that the j-shaped part of the curve becomes invisible or close to invisible in the plot, unless one focuses in on only these scales. The alternative is to plot the data and the curve in log–log space, as done in Figure 7.3. However, in log–log space the curve no longer can be inspected for its sigmoid shape by eye. Therefore, the actual inflection point is indicated with a circle.

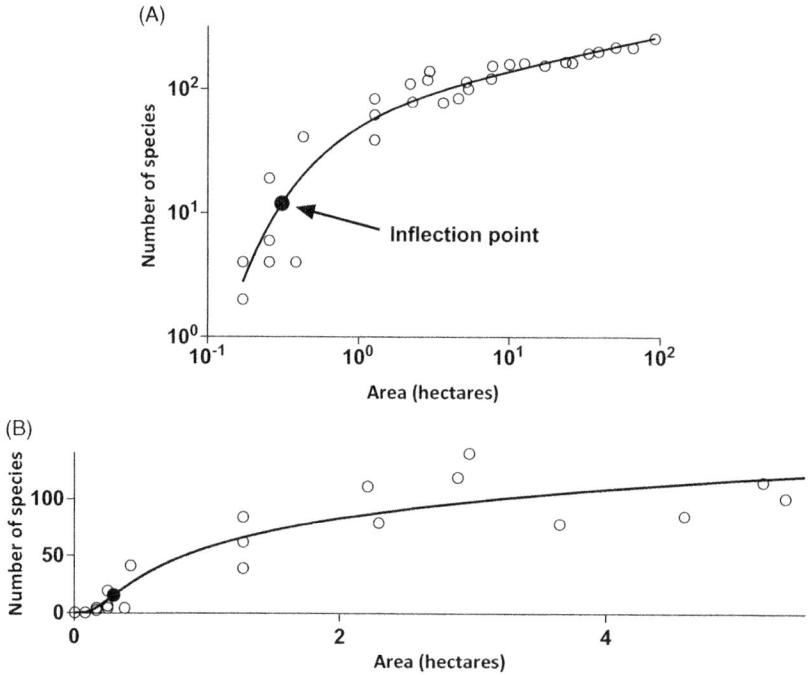

Figure 7.3 Deshaye and Morisset's (1988) dataset of plants on Canadian islands in the Richmond Gulf. The second persistence (P2) model was fitted to the data with a log-transformed dependent variable (number of species). (A) The curve in log–log space, (B) the curve in arithmetic space, although here the curve above six hectares is left out. The curves illustrate well how a sigmoid SAR can become convex upward in log–log space. The inflection point of the P2 model is shown in both plots by the solid black point.

7.5.3 The Shortlist

If the purpose of fitting a curve is to find a good fit to data or to compare parameter values between datasets, both the power and logarithmic models can be considered good heuristic devices, not only for saSARs but also for ISARs. The Tjørve hybrid model may potentially be more useful for saSARs when extrapolating above or below the surveyed scale window. Both the power model and the logarithmic model (as well as Tjørve's hybrid model) return decelerating non-asymptotic curves for the parameter values we wish to consider. However, if we want to discuss complete ISARs, including the smallest scales, a sigmoid model with no asymptote, such as the P2 model or the quadratic logarithmic model, may be better choices (where the P2 model intersects the origin of axes and the quadratic logarithmic has a starting point on the x-axis). At this point, we note that even the P2 and the quadratic models are not necessarily sufficiently flexible to trace the data points well at the lower j-shaped part of an ISAR, resulting in bias or a bad fit. Nonetheless, this provides us with a shortlist of five models, the: 1) power model, 2) logarithmic model, 3) Tjørve hybrid model, 4) P2 model and 5) quadratic logarithmic model.

The large SAR meta-analyses, where a number of models have been fitted to many datasets, have shown us that we do not need all the models that have been proposed and we might even want to focus on models that follow curve shapes we can expect from known patterns in nature rather than those that most often return the best fit. In addition, there may be other useful models that have not yet been proposed.

7.6 Conclusion

A large number of mathematical models have been proposed to describe SARs, either saSARs or ISARs. Here, we have reviewed thirty models that have been fitted to, or suggested for fitting to, SAR data. With such a melange of models it seems difficult to decide which model to choose. Therefore, we have proposed a shortlist of five models: the power model, the logarithmic model, the Tjørve hybrid, the P2-model and the quadratic logistic model. The selection of these models is based on the assumption that both saSARs and ISARs are monotonically increasing, and that ISARs may be sigmoid if the finest scales are included, although when restricted to narrower scale windows most ISAR data exhibit upward convex curve shapes, as do saSAR data.

It may be argued that the model chosen should not only fit the data, but behave according to the expected curve shape outside the sampled scale window. It is important to ask both whether the curve is expected to be decelerating (convex) or sigmoid and whether it is expected to have an asymptote or not. It should also be noted that in species–area datasets, especially ISAR data, the variance (or scatter) is typically large, as there are many confounding factors. This may provide a reason for why simple decelerating models, such as the power curve and the logarithmic curve, often provide better fits than more complex models that are thought to better resemble expected shapes. If the ISAR is inherently sigmoid, then another reason that simple decelerating models (without asymptotes) produce good fits could be that the j-shaped lower part of an ISAR lies outside the sampled scale window.

If we want to look for sigmoid datasets, inclusion of the very smallest scales is more important than a very large scale ratio. For example, the datasets of Deshaye and Morriset (1988) (see Figure 7.3) not only have a large scale window (with a ratio of 1/18,428 between the finest and coarsest scale), but also contain very small islands; down to 0.005 ha. Very few datasets contain such small isolates (or islands). Moreover, it has become generally accepted that saSARs that cover large scale windows are triphasic in log–log space (e.g. Fridley et al., 2005; Storch, 2016). Such a triphasic pattern cannot be fitted successfully with a power law curve with a single z-value segment. Instead, today a piecewise linear approximation (in log–log space), each piece with a different z-value (read: slope in log–log space) is recommended (e.g. Delsol et al., 2018).

It is therefore likely, we would argue, that the success of the power and logarithmic models is largely due to the fact that they are used for data within a restricted scale range, which means that the resultant curve shape is likely simpler than that generated using a larger scale window. This may be the case both for ISARs, which we believe are sigmoid in arithmetic space if the finest scales are included, and for saSARs, which are generally believed to be triphasic. Consequently, depending on how we measure best fit, simpler models will be preferred within narrower scale windows. Despite the large number of proposed models, few models meet the criteria of no upper asymptote, especially when considering the sigmoid models, which are also somewhat restricted in their behaviour. This should prompt us to keep up the search for alternative sigmoid regression models. Moreover, as the power and logarithmic models cannot accommodate triphasic SARs without the use of piecewise regression, the Tjørve hybrid model represents a useful alternative and there is a need for future comparative SAR studies and meta-analyses that incorporate this model.

References

Arrhenius, O. (1920a) Distribution of the species over the area. *Meddelanden från Kungliga Vetenskapsakademiens Nobelinstitut*, **4**, 1–6.

Arrhenius, O. (1920b) Yta och arter. I. *Svensk Botanisk Tidsskrift*, **14**, 327–329.

Arrhenius, O. (1921) Species and area. *Journal of Ecology*, **9**, 95–99.

Arrhenius, O. (1923) On the relation between species and area – A reply. *Ecology*, **4**, 90–91.

Barrett, K., Wait, D. A. & Anderson, W. B. (2003) Small island biogeography in the Gulf of California: Lizards, the subsidized island biogeography hypothesis, and the small island effect. *Journal of Biogeography*, **30**, 1575–1581.

Braun-Blanquet, J. & Jenny, H. (1926) Vegetations-entwicklung und bodenbildung in der alpine stufe der Zentralalpen (Klimaxgebiet des Caricion curvulae). *Denkschriften Schweizerische Naturforsch Gesellschaft*, **63**, 183–344.

Brenner, W. (1921) Växtgeografiska studier i Barösunds skärgård. *Acta Sociatatis pro Fauna et Flora Fennica*, **49**, 1–151.

Brown, J. H. (1971) Mammals on mountaintops: Nonequilibrium insular biogeography. *The American Naturalist*, **105**, 467–478.

Cain, S. A. (1934) Studies of virgin hardwood forest: II, A comparison of quadrat sizes in a quantitative phytosociological study of Nash's Woods, Posey County, Indiana. *The American Midland Naturalist*, **15**, 529–566.

Cain, S. A. (1938) The species–area curve. *The American Midland Naturalist*, **19**, 573–581.

Clench, H. K. (1979) How to make regional lists of butterflies: Some thoughts. *Journal of the Lepidopterists' Society*, **33**, 216–231.

Cody, M. L., Moran, R., Rebman, J. & Thompson, H. (2002) *Plants. A new island biogeography of the Sea of Cortés* (ed. by T. J. Case, M. L. Cody and E. Ezcurra), pp. 63–111. New York: Oxford University Press.

Colwell, R. K. & Coddington, J. A. (1994) Estimating terrestrial biodiversity through extrapolation. *Philosophical Transactions of the Royal Society B: Biological Sciences*, **345**, 101–118.

Colwell, R. K., Chang, X. M. & Chang, J. (2004) Interpolating, extrapolating, and comparing incidence-based species accumulation curves. *Ecology*, **85**, 2717–2727.

Condit, R., Hubbell, S. P., LaFrankie, J. V., Sukumar, R., Monokaran, N., Foster, R. B. & Ashton, P. S. (1996) Species–area and species–individual relationships for tropical trees: A comparison of three 50-ha plots. *Journal of Ecology*, **84**, 549–562.

Connor, E. F. & McCoy, E. D. (1979) The statistics and biology of the species–area relationship. *The American Naturalist*, **113**, 791–833.

Crawley, M. J. & Harral, J. E. (2001) Scale dependence in plant biodiversity. *Science*, **291**, 864–868.

Darlington, P. J. (1957) *Zoogeography: The geographical distribution of animals*. New York: John Wiley.

Delsol, R., Loreau, M. & Haegeman, B. (2018) The relationship between spatial scaling of biodiversity and ecosystem stability. *Global Ecology & Biogeography*, **27**, 439–449.

Dengler, J. (2008) Pitfalls in small-scale species–area sampling and analysis. *Folia Geobotanica*, **43**, 269–287.

Dengler, J. (2009) Which function describes the species–area relationship best? A review and empirical evaluation. *Journal of Biogeography*, **36**, 728–744.

Deshaye, J. & Morisset, P. (1988) Floristic richness, area and habitat diversity in a hemiarctic archipelago. *Journal of Biogeography*, **15**, 747–757.

Diamond, J.M. (1975) Assembly of species communities. *Ecology and evolution of communities* (ed. by M. L. Cody and J. M. Diamond), pp. 342–444. Cambridge, MA: Belknap Press.

Dony, J. G. (1977) Species–area relationships in an area of intermediate size. *Journal of Ecology*, **65**, 475–484.

Drakare, S., Lennon, J. J. & Hillebrand, H. (2006) The imprint of geographical, evolutionary and ecological context on species–area relationships. *Ecology Letters*, **9**, 215–227.

Engen, S. (1978) *Stochastic abundance models*. London: Chapman and Hall.

Fattorini, S. (2006) Detecting biodiversity hotspots by species–area relationships: A case study of Mediterranean beetles. *Conservation Biology*, **20**, 1169–1180.

Fattorini, S. (2007a) Levels of endemism are not necessarily biased by the co-presence of species with different range sizes: A case study of Vilekin and Chikatunov's models. *Journal of Biogeography*, **34**, 994–1007.

Fattorini, S. (2007b) To fit or not to fit? A poorly fitting procedure produces inconsistent results when the species–area relationship is used to locate hotspots. *Biological Conservation*, **16**, 2531–2538.

Fattorini, S. & Fowles, A. P. (2005) A biogeographical analysis of the tenebrionid beetles (Coleoptera, Tenebrionidae) of the island of Thasos in the context of the Aegean Islands (Greece). *Journal of Natural History*, **39**, 3919–3949.

Fisher, R. A., Corbet, A. S. & Williams, C. B. (1943) The relation between the number of species and the number of individuals in a random sample of an animal population. *Journal of Animal Ecology*, **12**, 42–58.

Flather, C. H. (1996) Fitting species-accumulation functions and assessing regional land use impacts on avian diversity. *Journal of Biogeography*, **23**, 155–168.

Fridley, J. D., Peet, R. K., Wentworth, T. R. & White, P. S. (2005) Connection fine- and broad-scale species–area relationships of southeastern U.S. flora. *Ecology*, **86**, 1172–1177.

Gentile, G. & Argano, R. (2005) Island biogeography of the Mediterranean Sea: The species–area relationship for terrestrial isopods. *Journal of Biogeography*, **32**, 1715–1726.

Gitay, H., Roxburgh, S. H. & Wilson, J. B. (1991) Species–area relations in a New-Zealand tussock grassland, with implications for nature-reserve design and for community structure. *Journal of Vegetation Science*, **2**, 113–118.

Gleason, H. A. (1922) On the relation between species and area. *Ecology*, **3**, 158–162.

Gleason, H. A. (1925) Species and area. *Ecology*, **6**, 66–74.

Goodall, D. W. (1952) Quantitative aspects of plant distribution. *Biological Reviews*, **27**, 194–242.

Gotelli, N. & Colwell, R. K. (2001) Quantifying biodiversity: Procedures and pitfalls in the measurement and comparison of species richness. *Ecology Letters*, **4**, 379–391.

Green, J. L. & Ostling, A. (2003) Endemics–area relationships: The influence of species dominance and spatial aggregation. *Ecology*, **84**, 3090–3097.

Green, J. L. & Plotkin, J. B. (2007) A statistical theory for sampling species abundances. *Ecology Letters*, **10**, 1037–1045.

Guilhaumon, F., Gimenez, O., Gaston, K. J. & Mouillot, D. (2008) Taxonomic and regional uncertainty in species–area relationships and the identification of richness hotspots. *Proceedings of the National Academy of Sciences USA*, **105**, 15458–15463.

Hanski, I., Zurita, G. A., Bellocq, M. I. & Rybicki, J. (2013) Species–fragmented area relationship. *Proceedings of the National Academy of Sciences USA*, **110**, 12715–12720.

Harte, J., Blackburn, T. & Ostling, A. (2001) Self-similarity and the relationship between abundance and range size. *The American Naturalist*, **157**, 374–386.

Harte, J., Kinzig, A. & Green, J. (1999) Self-similarity in the distribution and abundance of species. *Science*, **284**, 334–336.

He, F. & Hubbell, S. P. (2011) Species–area relationships always overestimate extinction rates from habitat loss. *Nature*, **473**, 368–371.

He, F. & Hubbell, S. P. (2013) Estimating extinction from species–area relationships: Why the numbers do not add up. *Ecology*, **94**, 1905–1912.

He, F. & Legendre, P. (1996) On species–area relations. *The American Naturalist*, **148**, 719–737.

He, F. & Legendre, P. (2002) Species diversity patterns derived from species–area models. *Ecology*, **83**, 1185–1198.

Hopkins, B. (1955) The species–area relations of plant communities. *Journal of Ecology*, **43**, 409–426.

Hopkins, B. & Skellam, J. G. (1954) A new method for determining the type of distribution of plant individuals. *Annals of Botany*, **18**, 213–217.

Hubbell, S. P. (2001) *The unified neutral theory of biodiversity and biogeography*. Princeton, NJ: Princeton University Press.

Irie, H. & Tokita, K. (2012) Species–area relationship for power-law species abundance distribution. *International Journal of Biomathematics*, **5**, 1260014.

Kangas, P. (1987) On the use of species area curves to predict extinctions. *Bulletin of the Ecological Society of America*, **68**, 158–162.

Keeley, J. E. & Fotheringham, C. J. (2005) Plot shape effects on plant species diversity measurements. *Journal of Vegetation Science*, **16**, 249–256.

Keil, P., Pereira, H. M., Cabral, J. S., Chase, J. M., May, F., Martins, I. S. & Winter, M. (2018) Spatial scaling of extinction rates: Theory and data reveal nonlinearity and a major upscaling and downscaling challenge. *Global Ecology & Biogeography*, **27**, 2–13.

Keil, P., Storch, D. & Jetz, W. (2015) On the decline of biodiversity due to area loss. *Nature Communications*, **6**, 8837.

Kilburn, P. D. (1966) Analysis of the species–area relation. *Ecology*, **47**, 831–843.

Kobayashi, S. (1974) The species–area relation I. A model for discrete sampling. *Researches on Population Ecology*, **15**, 223–237.

Kobayashi, S. (1975) The species–area relation II. A second model for continuous sampling. *Researches on Population Ecology*, **16**, 265–280.

Krishnamari, R., Kumar, A. & Harte, J. (2004) Estimating species richness at large spatial scales using data from small discrete plots. *Ecography*, **27**, 637–642.

Kunin, W. E., Harte, J., He, F., Hui, C., Jobe, R. T., Ostling, A., Polce, C., Šizling, A., Smith, A. B., Smith, K., Smart, S. M., Storch, D., Tjørve, E., Ugland, K.-I., Ulrich, W. & Varma, V. (2018) Upscaling biodiversity: Estimating the species–area relationship from small samples. *Ecological Monographs*, **88**, 170–187.

Kylin, H. (1923) Växtsociologiska randanmärkningar. *Botaniska Notiser*, **1923**, 161–234.

Lennon, J. J., Kunin, W. E. & Hartley, S. (2002) Fractal species distributions do not produce power-law species–area relationships. *Oikos*, **97**, 378–386.

Lennon, J. J., Kunin, W. E., Hartley, S. & Gaston, K. J. (2007) Species distribution patterns, diversity scaling and testing for fractals in Southern African birds. *Scaling biodiversity* (ed. by D. Storch, P. Marquet and J. Brown), pp. 51–76. Cambridge: Cambridge University Press.

Lomolino, M. V. (2000) Ecology's most general, yet protean pattern: The species–area relationship. *Journal of Biogeography*, **27**, 17–26.

Lomolino, M. V. (2001) The species–area relationship: New challenges for an old pattern. *Progress in Physical Geography*, **25**, 1–21.

Lomolino, M. V. (2002) '... there are areas too small, and areas too large to show clear diversity patterns...' R. H. MacArthur (1972: 191). *Journal of Biogeography*, **29**, 555–557.

Lomolino, M. V. & Weiser, M. D. (2001) Towards a more general species–area relationship: Diversity of all islands, great and small. *Journal of Biogeography*, **28**, 431–445.

MacArthur, R. H. & Wilson, E. O. (1963) An equilibrium theory of insular zoogeography. *Evolution*, **17**, 373–387.

MacArthur, R. H. & Wilson, E. O. (1967) *The theory of island biogeography*. Princeton, NJ: Princeton University Press.

Maddux, R. D. (2004) Self-similarity and the species–area relationship. *The American Naturalist*, **163**, 616–626.

Malyshev, L. I. (1991) Some quantitative approaches to problems of comparative floristics. *Quantitative approaches in phytogeography* (ed. by P. L. Nimis and T. J. Crovello), pp. 15–33. Dordrecht: Kluwer Academic Publishers.

Martín, H. G. & Goldenfeld, N. (2006) On the origin and robustness of power-law species–area relationships in ecology. *Proceedings of the National Academy of Sciences USA*, **103**, 10310–10315.

Matthews, T. J., Borregaard, M. K., Guilhaumon, F., Triantis, K. A. & Whittaker, R. J. (2016a) On the form of species–area relationships in habitat islands and true islands. *Global Ecology & Biogeography*, **25**, 847–858.

Matthews, T. J., Steinbauer, M., Tzirkalli, E., Triantis, K. A. & Whittaker, R. J. (2014) Thresholds and the species–area relationship: A synthetic analysis of habitat island datasets. *Journal of Biogeography*, **41**, 1018–1028.

Matthews, T. J., Triantis, K. A., Rigal, F., Borregaard, M. K., Guilhaumon, F. & Whittaker, R. J. (2016b) Island species–area relationships and species accumulation curves are not equivalent: An analysis of habitat island datasets. *Global Ecology & Biogeography*, **25**, 607–618.

Miller, R. I. & Wiegert, R. G. (1989) Documenting completeness, species–area relations, and the species-abundance distribution of a regional flora. *Ecology*, **70**, 16–22.

Olszewski, T. D. (2004) A unified mathematical framework for the measurement of richness and evenness within and among multiple communities. *Oikos*, **104**, 377–387.

Palmer, M. W. (1990) The estimation of species richness by extrapolation. *Ecology*, **71**, 1195–1198.

Pereira, M. & Daily, G. C. (2006) Biodiversity dynamics in countryside landscapes. *Ecology*, **87**, 1877–1885.

Picard, N., Karambé, M. & Birnbaum, P. (2004) Species–area curve and spatial pattern. *Écoscience*, **11**, 45–54.

Pimm, S. L. & Askins, R. A. (1995) Forest loss predict bird extinctions in eastern North America. *Proceedings of the National Academy of Sciences USA*, **92**, 9343–9347.

Pimm, S. L. & Raven, P. (2000) Extinction by numbers. *Nature*, **403**, 843–845.

Plotkin, J. B., Potts, M. D., Leslie, N., Manokaran, N., LaFrankie, J. & Ashton, P. S. (2000) Species–area curves, spatial aggregation, and habitat specialization in tropical forests. *Journal of Theoretical Biology*, **207**, 81–99.

Preston, F. W. (1960) Time and space and the variation of species. *Ecology*, **41**, 611–627.

Preston, F. W. (1962) The canonical distribution of commonness and rarity: Part I & II. *Ecology*, **43**, 185–215, 410–432.

Rosenzweig, M. L. (1995) *Species diversity in space and time*. Cambridge: Cambridge University Press.

Scheiner, S. M. (2003) Six types of species–area curves. *Global Ecology & Biogeography*, **12**, 441–447.

Scheiner, S. M. (2004) A mélange of curves – further dialogue about species–area curves. *Global Ecology & Biogeography*, **13**, 479–484.

Simberloff, D. (1992) Do species–area curves predict extinction in fragmented forests? *Tropical deforestation and species extinction* (ed. by T. C. Whitmore and J. A. Sayer), pp. 75–89. London: Chapman and Hall.

Šizling, A. L. & Storch, D. (2004) Power-law species–area relationships and self-similar species distributions within finite areas. *Ecology Letters*, **7**, 60–68.

Šizling, A. L. & Storch, D. (2007) Geometry of species distributions: Random clustering and scale invariance. *Scaling biodiversity* (ed. by D. Storch, P. A. Marquet and J. H. Brown), pp. 77–100. Cambridge: Cambridge University Press.

Šizling, A. L., Šizlingová, E., Tjørve, E., Tjørve, K. M. C. & Kunin, W. E. (2017) How to allow SAR collapse across local and continental scales: A resolution of the controversy between Storch et al. (2012) and Lazarina et al. (2013). *Ecography*, **40**, 971–981.

Storch, D. (2016) The theory of the nested species–area relationship: Geometric foundations of biodiversity scaling. *Journal of Vegetation Science*, **27**, 880–891.

Storch, D., Šizling, A. L. & Gaston, K. J. (2003) Geometry of the species–area relationship in central European birds: Testing the mechanism. *Journal of Animal Ecology*, **72**, 509–519.

Taylor, L. R., Woiwood, I. P. & Perry, J. N. (1978) The density-dependence of spatial behaviour and the rarity of randomness. *Journal of Animal Ecology*, **47**, 383–406.

Tjørve, E. (2002) Habitat size and number in multi-habitat landscapes: A model approach based on species–area curves. *Ecography*, **25**, 17–24.

Tjørve, E. (2003) Shapes and functions of species–area curves: A review of possible models. *Journal of Biogeography*, **30**, 827–835.

Tjørve, E. (2009) Shapes and functions of species–area curves (II): A review of new models and parameterizations. *Journal of Biogeography*, **36**, 1435–1445.

Tjørve, E. (2010) How to resolve the SLOSS debate: Lessons from species-diversity models. *Journal of Theoretical Biology*, **264**, 604–612.

Tjørve, E. (2012) Arrhenius and Gleason revisited: New hybrid models resolve an old controversy. *Journal of Biogeography*, **39**, 629–639.

Tjørve, E. & Tjørve, K. M. C. (2008) The species–area relationship, self-similarity, and the true meaning of the *z*-value. *Ecology*, **89**, 3528–3533.

Tjørve, E. & Tjørve, K. M. C. (2011) Subjecting the theory of the small-island effect to Ockham's razor. *Journal of Biogeography*, **38**, 1834–1839.

Tjørve, E. & Tjørve, K. M. C. (2017) Species–area relationship. *eLS* (*Encyclopedia of Life Sciences Online*), pp. 1–9. Chichester: John Wiley & Sons.

Tjørve, E. & Turner, W. R. (2009) The importance of samples and isolates for species–area relationships. *Ecography*, **32**, 391–400.

Tjørve, E., Kunin, W. E., Polce, C. & Tjørve, K. M. C. (2008) The species–area relationship: Separating the effects of species-abundance and spatial distribution. *Journal of Ecology*, **96**, 1141–1151.

Tjørve, E., Tjørve, K. C. M., Šizlingová, E. & Šizling, A. L. (2018) Great theories of species diversity in space and why they were forgotten: The beginnings of a spatial ecology and the Nordic early 20th-century botanists. *Journal of Biogeography*, **45**, 530–540.

Triantis, K. A., Guilhaumon, F. & Whittaker, R. J. (2012) The island species–area relationship: Biology and statistics. *Journal of Biogeography*, **39**, 215–231.

Triantis, K. A., Mylonas, M., Lika, K. & Vardinoyannis, K. (2003) A model for the species–area–habitat relationship. *Journal of Biogeography*, **30**, 19–27.

Turner, W. R. & Tjørve, E. (2005) Scale-dependence in species–area relationships. *Ecography*, **28**, 721–730.

Ulrich, W. & Buszko, J. (2003) Self-similarity and the species–area relation of Polish butterflies. *Basic and Applied Ecology*, **4**, 263–270.

Ulrich, W. & Buszko, J. (2005) Detecting biodiversity hotspots using species–area and endemics–area relationships. *Biodiversity and Conservation*, **14**, 1977–1988.

Veech, J. A. (2000) Choice of species–area function affects identification of hotspots. *Conservation Biology*, **14**, 140–147.

Veech, J. A., Crist, T. O. & Summerville, K. S. (2003) Intraspecific aggregation decreases local species diversity of arthropods. *Ecology*, **84**, 3376–3383.

Williams, C. B. (1950) The application of the logarithmic series to the frequency of occurrence of plant species in quadrats. *Journal of Ecology*, **38**, 107–138.

Williams, C. B. (1964) *Patterns in the balance of nature and related problems in quantitative ecology*. London: Academic Press.

Williams, M. R. (1995) An extreme-value function model of the species incidence and species–area relations. *Ecology*, **76**, 2607–2616.

Williams, M. R., Lamont, B. B. & Hestridge, J. D. (2009) Species–area functions revisited. *Journal of Biogeography*, **36**, 1994–2004.

Williamson, M., Gaston, K. J. & Lonsdale, W. M. (2001) The species–area relationship does not have an asymptote! *Journal of Biogeography*, **28**, 827–830.

Williamson, M., Gaston, K. J. & Lonsdale, W. M. (2002) An asymptote is an asymptote and not found in species–area relationships. *Journal of Biogeography*, **29**, 1713.

Wilson, E. O. (1961) The nature of the taxon cycle in the Melanesian ant fauna. *The American Naturalist*, **95**, 169–193.

8 · Biodiversity Scaling on a Continuous Plane: Geometric Underpinnings of the Nested Species–Area Relationship

ARNOŠT L. ŠIZLING AND DAVID STORCH

8.1 Introduction

The species–area relationship (SAR) is considered to be one of the most universal ecological patterns. Indeed, the notion that the number of species recorded within a region should increase as the area of the region increases seems intuitive. No study of diversity patterns in space or time makes sense without accounting for this simple fact (Rosenzweig, 1995). To account for the increase in species richness with area, one needs to know the SAR's properties (e.g. its shape and slope) and the factors that affect them. There is a considerable history of research on this topic (e.g. Arrhenius, 1921; Gleason, 1922; Preston, 1960; MacArthur & Wilson, 1967; May, 1975; Williamson, 1988; Rosenzweig, 1995; He & Legendre, 1996; Hubbell, 2001; Šizling & Storch, 2004; Drakare et al., 2006; Rosindell & Cornell, 2007, 2009; Storch et al., 2007, 2008, 2012; Harte et al., 2009; Lazarina et al., 2013; Storch, 2016; Šizling et al., 2017; see Chapters 3 and 4), but a mechanistic understanding of the underlying causes of SAR properties has emerged only recently. An important component of this understanding is that the drivers of the SAR can be seen as being organized across two hierarchical levels: the geometric and the biological. Each SAR is shaped by biological drivers within the constraints given by geometric rules. Geometry, unlike biology, cannot directly determine actual species richness, but it determines the constraints for the differences in species richness among sites and scales. Geometric rules provide links between the SAR and other macroecological patterns, namely the frequency distribution of species abundances (species abundance distribution, hereafter SAD), species spatial turnover (beta diversity) and, in particular, the spatial distribution patterns of individual species.

8.2 A Typology of SARs

The properties of the SAR and its relationships to other macroecological patterns depend on the way the SAR is constructed (Scheiner, 2003,

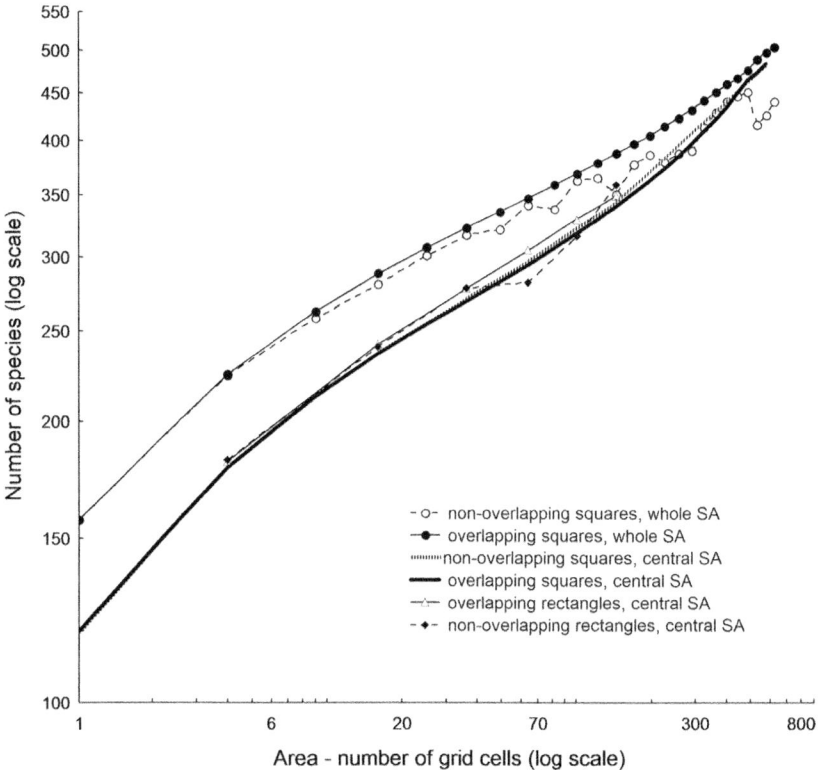

Figure 8.1 An example of the nested SAR constructed using different designs. The nested SAR here is based on the South African Bird Atlas Project (Harrison et al., 1997) in which the area of each grid cell is c. 676 km². The construction depends on 1) the shape of the plots (squares versus rectangles 1 × 4), 2) whether an overlapping or non-overlapping design was used and 3) whether the whole area of South Africa (SA) was sampled (which leads to the situation in which peripheral regions are sampled only by smaller plots) or only a square in the central region (leading to equally intensive sampling for all areas, at the expense of avoiding peripheral areas). These peripheral areas are the most species rich areas in SA and so the nested SARs for the central region lie below those for the whole area. Note that the non-overlapping design for the whole area leads to a fluctuating nested SAR, which can be attributed simply to the incomplete sampling of equal-sized plots that cannot piece together the whole arena. The non-overlapping design leads to a smooth, triphasic nested SAR only in the case of complete sampling, as in the central SA sample area.

2004; Chapter 4). One classical distinction focuses on the differences between island and mainland SARs (Rosenzweig, 1995), but there are actually two mutually independent aspects of this distinction. First, SARs for isolated areas may differ from those for areas that are interconnected by intervening habitat (which facilitates migration), because migration increases the number of species inhabiting smaller areas within which the species could not otherwise have viable populations. However, conceptually more important is the second distinction: whether we examine smaller areas that are nested within larger studied areas or, alternatively, compare mutually disjunct areas differing in size (Tjørve & Turner, 2009). A nested sample design implies strict constraints on species richness at larger spatial scales, as the species richness of larger areas encompasses all the species of the smaller plots within them. In such nested surveys, the species richness of small areas can never be higher than the richness of the larger areas within which they sit. This interdependence of species richness across scales in nested SARs, however, depends on the exact design of the nested sampling (Figure 8.1, Box 8.1). Here, we will focus on nested mainland SARs, as this type of SAR is the most affected by geometric constraints.

Box 8.1 *Designing a nested sampling strategy*

When constructing nested SARs, the aim is to produce a relationship between area and (mean) species richness that accurately represents the situation in a given region and for a focal taxonomic group. However, ultimately this aim can never be attained due to data limitations and the available techniques for plotting SARs. The major limitations and problems are as follows:

Incomplete sampling. Almost no taxonomic group can be sampled completely within an area; there are always unobserved individuals. Even in apparently complete samples, as is the case of some forest tree plots (Condit et al., 1996), the sample is typically limited by a priori criteria (e.g. minimum diameter at breast height). This leads to lower observed species occupancies and consequently to overestimating nested SAR slopes.

Temporal dynamics. Species populations and communities are not static and the longer we observe a given area, the more species we record. The nested SAR thus interacts with the species–time relationship (Adler & Lauenroth, 2003; Chapter 19), so that there is a negative interaction between time and area: the longer we observe a community, the lower is the nested SAR slope and vice versa, and smaller areas have steeper species–time relationships (Adler et al., 2005; White, 2007). As in the previous case, this can be attributed to the increase in species occupancy with increasing time.

Finite sampled area. In the case of a nested design, all the sample plots are located within the arena. This has several consequences. Larger plots must either overlap each other or their number has to be limited; both of these effects decrease variation in species richness with sampling area and make our estimation of species richness for larger plots sensitive to the particular placement of the whole arena within the larger region.

Discrete increment of sample sizes. Nested designs are mostly based on finite grids, which means that there is a smallest sample area (one grid cell) and a minimum distance by which a moving window can be shifted. Therefore, it is not possible to estimate mean species richness across all possible plots. If species distributions have a distinctive spatial structure due to some landscape (ir)regularities, this structure may be blurred by using a particular grid. Species occupancies (and thus nested SARs) may thus differ according to the exact position and resolution of the grid.

Shape of sample plots. In the case of a gridded design, squares are the most often used shapes of sample plots. This is obviously not the only possibility. In the case of other designs, the shape of plots may be different and fixed (e.g. individual nature reserves). This can affect the number of sampled species. The role of the shape of a sample plot on plot species richness is poorly explored, although Kunin (1997) showed that elongated samples generally capture more species than square samples of equal area. This is reasonable, given that there is distance decay in similarity in species composition (Nekola & White, 1999) and elongated areas incorporate more distant sites.

Irregular arena. If a nested sampling design is to be used within a region of irregular shape (i.e. one that is not a square or rectangle) and we still want to use a gridded design, larger sample plots cannot cover the whole area, as they cannot fit into the irregular shape at the periphery of the whole arena (e.g. peninsulas). The areas close to the edge are thus under-represented, which may affect estimated species richness of large areas if peripheral regions differ in species richness from more central regions. It is thus recommended to avoid such incompletely nested designs and restrict the study to arenas where plots of all sizes can be used to sample the whole arena. It is also reasonable to exclude coastal species from sampled islands and/or continents. An alternative approach may be to relax the constraints on sample shape at coarse scales, allowing samples to be fit into the available space or to use the shape of the whole region as the basic shape for all sample areas (Storch et al., 2012).

Since samples of similar area may have different numbers of species, if we plot raw data of species richness against the areas of each sample separately, the relationship may be quite scattered. In most cases, we need a representation of this relationship by a curve for which there is only one value of species richness for a given area. It is thus necessary either to divide the samples into particular size (area) classes and esti-mate mean species richness across all the plots within each class or to find some mathematical function which best approximates the scattered relationship, typically fit to the data by some form of regression (Chapter 7). If the smoothing function is well chosen, both approaches lead to similar results, as the fitted function approximates the mean values for each area (arithmetic means if we perform a least squares regression in arithmetic space and geometric means if the regression is performed for logarithmically transformed values of species richness). The nested SAR is thereafter a relationship between area and the mean species richness of that area. This is the most obvious way to plot nested SARs and particularly SARs based on grid or transect data. In such a case, it is possible to calculate mean species richness across the windows of given size in the nested design. Constructed in this way, there is a straightforward relationship between the SAR and the spatial distribu-tion patterns of the surveyed species.

8.3 Constraints Imposed by Species Occupancies

Here, we will show how the shape of the nested SAR is determined by the spatial distribution patterns of individual species and demonstrate that the slope of the SAR is driven by patterns of rarity and commonness.

8.3.1 Independence from Interspecific Interactions

Species distributions are spatially complex, with patches and gaps in a species' geographic range across a wide range of scales. One could take a finite set of plots of a given size and calculate the fraction that is occupied by a species, obtaining *relative occupancy* in the case of non-overlapping plots or *probability of occurrence* in the limiting case of an infinite number of mutually overlapping plots. As we cannot have an infinite number of plots in real data, we usually calculate *relative frequency of occurrence*, employing as many overlapping sampling plots as possible. The relative frequency of occurrence is a good estimate of the probability of occurrence and, for the sake of simplicity, most of the literature calls this estimate the probability of occupancy. The relative occupancy is also a good, although slightly worse, approximation of the probability of occurrence. Probability of occurrence tends to increase with the area of the plot. The expected species richness in a random plot of a given area is identical to the sum of the probabilities of occurrence across all the species. Technically,

$$\bar{S}[A] \equiv \sum_{i=1}^{S_{\text{tot}}} p_i[A], \tag{8.1}$$

where A is the area, $\bar{S}$ is the mean species richness, S_{tot} is the total number of species in the whole studied area and p_i is the probability of occupancy attributed to the i-th species (see Coleman, 1981, for relative occupancy, and Šizling et al., 2017 for probability of occurrence). Equation (8.1) implies that species occupancies across the scales considered provide complete information on the nested SAR, which means that no additional information such as specific placement of occupied and empty grid cells for different species can improve the estimate and all relevant biological causes are mediated via the occupancy patterns. This implies that, even though interspecific interactions can affect the co-occurrence patterns of different species (e.g. negative correlations in the case of competition or positive associations in the case of facilitation or other positive interactions), they cannot affect the nested SAR except insofar as they affect the species occupancies themselves. Such correlations may affect variation around the SAR (i.e. with two positively correlated species, samples would tend to capture either both or neither; whereas for negatively correlated species they would tend to capture either one or the other), but not the mean richness determining the nested, mainland SAR.

8.3.2 Shape of the Relative-Occupancy–Area Relationship

The function which relates probabilities of species occurrence to area can be extracted from occupancy data. The resulting functions are variously referred to as scale–area curves (Kunin, 1998), incidence functions (Ovaskainen & Hanski, 2003), occupancy–area relationships (He & Condit, 2007), range–area relationships (Harte et al., 2005) and P–area relationships (Storch et al., 2008). Although these relationships can be fitted using various mathematical functions (Nachman, 1981; Wright, 1991; Hanski & Gyllenberg, 1997; Leitner & Rosenzweig, 1997; Kunin, 1998; He & Gaston, 2000; for review see He & Condit, 2007), they are generally more or less linearly increasing when plotted on log–log axes (Kunin, 1998), as would be expected for a fractal (self-similar) distribution. However, when examined closely most appear somewhat decelerating across relatively fine scales and accelerating over very coarse scales (Kunin, 1998; He & Gaston, 2000), rather than being precisely linear. At coarser scales, these curves generally reach saturation at a particular area (Šizling & Storch, 2004; He & Condit, 2007). The point at which the curve reaches saturation depends on the relative rarity of the species, that is, on occupancy at the basal scale corresponding to unit grid cells (appendix by Šizling and Storch in Kunin et al., 2018). The probabilities of occurrence and, thus, relative occupancies of rare species reach saturation later than is the case for more common species, depending on the species placement within the total arena.

The finding that a species' occupancy at fine scales (i.e. species rarity) affects the slope of that species' occupancy curve suggests that the overall shape and slope of the nested SAR depends on the proportion of common and rare species in a given assemblage. When the majority of species are widely dispersed, even small sampling areas contain most of the species, as only small gaps in the spatial distributions of species are likely; sampling plots will only rarely fall within these gaps. Further increase in area does not lead to a substantial additional increase in species richness (as most species have already been sampled), which results in decelerating SARs. In contrast, if rare species predominate, most species will not occur in smaller sampling areas. Species number thus increases quite slowly with increasing area at the beginning, before increasing more rapidly when the sample area approaches the size of the total focal area and the sampling areas have become large enough to sample even the quite rare species. The SAR is consequently upward accelerating in such a case. Natural communities are typically made up of a mixture of common and rare species, and the shape of the SAR at fine scales is more affected by common species, while rare species are responsible for the continuing rise of the curve at large scales (Tjørve et al., 2008). The exact shape, however, depends on the occupancy–area relationships of all species, as the nested SAR is determined by summing these relationships.

8.3.3 Overall Slope of the Nested SAR

One important consequence of the fact that the mean species richness of an area is given by the relative occupancies of all species is that the overall slope of the SAR of a grid, when plotted in log–log space (usually denoted as z), is determined by the mean species occupancy of the unit grid cells in combination with the total number of these unit grid cells (Šizling & Storch, 2004). The rarer the species are, the steeper the SAR will be. This follows from the fact that the overall slope of the log-transformed nested SAR can be estimated from the two endpoints of the relationship: at the upper right end, the total species richness S_{tot}, corresponding to the total focal area (measured in number of unit grid cells) A_{tot}; and at the lower left the minimum area considered, that of the unit grid cell ($A = 1$) and the mean species richness within the unit grid cell, which is equal to the sum of the species' relative occupancies at this finest scale (see Equation 8.1) (Figure 8.2). The slope of the line defined by these two extreme points in log–log space is then

$$ Z = \frac{\ln\left(\dfrac{S_{tot}}{\sum p_i[1]}\right)}{\ln\left(A_{tot}\right)}. \tag{8.2} $$

Since the mean species occupancy at one, $\bar{p}[1]$, equals $\sum_{i=1}^{S_{tot}} p_i[1]/S_{tot}$, then $z = -\ln\left(\bar{p}[1]\right)/\ln\left(A_{tot}\right)$. As a consequence, any external or internal factor that affects mean relative species occupancy necessarily also affects the nested SAR slope, so that higher mean relative occupancy leads to a lower nested SAR slope. This is the reason why more productive areas that

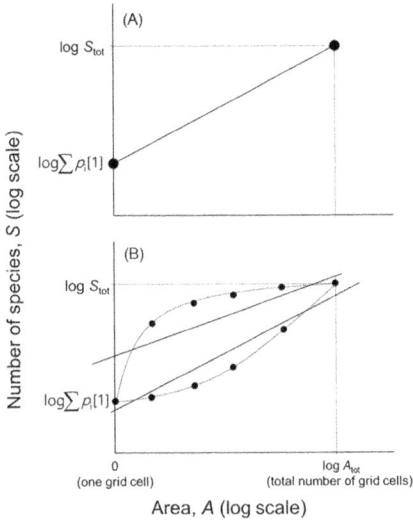

Figure 8.2 The overall slope of the nested SAR and its relationship with mean species relative occupancy for a complete nested design. (A) Assuming that the overall slope of the nested SAR plotted on logarithmic axes is given by the two endpoints of the relationship, the slope is thus determined by mean species occupancy (i.e. the number of occupied cells for each species), total grid size (total number of basic grid cells) and total number of species. (B) If the nested SAR is not linear in a log–log plot, however, the overall slope and the slope estimated using linear least squares regression (solid lines) may deviate from each other, depending on the exact shape of the nested SAR (dotted curved lines) and the differences in the size of the sampling areas (i.e. the distances along the x-axis).

are characterized by higher bird occupancies (Bonn et al., 2004) have lower nested SAR slopes (Storch et al., 2005) and also why higher nested SAR slopes are typical for higher trophic levels which typically have low fine scale occupancies (for a review see Drakare et al., 2006). In line with this finding, lower nested SAR slopes have been reported for plant communities with higher biomass (Chiarucci et al., 2006), later successional stage communities (Lepš & Štursa, 1989; Carey et al., 2006) and for taxa with small body sizes (Azovsky, 2002; Finlay, 2002).

8.4 Constraints Imposed by the Fractal Geometry of Species Ranges

Many mathematical functions have been proposed to model the SAR (Tjørve, 2003, 2009; Dengler, 2009; Chapter 7). The most commonly used expression is the power law, which is given by $S = cA^z$, where S is the number of species, A is area, c is a constant related to the number of species in a plot of unit area ($A = 1$) and z is the slope of the SAR on a log–log scale (Rosenzweig, 1995). Thus, here z represents the multiplicative increase in species for each multiplicative increase in area (e.g. 'Darlington's rule' that a 10-fold increase in area brings a doubling in species richness is equivalent to a z of $\log(2) \approx 0.3$). The power law was first suggested as a good model for the SAR by Arrhenius (1921; see also Chapter 2) in what was the first formal description of the SAR. It is an advantageous model for several reasons. First, the SAR has been repeatedly reported to be close to linear in log–log space, which means that the power law is indeed an appropriate approximation (Connor & McCoy, 1979; Rosenzweig, 1995; Dengler, 2009). In addition, the power law has been shown to generally provide a better description of the SAR than other common alternatives (e.g. than a logarithmic function of the form $S = k + m \log A$, where k and m are constants, Gleason, 1922), although this may not apply across all spatial scales. Second, and perhaps more importantly, the slope z is a dimensionless number which allows comparison between SARs from different assemblages using a single variable. Third, the z-value of the nested SAR is related to several metrics of species spatial turnover (beta diversity), as we will show in Section 8.9.

That said, it is important to note that there is no reason why the power law should be the universal and invariantly proper description of the SAR. In fact, the power law cannot be a universal description of the SAR across all scales and for all taxa (Storch & Šizling, 2008; Figures 8.3 and 8.4). As mentioned in Section 8.3.1, the nested SAR is determined by summing individual species occupancy–area curves, which in turn are dependent on species' distribution patterns. The nested SAR can thus potentially have various shapes depending on the proportion of rare and widely dispersed species in an assemblage, and on the occupancy–area relationship of each species. Indeed, insofar as occupancy–area relationships vary in their parameters, there is no reason to expect any simple mathematical function to be the proper and universal description of the nested SAR (Šizling & Storch, 2004), unless the frequency distribution of these parameters was universal across taxa and biomes.

Nonetheless, it does make sense to ask which types of species spatial distributions and which corresponding types of occupancy–area curves lead to particular shapes and slopes of the nested SAR. Is there a particular set of occupancy–area relationships whose sum approaches the power law SAR? It was initially suggested (Harte et al., 1999) that fractal species spatial distributions (i.e. ones that are self-similar across scales), which produce power law occupancy–area curves, would produce power law SARs. However, this idea was later proved not to be the case (Lennon et al., 2002), as the sum of multiple power laws with different slopes is not itself a power law, but instead tends to accelerate on logarithmic axes. This is a seldom seen pattern in measured nested SARs except at extremely coarse (e.g. continental) scales. However, Šizling and Storch (2004) have shown that the nested SAR can statistically approach the power law if individual species have spatial distributions that are effectively close to fractal and if most species have relatively low occupancies, so that their occupancy–area relationships

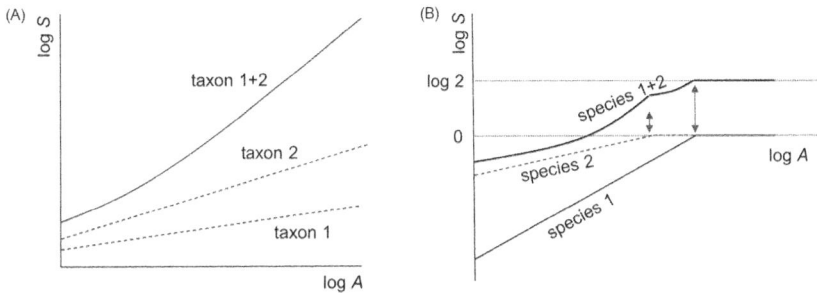

Figure 8.3 The power law approach to the nested SAR cannot be universal across taxa. (A) Let us assume we have two taxa, t1 and t2, which together comprise the higher taxon t(1+2). If the nested SARs for both taxa t1 and t2 are power laws with different slopes, the nested SAR for the higher taxon t(1+2) cannot follow a power law, as two power laws form a power law only if both have the same parameters (Lennon et al., 2002; Storch & Šizling, 2008). The power law nested SAR thus cannot be universal across all taxa simultaneously. This can be extended even up to the species level: if the occupancy–area relationships for individual species are power laws with different slopes, the resulting nested SAR is an upward-accelerating curve on a log–log scale (Lennon et al., 2002). (B) A nested SAR approaches a power law (bold line for s(1+2)) if the approximate power law occupancy–area curves reveal saturation for some area (arrows) instead of increasing across all scales. This must indeed be the case for any nested SARs measured in a finite area (Šizling & Storch, 2004) (species richness, S, for one species is equivalent to its probability of occurrence within a plot of a given area, which can be estimated using relative occupancy at the respective scale, and S of a single species is thus between zero and one, that is, its logarithm is between zero and minus infinity). In such a case, the upward-accelerating tendency of the nested SAR, driven by summing the power law occupancy–area curves with different exponents, is compensated for by saturation of these curves, which is related to the fact that relative occupancy or probability of species occurrence cannot be higher than 1.

saturate to a value of one at relatively large sampling areas. The tendency for both downward and upward acceleration of the SAR, mentioned above, has a compensatory effect, generating power law like SARs. The fact that occupancy–area relationships are effectively almost linear up to the point of saturation has indeed been commonly reported (Virkkala, 1993; Kunin, 1998; Ulrich & Buszko, 2003; Šizling & Storch, 2004). However, even though self-similar properties of species spatial distributions have been empirically reported, the cause of this pattern was unclear until recently. To explain this observation, Šizling and Storch (2007) and Storch et al. (2008) built a theory of *generalized fractals* to show that patterns which are effectively indistinguishable from fractals can emerge due to any process comprising intraspecific aggregation at multiple spatial scales, for example due to a hierarchy of habitat resolution (so that finely defined habitat patches are nested within habitats defined more broadly). Thus, this theory represents a null model of aggregated species spatial distributions and consequent macroecological patterns.

8.5 Triphasic SARs and Constraints of the Finite-Area Effect

As mentioned above, nested SARs are not properly characterized by the power law across all spatial scales (Hubbell, 2001; Fridley et al., 2005; Storch & Šizling, 2008; Harte et al., 2009; Storch, 2016). First, if we examine sufficiently fine scales, in log–log space the nested SAR typically becomes flatter as the sampling area gets larger, so that the logarithmic function may represent a better approximation than the power law at small scales (Rosenzweig, 1995). This has been attributed by some authors (e.g. Rosenzweig, 1995; Hubbell, 2001) to sampling effects; the number of individuals in such small sample plots is comparatively low and approaches the number of species. As there cannot be more species than individuals, a further decrease in sampling area (and thus the number of individuals) must lead to an increasingly faster decrease of species number. This issue is not trivial and we will deal with it later in the section devoted to the relationship between the SAR and abundance patterns.

A different deviation from the power law has been described as occurring at very large scales (i.e. larger than the extent of whole species ranges), whereby the slope of the logarithmically transformed SAR again increases (Shmida & Wilson, 1985; Storch et al., 2012). This leads to the notion of a generally triphasic nested SAR (i.e. with downward decelerating, linear and upward accelerating phases in log–log space; Fridley et al., 2005). The upward trajectory of the SAR at very coarse spatial scales is consistent with the behaviour of the individual species occupancy curves discussed above, which bend upwards before reaching saturation. At any scale (resolution) coarse enough that the full range of a species is contained within a single sample area, its occupancy behaves like a single point, resulting in an occupancy curve with a slope that approaches one as area approaches the area of saturation. Allen and White (2003) have shown that the triphasic SAR emerges whenever the distribution of individual species is represented by distinct clumps that are generally smaller than the whole studied area (for instance, if the area comprises whole species ranges; see also McGill & Collins, 2003). A triphasic SAR is also predicted by the neutral model of biodiversity dynamics (Hubbell, 2001; Rosindell & Cornell, 2007, 2009; O'Dwyer & Green, 2010; see Chapter 11), its properties depending on several parameters of neutral dynamics, namely dispersal kernels and speciation rate. Finally, the points that separate the three phases of the triphasic SAR have been shown to depend on the distribution of species range sizes (Storch et al., 2012).

8.6 Constraints Imposed by the Species Abundance Distribution

There have been many attempts to derive the SAR from the species abundance distribution (hereafter SAD; see May, 1975; Williams, 1995; He & Legendre, 2002; Martín & Goldenfeld, 2006; Chapter 4). If individuals were distributed randomly in space, the SAR would effectively be equivalent to a *species accumulation curve* (Gray et al., 2004; Ugland et al., 2005), which is the relationship between mean species richness and the number of individuals drawn from a well-mixed pool. The shape of such a curve entirely depends on the SAD. Such a situation has never been observed, with species spatial distributions generally being more clustered than random (Kunin, 1998; Harte et al., 2005; Storch et al., 2008; McGill, 2010). A SAR derived strictly from random placement would rise relatively quickly to an asymptote once samples were large enough to capture even the rare species and level off thereafter, resulting in a prediction of unrealistically low slopes (and unrealistically high predicted richness) of logarithmically transformed SARs at intermediate to coarser scales.

Further development of these ideas has combined the sampling effect with the spatial aggregation of individuals. Intraspecific aggregation causes larger gaps in species spatial distributions, assuming a given species abundance, which leads to lower mean species richness, as each species occurs in a lower number of quadrats of a given size than would be expected for a random spatial distribution. Evidently, the higher the degree of spatial aggregation of individuals, the higher the overall slope of the nested SAR (z; He & Legendre, 2002), which resonates with the previous findings described above, if we recognize that high aggregation leads to lower occupancy. According to this approach, therefore, a realistic nested SAR emerges due to the combination of the SAD and the spatial aggregation of individuals (Martín & Goldenfeld, 2006; Tjørve et al., 2008).

The problem with this approach, however, is the assumption that the SAD is given a priori for the whole arena and that SADs of smaller areas can be derived by aggregation (Green & Plotkin, 2007). This assumption would mean that the arena we are studying is ecologically meaningful, where an underlying mechanism produces the abundances of all species within the whole arena. If we choose arenas arbitrarily or such a mechanism does not exist, the SAD would not be general across space and taxa. The reason is that the SAD is not scale invariant and thus SAD form would vary across arenas. Recently, it has been demonstrated that the SAD is not an independent macroecological pattern but itself emerges due to species spatial turnover at multiple scales (Šizling et al., 2009a, b; Kůrka et al., 2010), which is the same pattern that drives the nested SAR. We thus do not need the assumption of a mechanism that centrally controls abundances in large arenas. Generally, there is good reason to assume that the SAD is actually a derived pattern and of limited value as an explanation of the nested SAR.

8.7 Constraints Imposed by the Mean Number of Individuals per Species

Although the nested SAR cannot simply be viewed as being determined by the SAD (in combination with aggregation), species abundances still affect the shape of the nested SAR. Just as we can plot the relationship between mean (expected) species richness and sampling area, we can construct the relationship between the expected number of individuals and the sampling area. Let us call it the 'individuals–area relationship' or IAR. Such a relationship must be necessarily linear in a nested design (so that doubling the area of a sample doubles the mean number of individuals it contains), even in the case of a highly heterogeneous distribution of individuals in space. In a log–log plot (Figure 8.4), the SAR will lie below the IAR, since there can never be more species than individuals. The vertical distance between both curves then equals the logarithm of the mean number of individuals per species (as $\log(I) - \log(S) = \log(I/S)$). The IAR follows a line with slope of one in a log–log plot and thus only the first statistical moment of the SAD, namely mean abundance per species, uniquely determines the shape and slope of the IAR. Such plots make it quite clear that the nested SAR cannot be linear across all scales in log–log space; it thus cannot follow the power law universally. This is because, on one hand, the slope of the nested SAR must be lower than that of the IAR (otherwise doubling the area would lead to doubling the number of species, which can only occur if there were no overlaps in species between sites) and at the same time the nested SAR cannot cross

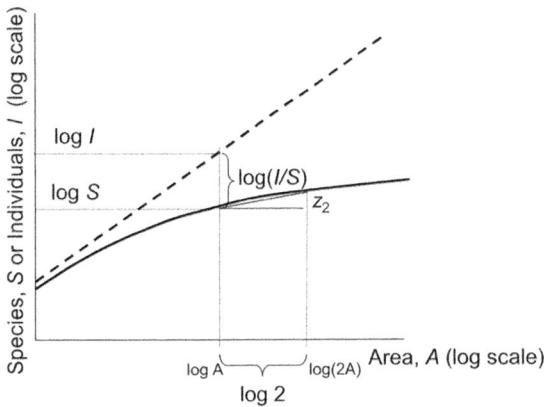

Figure 8.4 The relationship between the nested SAR (bold) and the increase in the number of individuals with area (IAR, dashed). The total number of individuals should increase with area linearly in a completely nested design, which means that it is a straight line with slope equal to one on a log–log scale. Species richness must be lower than the number of individuals for each area, so that the distance between the individuals–area relationship (IAR) and the nested SAR is $\log(I) - \log(S)$, which is equal to $\log(I/S)$, that is, to the logarithm of mean species abundance for a given area. Clearly, if the slope of the nested SAR is (labeled as z_2, that is, the rate of species richness increase when doubling area) lower than the slope of the SAR for large areas, the same slope cannot be maintained for smaller areas, otherwise the nested SAR would cross the IAR. If the nested SAR is close to the IAR (and thus the mean species abundance is low), its slope must approach the slope of the IAR, that is, it must approach one (modified from Šizling et al., 2011).

the IAR, as the number of species must be lower or equal to the number of individuals (Figure 8.4). This constraint causes the nested SAR to rise steeply (approaching the slope of the IAR, $z = 1$) at very fine scales, where species richness is limited by the number of individuals sampled, but then to decelerate as we move to somewhat larger areas, where the nested SAR and the IAR pull apart from each other. This corresponds to the abovementioned curvature of the nested SAR for small areas, where the number of individuals is too low to encompass a higher number of species. In the limit, when each species is represented by just one individual, species richness increases proportionally to area, that is, the nested SAR slope is 1 and is equivalent to the IAR.

Harte et al. (2009) have argued that the relationship between mean abundance per species and the local slope of the nested SAR (i.e. the derivative of the log–log nested SAR) is a universal function across space and taxa. They have based their approach on Maximum Entropy machinery, which calculates the most likely distribution within particular constraints (see Jaynes, 1957, 1982). Harte et al. (2009) named their approach the 'Maximum Entropy Theory for Ecology' (METE). Harte et al. assumed that the total number of species, the total number of individuals and the total energy consumed are conserved at a given area and then derived a one-to-one relationship between the local slope of the SAR and the ratio between the total number of individuals and number of species (i.e. mean species abundance; see Chapter 10 for a discussion of METE and SARs). Although the assumptions may be questionable and the METE approach has several problems (Haegeman & Etienne, 2010) and the discussed relationship cannot apply universally to all taxa (Šizling et al., 2011), it follows from planar geometry of the log–log graph (Figure 8.4) that the local slope of the SAR should indeed be constrained by mean population size of a given area. This can be derived from simple considerations regarding an autocorrelation of the Jaccard index (a proxy for species spatial turnover) across scales (Šizling et al., 2011). The Jaccard index between large adjacent plots is constrained by the Jaccard index for small adjacent subplots. Because the Jaccard index of two adjacent plots scales with z in a one-to-one manner, the constraints upon the Jaccard index determine the constraints for the upper and lower value of z (Šizling et al., 2011). The METE prediction of the relationship runs in the middle between these constraints, but relationships close to the limits are also likely and observed (Šizling et al., 2011). For example, this was found to be the case for an analysis of a British plant dataset, where the constraints of fractal geometry predicted SAR shape better than METE (Kunin et al., 2018). Note that species abundances are positively related to species occupancies (Gaston et al., 1997) and this finding is thus consistent with the abovementioned relationship between the overall slope of the nested SAR and mean species occupancy (Figure 8.2). This cross-validation of the METE is important for its consistency.

8.8 The Generality of the Nested SAR

Although the claim of generality by the METE has been challenged, there have been several other attempts to find a universal pattern beyond the SAR. Allen and White (2003) showed, using computer-based simulations, that SARs measured on an infinite plane tend to collapse into a universal curve when rescaled by mean species range size and mean richness. Storch et al. (2012) further demonstrated that rescaled SARs extracted from data of five taxa across five continents collapse into a universal curve. Lazarina et al. (2013), however, explored a huge data set on SARs where the arenas were smaller than continents, varying between 0.004 and 27,000 km^2 and reported considerable deviations from the expected collapse at small and intermediate scales. The deviation from the expected collapse can be attributed to the finite area effect (Šizling et al., 2017): the universal expected collapse fails where species ranges are comparable in size with the arena. This is consistent with the conclusion that the nested SAR is upward accelerating at large scales where ranges are small compared to the whole sampled arena. At small scales, however, where ranges are comparable in size to the sampled arena, SARs vary depending on the frequency distribution of occupancies and areas of gaps within the ranges; both these factors influence the relationship between probability of occurrence and area.

8.9 The Relationship between the Nested SAR and Beta Diversity and Other Macroecological Patterns

The close relationship between measures of species spatial turnover and the slope of the nested SAR in log–log plots (i.e. z in $S = cA^z$) has been repeatedly recognized (Harte & Kinzig, 1997; Arita & Rodríguez, 2002; Koleff et al., 2003; Gaston et al., 2007; Tjørve & Tjørve, 2008) and is in fact quite obvious. If there is no species turnover, an increase in area does not lead to an increase in the number of species, whereas, if species composition changes rapidly between different plots, any increase in area is followed by a considerable increase in species richness as new areas containing new species are encountered. However, for most formulations developed to date this strictly applies only for measures of species turnover between two equal-area adjacent plots, because in this case the total species richness concerns the contiguous area which is twice the size of each of the original two plots. For instance, Whittaker's (1960) beta diversity index is computed as the ratio between total species richness and the mean species richness of each subplot, $\beta_w = \frac{S_{tot}}{S_{sample}}$, and as such it clearly relates to the nested SAR, such that $z = \ln \beta_w / \ln \left(A_{tot} / A_{sample} \right)$, where A_{tot} is the area of the sampled arena and A_{sample} is the area of the sampling plot. Similar mathematical relationships exist for all the beta diversity measures that scale in a one-to-one manner with Whittaker's beta, such as the Jaccard index: the slope z_2 of the logarithmically transformed nested SAR between an area, and an area twice as large can be estimated as $z_2 = \ln \beta_w / \ln 2 = 1 - \ln (1 + J) / \ln 2$, and thus $J = 2^{1 - z_2} - 1$ (Šizling et al., 2017).

Much more problematic is the relationship between the nested SAR and beta diversity for non-adjacent areas. Harte and Kinzig (1997) have derived several macroecological measures in this context, including species spatial turnover and its distance-dependence using the assumption of an exact power law SAR (see also Harte et al., 1999). However, these relationships are only approximate, as they implicitly assume that species turnover between the nonadjacent samples estimates turnover between adjacent samples. These relationships therefore provide realistic predictions only in some cases. So far, the mathematical connections between the nested SAR and the scaling of species turnover (its distance decay; Nekola & White, 1999) are poorly explored (but see Azaele et al., 2008) and represent an exciting area of future research.

The nested SAR is also related to various other macroecological patterns (Storch et al., 2008; McGill, 2010). An obvious example is the regional–local richness relationship (Caley & Schluter, 1997), because regional species richness is simply the number of species in an area larger than that used to calculate local richness, that is, two points in the SAR (Rosenzweig & Ziv, 1999; Bartha & Ittzés, 2001). Another pattern which has been related to the SAR is the density–area relationship (e.g. Pautasso & Weisberg, 2008), that is, the observation that population densities are lower if they are measured over larger areas. According to Nee and Cotgreave (2002), this is a simple consequence of the fact that species richness increases with area, whereas total density (of all species combined) tends to remain constant, so that larger areas are necessarily characterized by lower mean per-species density. As argued above, the increase in richness with area is due to species spatial turnover and thus density–area relationships and community turnover patterns are also mutually dependent.

It is important to note that, in all these cases, it is not always clear what is the primary pattern and what is derived. As we have noted above, all the major macroecological patterns are related to species spatial aggregation at multiple scales (Storch et al., 2008), which can then be considered as the pattern which drives all the other macroecological patterns (McGill, 2010). However, in contrast to the traditional view that

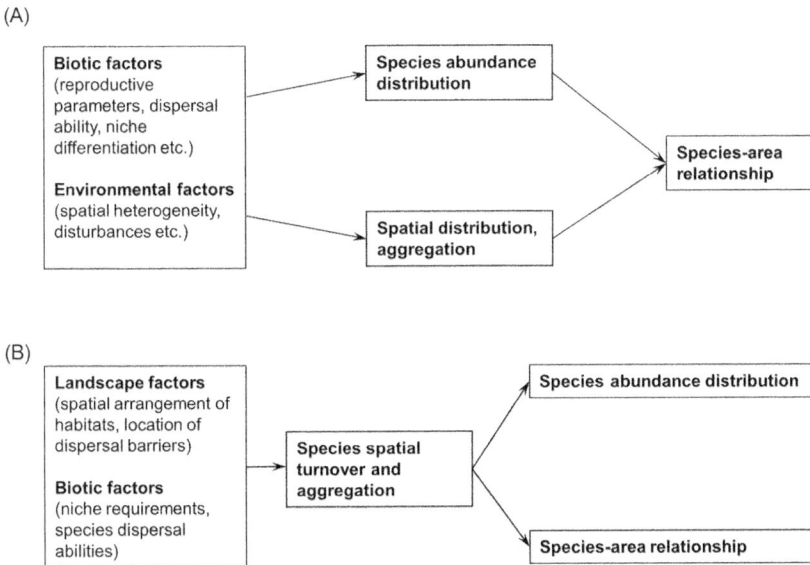

Figure 8.5 The problem of primacy of macroecological patterns which are naturally connected to each other. (A) Many approaches (e.g. He & Legendre, 2002; Martín & Goldenfeld, 2006) assume that the nested SAR can be derived from the SAD in combination with species spatial aggregation, with both of these patterns being determined by different biological factors and processes. (B) However, current findings (Šizling et al., 2009b) indicate that all the patterns are primarily affected by the factors affecting the spatial correlation structure of species distributions, which leads to species spatial turnover and spatial autocorrelation of abundances resulting from aggregation patterns of species distributions. This then leads to many other macroecological patterns, including the SAD.

combines spatial aggregation with the SAD as two independent effects (He & Legendre, 2002), there are good reasons to believe that the truly basic 'master' patterns comprise species spatial turnover between neighbouring sites and the spatial autocorrelation of species abundances (Šizling et al., 2009a, b), that is, phenomena that are very familiar to field ecologists (Figure 8.5); the nested SAR and the SAD are then derived from these 'master' patterns.

8.10 Conclusions and Perspectives

There is no universal nested SAR that is observed in all situations. However, although empirical nested SARs have various slopes and shapes, these properties are tightly related to patterns in species spatial distributions, such that nested SAR properties can be at least partially predicted on the basis of knowledge of these patterns or the biological factors that affect them. This particularly applies to SARs constructed using a fully nested design, where the species richness of larger areas is constrained by the richness of smaller areas and vice versa. Recent research on the relationships of the nested SAR to other patterns has led to several generalizations.

First, the shape and slope of nested SARs are given exactly by the spatial scaling of species occupancies, that is, they are determined by the occupancy–area relationships of the focal species. Species interactions thus affect the nested SAR only if they affect species occupancies, whereas the exact locations and correlations of species presences are not important for the mean species richness of an area. This also implies that the overall slope of the nested SAR measured on a grid is given by the size of the grid (measured in number of grid cells) and mean species occupancy across those finest-scale cells. Therefore, any biological factors that affect mean species occupancy of the studied taxa necessarily affect the overall nested SAR slope.

Second, there is no a priori reason to expect that the nested SAR follows a simple mathematical function. It is quite practical, however, to approach nested SARs across a range of intermediate scales using the power law, which allows comparison between different nested SARs using two parameters: the overall or local slope of the log-transformed relationship, z, and the intercept c. For this reason, it is useful to plot the nested SAR using log–log plots; nonlinearities in these plots reflect deviations from the power law. When examined across a sufficiently wide range of scales, the nested SAR cannot follow the power law and there are good reasons to expect generally triphasic SARs. Triphasic SARs are actually produced by several models of biodiversity dynamics. However, realistic nested SARs can be produced by any model that incorporates species aggregation at multiple scales, regardless of whether it is habitat heterogeneity or dispersal limitation (or indeed some other property) which is deemed responsible for the spatial clumping of individuals.

Third, there is a tight relationship between the nested SAR and SADs, species accumulation curves and species spatial turnover. We have argued, however, that attempts to derive the nested SAR using sampling from a distribution of abundances may be misleading, as these distributions are themselves scale dependent and are actually determined by the same pattern as the nested SAR, that is, the clumped spatial distribution of individuals and species spatial turnover. Indeed, species spatial turnover is directly related to the slope of the log–log nested SAR at each particular scale and appears to be the underlying factor driving the pattern. Species spatial turnover increases with the level of spatial aggregation of species and decreases with mean species abundance (and also occupancy, see Section 8.3.1), so that it is natural that these factors are directly related to the nested SAR slope.

Regardless of the progress that has been made in understanding the abovementioned patterns and relationships, several issues remain unresolved. The relationship between the nested SAR and the scaling of species turnover (its distance decay) is still poorly understood, which limits our ability to up-scale regional species richness from scattered local samples (e.g. Ugland et al., 2003; Kunin et al., 2018).

The patterns we have described obtain their universality on the basis of geometric constraints and mathematical logic. However, biological processes, such as the spatial aggregation of conspecifics, are crucial in driving finer and more quantitative features of spatial patterns. Observed nested SARs, as well as their proximate drivers (i.e. the spatial distribution of individual species), therefore emerge due to the interplay between geometry and biology.

Acknowledgements

The work was supported by the EU FP7 SCALES project ('Securing the Conservation of biodiversity across Administrative Levels and spatial, temporal and Ecological Scales'; project No. 26852) and by the Czech Science Foundation (project No. 20-29554X).

References

Adler, P. B. & Lauenroth, W. K. (2003) The power of time: Spatiotemporal scaling of species diversity. *Ecology Letters*, **6**, 749–756.

Adler, P. B., White, E. P., Lauenroth, W. K., Kaufman, D. M., Rassweiler, A. & Rusak, J. A. (2005) Evidence for a general species–time–area relationship. *Ecology*, **86**, 2032–2039.

Allen, A. P. & White, E. P. (2003) Effects of range size on species–area relationships. *Evolutionary Ecology Research*, **5**, 493–499.

Arita, H. T. & Rodríguez, P. (2002) Geographic range, turnover rate and the scaling of species diversity. *Ecography*, **25**, 541–550.

Arrhenius, O. (1921) Species and area. *Journal of Ecology*, **9**, 95–99.

Azaele, S., Muneepeerakul, R., Maritan, A., Rinaldo, A. & Rodriguez-Iturbea, I. (2008) Predicting spatial similarity of freshwater fish biodiversity. *Proceedings of the National Academy of Sciences USA*, **106**, 7058–7062.

Azovsky, A. I. (2002) Size-dependent species–area relationship in benthos: Is the world more diverse for microbes? *Ecography*, **25**, 273–282.

Bartha, S. & Ittzés, P. (2001) Local richness-species pool ratio: A consequence of the species–area relationship. *Folia Geobotanica*, **36**, 9–23.

Bonn, A., Storch, D. & Gaston, K. J. (2004) Structure of the species–energy relationship. *Proceedings of the Royal Society B: Biological Sciences*, **271**, 1685–1691.

Caley, M. J. & Schluter, D. (1997) The relationship between local and regional diversity. *Ecology*, **78**, 70–80.

Carey, S., Harte, J. & delMoral, R. (2006) Effect of community assembly and primary succession on the species–area relationship in disturbed ecosystems. *Ecography*, **29**, 866–872.

Chiarucci, A., Viciani, D., Winter, C. & Diekmann, M. (2006) Effects of productivity on species–area curves in herbaceous vegetation: Evidence from experimental and observational data. *Oikos*, **115**, 475–483.

Coleman, D. B. (1981) On random placement and species–area relations. *Mathematical Biosciences*, **54**, 191–215.

Condit, R., Hubbell, S. P., Lafrankie, J. V., Sukumar, R., Manokaran, N., Foster, R. B. & Ashton, P. S. (1996) Species–area and species–individual relationships for tropical trees: A comparison of three 50-ha plots. *Journal of Ecology*, **84**, 549–562.

Connor, E. F. & McCoy, E. D. (1979) The statistics and biology of the species–area relationship. *The American Naturalist*, **113**, 791–833.

Dengler, J. (2009) Which function describes the species–area relationship best? A review and empirical evaluation. *Journal of Biogeography*, **36**, 728–744.

Drakare, S., Lennon, J. L. & Hillebrand, H. (2006) The imprint of the geographical, evolutionary and ecological context on species–area relationships. *Ecology Letters*, **9**, 215–227.

Finlay, B. J. (2002) Global dispersal of free-living microbial eukaryote species. *Science*, **296**, 1061–1063.

Fridley, J. D., Peet, R. K., Wentworth, T. R. & White, P. S. (2005) Connecting fine- and broad-scale species–area relationships of southeastern U.S. flora. *Ecology*, **86**, 1172–1177.

Gaston, K. J., Blackburn, T. M. & Lawton, J. H. (1997) Interspecific abundance–range size relationships: An appraisal of mechanisms. *Journal of Animal Ecology*, **66**, 579–601.

Gaston, K. J., Evans, K. L. & Lennon, J. J. (2007) The scaling of spatial turnover: Pruning the thicket. *Scaling Biodiversity*. (ed. by D. Storch, P. A. Marquet & J. H. Brown), pp. 181–214. Cambridge: Cambridge University Press.

Gleason, H. A. (1922) On the relation between species and area. *Ecology*, **3**, 158–162.

Gray, J. S., Ugland, K. I. & Lambshead, J. (2004) On species accumulation and species–area curves. *Global Ecology & Biogeography*, **13**, 567–568.

Green, J. L. & Plotkin, J. B. (2007) A statistical theory for sampling species abundances. *Ecology Letters*, **10**, 1037–1045.

Haegeman, B. & Etienne, R. S. (2010) Entropy maximization and the spatial distribution of species. *The American Naturalist*, **175**, E74–E90.

Hanski, I. & Gyllenberg, M. (1997) Uniting two general patterns in the distribution of species. *Science*, **275**, 397–400.

Harrison, J. A., Allan, D. G., Underhill, L. G., Herremans, M., Tree, A. J., Parker, V. & Brown, C. J. (1997) *The atlas of southern African birds. Vol I & II.* Johannesburg: Bird Life South Africa.

Harte, J. & Kinzig, A. P. (1997) On the implications of species–area relationships for endemism, spatial turnover, and food web patterns. *Oikos*, **80**, 417–427.

Harte, J., Conlisk, E., Ostling, A., Green, J. L. & Smith, A. B. (2005) A theory of spatial structure in ecological communities at multiple spatial scales. *Ecological Monographs*, **75**, 179–197.

Harte, J., Kinzig, A. & Green, J. (1999) Self-similarity in the distribution and abundance of species. *Science*, **284**, 334–336.

Harte, J., Smith, A. B. & Storch, D. (2009) Biodiversity scales from plots to biomes with a universal species–area curve. *Ecology Letters*, **12**, 789–797.

He, F. & Condit, R. (2007) The distribution of species: Occupancy, scale, and rarity. *Scaling biodiversity* (ed. by D. Storch, P. A. Marquet and J. H. Brown), pp. 32–50. Cambridge: Cambridge University Press.

He, F. & Gaston, K. J. (2000) Estimating species abundance from occurrence. *The American Naturalist*, **156**, 553–559.

He, F. & Legendre, P. (1996) On species–area relations. *The American Naturalist*, **148**, 719–737.

He, F. & Legendre, P. (2002) Species diversity patterns derived from species–area models. *Ecology*, **85**, 1185–1198.

Hubbell, S. P. (2001) *The unified theory of biodiversity and biogeography*. Princeton, NJ: Princeton University Press.

Jaynes, E. T. (1957) Information theory and statistical mechanics. *Physical Review*, **106**, 620–630.

Jaynes, E. T. (1982) On the rationale of maximum entropy methods. *Proceedings of the IEEE*, **70**, 939–952.

Koleff, P., Gaston, K. J. & Lennon, J. J. (2003) Measuring beta diversity for presence-absence data. *Journal of Animal Ecology*, **72**, 367–382.

Kunin, W. E. (1997) Sample shape, spatial scale and species counts: Implications for reserve design. *Biological Conservation*, **82**, 369–377.

Kunin, W. E. (1998) Extrapolating species abundances across spatial scales. *Science*, **281**, 1513–1515.

Kunin, W. E., Harte, J., He, F., Hui, C., Jobe, R. T., Ostling, A., Polce, C., Šizling, A. L., Smith, A. B., Smith, K., Smart, S. M., Storch, D., Tjørve, E., Ugland, K.-I., Ulrich, W. & Varma, V. (2018) Up-scaling biodiversity: Estimating the species–area relationship from small samples. *Ecological Monographs*, **88**, 170–187.

Kůrka, P., Šizling, A. L. & Rosindell, J. (2010) Analytical evidence for scale-invariance in the shape of species abundance distributions. *Mathematical Biosciences*, **223**, 151–159.

Lazarina, M., Kallimanis, A. S. & Sgardelis, S. (2013) Does the universality of the species–area relationship apply to smaller scales and across taxonomic groups? *Ecography*, **36**, 965–970.

Leitner, W. A. & Rosenzweig, M. L. (1997) Nested species–area curves and stochastic sampling: A new theory. *Oikos*, **79**, 503–512.

Lennon, J. J., Kunin, W. E. & Hartley, S. (2002) Fractal species distributions do not produce power-law species area distribution. *Oikos*, **97**, 378–386.

Lepš, J. & Štursa, J. (1989) Species–area curve, life history strategies, and succession: A field test of relationships. *Vegetatio*, **83**, 249–257.

MacArthur, R. H. & Wilson, E. O. (1967) *The theory of island biogeography*. Princeton, NJ: Princeton University Press.

Martín, H. G. & Goldenfeld, N. (2006) On the origin and robustness of power-law species–area relationships in ecology. *Proceedings of the National Academy of Sciences USA*, **103**, 10310–10315.

May, R. (1975) Patterns of species abundance and diversity. *Ecology and evolution of communities* (ed. by M. L. Cody and J. M. Diamond), pp. 81–120. Cambridge, MA: Belknap Press.

McGill, B. J. (2010) Towards a unification of unified theories of biodiversity. *Ecology Letters*, **13**, 627–642.

McGill, B. J. & Collins, C. (2003) A unified theory for macroecology based on spatial patterns of abundance. *Evolutionary Ecology Research*, **5**, 469–492.

Nachman, G. (1981) A mathematical model of the functional relationship between density and spatial distribution of a population. *Journal of Animal Ecology*, **50**, 453–460.

Nee, S. & Cotgreave, P. (2002) Does the species–area relationship account for the density–area relationship? *Oikos*, **99**, 545–551.

Nekola, J. C. & White, P. S. (1999) Distance decay of similarity in biogeography and ecology. *Journal of Biogeography*, **26**, 867–878.

O'Dwyer, J. P. & Green, J. L. (2010) Field theory for biogeography: A spatially explicit model for predicting patterns of biodiversity. *Ecology Letters*, **13**, 87–95.

Ovaskainen, O. & Hanski, I. (2003) The species–area relationship derived from species-specific incidence functions. *Ecology Letters*, **6**, 903–909.

Pautasso, M. & Weisberg, P. J. (2008) Negative density–area relationship: The importance of zeros. *Global Ecology & Biogeography*, **17**, 203–210.

Preston, F. V. (1960) Time and space and the variation of species. *Ecology*, **41**, 611–627.

Rosenzweig, M. L. (1995) *Species diversity in space and time*. Cambridge: Cambridge University Press.

Rosenzweig, M. L. & Ziv, Y. (1999) The echo pattern of species diversity: Pattern and processes. *Ecography*, **22**, 614–628.

Rosindell, J. & Cornell, S. J. (2007) Species–area relationships from a spatially explicit neutral model in an infinite landscape. *Ecology Letters*, **10**, 586–595.

Rosindell, J. & Cornell, S. J. (2009) Species–area curves, neutral models, and long-distance dispersal. *Ecology*, **90**, 1743–1750.

Scheiner, S. M. (2003) Six types of species–area curves. *Global Ecology & Biogeography*, **12**, 441–447.

Scheiner, S. M. (2004) A mélange of curves – Further dialogue about species–area relationships. *Global Ecology & Biogeography*, **13**, 479–484.

Shmida, A. & Wilson, M. V. (1985) Biological determinants of species diversity. *Journal of Biogeography*, **12**, 1–20.

Šizling, A. L. & Storch, D. (2004) Power-law species–area relationships and self-similar species distributions within finite areas. *Ecology Letters*, **7**, 60–68.

Šizling, A. L. & Storch, D. (2007) Geometry of species distributions: Random clustering and scale invariance. *Scaling Biodiversity* (ed. by D. Storch, P. A. Marquet & J. H. Brown), pp. 77–99. Cambridge: Cambridge University Press.

Šizling, A. L., Kunin, W. E., Šizlingová, E., Reif, J. & Storch, D. (2011) Between geometry and biology: The problem of universality of the species–area relationship. *The American Naturalist*, **178**, 602–611.

Šizling, A. L., Šizlingová, E., Tjørve, E., Tjørve, K. M. C. & Kunin, W. E. (2017) How to allow SAR collapse across local and continental scales: A resolution of the controversy between Storch et al. (2012) and Lazarina et al. (2013). *Ecography*, **40**, 971–981.

Šizling, A. L., Storch, D., Reif, J. & Gaston, K. J. (2009b) Invariance in species-abundance distributions. *Theoretical Ecology*, **2**, 89–103.

Šizling, A. L., Storch, D., Šizlingová, E., Reif, J. & Gaston, K. J. (2009a) Species abundance distribution results from a spatial analogy of central limit theorem. *Proceedings of the National Academy of Sciences USA*, **106**, 6691–6695.

Storch, D. (2016) The theory of the nested species–area relationship: Geometric foundations of biodiversity scaling. *Journal of Vegetation Science*, **27**, 880–891.

Storch, D. & Šizling, A. L. (2008) The concept of taxon invariance in ecology: Do diversity patterns vary with changes in taxonomic resolution? *Folia Geobotanica*, **43**, 329–344.

Storch, D., Evans, K. L. & Gaston, K. J. (2005) The species–area–energy relationship. *Ecology Letters*, **8**, 487–492.

Storch, D., Keil, P. & Jetz, W. (2012) Universal species–area and endemics–area relationships at continental scales. *Nature*, **488**, 78–81.

Storch, D., Marquet, P. A. & Brown J. H. (eds.) (2007) *Scaling biodiversity*. Cambridge: Cambridge University Press.

Storch, D., Šizling, A. L., Reif, J., Polechová, J., Šizlingová, E. & Gaston, K. J. (2008) The quest for a null model for macroecological patterns: Geometry of species distributions at multiple spatial scales. *Ecology Letters*, **11**, 771–784.

Tjørve, E. (2003) Shapes and functions of species–area curves: A review of possible models. *Journal of Biogeography*, **30**, 827–835.

Tjørve, E. (2009) Shapes and functions of species–area curves (II): A review of new models and parameterizations. *Journal of Biogeography*, **36**, 1435–1445.

Tjørve, E. & Tjørve, K. M. C. (2008) The species–area relationship, self-similarity, and the true meaning of the z-value. *Ecology*, **89**, 3528–3533.

Tjørve, E. & Turner, W. R. (2009) The importance of samples and isolates for species–area relationships. *Ecography*, **32**, 391–400.

Tjørve, E., Kunin, W. E., Polce, C. & Tjørve, K. M. C. (2008) Species–area relationship: Separating the effects of species abundance and spatial distribution. *Journal of Ecology*, **96**, 1141–1151.

Ugland, K. I., Gray, J. S. & Ellingsen, K. E. (2003) The species-accumulation curve and estimation of species richness. *Journal of Animal Ecology*, **72**, 888–897.

Ugland, K. I., Gray, J. S. & Lambshead, J. D. (2005) Species accumulation curves analysed by a class of null models discovered by Arrhenius. *Oikos*, **108**, 263–274.

Ulrich, W. & Buszko, J. (2003) Self-similarity and the species–area relation of Polish butterflies. *Basic and Applied Ecology*, **4**, 263–270.

Virkkala, R. (1993) Ranges of northern forest passerines: A fractal analysis. *Oikos*, **67**, 218–226.

White, E. P. (2007) Spatiotemporal scaling of species richness: Patterns, processes, and implications. *Scaling Biodiversity* (ed. by D. Storch, P. A. Marquet and J. H. Brown), pp. 325–346. Cambridge: Cambridge University Press.

Whittaker, R. H. (1960) Vegetation of the Siskiyou Mountains, Oregon and California. *Ecological Monographs*, **30**, 279–338.

Williams, M. R. (1995) An extreme-value function model of the species incidence and species–area relationship. *Ecology*, **76**, 2607–2616.

Williamson, M. H. (1988) Relationship of species number to area, distance and other variables. *Analytical biogeography* (ed. by A. A. Myers & P. S. Giller), pp. 91–115. London: Chapman & Hall.

Wright, D. H. (1991) Correlations between incidence and abundance are expected by chance. *Journal of Biogeography*, **18**, 463–466.

9 · *Species Accumulation Curves and Extreme Value Theory*

LUÍS BORDA-DE-ÁGUA, SAEID
ALIREZAZADEH, MANUELA NEVES,
STEPHEN P. HUBBELL, PAULO A. V.
BORGES, PEDRO CARDOSO, FRANCISCO
DIONÍSIO AND HENRIQUE M. PEREIRA

9.1 Introduction

The quantitative study of how the number of species increases with the size of the area where they are observed, the species–area relationship (hereafter SAR), has a venerable history that started in the nineteenth century (reviewed in Rosenzweig, 1995). However, different sampling circumstances can give rise to different SARs. For instance, Scheiner (2003) and Matthews et al. (2016) identified different types of species–area curve. One common form of the species–area curve results from plotting the number of species in each island of an archipelago as a function of the areas of the islands (hereafter ISAR). Here, on the other hand, we are solely concerned with species–area curves resulting from the relationship between the accumulation of species and the increasing size of the areas sampled in a nested fashion. We will call this type of species–area curve a 'species accumulation curve' (hereafter SAC), but note that other authors use the same terminology to denote slightly different types of species accumulation curves (e.g. Matthews et al., 2016; Chapter 1). Although we only deal with a specific type of SAC, we can say that it corresponds to the answer to the basic question that ecologists are often interested in: how does the number of species change when we sample nested areas of increasing size?

When plotted on linear scales SACs are, as expected, increasing curves, but with a decelerating rate of growth. This simply means that the rate at which we find new species decreases as our sample size increases. An important question is whether this rate of increase of species can be described with generality by a single mathematical expression or, in other words, if there is a quantifiable pattern associated with SACs. Arrhenius

(1921) suggested a power law, $S = cA^z$, to fit these curves, where S is the number of species, A is the area, and c and z (<1) are parameters to be estimated, a formulation that has remained a favourite among theoreticians and empiricists in ecology. For example, Preston (1962) and later May (1975) derived the power law relationship under certain assumptions (that we now know to be rarely observed) between the number of species and their relative abundances. Interestingly, the power law has also been suggested to fit ISARs. However, the power law has not been free of criticism. For instance, soon after Arrhenius published his work, Gleason (1922) suggested that the logarithmic function $S = \log(cA^z)$ provided a better fit (see Chapter 2). Since then, the number of alternative functions has proliferated (Chapter 7) and, recently, Guilhaumon et al. (2008) and Triantis et al. (2012), dealing with ISARs, compared the ability of several different candidate models to fit different datasets, with the latter paper revealing that a power law provided the best fit to the most datasets. As we will see, among these alternative functions, the Weibull (e.g. Tjørve, 2003; Williams et al., 2009; Chapter 7) is especially relevant to the present work. Notice, however, that the shape of the SAC depends on the range of the spatial scales involved (Chapter 4). For example, the neutral theory of biodiversity and biogeography predicts a triphasic SAC (Hubbell, 2001; Chapter 11): for very small areas it is a concave downward function when plotted in a double logarithmic plot, then for some range of (intermediate) area sizes it is well described by a power law and, finally, for very large areas the slope of the power law approaches one. The latter arises because samples come from different metacommunities or independent evolutionary biogeographic units.

The primary purpose of this work is to provide a theoretical framework by which to analyse SACs and not simply to introduce another function to the long list of functions already used to fit the SAC for the sake of it. Specifically, we use a result from a branch of probability theory called extreme value theory (hereafter EVT), also known as the statistics of extremes (e.g. Coles, 2001; Castillo et al., 2005) that deals with the extreme values of a distribution. The main result of the EVT that we apply to this work is that the curve of the SAC is well fitted by the generalized extreme value cumulative distribution function. Thus, although we do suggest another function to fit the SAC, we do it based on a theoretical result from probability theory. In doing so, we are not seeking a biological 'meaning' for the shape of a SAC. Rather, the shape is 'explained' by a probability theorem. Nevertheless, we recognize that the numerical values of the parameters estimated to fit the data result from biological processes and, hence, comparison among different sets of parameters is bound to involve a biological interpretation.

EVT addresses the non-degenerate distribution (i.e. a non-deterministic distribution, meaning that when only a single random variable is involved it does not take only a single value) of the maxima or minima values of a sequence of samples, after a suitable normalization. For instance, these can be the lowest or the highest temperatures, the largest dispersal distances of seeds or any other extreme value. The argument for using EVT is that to obtain a SAC we look for the first individual of a given species to be found in the sample area, which leads to the minimum distances to the first encounter. For instance, if we start from an arbitrary point, as exemplified in Figure 9.1, we look for the nearest individual, that is, the closest, and thus the *minimum* distance among all individuals of a given species. After this individual has been found, all the remaining individuals of that species are ignored (abundance is not relevant for the determination of the SAC). After applying this procedure to all the species present in a given area we end up with the distribution of the *minimum* distances. The cumulative distribution of these distances is a SAC. Indeed, this was previously recognized by He and Hubbell (2011, see their figure 1) and Martín and Goldenfeld (2006). The latter authors recognized the link with EVT but dismissed its use for reasons with which we are not in accordance, as we outline in the

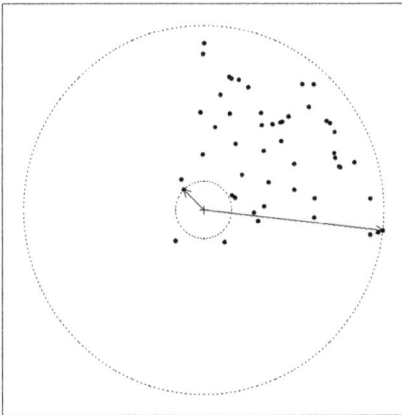

Figure 9.1 The distribution of the individuals of a hypothetical species. The arrows show the minimum and maximum distances of the individuals to the central point identified with a cross and the circles delineate the areas of the two circumferences delimiting the areas of the first and last encounters.

discussion. Notice, however, that the nature of our work is exploratory because, as we will elaborate in the discussion, the validity of the assumptions of the models when applied to our datasets needs to be further verified.

9.2 Extreme Value Theory and the Block Minima Approach

This section is admittedly technical, as it describes the main results of EVT. However, we end it with a summary of what is required to apply EVT to a practical situation.

The statistics of extremes parametric approach mainly involves two procedures. One is called the 'Block Maxima or Minima' procedure and focuses on the largest (or smallest) value in repeated same-sized samples from a distribution. The other is called the 'Peaks-Over-Threshold' procedure and deals with the values larger (or smaller) than a predefined threshold. The latter is not appropriate for the SAC problem, thus we will not deal with it here. For an application in ecology of these two procedures, but applied to the maxima instead of the minima, see for example García and Borda-de-Água (2017). To exemplify the 'Block Minima approach' consider, for instance, the daily temperature data for the last 100 years from a northern hemisphere location and assume that we are interested in the lowest temperature in each year, presumably occurring in the winter. If we divide the data into non-overlapping 'blocks', each starting on the 1st July (one year) and ending on the 30th June (the following year) and for each of these periods we identify the temperature of the coldest day, we will have a dataset of approximately 100 points of the recorded temperature of the coldest day of each block. The resulting histogram of these points is the beginning of the subject of the EVT analysis.

For the present chapter, species are the 'blocks' and we focus solely on the distance of the nearest individual of every species to a given point in space (see Figure 9.1). This implies that we compute the distance of all individuals of a species to a given focal point and select the minimum distance. The distribution from all the minimum distances (one for every species) can be plotted as a histogram, from which we can construct a cumulative distribution function, that becomes a SAC, after renormalizing it considering the total number of species. Once we have identified the closest individual of a given species, all the other individuals of the same species are irrelevant for the construction of the SAC. Notice that under this scheme we obtain one single SAC because we based its construction on a single focal point. An alternative is to consider different starting points and obtain for each one a SAC and estimate the corresponding parameters. Then, assuming a large number of starting points, draw the histograms of the estimated parameters and calculate the appropriate statistics, such as the mean or the standard deviation. Alternatively, one could even consider a Bayesian hierarchical approach where the parameters of each SAC are derived from common prior distributions (e.g. Gelman et al., 2014), but we will leave this to future work. In the next paragraphs we give the most important theoretical results needed to apply the EVT theory.

In EVT, in contrast to the Central Limit Theory wherein the focus is on the values around the mean of the data, the aim is to describe the tails of the underlying distribution with a view to modelling extreme or even rare events. Classic theory of extremes is concerned with the limiting behaviour of the maximum $M_n = \max(X_1, \ldots, X_n)$ or the minimum $m_n = \min(X_1, \ldots, X_n)$, as $n \to \infty$, of a sample $(X_1, \ldots, X_n)$ of independent and identically distributed (i.i.d.) or possibly stationary, weakly dependent, random variables with unknown distribution function F.

The first steps in EVT dealing with the problem of obtaining a non-degenerate limiting behaviour for the maximum (or minimum) date back to Fréchet (1927), Fisher and Tippett (1928), Gumbel (1935) and von Mises (1936). Gnedenko (1943) and de Haan (1970) finally established the theory of the asymptotic behaviour of statistical extremes, by giving conditions for the existence of sequences $\{a_n\} \in \mathbb{R}^+$ and $\{b_n\} \in \mathbb{R}$ such that, for minima,

$$\lim_{n\to\infty} P\left(\frac{m_n - b_n}{a_n} \leq x\right) = \lim_{n\to\infty}\left(1 - [1 - F(a_n x + b_n)]^n\right) = G_\xi(x) \quad \forall x \in \mathbb{R},$$

(9.1)

where G_ξ is a nondegenerate distribution function, belonging to one of the following families:

- Type I (reversed Gumbel, $\Lambda^r(x)$)

$$\Lambda^r(x) = 1 - \exp\left[-\exp(x)\right], \qquad x \in \mathbb{R},$$

(9.2a)

- Type II (reversed Fréchet, $\Phi^r(x)$)

$$\Phi^r(x) = \begin{cases} 1 - \exp\left[-(-x)^{-\alpha}\right] & x \leq 0, \quad \alpha > 0 \\ 1 & x > 0, \end{cases}$$

(9.2b)

- Type III (Weibull, $\Psi(x)$)

$$\Psi(x) = \begin{cases} 0 & x < 0 \\ 1 - \exp\left[-x^{\alpha}\right] & x \geq 0, \;\; \alpha > 0. \end{cases} \tag{9.2c}$$

The constants a_n and b_n defined in Equation (9.1) are called *stabilizing* or *normalizing constants* or *attraction coefficients*. These constants are defined in terms of F or the quantiles of F but, as we will see, we do not need to know them in practical applications.

The previous paragraph stated a major result in EVT that shows which analytical distributions should be used when fitting data on the distribution of minima (there are similar formulas for the maxima, e.g. Coles, 2001). This can be further simplified by noticing that the distributions in Equations (9.2a–c) can be combined into a single family of models called the *generalized extreme value* distribution function, usually denoted by GEV, given by

$$GEV_m(x) = \begin{cases} 1 - \exp\left\{-[1 - \xi x]^{-\frac{1}{\xi}}\right\}, & 1 - \xi x > 0, \quad \text{if } \xi \neq 0 \\ 1 - \exp\left\{-\exp\left[x\right]\right\}, & x \in \mathbb{R}, \quad \text{if } \xi = 0, \end{cases} \tag{9.3}$$

where we add the subscript m to explicitly state that it is for minima, and $\xi \in \mathbb{R}$ is the shape parameter, the so called *extreme value index*, the primary parameter in EVT. In fact, depending on the value of ξ, we can recover the distributions given by Equations (9.2a–c) from the Equation (9.3). Thus, the reversed Gumbel distribution (Equation 9.2a) is the particular case of Equation (9.3) when $\xi = 0$ is interpreted as the limit $\xi \to 0$, the reversed Fréchet distribution (Equation 9.2b) when $\xi > 0$ and the Weibull distribution (Equation 9.2c) when $\xi < 0$.

As a consequence of the existence of the limit in Equation (9.1), when $n \to \infty$ we may consider the approximation

$$P[m_n \leq x] = 1 - [1 - F(x)]^n \approx GEV_m((x - \lambda)/\delta), \tag{9.4}$$

where $(\lambda, \delta) \in (\mathbb{R}, \mathbb{R}^+)$ are the unknown location and scale parameters, that incorporate the attraction coefficients (a_n, b_n). As mentioned before, in practice we do not need to know the constants a_n and b_n because if $\text{Prob}\{(m_n - b_n)/a_n \leq x\} \to GEV_m(x)$, when $n \to \infty$, then it can be shown that $\text{Prob}\{m_n \leq x\} \to GEV_m^\star(x)$, where $GEV_m^\star(x)$ also belongs to the generalized extreme value family (e.g. Coles, 2001). Therefore, in real applications, $GEV_m^\star(x)$ is the distribution for which we estimate the parameters. Observe that the theorems stated above require the data to come from independent and identically distributed distributions. However, an important result from EVT states that the distribution given in Equation (9.3) still applies to data that are near-independent if the underlying process is stationary (Coles, 2001; Castillo et al., 2005); we will return to this point in Section 9.4.

In summary, assuming that the data obey the assumptions of the EVT theorems, then the distribution of minima is given by the GEV_m (Equation 9.3). In addition, although this may not be essential when analysing data, based on the value of the estimated parameter ξ we can also decide which of the distributions (Equations 9.2a–c) is the most likely; there are also statistical tests for deciding which of the distributions fits the data better. The application of EVT to the SAC suggested here requires that we have full information on the location of all individuals of all the species. Although we recognize that not many datasets have such detailed information, we hope that more will become available in the future. In Section 9.3 we apply the EVT theory to a dataset on tropical tree species.

We performed all analyses in R version 3.5.1 (R Core Team, 2018) with the package extRemes 2.0–9 (Gilleland & Katz, 2016). This package implements solely the block maxima approach, but it can still be used for the block minima approach, given the 'quasi-symmetric' relationship, $\min(X_1, \ldots, X_n) = -\max(-X_1, \ldots, -X_n)$. Then, given a dataset, X, we first make $Y = -X$, estimate the parameters for Y and then convert the estimated location parameter, $\hat{\lambda}$, to $-\hat{\lambda}$ (Coles, 2001, p. 53).

9.3 Case Study: Application of the EVT to Data on Tree Species from Barro Colorado Island

We illustrate the application of the block minima approach to data on mapped populations of woody plant species (except lianas) located in a 50 ha (500 × 1,000 m) plot of old-growth tropical moist forest on Barro Colorado Island (BCI), Panama (Condit, 1998). This dataset contains information on the species and spatial location of all individuals with a stem diameter ≥ 1 cm diameter at breast height, a total of 242,077 individuals corresponding to 305 species (data from the 1985 census).

To stay close to the suggested sampling scheme of Figure 9.1, we use the central point of the 50 ha plot and determine the radii of the circumferences determined by the nearest individual of every species with at least one individual within the largest possible radius of 250 m.

Because this scheme leaves parts of the plot unsampled, the number of species present within this circle is 268.

The spatial distribution of the nearest individual of each species to the central point of the plot is shown in Figure 9.2 and the corresponding species accumulation curve is depicted in Figure 9.3A and B. As expected, the rate at which new species are found tends to decrease as the area increases. The solid curve in Figure 9.3A and B corresponds to the fitting obtained with the GEV_m, where the estimated parameters are $(\hat{\lambda}, \hat{\delta}, \hat{\xi}) = (1.13, 2.29, -2.03)$ and the corresponding standard errors are $SE(\hat{\lambda}, \hat{\delta}, \hat{\xi}) = (0.07, 0.08, 0.13)$. The values of the shape parameter $\hat{\xi} = -2.03$ and of the associated standard error $SE(\hat{\xi}) = 0.13$ clearly restrict the plausible values of $\hat{\xi}$ to negative values. This gives strong evidence that, among the three possible extreme value distributions (Equations 9.2a–c), the Weibull (Equation 9.2c) is the best candidate. This result corroborates Tjørve (2003) and Williams et al. (2009), who found that the Weibull distribution provides a good fit to the SAC and, in general, to the SAR.

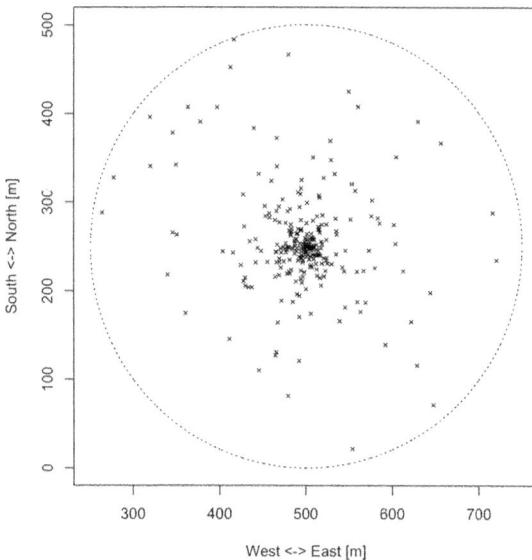

Figure 9.2 The spatial location of the nearest individuals for all species of trees relative to the central point of the 50 ha rectangular plot in Barro Colorado Island, Panama (e.g. Condit, 1998), within a 250 m radius (dotted circumference). The axes show the distances in metres to the south-west corner of the plot; thus, the central point is located at coordinates (500 m, 250 m).

It is clear from Figure 9.3B that the SAC is not a straight line when plotted in a log–log plot. This indicates that, at least on these local spatial scales, a power law gives a poor fit in the present case. This result is not surprising given previous analyses of these data (Condit et al., 1996; Borda-de-Água et al., 2002). Often, the SAC is fitted only within the range of values of sample area within which a straight line is observed in the log–log plot. For example, Borda-de-Água et al. (2002) fitted the SAC to the BCI tree species data using only the points above 1 ha. Using the present sampling scheme, we also observe an approximately straight

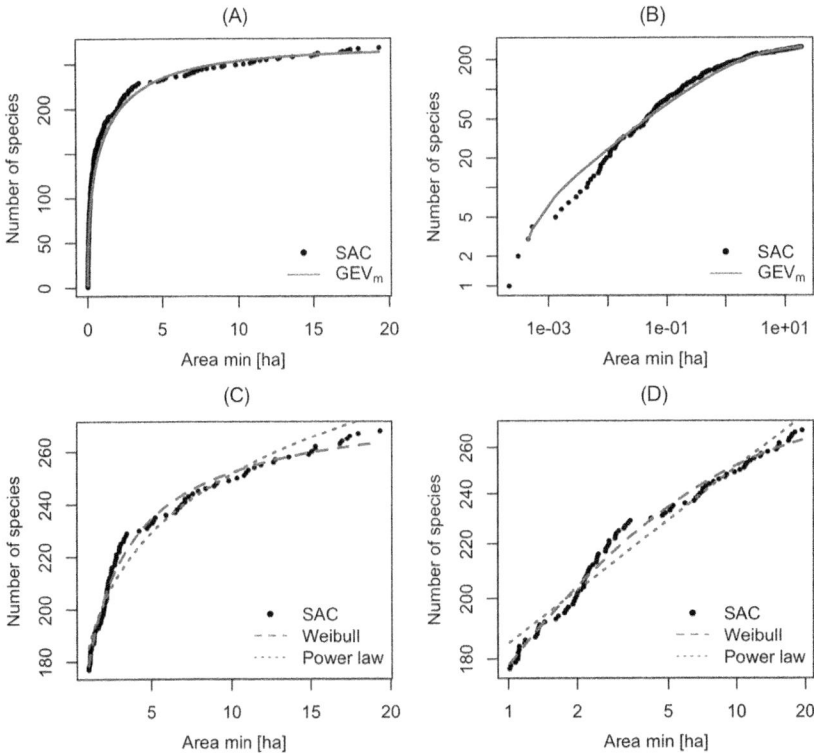

Figure 9.3 The species accumulation curves (SAC) for the tree species of the Barro Colorado Island 50 ha plot obtained using the sampling scheme illustrated in Figure 9.1. The plots on the left have axes with linear scales, while the ones on the right have logarithmic ones. The plots on the top row depict the SAC for the entire range of areas observed and the empirical curves were fitted using the GEV_m distribution, while the plots on the bottom row depict the SAC only for areas above 1 ha and the empirical curve was fitted for this range of areas only, using the Weibull and power law functions.

line for areas above 1 ha (cf. Figure 9.3C and D). Notice, however, that, by fitting only the points above 1 ha, we disregard the information provided by 65 per cent of the species, those that were first encountered at spatial scales below 1 ha and, if we use a more stringent threshold, such as only areas above 4 ha, then we would be disregarding 85 per cent of the species. In fact, in both of these cases we would be fitting only the tail of the distribution. Still, given that this is common practice, we fitted the data above 1 ha using a three parameter Weibull cumulative distribution function, $S = C\left(1 - \exp\left[-\left(\frac{A}{a}\right)^{\theta}\right]\right)$, and a power law function, $S = cA^z$. The estimated parameters for the Weibull distribution are $(\hat{C}, \hat{a}, \hat{\theta}) = (274.2, 0.885, 0.381)$, with standard errors $\mathrm{SE}(\hat{C}, \hat{a}, \hat{\theta}) = (2.876, 0.032, 0.015)$ and $R^2 = 0.987$, and for the power law $(\log(\hat{C}), \hat{z}) = (5.22, 0.134)$, with standard errors $\mathrm{SE}(\log(\hat{C}), \hat{z}) = (0.004, 0.003)$, and $R^2 = 0.961$. The resulting curves are shown in Figure 9.3C and D and both give a very good fit to the observed SAC.

9.4 Discussion

By definition, the species accumulation curve is the result of finding the nearest (first) individual of each species as the size of nested areas increases. Therefore, we argue that Extreme Value Theory is an appropriate theoretical framework within which to model SACs. However, the degree to which the SAC obeys or violates the assumptions of the EVT's classical theorems must be considered. The argument for establishing the connection between the EVT and the Endemics−Area Relationship (He & Hubbell, 2011) is similar to that developed here between the EVT and SAC. While the SAC is concerned with the occurrence of the first (nearest) observed individual of a species, hence of the minimum distance, the endemics−area relationship is concerned with the occurrence of the last (furthest) individual of a species, hence the maximum distance. Again, the problems associated with the applicability of the EVT to the SAC applies to the Endemics−Area Relationship.

Other authors have used EVT distributions. For instance, Williams (1995) used the reversed Gumbel distribution (Equation 9.2a) and Tjørve (2003) and Williams et al. (2009) used the Weibull distribution (Equation 9.2c). However, they did not frame the problem of the SAC in terms of the statistics of extremes. To the best of our knowledge, the only authors who explicitly acknowledged the connection between the EVT and the

SAC were Martín and Goldenfeld (2006). However, they rejected the use of the classical results of extreme value theory because:

[t]he distances from a point in the area Ω to each of the individuals are not independent and identically distributed because they are clustered. If an individual is at a distance R from $\vec{r}$, it is very likely that other individuals are at a distance similar to R: [t]hey are strongly correlated. In the same way, the set of distances from a point $\vec{r}$ is correlated with the set of distances from a nearby point $\vec{r}'$ (Martín & Goldenfeld, 2006, p. 10311).

Although we agree that individuals of a given species are clustered in several cases, we do not so readily dismiss the application of the classical results of statistics of extremes to the SAC. To start with, from the perspective of EVT, it is not relevant whether the individuals of a species are clustered or not. The requirement of independence is among the nearest (minima) individuals of the species present. Nevertheless, if this is not the case, but the distribution of individuals of a set of tree species is assumed to follow a stationary process then it can be shown that the results for the independent data (Equations 9.2a–c and 9.3) still apply for weakly dependent data. In practice, stationarity, applied to our case study, implies that the SAC obtained from an area of size A of the forest is similar to that that may be obtained from another area of equal size or, in other words, the forest is relatively homogeneous within the study region (an assumption that is often made). The parameters of the distributions of the independent and weak dependent cases are related, but this relationship is not a problem that should concern us, because the parameters have to be estimated in any case (Coles, 2001; pp. 97–98; see also chapter 8 in Castillo, 1988; and chapter 9 in Castillo et al., 2005).

We have, however, identified another problem that may hinder use of the classical distributions of the EVT. The theorem leading to Equation (9.1) assumes that the minima are sampled from identical distributions. Of course, in real applications this requirement has to be relaxed. The question is whether the distributions of the distances of the individuals of each species to the focal point of the plot, from where we then extract the minimum distance, are very different from each other. In order to address this concern, we computed these distributions for all species. Figure 9.4A–D illustrates some typical cases. Because the distributions of the distances for all individuals of each species are typically approximately bell-shaped (Figure 9.4A–D), we summarize the results by plotting the distributions of the means and standard deviations (Figure 9.4E and F). Figure 9.4E and F shows that species do have

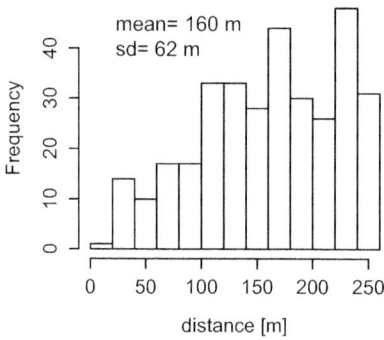

(A) - *Acalypha diversifolia*

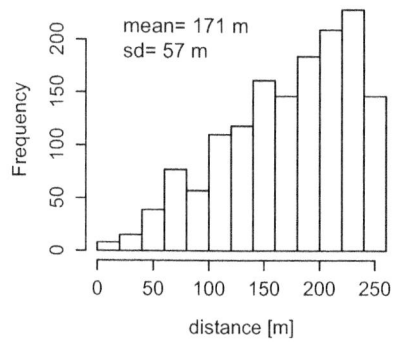

(B) - *Capparis frondosa*

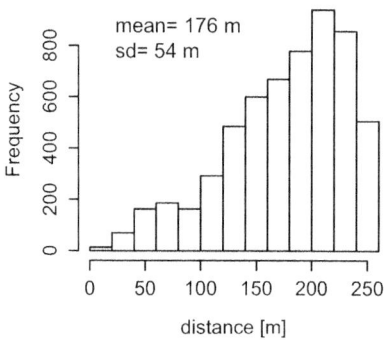

(C) - *Trichilia tuberculata*

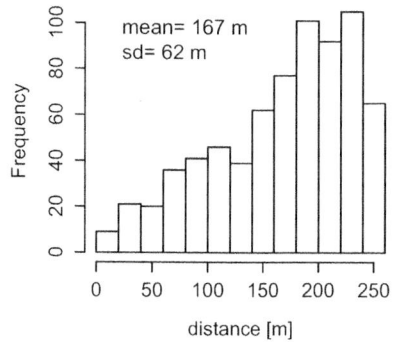

(D) - *Virola sebifera*

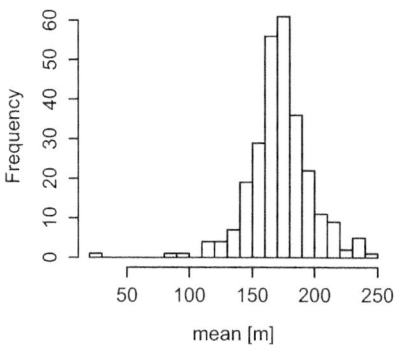

(E)

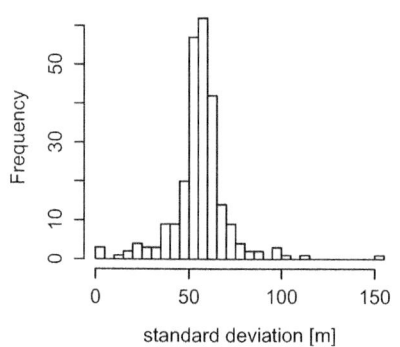

(F)

Figure 9.4 Distributions of the distances to the central point of the plot of the individuals of (A) *Acalypha diversifolia*, (B) *Capparis frondosa*, (C) *Trichilia tuberculata* and (D) *Virola sebifera*, for the 50 ha plot from Barro Colorado Island, Panama. Panels (E) and (F) are the histograms of the mean and standard deviation, respectively, of the distributions of all species (n = 305 species).

different distributions (different means and standard deviations) and, in addition, some species are rare, such as those with one or two individuals, cases where one cannot easily attribute a specific distribution. The implication of the two latter cases for the application of the EVT to the SAC should be addressed in future studies, taking advantage of the most recent developments in EVT concerning non-identical distributions (e.g. Falk et al., 2011).

The method we propose here is data demanding. It requires information on the location of the individuals of the species of the taxa we are interested in. This is undoubtedly the major problem for the application of EVT to SACs as there are not many datasets with such detailed information. We hope, nonetheless, that researchers will appreciate having the methods out on the shelf, so to speak, to apply in the future when more such datasets become available. The datasets on tropical tree species managed by the Smithsonian Tropical Research Institute, such as the one from BCI, have the right characteristics to apply the EVT as developed here. In particular, we expect that the development of technology, whether, for example, in genetics or remote sensing (e.g. Levy et al., 2014; Kellner & Hubbell, 2017; Kellner et al., 2019), may enable the collection of appropriate information on species composition at large spatial scales and for a wider range of taxa.

As stated in the introduction, we do not attempt to provide a biological interpretation for the shape of the SAC. Rather, we argue that the shape of the SAC is dictated by the appropriate extreme value distribution, as that of any other phenomenon that is generated by any extreme process. This said, it is potentially fruitful to search for biological interpretations of the numerical values of the parameters of the distribution. For example, a larger value of z in the power law formulation of the SAC may be associated with groups of species with low dispersal ability (e.g. Borda-de-Água et al., 2007).

9.5 Concluding Points and Future Work

Our main purpose herein was to anchor the study of a particular form of SAC, the one resulting from nested samples, on well-established probability results; specifically, those related to the classical theorems of extreme value theory. In a sense, we are doing 'theorem fitting' not 'curve fitting', though the latter still plays an important guiding role in assessing the validity of the applicability of a theorem. From this perspective, a good fit is a necessary condition, but not a sufficient one. If curve fitting were the sole purpose, then the fit provided by the GEV_m shown in Figure 9.3A and B would have been quite satisfactory. Nevertheless, we should not stop there because we know that the assumptions underlying the theorem that leads to the GEV_m are not fully satisfied by our data or other data from similar ecological communities. Thus, we emphasize the exploratory nature of this work. Future research will consist of analysing the relevance of the non-independent and non-identical nature of the natural distributions of species, aspects of the EVT that have already been developed; see, for example, Part III of the book by Falk et al. (2011). Whether the deviations from the assumptions of the theorems, which inevitably exist in real applications, may invalidate the application of the classical results of extreme value theory will be the focus of future research and we hope it will lead to a healthy collaboration between (statisticians) mathematicians and ecologists.

Acknowledgements

We thank R. J. Whittaker and an anonymous reviewer for comments and suggestions that considerably improved the chapter. This work is a result of the FCT project MOMENTOS (PTDC/BIA-BIC/5558/2014) and the project NORTE-01-0145-FEDER-000007, supported by Norte Portugal Regional Operational Programme (NORTE2020), under the PORTUGAL 2020 Partnership Agreement, through the European Regional Development Fund (ERDF). SA was funded by the FCT project MOMENTOS. MN was funded by FCT – Fundação para a Ciência e a Tecnologia, Portugal, through the project UID/MAT/00006/2019.

References

Arrhenius, O. (1921) Species and area. *Journal of Ecology*, **9**, 95–99.

Borda-de-Água, L., Hubbell, S. P. & He, F. (2007) Scaling biodiversity under neutrality. *Scaling Biodiversity* (ed. by D. Storch, P. A. Marquet and J. H. Brown), pp. 347–375. Cambridge: Cambridge University Press.

Borda-de-Água, L., Hubbell, S. P. & McAllister, M. (2002) Species–area curves, diversity indices, and species abundance distributions: A multifractal analysis. *The American Naturalist*, **159**, 138–155.

Castillo, E. (1988) *Extreme value theory in engineering*. San Diego, CA: Academic Press.

Castillo, E., Hadi, A. S., Balakrishnan, N. & Sarabia, J. M. (2005) *Extreme value and related models with applications in engineering and science*. Hoboken NJ: John Wiley & Sons.

Coles, S. (2001) *An introduction to statistical modeling of extreme values.* London: Springer.

Condit, R. (1998) *Tropical forest census plots: Methods and results from Barro Colorado Island, Panama and a comparison with other plots.* Berlin: Springer.

Condit, R., Hubbell, S. P., Lafrankie, J. V., Sukumar, R., Manokaran, N., Foster, R. B. & Ashton, P. S. (1996) Species–area and species–individual relationships for tropical trees: A comparison of three 50-ha plots. *Journal of Ecology*, **84**, 549–562.

de Haan, L. (1970) *On regular variation and its applications to the weak convergence of sample extremes.* Matematisch Centrum Amsterdam, Amsterdam.

Falk, M., Hüsler, J. & Reiss, R. D. (2011) *Laws of small numbers: Extremes and rare events*, 3rd ed. Berlin: Springer.

Fisher, R. A. & Tippett, L. H. C. (1928) Limiting forms of the frequency distribution of the largest or smallest member of a sample. *Proceedings of the Cambridge Philosophical Society*, **24**, 180–190.

Fréchet, M. (1927) Sur la loi de probabilité de l'écart maximum. *Annales de la Société Polonaise de Mathématique*, **6**, 93–116.

García, C. & Borda-de-Água, L. (2017) Extended dispersal kernels in a changing world: Insights from statistics of extremes. *Journal of Ecology*, **105**, 63–74.

Gelman, A., Carlin, J. B., Stern, H. S., Dunson, D. B., Vehtari, A. & Rubin, D. B. (2014) *Bayesian data analysis*, 3rd ed. Boca Raton, FL: Chapman & Hall/CRC.

Gilleland, E. & Katz, R. W. (2016) extRemes 2.0: An extreme value analysis package in R. *Journal of Statistical Software*, **72**, 1248–1287.

Gleason, H. A. (1922) On the relation between species and area. *Ecology*, **3**, 158–162.

Gnedenko, B. V. (1943) Sur la distribution limite d'une série aléatoire, *Annals of Mathematics*, **44**, 423–453.

Guilhaumon, F., Gimenez, O., Gaston, K. J. & Mouillot, D. (2008) Taxonomic and regional uncertainty in species–area relationships and the identification of richness hotspots. *Proceedings of the National Academy of Sciences USA*, **105**, 15458–15463.

Gumbel, E. J. (1935) Les valeurs extrêmes des distributions statistiques. *Annales de l' Institute Henri Poincaré*, **5**, 115–158.

He, F. & Hubbell, S. P. (2011) Species–area relationships always overestimate extinction rates from habitat loss. *Nature*, **473**, 368–371.

Hubbell, S. P. (2001) *The unified neutral theory of biodiversity and biogeography.* Princeton, NJ: Princeton University Press.

Kellner, J. R. & Hubbell, S. P. (2017) Adult mortality in a low-density tree population using high-resolution remote sensing. *Ecology*, **98**, 1700–1709.

Kellner, J. R., Armston, J., Birrer, M., Cushman, K. C., Duncanson, L., Eck, C., Falleger, C., Imbach, B., Král, K., Krůček, M., Trochta, J., Vrška, T. & Zgraggen, C. (2019) New opportunities for forest remote sensing through ultra-high-density drone lidar. *Surveys in Geophysics*, **40**, 959–977.

Levy, O., Ball, B. A., Bond-Lamberty, B., Cheruvelil, K. S., Finley, A. O., Lottig, N. R., Punyasena, S. W., Xiao, J., Zhou, J., Buckley, L. B., Filstrup, C. T., Keitt, T. H., Kellner, J. R., Knapp, A. K., Richardson, A. D., Tcheng, D., Toomey, M., Vargas, R., Voordeckers, J. W., Wagner, T. & Williams, J. W.

(2014) Approaches to advance scientific understanding of macrosystems ecology. *Frontiers in Ecology and the Environment*, **12**, 15–23.

Martín, H. G. & Goldenfeld, N. (2006) On the origin and robustness of power-law species–area relationships in ecology. *Proceedings of the National Academy of Sciences USA*, **103**, 10310–10315.

Matthews, T. J., Triantis, K. A., Rigal, F., Borregaard, M. K., Guilhaumon, F. & Whittaker, R. J. (2016) Island species–area relationships and species accumulation curves are not equivalent: An analysis of habitat island datasets. *Global Ecology & Biogeography*, **25**, 607–618.

May, R.M. (1975) *Patterns of species abundance and diversity. Ecology and evolution of communities* (ed. by M. L. Cody and J. M. Diamond), pp. 81–120. Harvard University Press, Cambridge Mass.

von Mises, R. (1936) La distribution de la plus grande de n valeurs. Reproduced, Selected papers of von Mises, R. (1964) *American Mathematical Society*, **2**, 271–294.

Preston, F. W. (1962) The canonical distribution of commonness and rarity: Part I. *Ecology*, **43**, 185–215.

R Core Team (2018) *R: A language and environment for statistical computing*. Vienna, Austria: R Foundation for Statistical Computing. www.R-project.org/.

Rosenzweig, M. L. (1995) *Species diversity in space and time*. Cambridge: Cambridge University Press.

Scheiner, S. M. (2003) Six types of species–area curves. *Global Ecology & Biogeography*, **12**, 441–447.

Tjørve, E. (2003) Shapes and functions of species–area curves: A review of possible models. *Journal of Biogeography*, **30**, 827–835.

Triantis, K. A., Guilhaumon, F. & Whittaker, R. J. (2012) The island species–area relationship: Biology and statistics. *Journal of Biogeography*, **39**, 215–231.

Williams, M. R. (1995) An extreme-value function model of the species incidence and species-area relations. *Ecology*, **76**, 2607–2616.

Williams, M. R., Lamont, B. B. & Henstridge, J. D. (2009) Species–area functions revisited. *Journal of Biogeography*, **36**, 1994–2004.

10 · *The Species–Area Relationship: Idiosyncratic or Produced by 'Laws Acting around Us'?*

JOHN HARTE

10.1 Introduction

> Happy families are all alike; every unhappy family is unhappy in its own way.
>
> Leo Tolstoy, *Anna Karenina*

Ecologists generally hold out hope that a unified understanding of ecosystems is possible because, in Darwin's words, they are governed by 'laws acting around us' (Darwin, 1859). But at the same time, ecologists take delight in the idiosyncrasies of nature, in the features that are unique to an organism or an ecosystem, in the phenomena that resist general theory. Sometimes this duality leads particularists to condemn the search for laws and universal theory and theorists to denigrate natural history as stamp collecting.

Such conflicts are foolish. Here I demonstrate that the search to understand the species–area relationship (SAR) and the other patterns studied by macroecologists as well, exemplify how a well-defined boundary can be drawn between two legitimate domains: the phenomena that are unified by theory and those that are idiosyncratic. The focus in this chapter is on nested species–area relationships, although the insights reached here extend more broadly to all of macroecology.

10.1.1 Macroecological Theory: What Are We Looking For?

Theories that predict accurately a wide variety of phenomena are more satisfying than those that explain only a few things. Moreover, predictions based on few assumptions or fitting parameters are more satisfying than those based on many. Theories that do both have been termed 'efficient theories' (Marquet et al., 2014).

While the SAR is among the most prominent and longest studied (see e.g. Rosenzweig, 1995, for a review; see also Chapter 2) phenomena in ecology, it is just one of many relationships and distributions studied by macroecologists. Others of interest include the distribution of abundances over species, the distribution of body sizes or metabolic rates over individuals, the relationship between the abundances of species and the metabolic rate of the individuals within those species, the distribution over space of individuals within species, the structure of trophic networks and both the structure of taxonomic trees and the dependence of the distributions and relationships listed above on that structure.

With such a rich array of phenomena for which data are available, ecologists have a marvellous opportunity to attempt to develop and test comprehensive, unifying macroecological theory. In this chapter, I describe such a theory, its domain of applicability, and the insights it provides into the form of the SAR and other macroecological metrics. The larger goal of this review is to illuminate the role of mechanism in ecological theory, and the boundary between the idiosyncratic and the law–abiding facets of nature.

To a considerable extent, two constraints have been self-imposed on macroecology: a focus on relatively undisturbed and quasi-static ecosystems (or at least on theories that, in effect, assume the systems studied are undisturbed and static), and a focus on species, as opposed to other taxonomic, phylogenetic, or trait based definitions of the categories of analysis. I emphasize here the multiple insights that can be obtained by liberating macroecological theory from those constraints.

10.2 MaxEnt: A Foundation for Macroecological Theory

The maximum information entropy principle (MaxEnt) provides a powerful framework for theory construction (Harte & Newman, 2014). The motivation for use of the principle arises from the fact that science is an inference process. Suppose we wish to infer the probability distribution describing some phenomena. Our goal, as scientists, is to attempt to make that inference as free from bias as possible. In doing so, we should of course make use of our prior knowledge of the phenomenon. MaxEnt is a procedure for making inferences about the form of probability distributions that are compatible with prior knowledge and makes no further implicit or explicit assumptions. In that sense, it yields unbiased inferences.

The core idea builds upon the Shannon information entropy, which is a property of a probability distribution. Given a probability distribution, $p(n)$, its Shannon information entropy, I, is:

$$I = -c \sum_n p(n) \log (p(n)). \tag{10.1}$$

The constant, c, need not concern us.

Confusion about Equation (10.1) arises because entropy and information can, in an intuitive sense, be thought of as opposing concepts. So, let us be clear about the meaning of information entropy. Suppose your knowledge of a system consists in knowing the shape of a probability distribution, $p(n)$. If you subsequently make a measurement of n, you will not be very surprised about the outcome if the distribution is sharply peaked. You already know enough (from the shape of the distribution) to know what outcome to expect. In contrast, a draw from a relatively flat distribution provides unexpected information. I is a relative measure of how much your uncertainty is reduced as a result of a draw from a distribution of known shape. Equation (10.1) informs us that the flat distribution has high information entropy while the sharply peaked distribution has low information entropy.

In fact, Shannon information entropy has the mathematical property that it is maximum when the distribution, $p(n)$, is as smooth and flat as is possible. One possible way to maximize I, then, is to set $p(n)$ equal to a constant. But that is where prior knowledge enters the picture. Our prior knowledge may be, and usually is, incompatible with a uniform distribution. By maximizing I subject to the constraints imposed by our prior knowledge, we are making it as smooth and flat as possible, without contradicting our prior knowledge. Any distribution $p(n)$ that was less smooth and flat would, in effect, incorporate hidden assumptions and so maximizing I subject to the constraints meets the criterion for least biased inference. The task of inferring probability distributions is thus the task of maximizing information entropy. The mathematical procedure to accomplish that task is the method of Lagrange multipliers (Arfken & Weber, 2005).

The first formal derivation and application of the MaxEnt procedure was made by Jaynes (1957a, b, 1963, 1968, 1979, 1982). In a series of publications, he showed how MaxEnt could be used to derive all the basic findings of equilibrium thermodynamics and statistical mechanics. For a thermodynamic system consisting of a container of gas molecules, the values of certain state variables such as pressure and volume comprise the prior knowledge. With those state variables as constraints, Jaynes maximized the Shannon information entropy of the probability distribution describing molecular kinetic energies and derived the Boltzmann distribution. The rest of the apparatus of thermodynamics then fell into place, including the identification of temperature as the inverse of a Lagrange multiplier, a derivation of the second law of thermodynamics and much more.

Ever since Jaynes, applications of the method have been accumulating, including image reconstruction in medicine and forensics (Frieden, 1972; Skilling, 1984; Gull & Newton, 1986; Roussev, 2010), neural net firing patterns (Meshulam et al., 2017), protein folding (Steinbach et al., 2002; Mora et al., 2010), reconstruction of incomplete input–output data and other applications in economics (Golan et al., 1996; Golan, 2018) and macroecology (Phillips et al., 2006; Shipley et al., 2006; Dewar & Porté, 2008; Harte et al., 2008, 2009, 2015; Williams, 2010; Harte, 2011; Elith et al., 2011). The procedure is powerful and is increasingly influencing how scientific inference is carried out.

Early applications of MaxEnt in ecology (Phillips et al., 2006; Elith et al., 2011) made use of the procedure to infer the climate parameters associated with the ranges of species and, thus, to be able to predict how species ranges would shift under a changing climate. In such applications, MaxEnt is a tool, in the same sense that regression techniques are a tool, for inferring relationships between two sets of data (climate and occupancy data). The application I describe here is conceptually quite different; in the Maximum Entropy Theory of Ecology, MaxEnt is the foundation for the construction of efficient (*sensu* Marquet et al., 2014) ecological theory.

10.2.1 The Macro and the Micro

It is instructive to contrast the widely-deployed strategy of process-based modelling in science with the MaxEnt inference strategy. Typically, process-based modelling seeks to understand macro-level phenomena by assuming certain mechanisms operate at the micro-level and then modelling the macro-level consequences of those mechanisms. For example, a grid-based spatially-explicit individuals-based model that incorporates assumed mathematical expressions describing the processes of birth, death, dispersal, and competition at the level of individual organisms could be used to predict community-level properties such as species richness, net primary productivity (NPP), and the shape of the SAR and the species abundance distribution (SAD). The key here is that

MaxEnt is a top-down (macro → micro) inference procedure

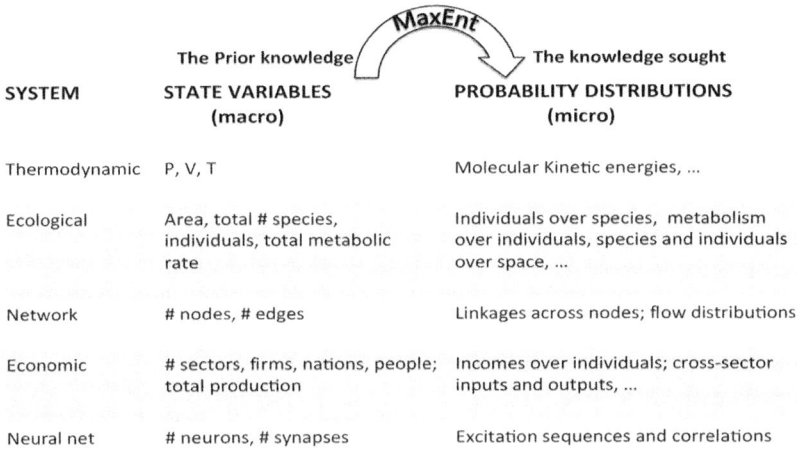

SYSTEM	The Prior knowledge STATE VARIABLES (macro)	The knowledge sought PROBABILITY DISTRIBUTIONS (micro)
Thermodynamic	P, V, T	Molecular Kinetic energies, …
Ecological	Area, total # species, individuals, total metabolic rate	Individuals over species, metabolism over individuals, species and individuals over space, …
Network	# nodes, # edges	Linkages across nodes; flow distributions
Economic	# sectors, firms, nations, people; total production	Incomes over individuals; cross-sector inputs and outputs, …
Neural net	# neurons, # synapses	Excitation sequences and correlations

Figure 10.1 A schematic illustrating the top-down logic underlying the MaxEnt inference procedure.

knowledge at some larger scale (e.g. landscape level) is derived from assumed mechanisms operating among agents (e.g. individual organisms) at a much smaller level.

The flow of insight in such process-based modelling is thus from the micro-level to the macro-level. In contrast, in MaxEnt-based theories, the flow is from the macro-level to the micro-level in the following sense: probability distributions describing phenomena at the micro scale, such as how metabolism is distributed over individuals, how individuals are distributed over space or how abundances are distributed over species are derived from knowledge of macro-scale quantities such as the total number of species and individuals and the total metabolic rate of the ecosystem (Figure 10.1). Moreover, no explicit information about mechanisms is assumed in the inference procedure, a topic to which I return in Sections 10.7 and 10.8.

10.3 The Maximum Entropy Theory of Ecology (METE): The Essential Idea

Any application of MaxEnt to construct scientific theory entails three stages.

In stage one, the categories of analysis and the phenomena that one wishes to predict are identified. The categories of analysis in METE include, for example, individuals and species; the phenomena to be predicted are listed in the left hand column of Table 10.1. Note that some of the items in the table are probability distributions, while others are functional relationships that can be derived from probability distributions. In the latter case, a typical relationship is that between the mean of a distribution and a measurable parameter influencing the shape of the distribution or the relationship between two ecological variables such as abundance and total metabolism of each of the species within a prescribed area.

In stage two, certain core probability distributions are defined and their independent variables identified. In Table 10.1 the independent variables are abundances, n, of species, metabolic rates, ε, of individuals, and species richnesses, m, of higher taxonomic categories. Intuition and guesswork are needed to define the core distributions; there are no formal prescriptions for how to do this, yet it is at the heart of theory construction. From these core distributions, if chosen wisely, the shapes of all the metrics in stage one (e.g. the items in Table 10.1) can be derived. In the case of METE, those core distributions include what are termed 'the ecological structure function', R, and a species-level spatial distribution, Π, of individuals within species.

The original version of METE (Harte et al., 2008) sought only to predict the entries in the first five rows in the table, grouped under the heading of ASNE, along with the spatial Π distribution and the SAR (the remaining entries in the table are discussed in Section 10.5). The acronym ASNE refers to the fact that in this version of METE the state variables are the area of the ecosystem (A_0), the total number of species (S_0) in that area, the total number of individuals (N_0) comprising those species and the total metabolic rate (E_0) of all those individuals. In the ASNE version of METE, the structure function R is defined by a two-step process. In step one, choose a species at random from the species pool. In practical applications, it is understood that the species are those found in some defined taxonomic group such as arthropods, plants or birds. In step two, choose an individual at random from the pool of all individuals that are found in species with abundance n. The structure function, $R(n,\ \varepsilon\,|\,S_0,\ N_0,\ E_0)$, is the joint probability that the chosen species has abundance n and the chosen individual has metabolic rate between ε and $\varepsilon + \mathrm{d}\varepsilon$, given values of the state variables S_0, N_0, and E_0. The spatial distribution $\Pi(n\,|\,n_0,\ A,\ A_0)$ is the probability that if there are n_0 individuals of a species in some area A_0, then there are n individuals of that species in a randomly selected area A contained within A_0.

In stage three, constraints on the core distributions in stage two are characterized. These constraints comprise the prior knowledge and appear as the 'conditionals' in the core distributions. As in thermodynamics where pressure, volume, etc. are considered as 'thermo-dynamic state variables', the constraints in METE are called the eco-logical state variables.

Table 10.1 *The non-spatial macroecological metrics predicted in ASNE and the additional ones predicted in AFSNE (see section 10.5.1). The λ_i are Lagrange Multipliers, which are numerically determined from the state variables S_0, N_0, E_0, F_0; $\beta = \lambda_1 + \lambda_2$ and $\gamma(\varepsilon) = \lambda_1 + \lambda_2\varepsilon$. In addition to these predictions, both ASNE and AFSNE predict the distribution $\Pi(n \mid n_0, A, A_0)$ discussed in the text. The $\approx$signs, as opposed to = signs, in the table, reflect the neglect of small terms that go to zero as E/N, N/S and S all go to infinity.*

Metric	Predicted form in ASNE version of METE
Distribution of abundances across species (SAD)	$\varphi(n) \approx \dfrac{e^{-\beta n}}{n \ln\left(\frac{1}{\beta}\right)}$
Distribution of metabolic rates across all individuals in community	$\Psi(\varepsilon) \approx \dfrac{\beta\lambda_2 e^{-\gamma(\varepsilon)}}{\left(1-e^{-\gamma(\varepsilon)}\right)^2}$
Distribution of metabolic rates over the individuals in a species, selected at random from the pool of all species	$\nu(\varepsilon) \approx \dfrac{\lambda_2}{\ln(1/\beta)} \dfrac{e^{-(\lambda_2(\varepsilon-1)+\beta)}}{(\lambda_2(\varepsilon-1)+\beta)}$
Distribution of metabolic rates across individuals in a species with n individuals	$\Theta(\varepsilon \mid n) \approx \lambda_2 n e^{\lambda_2 n(\varepsilon-1)}$
Dependence on species abundance of metabolic rates averaged over individuals within species	$<\varepsilon \mid n> \approx 1 + \dfrac{1}{n\lambda_2}$

Metric	Predicted form in AFSNE version of METE
Distribution of species richness over families: $\Gamma(m)$. m = # species/family	$\Gamma(m) \approx \dfrac{e^{-\lambda_1 m}}{m \ln(1/\lambda)}$
Distribution of species abundances for all the species in a family with m species	$\varphi(n \mid m) \approx \dfrac{e^{-\beta m n}}{n \ln(1/\beta)}$
Distribution of metabolic rates across individuals in the species with n individuals that are in families with m species	$\Theta(\varepsilon \mid m, n) \approx \lambda_3 m n e^{-\lambda_3 m n(\varepsilon-1)}$
Dependence on species abundance of metabolic rates averaged over individuals within species in a family with m species	$<\varepsilon \mid m, n> \approx 1 + \dfrac{1}{n m \lambda_3}$

The actual constraints on R are ratios of the values of the State Variables: the average abundance per species, $<n> = N_0/S_0$ and the average metabolic rate per species, $<n\varepsilon> = E_0/S_0$, where $<n>$ and $<n\varepsilon>$ are calculated directly from R. The constraint on Π is $<n> = n_0 A/A_0$.

The information entropy of R and Π are then maximized subject to those constraints. The MaxEnt inference procedure predicts the form of R and Π. If the state variables are measured, then there are no adjustable parameters. From those distributions the predicted forms of the ecological metrics listed under ASNE in Table 10.1 follow using straightforward calculus (Harte, 2011). In both Python$^{©}$ (Kitzes & Wilber, 2016) and in R (Rominger & Merow, 2016), software now exists to perform empirical tests of METE's predictions.

By introducing additional state variables, more complex forms of the structure function can be derived; it will depend on additional independent variables and be conditional on a larger set of state variables. This will be discussed further in Section 10.5.

10.4 Scale Collapse and the 'Universal SAR'

If all available SAR data are placed on a single graph, with log(species richness) on the ordinate and log(area) on the abscissa, the totality of data would comprise a large formless blob; the curves for each SAR will exhibit different slopes (local tangents to the curve) at any fixed area and of course intercepts will differ from dataset to dataset. No underlying simplicity would be apparent. In contrast, METE informs us that if the data from all the datasets are replotted in a certain way predicted by the theory, then an underlying simplicity emerges.

The SAR can be predicted from knowledge of the SAD, $\phi(n \,|\, S_0, N_0, E_0)$, and the individuals spatial distribution function, $\Pi(n \,|\, n_0, A, A_0)$. The formula for the number of species in a subplot area A, given state variables S_0, N_0, E_0 in area A_0, is:

$$S(A \,|\, S_0, N_0, E_0, A_0) = \sum_{n_0=1}^{N_0} \phi(n_0 \,|\, S_0, N_0, E_0)[1 - \Pi(0 \,|\, n_0, A, A_0)].$$

(10.2)

The first term under the summation in Equation (10.2) is the probability that a species has abundance n_0 in A_0 and the second is the probability that the species with abundance n_0 in A_0 is present in area A.

METE predicts (Harte et al., 2008) that ϕ is a log series distribution and Π is a truncated geometric distribution. Evaluation of Equation

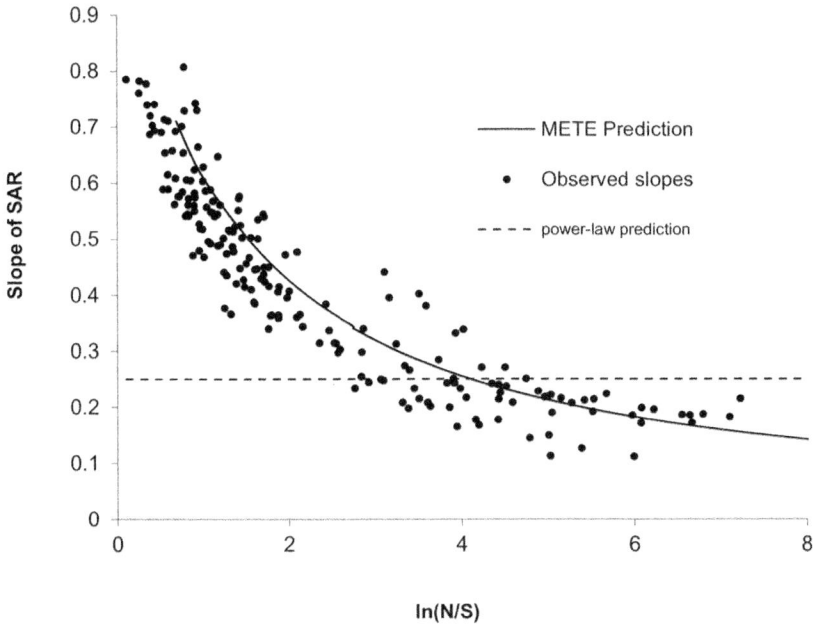

Figure 10.2 Predicted and observed scale collapse of the SAR. The local (i.e. at a specified spatial scale) slope of the log–log SAR is plotted on the vertical axis and the log of the average abundance per species at that same scale is plotted on the horizontal axis. The solid line is the METE prediction and filled circles are empirical values for plant, bird and arthropod data from approximately twenty distinct habitats and a wide range of spatial scales (data sources provided in Harte et al., 2009). The dashed line is the $z = \frac{1}{4}$ power law prediction.

(10.2) then results in the following unexpected prediction (Harte et al., 2009; Figure 10.2): plot on the vertical axis the local slope, z (tangent to the curve on a log–log graph), at area A; on the horizontal axis, plot the ratio log (total abundance at scale A/species richness at scale A). If you do that for every scale, A, for which there are data and for every SAR dataset, then all the data points, across habitats, taxonomic groups and spatial scales, should collapse onto a universal curve whose explicit shape is predicted.

Although an analytical expression for the predicted functional dependence of species richness on area is not obtainable, as long as $\exp(-S) \ll 1$, an analytical expression for the value of the area-dependent log–log slope, z, is (Harte et al., 2009):

$$z \approx [\ln(2) \bullet \ln\left(1/\left(1 - e^{-\beta(A)}\right)\right)]^{-1}, \qquad (10.3)$$

where an infinitely nested expression relates $\beta(A)$ to $N(A)/S(A)$:

$$\left(1 - e^{-\beta}\right)^{-1} = \frac{N(A)}{S(A)} \cdot \ln\left(\frac{N(A)}{S(A)} \cdot \ln\left(\frac{N(A)}{S(A)} \cdot \ln\left(\frac{N(A)}{S(A)} \cdots\right)\right)\right). \tag{10.4}$$

Typically, four or five iterations of the ln terms in Equation (10.4) give an excellent numerical approximation to $(1 - e^{-\beta})^{-1}$ and hence to z. Because Equation (10.4) provides only an implicit relationship for z (it depends on $S(A)$), a numerical solution is necessary to actually predict the shape of the SAR. A fit to that solution gives the approximate form for $S(A)$ for $N(A) \gg S(A) \gg 1$:

$$S(A) \approx c \cdot \ln\left(N(A)\right) \cdot \ln\left(\ln(N(A))\right), \tag{10.5}$$

where c is an intercept that distinguishes systems of differing overall diversity at a fixed scale. For a complete nested SAR, $N(A)$ is proportional to A, and hence the predicted SAR rises faster than the Gleason form: $S \sim \ln(A)$, but asymptotically slower than a fixed power law.

For comparison, the widely-assumed power law SAR with a slope parameter $z = 0.25$ is shown in Figure 10.2 as the horizontal dotted line. Available data clearly deviate from any fixed-z model. The standard practice of fitting a straight line through log(species richness) versus log

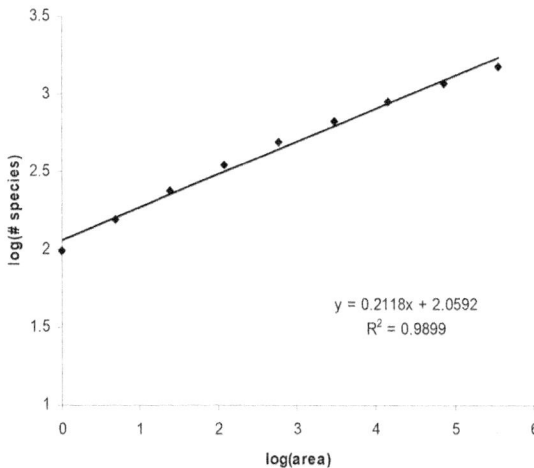

$$y = 0.2118x + 2.0592$$
$$R^2 = 0.9899$$

Figure 10.3 Illustration of the hazard of inferring power law SAR behaviour from 'good' fits. Although the R^2 is ~ 0.99, the local slope at the two smallest areas is roughly twice that at the two largest areas.

(area) graphs is, in fact, notoriously dangerous. This is illustrated in Figure 10.3, where a seemingly excellent ($R^2 = 0.99$) fit of a straight line through some concocted data obscures the more relevant fact that the slope calculated for the first two data points is roughly twice the slope calculated from the last two data points.

10.4.1 A Note on Microorganisms

Nothing in the foundational assumptions or deployed methods should, in principle, restrict the domain of METE to macroorganisms. Thus, the question arises as to whether microbial SARs follow the universal scale collapse prediction. Available evidence, though sparse, suggests an affirmative answer. Consider microbial taxon richness and abundance in a patch of soil. In areas ranging from square metres to hectares, data from multiple small samples indicate that typical taxon numbers are on the order of 10^3–10^4, while total abundances range from 10^{12} to 10^{16}. Because the predicted local slope of the SAR depends on the logarithm of N/S, these wide ranges of values are not going to have a huge influence on the predicted slope. Indeed, using Equations (10.3) and (10.4), I predict z-values that range from $1/(\ln(2) \star \ln(10^{15})) = 0.04$ to $1/(\ln(2) \star \ln(10^{10})) = 0.06$. Thus, despite the huge uncertainty in both taxon richness and abundance, fitted power law SAR slopes for microorganisms at field scale should be ~ 0.05, not 0.01 or 0.1. Graphically, if we refer to Figure 10.2, microorganism data points should have abscissas that lie far out to the right, beyond the range of data shown, and thus z should be very small.

Two studies (Green et al., 2004; Horner-Devine et al., 2004) that measured SARs for microorganisms both concluded that, when fitted with a power law function, very small but non-zero slopes (z ~ 0.05) were obtained. There is an important caveat, however. The shape of the SAR is much influenced by the species with the lowest abundances. But in field (and in human microbiome) measurements of the abundances of the microorganisms, techniques of sampling and measurement result in abundance cutoffs and hence the non-detection of the rarer taxa. Thus, tests of the abundance distribution predicted by METE and to some extent tests of the SAR are at best suggestive.

10.4.2 Upscaling Species Richness

Imposing a power law fit to SAR data is particularly dangerous when the power law is then extrapolated to large spatial scales, as is sometimes done in calculations of estimated extinction rates under landscape-scale habitat losses (see Chapters 14 and 16) or in estimations of species richness at large spatial scales based on extrapolations from small plot or single canopy census data.

Tree species richness in the Western Ghats Preserve in southern India provides a useful example (Figure 10.4). Using measured species richness values (trees with >10 cm dbh) in forty-eight quarter-hectare plots scattered throughout the Preserve, the universal SAR predicts a total of 1,070 tree species in the 60,000 km^2 Preserve (Harte et al., 2009). Currently, a total of about 1,000 tree species have been recorded in the Ghats, with several new ones being discovered each year in recent years. In contrast, extrapolating up to the scale of the entire Preserve, a power law SAR, using the average z-value obtained from power law fits to the SARs within each of the forty-eight quarter-hectare census plots, predicts 49,000 tree species in the Preserve (Krishnamani et al., 2004). Similar, impossibly large, estimates for tree species richness arise if power

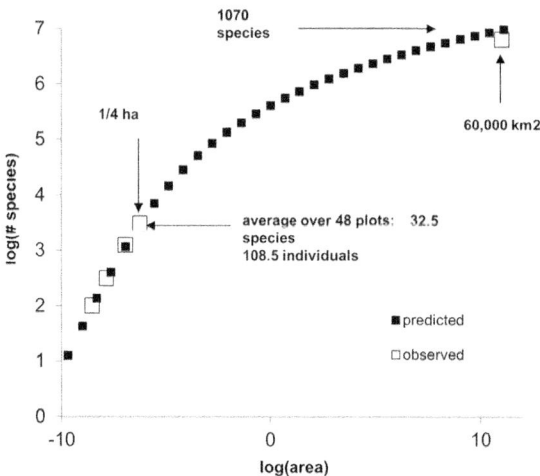

Figure 10.4 Test of upscaling and downscaling predictions for tree species richness in the Western Ghats. The input information for all the predicted values of species richness shown in the graph are the averages of total abundance and species richness in forty-eight quarter-hectare plots scattered throughout the 60,000 km^2 preserve (Krishnamani et al., 2004). In those small plots, abundance and spatial location data for every tree species with ≥30 mm dbh were obtained and used to construct forty-eight species–area relationships spanning nested areas within each plot of 0.25, 0.125, 0.05 and 0.025 ha (average species richness values shown by open squares in the graph). Area is in units of square kilometres and axes show natural logs.

law fits are extrapolated from small tropical forest plots to the entire Amazonian rain forest, whereas predictions based on the METE SAR are remarkably realistic, yielding an estimated 16,000 tree species (Harte & Kitzes, 2014).

The UK provides a clear example of the failure of METE to accurately upscale plant species richness from small plots located within a variety of distinct habitat types (heath, forest, grassland, etc.) to very large spatial scales within which multiple largely-non-overlapping habitats exist: namely the entire UK (Kunin et al., 2018). From the values of S and N within small plots, averaged over all the habitats within the UK, extrapolated species richness at the scale of the entire UK is greatly underestimated. At best METE should be able to upscale species richness from small-plot data within any specific habitat type up to the entirety of such contiguous habitat, with the UK species richness value then given by the sum of the separately up-scaled values. Indeed, this is the approach taken in Harte and Kitzes (2014) for estimating tree species richness in the Amazon, with six bioregions (ter Steege et al., 2013) comprising the entire Amazon assumed to be non-overlapping.

We note that the method proposed and tested in Harte et al. (2009) and Harte and Kitzes (2014) for extrapolating species richness from small plots to large spatial scales does not require abundance data, despite statements to the contrary in the literature. The extrapolation procedure can be carried out either with prior knowledge of the two state variables, S and N, at one spatial scale, in which case abundance data are obviously required, or with knowledge of the species richnesses, S_1 and S_2, at two spatial scales. In the latter case, the extrapolation procedure allows estimation of total abundance N at each of the scales from knowledge of the species richness at both of the scales.

10.4.3 A Note on the Triphasic SAR

METE predicts that the slope of the log–log nested SAR at *any* specified area, A, is a monotonically decreasing function of the ratio of total abundance, $N(A)$, to species richness, $S(A)$ (Figure 10.2). Because N increases linearly with A while S usually increases more slowly, this implies that slope, $z(A)$ is usually a monotonically decreasing function of area.

In contrast, there are clearly situations in which, at large spatial scales, the slope of the nested SAR increases with area. As area increases, distinctly different habitats, with largely non-overlapping lists of species present, will inevitably be encountered and this leads to an increase in the slope of the SAR. Thus, a triphasic (slope first decreases, then increases, then decreases as area increases) nested SAR is expected, although available nested SAR data are not sufficient to give us quantitative insight into its detailed form. Within a MaxEnt framework, the triphasic SAR is simply a consequence of additional biogeographic constraints.

But the application of METE to landscape- or continental-scale census data, in which different habitats are encountered as area increases, does pose an interesting challenge. In an extreme case (Figure 10.5A), the procedure for applying METE is straightforward: within each of the taxonomically non-overlapping habitats, each of area $A_0/2$ (Figure 10.5A), there are a set of state variables, S, N and E, and thus the SAR within each of the habitats is predicted by METE. At the scale A_0, the number of species is the sum of the numbers in the two habitats of area $A_0/2$ and, thus, the slope will increase at sufficiently large area. This situation might correspond to a census tract that includes subalpine forest in the left half of the tract and subalpine meadow in the other half.

More challenging is identifying the correct procedure for the case shown in Figure 10.5B. Here there is a continuous gradation of habitat such that there might be no overlap between the far left and far right of the combined habitats, but some overlap between the right and left halves. An example might be closed forest gradually transitioning to savanna, which in turn gradually transitions to grass-lands. Although complete nested SAR data for such transitional landscapes are not available to my knowledge, a plausible inference is that METE would probably fail at the scale of the entire landscape, but would accurately predict SARs within the predominantly forest, within the predominantly savanna and within the predominantly grassland habitat.

The failure of METE to accurately upscale plant species richness across the entirety of the UK stems from the existence of distinct (with nearly non-overlapping species) habitat types within the UK. In Section 10.6, I discuss a different kind of failure of the theory: in ecosystems with rapidly changing state variables. While the former source of discrepancy is conceptually simple and readily dealt with by confining analysis to relatively homogenous habitat types, we shall see that the latter likely necessitates a major structural change in the theoretical framework.

(A)

(B)

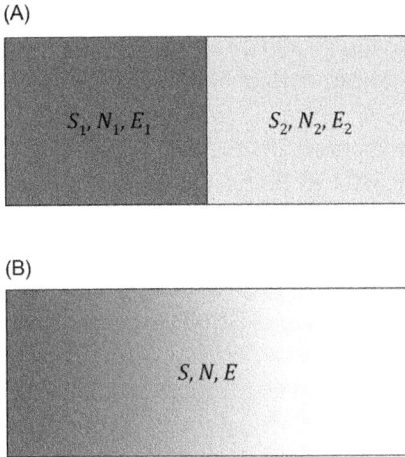

Figure 10.5 Depiction of a plot containing more than one habitat, with the darkest edge of the plot signifying closed forest and the lightest edge signifying open meadow. (A) A plot of area A_0 containing two spatially separated discrete habitats, each of area $A_0/2$. (B) A similar-sized plot containing a gradual transition from closed forest to open meadow.

10.5 Liberating Macroecology from Its Species Fixation

10.5.1 Climbing down the Taxonomic Tree

Can MaxEnt allow us to extend theory to populations, families, functional groups, etc. and what do we learn by doing so? The ASNE version of METE assumed prior knowledge of the number of species in the community and as a consequence it predicted the shape of the species–area relationship. But, if prior knowledge includes the number of, say, families or other taxonomic categories, they can also be included as constraints, in addition to species richness. The result of doing so is informative (Harte et al., 2015). Consider, for example, the AFSNE version of METE, in which an additional state variable, F, for number of families in the community, is included. Now additional metrics are predicted, most notably the distribution of species over families (see Table 10.1).

In the AFSNE version of METE, in which the number of families represented among the species in the ecosystem is included as a state variable and the variable m refers to the number of species in a family, the modified structure function, $Q(m, n, \varepsilon \mid F_0, S_0, N_0, E_0)$, is defined as follows. Pick a family at random from the pool of families; then $Qd\varepsilon$ is the probability that it has m species and that, if you pick a species from a family with m species, it has n individuals and that, if you pick an individual from a species with n individuals, then it has a metabolic rate in the interval $(\varepsilon, \varepsilon + d\varepsilon)$.

Interestingly, most of the ASNE predictions are unaltered. In particular, AFSNE predicts the SAD and the SAR to be essentially identical to the ASNE predictions. Very small predicted deviations would be impossible to detect empirically. The predicted distribution of metabolic rates over all individuals differs only slightly and only for the largest individuals. But even more interestingly, one prediction of ASNE is significantly modified: the prediction of energy equivalence.

The energy equivalence principle, in its original form, asserts that each species in the community takes an equal share of total community metabolism. In particular, ASNE predicts that the mean metabolic rate of the individuals within a species with n individuals is given by:

$$\langle \varepsilon \,|\, n,\, S_0,\, N_0,\, E_0 \rangle = 1 + ((E_0 - N_0)/S_0)/n. \tag{10.6}$$

In most ecosystems, the state variables obey the inequalities $S_0 \ll N_0 \ll E_0$ and the abundances of nearly all species are less than E_0/S_0. Thus, for nearly all species the second term on the right side of Equation (10.6) dominates. Hence, $<\varepsilon\,|\,n>n = $ constant, which is just the statement of energy equivalence: nearly all species have equal rates of metabolism.

In AFSNE, the prediction is modified:

$$\langle \varepsilon \,|\, n,\, m,\, F_0,\, S_0,\, N_0,\, E_0 \rangle = 1 + ((E_0 \quad N_0)/F_0)/(nm). \tag{10.7}$$

Here m is the number of species in the family that the species with abundance n is in. In words, Equation (10.7) states that species in species-rich families (large m) take a smaller share of total metabolism than do species in species-poor families. Families, not species, get equal shares of total metabolism.

As shown in Harte et al. (2015), this extension of METE to higher taxonomic categories explains a number of patterns observed in macroecology. In particular, systematic relationships between the species richness of higher taxonomic categories and the macroecological patterns exhibited by the species in those categories have been documented. For example, ter Steege et al. (2013) show that the most abundant Amazonian tree species belong to genera that contain relatively few species. Schwartz and Simberloff (2001) and Lozano and Schwartz (2005) show that rare vascular plant species are over-represented in species-rich families. Smith et al. (2004) show that species of mammals with the largest body sizes and, therefore, largest metabolic rates of individuals, belong to genera with the fewest species. Moreover, the variance of body size across species is also greatest in mammalian genera with the fewest species (Smith et al., 2004). All of these general trends are predicted by the taxonomically extended version of METE (Harte et al., 2015).

Additionally, I note that the extended METE also predicts the distributions of species richness values across genera or families, as well as the shapes of family–area or genus–area relationships (Harte et al., 2015). The former, but not the latter, have been extensively and successfully tested.

10.5.2 A Note on Ambiguities in Taxonomic Resolution

Macroecologists assume prevailing criteria for assigning individuals to species (e.g. accepting somewhat arbitrary genetic distance criteria in microbiology or whatever the current taxonomies happen to be for macrospecies). Moreover, somewhat arbitrary decisions about what constitutes an individual (e.g. ramets versus genets) sometimes have to be made. In Harte et al. (2013) we showed that the success of METE's predictions is remarkably robust to such ambiguities. Moreover, the prediction it makes for the SAD for, say, arthropods in undisturbed habitats resembles observed patterns in nature regardless of whether the data are examined separately order by order or with all orders lumped together (Harte & Kitzes, 2014).

That the success of the METE predictions might be resilient to the various types of changes in taxonomic resolution and definition can be qualitatively understood by recalling that the predicted METE structure function is a distribution over abundance, n, and metabolic rate, ε, and the shape of the function is conditional on the values of the state variables. A change in taxonomic resolution, or restriction to a subset of species present in a community (e.g. only those in a particular order), generally does result in a change in the detailed shape of the metrics of macroecology, such as the SAR and the SAD, but it will also result in a change in the values of the state variables and thus in the predicted shape of the metrics. These changes appear to be remarkably compensatory.

The likelihood that taxonomy will forever be in flux or at least always possess ambiguity, reinforces the case for developing theories like METE that have the potential to be resilient to taxonomic choices.

10.6 Liberating Macroecology from its Quasi-static Fixation

There can be ecological value even in a habitat that has been mangled by humans. Thus, a hectare of recently clear-cut mountainside temperate forest in an early successional stage of return to forest can harbour more avian diversity than a hectare of the surrounding intact forest. Yet, macroecologists generally choose for their census sites ecosystems that are relatively pristine, and not undergoing rapid response to recent disturbance. In standard classifications of habitat types, categories such as lowland tropical forest, cloud forest, taiga, tundra and alpine meadow appear, but rarely do categories such as soil subsidence scars, recently clear-cut temperate forest, heavily grazed subalpine meadow, dried up lake bed or suburbanized former prairie.

With ever-growing human numbers and ever-growing per capita rates of consumption of energy and material resources, the habitats on our planet are increasingly looking less and less like those traditionally studied. Even during the course of a three-year research grant, study sites may be changing dramatically, above and beyond changes driven by stochastic environmental factors such as weather, in response to prior or continuing disturbance. The question naturally emerges: how do macroecological patterns differ on quasi-static versus dynamic landscapes?

To be specific, I define a quasi-static ecosystem to be one in which the values of the state variables are unchanging, or only very slowly varying, in time. A dynamic system is one in which the state variables are rapidly changing. The boundary between 'slowly' and 'rapidly' will have to be determined from a combination of theory and empirics.

In ecosystems that clearly fall within a relatively pristine habitat classification, the predictions of METE, including the SAD, the universal scale collapse of the SAR, the distribution of metabolic rates over individuals and the influence of the shape of taxonomic trees on the energy-equivalence rule are fairly successful (Harte et al., 2008, 2009, 2015; Harte, 2011; White et al., 2012; Harte & Kitzes, 2014; Xiao et al., 2015).

However, discrepancies are often observed when METE's predictions are tested in ecosystems that are undergoing relatively rapid change. In particular, when state variables are changing as a consequence, for example, of succession in response to either natural or anthropogenic disturbance, the values of the state variables at any moment in time do not accurately predict the METE predictions for the shapes of the macroecological metrics at that same moment in time.

For example, at younger sites in the Hawaiian Islands where diversification is occurring more rapidly, both the distribution of abundances over species and metabolic rates over individuals show deviations from the METE predictions (Rominger et al., 2016). Similarly, moth censuses at Rothampsted reveal METE predicted log series abundance distributions at less disturbed, but lognormal distributions at more disturbed locations (Kempton & Taylor, 1974).

Across the Smithsonian tropical forest plots, systematic deviations from MaxEnt predictions appear prominently at the Barro Colorado Island (BCI) site in Panama, where the formation of Gatun Lake resulted in the semi isolation of the created island from its metacommunity and a steady decline in the number of tree species (E. Leigh, personal communication). When viewed as a log(abundance) versus rank graph, the shape of the SAD at BCI is currently intermediate between a Fisher logseries distribution and a lognormal distribution and fairly well described by the zero-sum multinomial distribution (Hubbell, 2001). In contrast, other Smithsonian tropical forest plots that are less disturbed, such as Cocoli and Bukit Timah, show closer agreement with the logseries abundance distribution predicted by static METE (Harte, 2011). Whether or not the deviation of the BCI SAD from the logseries distribution is a consequence of a decrease in immigration rates to the 50 ha plot remains to be seen.

The examples above pertain to systems subject to anthropogenic disturbance. We turn next to an ecosystem that is subject and adapted to recurring natural disturbance: serotinous, fire-adapted Bishop Pine forest (Harvey & Holzman, 2014). In the aftermath of a recent fire at Pt. Reyes National Seashore, CA, the SAR at a post-fire site deviates markedly from the METE prediction (Figure 10.6), whereas, in a comparison site that has not burned in at least four decades, the METE prediction is quite accurate (Newman et al., 2020). Thus, deviations from METE predictions arise under both anthropogenic and natural recurring disturbances (Newman et al., 2020). Whether there are systematic differences in the patterns of discrepancy in anthropogenic versus natural disturbance regimes remains to be seen.

Referring to Equation (10.2), the shape of the SAR is uniquely determined by the shape of the SAD and of the species-level spatial distributions, $\Pi(n \,|\, n_0, A, A_0)$. Thus, we can ask: if the SAR in a disturbed ecosystem deviates from the METE prediction, as in Figure 10.6, is that a consequence of the effect of disturbance on the SAD, on Π or on both of these distributions? A definitive answer is not in hand, but preliminary

(A)   (B)

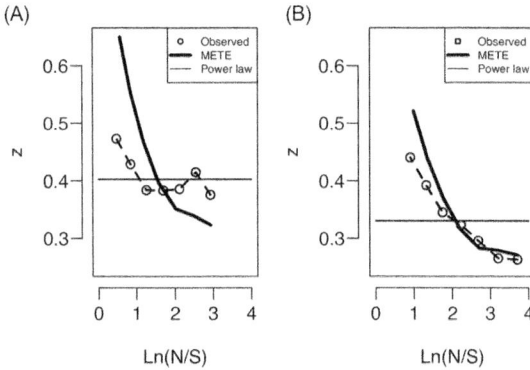

Figure 10.6 Universal scale collapse graphs (log–log SAR slope versus log(N/S)) are shown for the recently disturbed Bayview plot (A) and for the mature Mount Vision plot (B) at Pt. Reyes National Seashore, CA (Newman et al., 2020). The Bayview plot experienced a stand-replacing fire seventeen years prior to the survey, while the Mount Vision plot had no major disturbance for over forty years. Data are shown against METE predictions for each plot and best-fit power laws for comparison.

analysis (Newman et al., 2020) suggests that the deviation of the SAR from the METE prediction is driven by changes in both these distributions when the state variables are changing rapidly in time.

At the core of METE is the notion that a set of state variables provides a sufficient amount of ecological detail to determine the shapes of macroecological patterns using the MaxEnt inference procedure. METE, as heretofore developed, is static in the sense that the state variables are assumed to be constant in time or at least so slowly varying that the information to derive the distributions that the state variables constrain is adequately captured by the instantaneous values of those variables. The question remains as to whether a state variable approach will work in disturbed systems.

It is plausible that in a dynamic ecosystem, in which the state variables are rapidly changing, the existing theory might be inadequate. For an analogy, consider thermodynamics, where pressure, volume and temperature play the role of macroscopic state variables and from which the micro-level distributions such as the Boltzmann distribution of molecular kinetic energies can be derived using the MaxEnt inference procedure. In a 'disturbed' gas, such as one that is being heated from one side of a container, molecular motions are non-random across the temperature gradient in the container. These long-range correlations suggest that the container-averaged values of pressure, volume and temperature at any instant might not suffice to derive the instantaneous distribution of molecular kinetic energies.

On the other hand, support for a state-variable approach even in disturbed ecosystems emerged from an investigation of effects of disturbance on small-mammal communities (Supp et al., 2012). In that landmark study, the authors concluded 'Despite differences in species composition, the macroecological patterns showed no significant changes in response to altered seed predation, except when plant S and N were influenced by rodent removal'. If disturbances that do not change the state variables result in little change in the macroecological patterns, then the search for some sort of a link between macro-scale state variables and micro-scale probability distributions in disturbance regimes appears worth pursuing.

10.6.1 DynaMETE

To extend METE to the dynamic realm we are investigating several approaches that result in time-dependent state variables and time-dependent forms for the macroecological metrics. I refer to the dynamic theory of macroecology as DynaMETE.

One approach to formulating a dynamic theory is to view disturbance and the resulting time-dependent state variables as a source of additional constraints on the structure function, R, and the spatial Π function. In that approach, three additional Lagrange multipliers can be introduced, corresponding to the constraints imposed by the measured rates of changes of S, N and E. Expressing the state variable dynamics, incorporating explicit mechanisms such as birth, death, ontogenic growth, migration, speciation and extinction, in terms of averages of functions of the independent variables, n and ε, then these additional constraints result in altered functional forms for the shapes of the distributions R and Π.

In an alternative approach, a dynamic probability distribution, $P(S, N, E, t)$, for the state variables can be derived from a Markov-process, master equation incorporating those same mechanisms listed above. Using this distribution, the dynamic metrics of macroecology can be predicted using:

$$D(n, \varepsilon, t) = \sum_{S, N, E} P(S, N, E, t) R(n, \varepsilon \mid S, N, E), \qquad (10.8)$$

where R is the static structure function. From the dynamic structure function, $D(n, \varepsilon, t)$, the time-dependent metrics of macroecology can be derived in exactly the same way that static metrics are derived from R $(n, \varepsilon \mid S, N, E)$ in the static theory.

As of this writing, the detailed structure of a successful dynamic theory is not yet apparent. What does seem clear is that DynaMETE will be a hybrid theory in which one parent will be the incorporation of explicit mechanisms relevant to the disturbance into the theory; the other parent will be the MaxEnt inference method. I suspect that a successful dynamic theory will also build on the notion that disturbance alters the landscape in such a way as to void an assumption that, plausibly, is implicit in the success of static METE: fitness equality across taxonomic units, resulting in, among static METE's many other predictions, energy equivalence.

10.7 Mechanism: Where's the Meat in Static METE?

One of the advantages of a MaxEnt approach to theory construction is that it allows the sequential addition of ever more knowledge-based constraints, thereby providing a means of determining which state variables are more or less important drivers of patterns at the micro-scale. In the ASNE version of METE, the SAR, the SAD and the distribution of metabolic rates over individuals in static systems are accurately predicted in relatively static ecosystems. Adding additional higher-taxonomic state variables did not affect those predictions. We infer that, in such systems, whatever the mechanisms are that determine the values of the state variables S, N and E, at a given site, they are sufficient to determine the shape of those macroecological metrics. Mechanism in METE is embedded implicitly in the values of the state variables.

On the other hand, the form of the energy equivalence rule was significantly improved by the addition of a constraint arising from an additional taxonomic state variable: the number of families, F_0, resulting in the AFSNE version of METE. Moreover, a number of other patterns in macroecology that depend on the species richness of higher taxonomic categories were accurately predicted. This provides a window into possible mechanisms driving the energy-equivalence rule. In particular, even though taxonomic trees are not precise representations of the full evolutionary history leading up to the extant community, the extended theory represents a first attempt at bridging ecological and evolutionary metrics within the maximum entropy framework. In the language of phylogenetics, we can speculate that the division of the total metabolic pie can be thought of as a sequential sharing process beginning in deep time, leading to each family, rather than each species, having an equal share of total metabolism, E_0.

Connections between evolution and ecology (Webb et al., 2002; Kraft et al., 2007; Cavender-Bares et al., 2009; Graham et al., 2009) have been insightful, but the proper null model against which to test alternate hypotheses remains unclear (Swenson et al., 2006; Cavender-Bares et al., 2009). The maximum entropy framework used here may be able to provide such a null model linking macroecology and phylogenetic community structure. In particular, the division of the metabolic pie might reflect a stabilizing mechanism that insures fitness equalization throughout evolutionary time scales.

For another example of a link among state variables, macroecological patterns and mechanisms, consider the consequence of adding another macrovariable, W, in addition to total metabolic rate, E, and thus creating the ASNEW model of METE. Thinking of W as a resource (such as water) to be allocated to individuals using the same formalism as used to allocate metabolism to individuals, the SAD resulting from the MaxEnt calculation shifts from the Fisher log series distribution, $ce^{-\beta n}/n$, to the function $ce^{-\beta n}/n^2$. The latter predicts relatively more rare species than the former and also alters the shape of the SAR. Every additional allocated state variable increases the exponent to which n in the denominator is raised by 1. This suggests that in ecosystems in which there are more rare species than predicted by the Fisher log series distribution, one or more additional limiting resources might be influential.

We conclude that METE provides insight into mechanism, even though it does not explicitly embody mechanisms. In particular, the identity of the state variables that provide the necessary and sufficient constraints to accurately predict the shapes of the metrics of macroecology yields useful information about dominant mechanisms. Although the insight thus obtained is by no means a substitute for mechanistic models, it can help mechanistic modellers focus on what is most relevant. There is meat in METE, but it is implicit at the macro-scale, rather than explicit at the micro-scale in the manner familiar to mechanistic modellers.

10.8 Further Thoughts on Mechanism in Science

The above considerations raise the questions: what constitutes scientific explanation? Is mechanistic reasoning a necessary foundation for scientific theory? Is it a possible foundation? Is there an alternative? Let us speculate.

In everyday usage, we say that something is explained by a mechanism when we have identified some other *thing* that causes it. While that sounds straightforward, unfortunately what is accepted by one person as an explanatory mechanism will often not be accepted as such by another. The search for mechanism can propel us into a seemingly infinite regress of explanations.

Tied in with the notion of mechanism is the notion of causation. And here, too, what we often take for granted is not as simple as it seems. In an essay titled 'On the notion of cause', Bertrand Russell (1912) dissected the concept of cause and effect with razor sharp logic and sliced it to pieces. He showed that every definition of the notion of causation advanced by scientists or philosophers inevitably led to logical inconsistencies, and concluded: 'Causality ... is a relic of a bygone age, surviving, like the monarchy, only because it is erroneously supposed to do no harm'.

To illustrate one of the many problems Russell identified, consider the following two propositions:

I) Causes precede in time their effects.
II) Atmospheric carbon dioxide causes the planet to warm.

Most would agree with both of these statements. But now consider ice core data from the Antarctic ice shelf. It reveals something very interesting about Earth's temperature and atmosphere over the past hundreds of thousands of years. Temperature and carbon dioxide concentration rise and fall in approximate synchrony, but there is a slight time lag in the carbon dioxide record relative to the temperature record: prehistoric rises in temperature occur slightly *before* rising carbon dioxide concentrations. That might suggest to some that either proposition I or II must be wrong and some climate science deniers have used this argument to support their notion that II must be wrong.

The deniers of course are wrong. The resolution involves the notion of feedback: carbon dioxide indeed causes warming, but warming also causes a rise in atmospheric carbon dioxide. This is called a positive feedback because it acts to strengthen the warming caused by carbon dioxide emissions from fossil fuel burning. Feedback describes two mutually causal phenomena, but of course the two can't both precede each other. Which one lags depends on details such as how fast heat penetrates into the oceans, how fast vegetation and soil respond to warming and how fast ice caps melt. When feedbacks are at play in a complex system, as they almost always are, temporal history can be a poor guide to separating cause and effect.

The curdling of causal connection can also arise when we consider a hierarchy of levels of organization in a system. It is commonplace in the sciences to come across an argument like the following: to understand the properties of a macroscopic material object, say a plastic spoon, you have to understand the properties of the plastic, which entails understanding the molecules that comprise the plastic; and to understand that you have to understand the atoms of the elements that comprise the molecules. We could continue down to quarks, perhaps to strings. This is the standard reductionist view of science. Had we started with a frog, rather than a spoon, a similar but even longer passage along the hierarchy could be traversed. Reductionism is tantamount to saying that mechanistic causation flows from the small to the big, from the parts to the whole.

Does this make sense when trying to understand an ecosystem? Does causation solely run up a hierarchical ladder, from the individual organisms, the nutrients, the microtopography of the land surface, the spatially and temporally explicit sunlight intensity and the myriad other fine-scale components of an ecosystem on up a hierarchy of scales to determine the organization and functions of an ecosystem? Of course not. The shape of a complex forest canopy influences the amount of light and water that reaches the soil, which influences the likelihood that a seed will germinate, which influences the future canopy. Causal loops are at work. The ice core data showed us that a unique direction of causation cannot be inferred naively from a temporal sequence and the forest shows why it cannot be inferred in a unique direction along hierarchical levels.

Suppose, nevertheless, we seek to formulate a mechanistically-based theory of ecology, capable say of predicting the patterns that METE predicts and that we see in nature, including the SAR, the spatial distributions of individuals within species, the distributions of abundances and metabolic rates across species, the relationship between metabolic rate and abundance, the architecture of food webs, the structure of taxonomic and phylogenetic trees and much more. We would like the theory to be applicable to all ecosystems, from the deserts to the tundra, from the rainforests to the oceans. And surely we don't want to have one theory for, say, birds, and a very different one for plants or mammals or microorganisms. What mechanisms might we select as ingredients?

Such a mechanistically-based theory would include descriptions of the mechanisms by which species and individuals partition biotic and abiotic resources and interact with each other in doing so. A list of governing mechanisms and processes could surely include predation, competition, mutualism, commensalism, birth, death, migration, dispersal, speciation, disease and resistance to disease, effects of crowding, effects of rarity on reproductive pairing and genetic diversity, vocalizations, effects of aggregation of individuals on predation, harvesting strategies, predator-colouration, escape strategies, plant signalling, nutrient cycling, leaf geometry, resource substitution, hibernation, plasticity, life cycle adaptations to fluctuations in weather and resource supply and an assortment of reproductive strategies ranging from lekking behaviour to broadcast insemination to male sparring to flower enticement of pollinators.

The above is a long, yet still incomplete, list of the many mechanisms, processes and traits that are observed to govern the interactions of millions of species comprised of billions of locally adapted populations, all with distinguishable genetic and trait differences among their many trillions of individual organisms. Interactions among individuals and populations play out on a stage with abiotic conditions that vary unpredictably in both space and time. To complicate things further, those abiotic conditions are, in turn, partially influenced by the organisms.

From a morass of mechanistic complexity, the ecological phenomena we observe in nature are generated. Few if any of the mechanisms are well enough understood that they can be incorporated into ecological theory without making arbitrary choices about parameter values to describe them quantitatively. Picking from the menu of options a defensible set of ingredients for a mechanistic theory is a daunting challenge. The question thus raised is: under what circumstances will mechanistic modelling prove useful? The answer, I suggest in Section 10.9, takes us back to the question I raised at the outset: what is the boundary between the idiosyncratic and the universal?

10.9 The Idiosyncratic and the Universal: Where Is the Boundary?

It is abundantly clear to any observant ecologist that mechanisms abound in ecosystems. These mechanisms are a delight to contemplate and a major reason we enjoy studying ecosystems. Certainly, the considerations in Section 10.8 do not imply that causal explanation has no useful role in science, including ecology. That knotty conceptual problems arise when simplistic causal and mechanistic thinking is applied to complex systems is certainly not a good reason to scuttle the efforts we all make to ponder the connections among events and to attribute consequences to actions. Many important problems in ecology, such as explaining how the construction of a dam on a river will lead to massive fish kills, are best approached with traditional mechanistic, causal thinking. Thus, we return to the question posed at the beginning of this chapter: where is the boundary between the domain of macroecology in which idiosyncratic phenomena and mechanistic explanations are most appropriate and the domain of phenomena that can be corralled under a statistical 'null theory', such as METE?

I speculate here (see Table 10.2) that the boundary has to do with temporal behaviour. The critical distinction I draw is between phenomena arising in systems with constant or slowly varying state variables versus in systems for which the state variables are in flux. The former are static or quasi-static (we avoid the phrase 'in equilibrium' because no organism and no ecosystem is in thermodynamic equilibrium), the latter are dynamic.

In this view, static ecosystems, like happy families in Tolstoy's epigram, are in some fundamental sense all alike. Their SARs fall on the universal SAR curve in Figure 10.2 and their other macroecological metrics also obey the predictions of METE. No explicit mechanisms are needed to predict the forms of these metrics, but efficient theory does predict them. Such ecosystems do of course differ in the numerical values of their state variables, but that will not influence the functional form of the macroecological metrics.

Ecosystems undergoing change in response to disturbance, on the other hand, are idiosyncratic in the sense that the time-dependent probability distributions describing their state variables depend on the set of mechanisms that have disrupted them. And those distributions of

Table 10.2 *Characteristics of quasi-static and dynamically disturbed ecosystems.*

Quasi-static Systems	Dynamically Disturbed Systems
Top-down: Myriad mechanisms that need not be specified determine state variables, whose empirical values allow statistical inference of micro-level stationary stochastic patterns	Bottom-up + Top-down: Explicit causal mechanisms operating at agent-level (i.e. micro-level) govern dynamic, stochastic state variables, whose distributions allow statistical inference of micro-level non-stationary stochastic patterns
Equal fitness across species or across higher taxonomic categories	Unequal fitness as a consequence of differing taxon response to disturbance
'Laws acting around us' are inferential, statistical and result in universal patterns	Idiosyncratic outcomes result from specific mechanisms unique to the disturbance regime

state variables govern the functional forms of their macroecological metrics. Hence there is unlikely to be any universality, across disturbed ecosystems, in their SARs. Each disturbed ecosystem is displaced from universal static METE predictions in its own way.

10.10 Beyond Macroecology?

It is noteworthy that no information about organismal traits, such as specific leaf area or rooting depth of plants, colour patterns of birds or running speeds of mammals, is needed as input to METE. As a consequence, METE makes no predictions about what trait values are associated with, say, rarity, or with those species that are most responsive to, say, the onset of drought. Predictions such as those are not the domain of macroecology; rather, they require organismal-level knowledge in combination with insights from other domains of ecology such as behavioural and physiological ecology. METE can tell us what fraction of the species of plants in a meadow have fewer than, say, ten individuals, but it cannot tell us if the rare species tend to be those with red or with yellow flowers.

The question then arises: is comprehensive, unified theory possible beyond macroecology? Can, for example, information theory in the form of the MaxEnt inference principle be extended to these trait-laden domains of ecology? Or, on the other hand, are trait-based phenomena intrinsically idiosyncratic, resistant to broad theoretical generalizations of the sort that we have seen is possible in macroecology?

If we separate the question into two, distinguishing the issue of predicting trait distributions in quasi-static systems from the issue of the relationship between trait values and responses to disturbance, then informed speculation is possible. As to the first issue, there do not appear to be any intrinsic barriers to using MaxEnt inference methods to predict trait distributions over species or individuals within an ecological community and, indeed, Shipley et al. (2006) provided an early, albeit controversial (Haegeman & Loreau, 2009), example of how such an inference can be made. Given that body size is often a useful trait and given the metabolism–mass scaling pattern (Brown et al., 2004), METE's prediction of the distribution of metabolic rates can be construed as a predicted trait distribution. There is no intrinsic reason why METE cannot be extended to predict many other trait distributions.

The second issue, the influence of trait values on dynamic responses of ecosystems to disturbance, is arguably intrinsically idiosyncratic. Just as a set of explicit mechanisms appears to be necessary input to determine the time evolution of the probability distribution of the state variables in DynaMETE, connecting dynamics with traits will probably necessitate a rather considerable amount of habitat- and taxon- and stress-specific data from field observations. And that seems like a reasonable definition of the idiosyncratic.

I conclude that Tolstoy's epigram on families may provide an apt metaphor for ecology, extending well beyond the coarse-grained vision of the macroecologist to the keen observations of the natural historian.

References

Arfken, G. B. & Weber, H. J. (2005) *Mathematical methods for physicists*, 6th ed. Burlington, MA: Elsevier Academic Press.

Brown, J., Gillooly, J., Allen, A., Savage, V. & West, G. (2004) Toward a metabolic theory of ecology. *Ecology*, **85**, 1771–1789.

Cavender-Bares, J., Kozak, K. H., Fine, P. V. & Kembel, S. W. (2009) The merging of community ecology and phylogenetic biology. *Ecology Letters*, **12**, 693–715.

Darwin, C. (1859) *On the origin of species by means of natural selection, or the preservation of races in the struggle for life*. London: John Murray.

Dewar, R. C. & Porté, A. (2008) Statistical mechanics unifies different ecological patterns. *Journal of Theoretical Biology*, **251**, 389–403.

Elith, J., Phillips, S. J., Hastie, T., Dudík, M., Chee, Y. E. & Yates, C. J. (2011) A statistical explanation of MaxEnt for ecologists. *Diversity and Distributions*, **17**, 43–57.

Frieden, B. R. (1972) Restoring with maximum likelihood and maximum entropy. *Journal of the Optical Society of America*, **62**, 511–518.

Golan, A. (2018) *Foundations of info-metrics: Modeling, inference, and imperfect information*. Oxford: Oxford University Press.

Golan, A., Judge, G. & Miller, D. (1996) *Maximum entropy econometrics: Robust estimation with limited data*. New York: Wiley.

Graham, C. H., Parra, J. L., Rahbek, C. & McGuire, J. A. (2009) Phylogenetic structure in tropical hummingbird communities. *Proceedings of the National Academy of Sciences USA*, **106**, 19673–19678.

Green, J. L., Holmes, A. J., Westoby, M., Oliver, I., Briscoe, D., Dangerfield, M., Gillings, M. & Beattie, A. (2004) Spatial scaling of microbial eukaryote diversity. *Nature*, **430**, 135–138.

Gull, S. F. & Newton, T. J. (1986) Maximum entropy tomography. *Applied Optics*, **25**, 156–160.

Haegeman, B. & Loreau, M. (2009) Trivial and nontrivial applications of entropy maximization in ecology: A reply to Shipley. *Oikos*, **118**, 1270–1278.

Harte, J. (2011) *Maximum entropy and ecology: A theory of abundance, distribution, and energetics*. Oxford: Oxford University Press.

Harte, J. & Kitzes, J. (2014) Inferring regional-scale species diversity from small-plot censuses. *PLoS One*, **10**, e0117527.

Harte, J. & Newman, E. A. (2014) Maximum information entropy: A foundation for ecological theory. *Trends in Ecology & Evolution*, **29**, 384–389.

Harte, J., Kitzes, J., Newman, E. & Rominger, A. (2013) Taxon categories and the universal species–area relationship. *The American Naturalist*, **181**, 282–287.

Harte, J., Rominger, A. & Zhang, Y. (2015) Extending the maximum entropy theory of ecology to higher taxonomic levels. *Ecology Letters*, **18**, 1068–1077.

Harte, J., Smith, A. & Storch, D. (2009) Biodiversity scales from plots to biomes with a universal species area curve. *Ecology Letters*, **12**, 789–797.

Harte, J., Zillio, T., Conlisk, E. & Smith, A. (2008) Maximum entropy and the state variable approach to macroecology. *Ecology*, **89**, 2700–2711.

Harvey, B. J. & Holzman, B. A. (2014) Divergent successional pathways of stand development following fire in a California closed-cone pine forest. *Journal of Vegetation Science*, **25**, 88–99.

Horner-Devine, M., Lage, M., Hughes, J. & Bohannan, B. J. M. (2004) A taxa–area relationship for bacteria. *Nature*, **432**, 750–753.

Hubbell, S. P. (2001) *The unified neutral theory of biodiversity and biogeography*. Princeton, NJ: Princeton University Press.

Jaynes, E. T. (1957a) Information theory and statistical mechanics: I. *Physical Review*, **106**, 620–630.

Jaynes, E. T. (1957b) Information theory and statistical mechanics: II. *Physical Review*, **108**, 171–191.

Jaynes, E. T. (1963) Information theory and statistical mechanics. *Brandeis Summer Institute 1962, statistical physics* (ed. by K. Ford), pp. 181–218. New York: Benjamin.

Jaynes, E. T. (1968) Prior probabilities. *IEEE Transactions on Systems Science and Cybernetics*, **4**, 227–240.

Jaynes, E. T. (1979) Where do we stand on maximum entropy. *The maximum entropy principle* (ed. by R. Levine and M. Tribus), pp. 15–118. Cambridge, MA: MIT Press.

Jaynes, E. T. (1982) On the rationale of maximum entropy methods. *Proceedings of the IEEE*, **70**, 939–952.

Kempton, R. A. & Taylor, L. R. (1974) Log-series and log-normal parameters as diversity discriminants for the Lepidoptera. *Journal of Animal Ecology*, **43**, 381–399.

Kitzes, J. & Wilber, M. (2016) macroeco: Reproducible ecological pattern analysis in Python. *Ecography*, **39**, 361–367.

Kraft, N. J., Cornwell, W. K., Webb, C. O. & Ackerly, D. D. (2007) Trait evolution, community assembly, and the phylogenetic structure of ecological communities. *The American Naturalist*, **170**, 271–283.

Krishnamani, R., Kumar, A. & Harte, J. (2004) Estimating species richness at large spatial scales using data from small discrete plots. *Ecography*, **27**, 637–642.

Kunin, W., Harte, J., He, F., Hui, C., Jobe, J., Ostling, A., Polce, C., Šizling, A., Smith, A., Smith, K., Smart, S., Storch, D., Tjørve, E., Ugland, K., Ulrich, W. & Varma, V. (2018) Upscaling biodiversity: Estimating the species–area relationship from small samples. *Ecological Monographs*, **88**, 170–187.

Lozano, F. & Schwartz, M. (2005) Patterns of rarity and taxonomic group size in plants. *Biological Conservation*, **126**, 146–154.

Marquet, P. A., Allen, A. P., Brown, J. H., Dunne, J. A., Enquist, B. J., Gillooly, J. F., Gowaty, P. A., Green, J. L., Harte, J., Hubbell, S. P., O'Dwyer, J., Okie, J. G., Ostling, A., Ritchie, M., Storch, D. & West, G. B. (2014) On theory in ecology. *BioScience*, **64**, 701–710.

Meshulam, L., Gauthier, J., Brody, C., Tank, D. & Bialek, W. (2017) Collective behavior of place and non-place neurons in the hippocampal network. *Neuron*, **96**, 1178–1191.

Mora, T., Walczak, A., Bialek, W. & Callan, C. (2010) Maximum entropy models for antibody diversity. *Proceedings of the National Academy of Sciences USA*, **107**, 5405–5410.

Newman, E. A., Wilber, M. Q., Kopper, K. E., Moritz, M. A., Falk, D. A., McKenzie, D. & Harte, J. (2020) A comparative study of community structure metrics in a high-severity disturbance regime. *Ecosphere*, **11**(1):e3022.10.1002/ecs2.3022.

Phillips, S. J., Anderson, R. P. & Schapire, R. E. (2006) Maximum entropy modeling of species geographic distributions. *Ecological Modelling*, **190**, 231–259.

Rominger, A. J. & Merow, C. (2016) meteR: An R package for testing the maximum entropy theory of ecology. *Methods in Ecology and Evolution*, **8**, 241–247.

Rominger, A. J., Goodman, K. R., Lim, J. Y., Armstrong, E. E., Becking, L. E., Bennett, G. M., Brewer, M. S., Cotoras, D. D., Ewing, C. P., Harte, J., Martinez, N. D., O'Grady, P. M., Percy, D. M., Price, D. K., Roderick, G. K., Shaw, K. L., Valdovinos, F. S., Gruner, D. S. & Gillespie, R. G. (2016) Community assembly on isolated islands. *Global Ecology & Biogeography*, **25,** 769–780.

Rosenzweig, M. L. (1995) *Species diversity in space and time*. Cambridge: Cambridge University Press.

Roussev, V. (2010) Data fingerprinting with similarity digests. *Advances in digital forensics VI* (ed. by K.-P. Chow and S. Shenoi), pp. 207–226. Heidelberg, Germany: Springer.

Russell, B. (1912) On the notion of cause. *Proceedings of the Aristotelian Society*, **13**, 1–26.

Schwartz, M. & Simberloff, D. (2001) Taxon size predicts rates of rarity in vascular plants. *Ecology Letters*, **4**, 464–469.

Shipley, B., Ville, D. & Garnier, E. (2006) From plant traits to plant communities: A statistical mechanics approach to biodiversity. *Science*, **314**, 812–814.

Skilling, J. (1984) Theory of maximum entropy image reconstruction. *Maximum entropy and Bayesian methods in applied statistics* (ed. by J. H. Justice), pp. 156–178. Cambridge: Cambridge University Press.

Smith, F. A., Brown, J. H., Haskell, J. P., Lyons, S. K., Alroy, J., Charnov, E. L., Dayan, T., Enquist, B. J., Morgan Ernest, S. K., Hadly, E. A. & Jones, K. E. (2004) Similarity of mammalian body size across the taxonomic hierarchy and across space and time. *The American Naturalist*, **163**, 672–691.

Steinbach, P. J., Ionescu, R. & Matthews, C. R. (2002) Analysis of kinetics using a hybrid maximum-entropy/nonlinear-least-squares method: Application to protein folding. *Biophysical Journal*, **82**, 2244–2255.

Supp, S. R., Xiao, X., Ernest, S. & White, E. (2012) An experimental test of the response of macroecological patterns to altered species interactions. *Ecology*, **93**, 2505–2511.

Swenson, N. G., Enquist, B. J., Pither, J., Thompson, J. & Zimmerman, J. K. (2006) The problem and promise of scale dependency in community phylogenetics. *Ecology*, **87**, 2418–2424.

ter Steege, H., Pitman, N. C. A., Sabatier, D., Baraloto, C., Salomão, R. P., Guevara, J. E., Phillips, O. L., Castilho, C. V., Magnusson, W. E., Molino, J.-F., Monteagudo, A., Núñez Vargas, P., Montero, J. C., Feldpausch, T. R., Coronado, E. N. H., Killeen, T. J., Mostacedo, B., Vasquez, R., Assis, R. L., Terborgh, J., Wittmann, F., Andrade, A., Laurance, W. F., Laurance, S. G. W., Marimon, B. S., Marimon, B.-H., Guimarães Vieira, I. C., Amaral, I. L., Brienen, R., Castellanos, H., Cárdenas López, D., Duivenvoorden, J. F., Mogollón, H. F., Matos, F. D. d. A., Dávila, N., García-Villacorta, R., Stevenson Diaz, P. R., Costa, F., Emilio, T., Levis, C., Schietti, J., Souza, P., Alonso, A., Dallmeier, F., Montoya, A. J. D., Fernandez Piedade, M. T., Araujo-Murakami, A., Arroyo, L., Gribel, R., Fine, P. V. A., Peres, C. A., Toledo, M., Aymard, C. G. A., Baker, T. R., Cerón, C., Engel, J., Henkel, T. W., Maas, P., Petronelli, P., Stropp, J., Zartman, C. E., Daly, D., Neill, D., Silveira, M., Paredes, M. R., Chave, J., Lima Filho, D. d. A., Jørgensen, P. M., Fuentes, A., Schöngart, J., Cornejo Valverde, F., Di Fiore, A., Jimenez, E. M., Peñuela Mora, M. C., Phillips, J. F., Rivas, G., van Andel, T. R., von Hildebrand, P., Hoffman, B., Zent, E. L., Malhi, Y., Prieto, A., Rudas, A., Ruschell, A. R., Silva, N., Vos, V., Zent, S., Oliveira, A. A., Schutz, A. C., Gonzales, T., Trindade Nascimento, M., Ramirez-Angulo, H., Sierra, R., Tirado, M., Umaña Medina, M. N., van der Heijden, G., Vela, C. I. A., Vilanova Torre, E., Vriesendorp, C., Wang, O., Young, K. R., Baider, C., Balslev, H., Ferreira, C., Mesones, I., Torres-Lezama, A., Urrego Giraldo, L. E., Zagt, R., Alexiades, M. N., Hernandez, L., Huamantupa-Chuquimaco, I., Milliken, W., Palacios Cuenca, W., Pauletto, D., Valderrama Sandoval, E., Valenzuela Gamarra, L., Dexter, K. G., Feeley, K., Lopez-Gonzalez, G. & Silman, M. R. (2013) Hyperdominance in the Amazonian tree flora. *Science*, **342**, 1243092.

Tolstoy, L. (1952) *Anna Karenina* (trans by C. Garnett), p. 1. New York: Heritage Press.

Webb, C. O., Ackerly, D. D., McPeek, M. A. & Donoghue, M. J. (2002) Phylogenies and community ecology. *Annual Review of Ecology and Systematics*, **33**, 475–505.

White, E. P., Thibault, K. & Xiao, X. (2012) Characterizing species abundance distributions across taxa and ecosystems using a simple maximum entropy model. *Ecology*, **93**, 1772–1778.

Williams, R. J. (2010) Simple MaxEnt models explain food web degree distributions. *Theoretical Ecology*, **3**, 45–52.

Xiao, X., McGlinn, D. & White, E. (2015) A strong test of the maximum entropy theory of ecology. *The American Naturalist*, **185**, E70–E80.

11 · *The Species–Area Relationships of Ecological Neutral Theory*

JAMES ROSINDELL AND
RYAN A. CHISHOLM

11.1 Introduction

Neutral models give a baseline expectation for ecological patterns in the absence of significant selection or niche processes (Hubbell, 1997, 2001; Bell, 2001). They are stochastic models based on the key processes of ecological drift, speciation and dispersal limitation. The 'neutrality assumption' of all neutral models is that the fitness of an individual (defined by its prospects of reproduction and death) is unconnected with its species identity. This rules out straightforward application to entire communities of species at different trophic levels (but see Chapter 12); neutral theory is most often applied to a single 'guild' of similar species, such as tropical forest trees. Nevertheless, the neutrality assumption is a strong assumption that has attracted criticism (Ricklefs, 2006; Clark, 2009, 2012). In our view it is most productive to take this at face value: it is just an assumption in a model and not a claim that the real world is truly neutral, even if the model does happen to fit some empirical data rather well (Rosindell et al., 2012a). The classic use case of neutral theory in ecology was related to commonness and rarity of species, as captured by species abundance distributions (Hubbell, 2001; McGill, 2003; Volkov et al., 2003; Etienne & Alonso, 2005). However, species–area relationships (SARs) have also formed an important part of the research around neutral models in ecology (Durrett & Levin, 1996; Zillio et al., 2005; Rosindell & Cornell, 2007; Pigolotti & Cencini, 2009). One advantage of the neutral theory approach, as compared to alternatives involving complicating factors such as niches and selection, is that many neutral models have analytical solutions (see Table 11.1). These solutions make it easier to fully explore the behaviour of the model and gain an intuition for how its predictions follow from its processes and assumptions. For example, we can understand how the shape of a SAR depends on the strength of dispersal and speciation, which are key parameters in many neutral models (Rosindell & Cornell, 2007; Pigolotti & Cencini, 2009).

Table 11.1 Types of neutral models and current progress in assessing the SARs they generate. Some have been solved analytically (light grey) others studied by simulation (dark grey) and others not studied at all (black). The rows correspond to increasing levels of complexity, going from top to bottom, in terms of speciation; the columns correspond to increasing levels of complexity, from left to right, in terms of spatial structure. This table serves the purpose of demonstrating the frontier of progress and where it stands today and, thus, also where future efforts could be most useful.

Spatial structure of neutral model		Non spatial (Section 11.2.1)	Spatially implicit (Section 11.2.2)	Spatially explicit with Gaussian dispersal (Section 11.2.3)	Spatially explicit with any general dispersal kernel (Section 11.2.5)
Relevance for studying which kinds of SAR (Figure 11.1)		Island SAR and continental SAR at large scales (Sections 11.2 and 11.3)	Island SAR and continental SAR at small scales (Sections 11.2 and 11.3)	Continental SAR, fragmented SAR, and fragmented sampled SAR (Sections 11.3–11.5)	Continental SAR, fragmented SAR and fragmented sampled SAR (Sections 11.3–11.5)
Level of SAR solution currently available for differing speciation modes	Point mutation speciation (Section 11.2)	Analytically solved (Section 11.2.1) (Hubbell, 2001; Vallade & Houchmandzadeh, 2003)	Analytically solved (Section 11.2.2) (Vallade & Houchmandzadeh, 2003; Etienne & Alonso, 2005)	Analytically solved (Sections 11.2.3, 11.2.4, 11.4 and 11.5) (O'Dwyer & Cornell, 2018)	Simulation only (Section 11.2.5) (Rosindell & Cornell, 2009)
	Protracted speciation (Section 11.3.2)	Analytically solved (Rosindell et al., 2010; Haegeman & Etienne, 2017)	Simulation only (Section 11.3.2) (Rosindell et al., 2010)	Simulation only (Alzate et al., 2019)	Simulation only (Alzate et al., 2019)
	Protracted speciation with gene flow (Section 11.3.2)	Simulation only (Section 11.3.2) (Rosindell & Phillimore, 2011)	Simulation only (Section 11.3.2) (Rosindell & Phillimore, 2011)	Not studied	Not studied

Neutral theory is now a rich topic with many published neutral models and many ensuing ecological and evolutionary predictions (Rosindell et al., 2011). In this chapter we focus on neutral theory's predicted SARs and their applications. We distinguish four types of SAR (see Figure 11.1): *mainland SARs* (built using sampled focal areas of different sizes that are all part of a large contiguous habitat), *island SARs* (based on the diversity of entire islands of different sizes), *fragmented SARs* (constructed on more complex fragmented landscapes) and *fragmented sampled SARs* (constructed from more complex spatially structured samples taken from a contiguous landscape). These are related to Scheiner's (2003) 'six types of species–area curves': types I, IIA (mainland SARs distinguished by approach to sampling and analysis), types IIB, IIIA, IIIB (various special cases and combinations of our fragmented sampled SAR) and type IV (island SAR); Scheiner has no type corresponding to our fragmented SARs. We discuss neutral predictions and applications of our four types of SARs. This includes examples of how linking neutral theory (and thus ecological mechanisms) to the SAR enables us to go much further than if SARs are treated phenomenologically, as is customary (e.g. with the power law SAR). We conclude the chapter with a discussion of what we consider to be the most exciting future areas for research in the field of neutral theory with respect to its predictions for SARs.

We begin with a review of the SAR shapes typically observed in nature. For mainland SARs constructed from different contiguous sample areas within a large landmass, the canonical shape is a triphasic curve (Preston, 1960; Figure 11.2A). The first *sampling* phase applies at very small scales (a handful of individuals) and is roughly linear with each additional individual included in the sample area likely representing a novel species. The second phase applies at intermediate scales (local communities to geographic regions) and is roughly a power law with exponent, z, typically in the range 0.1–0.3. The third and final phase applies at large continental scales and is roughly linear again, albeit with a shallower gradient than in the sampling phase (note that on log–log axes, as in Figure 11.2A, any linear relationship has a slope of one). The intermediate phase is the most well-known, with its characteristic power law being the default SAR model in much of the literature. Preston (1960) proposed the triphasic SAR on empirical grounds, after plotting data on bird species richness from scales of a few square metres to an entire continent; he also proposed mechanisms that could underlie the three phases. We are not aware of any subsequent studies that have

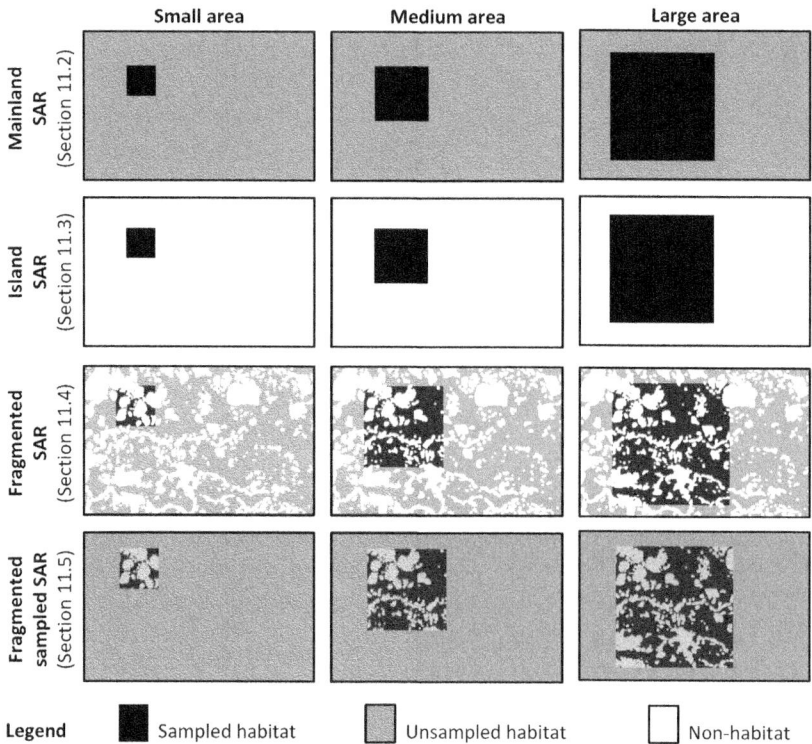

Figure 11.1 An illustration of different types of SAR. Areas of non-habitat (e.g. water or human-altered landscape) are white. Black corresponds to sampled areas of suitable habitat from which species richness, S, is to be calculated. Unsampled areas of suitable habitat are shown in grey. The three columns correspond to increasing sample area, A, from left to right (a SAR is typically plotted with S on the vertical axis and A on the horizontal axis). The four rows correspond to different types of SAR that are each covered in sections of this chapter. The fragmented landscapes here are based on image-processed satellite maps from Hansen et al. (2013), but are intended for illustration purposes only.

plotted empirical SARs over a range of scales broad enough to observe all three phases as Preston did. However, there are studies showing SARs exhibiting the transition from the second to the third phase for a variety of taxa (Storch et al., 2012) and small-scale studies often show the transition from the first to the second SAR phase (e.g. SARs for trees in the permanent forest plots of the ForestGEO network; Plotkin et al., 2000). For island SARs, the canonical shape is a generally increasing curve (Figure 11.2B; see Chapters 4 and 7). The exception is among

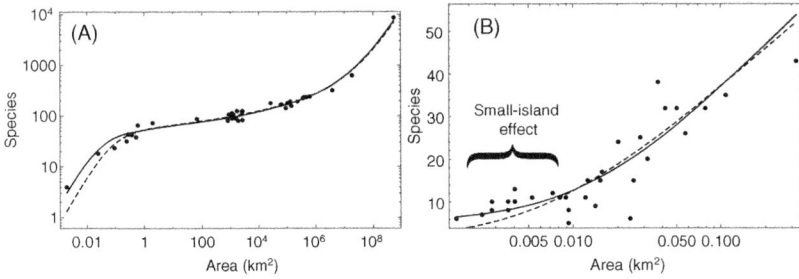

Figure 11.2 Two classic empirical SARs: (A) Preston's (1960) mainland SAR for breeding birds. Each point represents one contiguous sample area, from a small suburban plot in the north-eastern United States up to the world. The two curves show least squares fits of the theoretical SAR from a spatially explicit neutral model (Equation 11.11; Section 11.2.4): the solid curve has three free fitted parameters (standard deviation of dispersal kernel, $\sigma = 9.00$ km, per-capita speciation rate, $v = 6.12 \times 10^{-10}$, and individual density, $\rho = 1,650$ km^{-2}), the dashed curve has only two fitted parameters ($\sigma = 8.75$ km, $v = 2.70 \times 10^{-10}$, with individual density constrained at Preston's suggested empirical value of $\rho = 680$ km^{-2}). (B) Niering's (1963) island SAR for vascular plants on islands of the Kapingamarangi Atoll. Each point represents one island. The solid curve shows the fit of the theoretical SAR from the spatially implicit neutral model with two free fitted parameters (Equation 11.7; Section 11.2.2; fundamental biodiversity number, $\theta = 13.3$, per-capita immigration rate, $m = 2.35 \times 10^{-4}$, with fixed $\rho = 10^6$ km^{-2}). The dashed curve shows the fit of a niche-neutral model with three free fitted parameters (Section 11.3.1; $\theta = 15.3$, $m = 1.44 \times 10^{-4}$, and number of niches, $K = 5.3$, again with fixed $\rho = 10^6$ km^{-2}). Note the vertical axis is logarithmic in (A) and linear in (B), as is typical for mainland and island SARs, respectively.

small islands, where there is often no clear relationship between species richness and area – known as the *small-island effect* (Lomolino & Weiser, 2001; Chapter 19) (Figure 11.2B).

11.2 Mainland SARs

11.2.1 Non-spatial Neutral Model

The simplest neutral models are non-spatial, meaning there is no dispersal limitation and individuals disperse with equal ease to any location in a focal landscape. The most studied neutral models are *zero-sum*: there is a fixed community size and an increase in any one species' abundance must be exactly offset by a decrease in another species' abundance. Starting from any initial condition, the dynamics of the zero-sum model involve repeated birth–death events where one

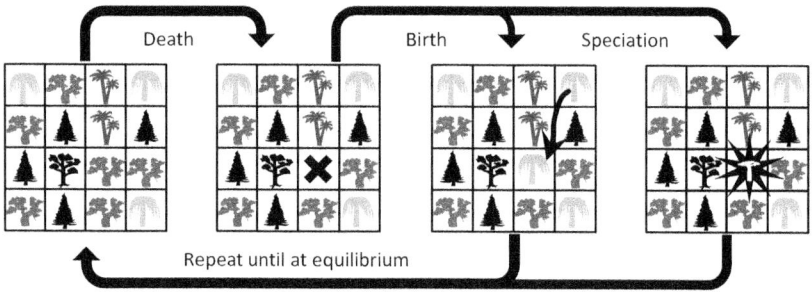

Figure 11.3 The rules for a non-spatial zero-sum neutral model with community size $J = 16$. First an individual chosen at random dies (black cross), leaving a gap in the habitat. Next, the gap is filled, either with the offspring of another individual (chosen at random according to a uniform distribution – black arrow) or by an entirely new species (speciation – black star). This process is repeated indefinitely. Eventually the system reaches a dynamic equilibrium in which speciation and extinction are on average balanced. A SAR can be constructed from equilibrium species richness versus community size (a proxy for area) across different values of the latter. Note the tree images are for illustration purposes only and are not intended to represent real world species.

individual is chosen at random to die and be replaced by the offspring of another randomly chosen individual (Figure 11.3). Species abundances thus drift randomly over time, with the eventual end state of this simplest model being extinction of all but one species. Thus, at equilibrium there is just one species in a sample area of any size, and we have the trivial species–area relationship $S(A) = 1$.

In more sophisticated versions of the zero-sum neutral model, the inevitable extinctions due to drift are balanced by occasional speciation events or by immigration events from a larger region (a *metacommunity*). The result in the long term is that the community exhibits a dynamic equilibrium in which species richness is maintained by a drift–speciation balance (Figure 11.3) or drift–immigration balance (Hubbell, 2001). We will deal with the speciation case first and the immigration case in Section 11.2.2.

In the case of a non-spatial neutral model incorporating speciation with a per-capita probability of v, the relationship between the community size (J) and the expected number of species at dynamic equilibrium (S) is given by (Etienne & Alonso, 2005):

$$S = \theta[\psi_0(\theta + J) - \psi_0(\theta)], \tag{11.1}$$

where $\theta = (J - 1)v/(1 - v)$ is a composite parameter termed the *fundamental biodiversity number* and $\psi_0(z)$ is the digamma function

(Abramowitz & Stegun, 1972). In the limit of high diversity (large θ), this becomes

$$S \sim \theta \log\left(\frac{1}{\nu}\right), \tag{11.2}$$

which is the species richness of a log–series distribution. The log–series also happens to be the exact solution to the non–zero–sum variant of the non–spatial neutral model, that is, a version of the model where death and birth events are not directly coupled (Vallade & Houchmandzadeh, 2003; Volkov et al., 2003). Equation (11.2) is a sufficiently accurate approximation to Equation (11.1) in most situations.

To develop a SAR for the non–spatial neutral model, we first relate the number of individuals to area via an individual density parameter ρ, such that $J = \rho A$. We then interpret the non–spatial model as a spatial community with no dispersal limitation (i.e. homogeneous dispersal), which means that Equation (11.1) applies. Substituting $J = \rho A$ into Equation (11.1) gives the SAR

$$S(A) = \theta(A)[\psi_0(\theta(A) + \rho A) - \psi_0(\theta(A))], \tag{11.3}$$

where $\theta(A) = (\rho A - 1)\nu/(1 - \nu)$. In the limit of high diversity, this reduces to (following Equation 11.2)

$$S(A) \sim \theta(A)\log\left(\frac{1}{\nu}\right) = \frac{(\rho A - 1)\nu}{1 - \nu}\log\left(\frac{1}{\nu}\right), \tag{11.4}$$

and, again, this is an accurate approximation of Equation (11.3) in most situations. These simple neutral SARs (Equations 11.3 and 11.4), from a non–spatial model where speciation is the source of all new diversity, can be applicable to real problems where the scale is large enough (e.g. continental scales) that most species are either entirely in a focal area or entirely outside it, so that dispersal limitation within the focal area can be ignored in the calculations. This is equivalent to assuming that speciation is a significantly more important source of new diversity than immigration. Equation (11.4) predicts a linear relationship between area and species richness, which is consistent with the third, large–scale, phase of the triphasic empirical SAR observed by Preston (1960).

11.2.2 Spatially Implicit Neutral Model

The model described in Section 11.2.1 is for a single well-mixed meta-community that ignores immigration and it is therefore most applicable at large scales. At smaller scales, where immigration from outside the focal community is more important than speciation as a source of diversity, the model can be adapted by introducing a new parameter, m, the per-capita immigration rate, which is the probability that a new individual is the offspring of a parent outside the focal community rather than of a parent within the focal community. The focal community in this model is termed the *local community* and the immigrants come from the much larger metacommunity, which is usually assumed to follow the speciation–extinction dynamics outlined in the previous section and to be static on time scales relevant to the local community (Hubbell, 2001). In the local community, speciation is assumed to be negligible in comparison to immigration and is thus ignored. This model is referred to as the *spatially implicit neutral model*, because the dispersal limitation implicit in the separation between the local community and the metacommunity introduces a weak notion of space.

Analytical solutions exist for the SAR in the spatially implicit neutral model. If it is assumed that the metacommunity species abundance distribution is log-series with parameters θ and ν (consistent with Equation 11.2) and the per-capita rate of immigration is m, with immigrants being taken at random from the metacommunity, then the expected species richness S of the local community of size J at the dynamic equilibrium is approximately (Chisholm et al., 2016):

$$S = \theta[\psi_0(\theta + \gamma(\psi_0(\gamma + J) - \psi_0(\gamma))) - \psi_0(\theta)], \qquad (11.5)$$

where $\gamma = (J - 1)m/(1 - m)$ and $\theta = (J_M - 1)\nu/(1 - \nu)$ (this is the same definition of θ as before, except now we denote the metacommunity size by J_M). In the limit of high diversity, Equation (11.5) becomes (Volkov et al., 2007):

$$S \sim \theta\left[\log\left(1 - \frac{\gamma}{\theta}\log(m)\right)\right]. \qquad (11.6)$$

We can convert Equation (11.5) into a SAR by again considering the area-dependence of its parameters. As in Section 11.2.1, we have $J = \rho A$, but now we must also consider the area-dependence of the per-capita immigration rate m, leading to the equation

$$S(A) = \theta[\psi_0(\theta + \gamma(A)(\psi_0(\gamma(A) + \rho A) - \psi_0(\gamma(A)))) - \psi_0(\theta)], \qquad (11.7)$$

where $\gamma(A) = (\rho A - 1)m(A)/(1 - m(A))$ and $\theta = (J_M - 1)\nu/(1 - \nu)$. The total average number of immigrants per generation is $J \times m(A) = \rho A m(A)$. Biologically justifiable choices for how the total number of immigrants scales with area include linear scaling with area (leading to $m(A) = k$ for some constant k) or linear scaling with the perimeter of the local community (leading to $m(A) = k/\sqrt{A}$) (Chisholm & Lichstein, 2009; Chisholm et al., 2016). Note that θ is not a function of area in Equation (11.7) because it represents metacommunity diversity and we are now treating the metacommunity as a fixed entity, only varying the area of the local community, unlike in Equation (11.4) where the metacommunity area was the subject of interest.

In the limit of high diversity, we have (from Equation 11.6)

$$S(A) \sim \theta \log \left(1 - \frac{\gamma(A)}{\theta} \log \left(m(A) \right) \right). \tag{11.8}$$

In the limit of small area A, Equation (11.7) tends to one, representing a single species, while Equation (11.8) tends to zero, emphasizing that Equation (11.8) is a valid limit only in the high-diversity case. For areas that are small but not trivially so, the behaviour of the SARs from both Equations (11.7) and (11.8) is approximately linear, consistent with the first *sampling* phase of Preston's (1960) classic empirical SAR. For still larger areas, the equations give decelerating SARs, qualitatively consistent with the second phase of Preston's empirical SAR, although the functional form of Equation (11.8) in this limit is not a power law.

An interesting special case of Equation (11.8) occurs when we take the limit $m \rightarrow 1$, representing either a local community experiencing very high immigration (i.e. almost all propagules coming from outside) or simply a random sample of individuals from the large metacommunity. Mathematically, in this limit we obtain

$$S(A) \sim \theta \log \left(1 + \frac{\rho A - 1}{\theta} \right), \tag{11.9}$$

which is consistent with the result obtained by He and Legendre (2002) for randomly sampling the log-series distribution.

Typical values for the immigration parameter m in the spatially implicit neutral model are on the order of 0.1 for a 50 ha forest plot, higher for smaller areas or better-dispersing species and lower for larger areas or more poorly dispersing species (Chisholm & Lichstein, 2009). Typical values for θ are on the order of 1–100, with larger values corresponding to more diverse metacommunities.

One limitation of the spatially implicit model is that the spatial structure of most real communities is not easily shoehorned into the model's local community versus metacommunity paradigm. An example is the 50 ha forest plot on Barro Colorado Island in Panama. In applications of neutral theory, the 50 ha plot is treated as the local community and some abstract larger region of forest outside the plot is treated as the meta-community (He, 2005). Barro Colorado Island itself is over 1,500 ha in size and the boundary of the 50 ha plot within it is arbitrary. Species see no change and experience no natural barriers as they cross the boundary of the 50 ha plot. As a result, the boundary is not ecologically meaningful and the distinction between the local community and metacommunity is made only for convenience rather than to capture something real. The fully spatially explicit neutral models we will discuss later do not suffer these limitations.

Summarizing this section and the previous one, we see that, despite the spatially implicit neutral model's simple structure, it yields SARs that arguably capture the essential mechanisms of real SARs at very large scales (the metacommunity SAR given by Equations 11.3 and 11.4) and very small scales (the local community SAR given by Equations 11.7 and 11.8). The reason we can get away without complex spatial structure at these scales is that i) at very small scales, as studied in this section, dispersal limitation is not important within the community because species can easily disperse throughout the whole community; and ii) at very large scales, as noted in Section 11.2.1, the vast majority of dispersal events occur within the community, without crossing the boundary, thus circumventing the need to model dispersal explicitly.

11.2.3 Spatially Explicit Neutral Models

As noted in Section 11.1, the canonical mainland SAR is a triphasic curve, with the middle phase being roughly a power law $S(A) = kA^z$ with z in the range 0.1–0.3. The non-spatial and spatially implicit models of Subsections 11.2.1 and 11.2.2 reproduce the third and first phases of the triphasic SAR, respectively, but do not capture the second phase where interaction between dispersal and the spatial scale of the focal area is important. Modelling the second phase is particularly important because it applies over many scales relevant to ecology and conservation. This points to the need for spatially explicit models.

Spatially explicit neutral models retain the key principle of neutral theory that fitness is independent of species labels. Unlike the non-spatial neutral models, however, each individual now has an explicit position in space, usually imagined as a single cell on a large square grid. Individuals die and are replaced as in the non-spatial model, but now remaining individuals are not equal in their probabilities of filling the space with their offspring; instead, the probabilities depend on spatial proximity to the vacated space and are described by a probability distribution known as a dispersal kernel. The most elementary case, known as the voter model (Holley & Liggett, 1975), allows dispersal only between nearest neighbour cells in the habitat. Though the voter model is a neutral model, it predates contemporary use of neutral theory in ecology and was originally conceived as a model of how political views (votes) might spread and compete in space. More germane to ecological settings, where dispersal is not typically so restrictive, is the case with a dispersal kernel given by a two-dimensional normal distribution (or fat tailed distribution; see Section 11.2.5). Such spatially explicit neutral models open up the possibility of capturing the complete triphasic SARs at all spatial scales and linking the three phases back to elementary processes such as dispersal limitation (Rosindell & Cornell, 2007).

There are two clearly distinct 'types' of area parameter that must be distinguished in order to study the fully spatially explicit model. First, there is the landscape area, which can be interpreted as the size of the whole continent or the 'world' of the model. Second, there is the (possibly much smaller) survey area describing the area of the region sampled from the model 'world'. The distinction between these two types of area is reflected in the contrast between island SARs (survey area = landscape area; landscape area is varied) and mainland SARs (survey area < landscape area; survey area is varied and landscape area is very large) (Figure 11.1). Spatially explicit, mechanistic models provide the tools necessary to separate landscape area from survey area more clearly, but a cost of this is the complexity brought about by having two area parameters. An elegant way to avoid this for studying mainland SARs is to assume the landscape area is sufficiently large that it can be approximated as infinite. The complicating effects of the boundaries of the model 'world' then disappear; instead, the effects of dispersal, speciation and survey area alone drive the SAR. An infinite-area model can be treated analytically. It can also be simulated with an approach known as coalescence (Kingman, 1982), which begins with a finite sample from the simulated world and works backwards in time (rather than forwards) tracing only the (finite) historic model components that can actually influence the outcome for the sample of interest set within an effectively infinite area. This coalescence method is well suited to neutral models, but does not transfer well to more complex models where myriad factors, other than distance, affect the probability that an individual will be the parent of a new individual at a particular location (Rosindell et al., 2008).

Now let us consider the mainland SARs predicted by the spatially explicit neutral model with a normal dispersal kernel. In general, the species richness S depends on five parameters: landscape area A_L, survey area A, density of individuals per unit area ρ, per-capita speciation rate ν and the variance of the dispersal kernel σ^2 (which governs dispersal limitation). Understanding the full behaviour of a many-parameter model can be challenging due to the size of parameter space requiring investigation. As described in the preceding paragraph, we can approximate a large landscape by an infinite landscape ($A_L \to \infty$) for the case of mainland SARs. Also, without loss of generality, we can assume units of distance and area are defined such that $\rho = 1$, removing another parameter and leaving a three-parameter function of interest $S(A, \nu, \sigma^2)$. Provided that dispersal is not very restricted, it turns out that the SAR predicted by this spatially explicit neutral model can be written in terms of a simpler two-parameter function (whose mathematical form we discuss later) (Rosindell & Cornell, 2007):

$$S\left(A, \nu, \sigma^2\right) \sim \sigma^2\, \Psi\left(\frac{A}{\sigma^2}, \nu\right). \tag{11.10}$$

This *scaling collapse* to a two-parameter function Ψ shows that what really matters for the shape of the SAR is not survey area alone, but the ratio of survey area to dispersal kernel variance: how far an individual's offspring travel in one generation compared to the distance across the survey area. To see intuitively why this scaling works, imagine the entire model dynamics taking place on a stretchable sheet of elastic. If we stretch the elastic and allow the species' dispersal ability to stretch proportionately, exactly the same dynamics occur but over a larger area A and with a larger dispersal distance σ^2, but without changing the ratio between them. Stretching does decrease the density of individuals, however. To get back to the original density of individuals we must increase the density by a factor equal to the stretching factor, which also increases species richness by the same factor. The effect of increasing the density to compensate for the stretching is captured by the multiplying factor of σ^2 in Equation (11.10). We stress that this result has been confirmed only for the specific spatially explicit neutral model described here; though it does apply more broadly (Rosindell & Cornell, 2009), it does not apply to every conceivable SAR model with these three parameters.

One of neutral theory's signature achievements was the discovery that the SARs produced by Equation (11.10) are triphasic and broadly consistent with Preston's observations from half a century earlier (Rosindell & Cornell, 2007). In particular, when the survey area is very large compared to the dispersal distance (specifically, $A \gg \sigma^2/\nu$) the spatially explicit SAR reduces to the non-spatial result in Equation (11.1), which captures the final phase of the triphasic SAR. This highlights how, in spatially explicit neutral theory, the global number of individuals representing a given species generally follows the same random walk dynamics as in the non-spatial model. However, in the spatially explicit model these individuals are distributed across space in a clustered way governed by the dispersal process; the full triphasic SAR emerges based on which individuals fall in the survey area. To the best of our knowledge this spatially explicit neutral model (including a range of minor variations on it) remains the only quantitative, mechanistic model that predicts a full triphasic SAR. Furthermore, the neutral model provides an excellent quantitative fit to the few empirical data that have been collated over sufficiently wide-ranging spatial scales (Figure 11.2A).

11.2.4 The Preston Function

In Equation (11.10), we defined a two-parameter function Ψ that characterizes the SAR of the spatially explicit model in a succinct manner. This function turns out to be enormously useful in spatial neutral ecology (Chisholm et al., 2018) and, thus, following a tradition in physics for naming *special functions*, we dub Ψ the *Preston function* in honour of that scientist's ground-breaking work in this field. The precise definition of the Preston function is the SAR for contiguous circular sample areas taken from an infinite world whose dynamics follow a non-zero-sum version of the spatial neutral model described in Section 11.2.3, with a bivariate normal dispersal kernel having $\sigma = 1$. In a non-zero-sum model births and deaths are not rigidly coupled, in contrast to the zero-sum model where a birth immediately follows a preceding death. Slight fluctuations in the local density of individuals are thus possible in non-zero-sum models, but do not meaningfully affect the resulting SAR. We choose to define the Preston function based on a non-zero-sum model, as this facilitates analytical treatment (Chisholm et al., 2018; O'Dwyer & Cornell, 2018).

Defining the Preston function enables the methods used for evaluating the function (simulation, analytical approximations, efficient algorithms, etc.) (Rosindell et al., 2008; O'Dwyer & Green, 2010; O'Dwyer & Cornell, 2018) to be abstracted away from the function's applications (Chisholm et al., 2018; Thompson et al., 2019). A mathematician or computer scientist could thus be motivated to study ways of evaluating the function faster and more accurately, knowing that it has applications in ecology, whilst an ecologist may study ways of using the function as a 'black box' without troubling to understand how it is evaluated; in much the same way as one does not typically trouble to understand what goes on inside a computer when taking the logarithm of a number, but one knows how to apply logarithms as useful tools without this knowledge. We will later return to further applications of the Preston function.

To evaluate the Preston function, one can simply use coalescence-based simulations of the spatial neutral model but this can be slow and, fortunately, an approximate analytical solution to the Preston function was recently published (O'Dwyer & Cornell, 2018):

$$
\Psi(A^*, \nu) \approx \nu_{eff} A^* + \cfrac{2\pi\sqrt{\dfrac{A^*}{\pi}}\left(1 - \nu_{eff}\right)I_1\left(\sqrt{\dfrac{A^*}{\pi}}\right)}{\dfrac{1}{\sqrt{\nu_{eff}}}I_1\left(\sqrt{\dfrac{A^*}{\pi}}\right)\dfrac{K_0\left(\sqrt{\dfrac{\nu_{eff}A^*}{\pi}}\right)}{K_1\left(\sqrt{\dfrac{\nu_{eff}A^*}{\pi}}\right)} + I_0\left(\sqrt{\dfrac{A^*}{\pi}}\right)},
$$

(11.11)

where $A^* = A/\sigma^2$, $\nu_{eff} = \nu \log(1/\nu)/(1 - \nu)$ and $I_i(z)$ and $K_i(z)$ are the modified Bessel functions of the first and second kinds. In most circumstances, this formula provides a way to compute the triphasic SAR in a mechanistic model quickly and accurately (Figure 11.2A).

11.2.5 Effects of Long-Distance Dispersal

All the results presented for spatially explicit models in Sections 11.2.3 and 11.2.4 assume a normally distributed dispersal kernel. Indeed, we specifically defined the Preston function to correspond to the case of a normal kernel. However, one can also consider spatially explicit neutral models with different dispersal kernels. The results involving

Preston functions such as Equation (11.10) are to some extent robust to dispersal kernels other than a normally distributed one. For example, a 'square' dispersal kernel where dispersal occurs according to a uniform distribution in a square region of a given width produces almost exactly the same SAR as the normally distributed kernel, meaning that Equation (11.10) can again be used for this case, providing that σ^2 is correctly calculated as the variance of the square kernel (Rosindell & Cornell, 2007).

In empirical studies, 'fat-tailed' dispersal kernels are often considered more realistic than normally distributed ones (Clark et al., 1999). Such dispersal kernels increase the probability of rare yet extreme dispersal events over very long distances. In the tails of the kernel, the probability of dispersal follows a power law with dispersal distance. For fat-tailed dispersal kernels, higher moments are not well defined and in the most extreme cases even the mean and variance are undefined, that is, the mean and variance of a sample of dispersal distances drawn from the kernel do not converge to a finite value as the sample size is increased to infinity. Fat-tailed dispersal kernels in spatially explicit neutral theory also result in triphasic SARs, but the second phase, corresponding to the classic power law SAR, is much broader (Rosindell & Cornell, 2009; Figure 11.4). The tendency of fat-tailed kernels to yield power law SARs over a broader range of spatial scales may partly explain the preponderance of power law SARs in empirical data.

At small spatial scales, the SARs produced by a spatially explicit neutral model with fat-tailed dispersal look almost identical to those produced with a normal dispersal kernel with a suitably elevated speciation rate. The ecological intuition for this result is that occasional and rare dispersal events over very long distances have a similar effect on a local scale community as would point speciation (Rosindell & Cornell, 2009; Figure 11.4). Indeed, a component of the point speciation process in neutral models could be interpreted as a surrogate for long-distance dispersal events rather than speciation per se. One consequence of this small-scale qualitative similarity of SARs is that a suitably rescaled Preston function could be used to predict SARs for systems with fat-tailed dispersal, but only at local scales. An analytical solution to the complete SAR resulting from a spatially explicit neutral model with fat-tailed dispersal remains an open problem and, though challenging, would represent a very worthwhile area for future study.

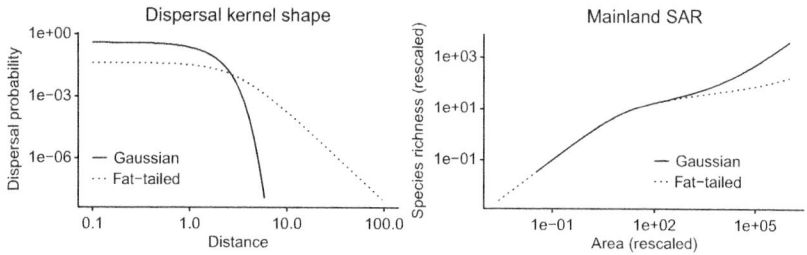

Figure 11.4 Comparison between dispersal kernels and their corresponding mainland SARs. Left: a Gaussian (normal distributed) dispersal kernel with variance $\sigma^2 = 1$ and a fat-tailed dispersal kernel given by $K(x) = \frac{2.4}{2\pi\sigma^2}\left(1 + \frac{x^2}{\sigma^2}\right)^{-2.2}$ with $\sigma^2 = 3$ after Clark et al. (1999). Note the behaviour around the tails of each dispersal kernel. Right: the SAR for a Gaussian dispersal kernel drawn with $\nu = 0.0004$ and the SAR for a fat-tailed dispersal kernel with $\nu = 0.000002$ rescaled to coincide with the Gaussian SAR during the first SAR phase. This figure incorporates a broad range of values for σ^2 rescaled by dividing both axes by σ^2 (see Equation 11.10). Based on the data in figure 3 of Rosindell and Cornell (2009)

11.3 Island SARs

In this section we are concerned with using neutral models to predict SARs for island archipelagos, where each data point comprises the area of an island and the species richness of a focal guild on that island. Such SARs have been extensively studied in island biogeography theory (MacArthur & Wilson, 1963, 1967; see Chapters 3 and 15).

11.3.1 Spatially Implicit Neutral Model and Island SARs

The spatially implicit model presented in Section 11.2.1 is particularly appropriate for island communities. Indeed, it is arguably more appropriate for island communities than for the mainland communities to which it was originally applied. This is because, while the local versus metacommunity separation is artificial for most mainland communities, it is completely natural for island communities: the community on the island itself becomes the local community, while the communities elsewhere become the metacommunity. The formulas for the spatially implicit neutral model in Section 11.2.1, presented in the context of mainland SARs, are therefore applicable to the case of islands too. When the model is applied to island data, the fitted immigration rates are typically orders of magnitude lower than for mainland communities, reflecting the greater isolation of islands (Volkov et al., 2007).

The SAR is thus again described by Equations (11.7) and (11.8), but with lower values of m. This results in a curve that predicts just one species (Equation 11.7) or zero species (Equation 11.8) for small islands, but for large islands the curve ticks upwards and produces an increasing SAR consistent with classic island biogeography theory (MacArthur & Wilson, 1967).

Island SARs are sometimes plotted on linear–log axes, with S on the vertical axis and $\log A$ on the horizontal axis, because variation in S is not as great as in mainland contexts. Plotted in this way, the gradient of the spatially implicit neutral SAR for large areas is proportional to the fundamental biodiversity number of the metacommunity, θ, that is,

$$\frac{dS(A)}{d \log A} \rightarrow (1 - c)\theta, \tag{11.12}$$

where c is the scaling exponent of the immigration rate with area, that is, $m(A) = k/A^c$ (with $c = 0$ and $c = 1/2$ being two biologically justifiable choices, as noted in Section 11.2.2).

Many empirical island SARs exhibit a phenomenon known as the *small-island effect*, whereby species richness for small islands is independent of area (Lomolino & Weiser, 2001; Chapter 19) due to a variety of possible reasons. This can be captured by adding a coarse niche structure to the spatially implicit model (after which the model is, by definition, no longer neutral). If an island has n independent non-overlapping equal-sized niches and the community within each niche undergoes neutral drift governed by the spatially implicit neutral model, then the island SAR tends to a species richness of $S = n$ (instead of $S = 1$) as $A \rightarrow 0$, reproducing the small-island effect and providing an excellent fit to archipelago data (Figure 11.2B). The biological interpretation of this result is that equilibrium diversity is maintained by niches on small islands but by an immigration–extinction balance – with many species in each niche – on large islands (Chisholm et al., 2016).

11.3.2 Speciation on Islands

Of particular interest in island biogeography are islands that support endemic species. The speciation process that gives rise to these endemic species is reflected in classic neutral theory, where it is essential to counterbalance inevitable extinctions of species over long timescales at large spatial scales (Hubbell, 2001). Thus, neutral theory may be suitable as a means to study broad patterns of endemic and non-endemic species on islands. Many endemic species arise through *anagenetic* speciation: an anagenetic species is one that immigrated to the island and then became distinct from the mainland population from which it originated. Anagenetic species do not strongly influence the overall species richness on an island because they represent species that would have been present anyway but have just become distinct from their mainland counterparts. However, the number of endemic species is of general interest from an evolutionary and also a conservation perspective. The SAR on islands could thus be considered in a disaggregated form in which the total species richness is separated into contributions from endemic species and non-endemic *immigrant* species that also exist in living mainland populations (Rosindell & Phillimore, 2011). On some islands there are *cladogenetic species*, endemics that evolve in situ on an island (or archipelago), with their close relatives also endemic to the island (Heaney, 2000). Cladogenetic species do increase the total species richness on islands, with endemic radiations sometimes producing hundreds of related species, all founded by a single immigration event from the mainland.

Neutral models can be applied to mechanistically predict SARs on islands, as well as how these SARs decompose into contributions from immigrant, anagenetic and cladogenetic species. In order to be suitable for this purpose, neutral theory requires a more complex view of speciation. In particular, neutral theory can be extended to include protracted speciation, where speciation takes time, rather than being an instantaneous event (Rosindell et al., 2010). It can also incorporate the idea that repeated immigration events of the same species from the mainland slow down the protracted speciation process in which immigrant species become distinct from their mainland cousins (Rosindell & Phillimore, 2011). The shape of the resulting SAR and its composition in terms of immigrant, anagenetic and cladogenetic species, is heavily dependent on the level of isolation of the island as well as its area (Figure 11.5). On islands close to the mainland with high levels of immigration, SARs take the form of species accumulation curves consisting of species sampled from the mainland. On islands further away from the mainland, SARs follow essentially the same pattern as described in Section 11.3.1 in terms of total species richness, but when we consider contributions to this from immigrant, anagenetic and cladogenetic species, there is a clear pattern of increasing endemism further away from the mainland. Anagenetic species in particular are only prevalent on islands with intermediate isolation, sufficiently far away to prevent too many repeated immigration events that impede speciation, but close enough that immigrants arrive

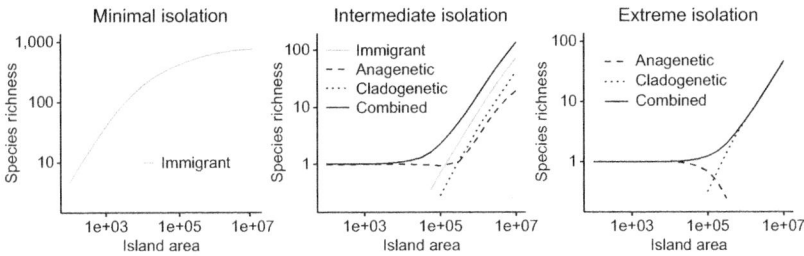

Figure 11.5 Species–area relationships (SARs) on islands with different levels of isolation. In all panels the lines show SARs considering just a subset of the species: immigrant species (grey), anagenetic species (dashed) and cladogenetic species (dotted). The black line shows the total SAR from immigrant, anagenetic and cladogenetic species combined. This figure was replotted using data from Rosindell and Phillimore (2011, figure 2). Note the combined curve is not shown in the minimal isolation case where all species are immigrant species.

comparatively often, each with the potential to become an anagenetic species (Rosindell & Phillimore, 2011).

11.4 Fragmented SARs

Fragmented SARs are more complex than both island SARs and mainland SARs because they involve spatially explicit patterns of underlying habitat. In fragmented habitats, we can again plot a SAR as species richness versus sampled habitat area (Figure 11.1), but there may now be substantial scatter in the graph if different sample areas have different levels of fragmentation. Extensions to the power law SAR incorporating a further parameter for fragmentation have been proposed and illustrated with metapopulation-based simulations (Hanski et al., 2013). The primary challenge, however, becomes one of developing appropriate fragmentation metrics (McGarigal & Marks, 1995) that can account for any variation in species richness that is not attributable to area.

The fragmented SAR produces an estimate of the species richness in a spatially explicit landscape that has been fragmented into regions of habitat and non-habitat for long enough to reach an equilibrium level of species richness that reflects the pattern of fragmentation and will remain stable in the long term provided the habitat is not altered further. The species richness of an arbitrary fragmented landscape in which species obey neutral dynamics can be simulated using coalescence, but can also be treated analytically. To do so requires making two assumptions: first, that the pattern of fragmentation surrounding the focal area is similar to that within the focal area; second, that there are no habitat fragments that are sufficiently isolated as to eventually develop their own endemic species. One might assume that any analytical solution for the general fragmented SAR would depend on multiple independent geometrical parameters of the fragmented landscape, for example average patch size, variance in patch size, edge-to-area ratio and potentially a large number of others. Surprisingly, an excellent analytical approximation can be obtained (Thompson et al., 2019) with just two compound parameters that can readily be calculated from the landscape fragmentation map: effective area and effective connectivity.

'Effective area' (A_e) is defined as the number of individual organisms of any species from the focal guild of interest that occupy the focal region. Thus, a given landscape will have a greater effective area for a guild whose component species live at higher densities. The 'effective connectivity of a cell' within the landscape is defined as the mean distance travelled per generation in the (possibly fragmented) landscape for an individual starting at that cell, taken as the mean over many successive hops in the landscape. If a single cell is empty of habitat, its effective connectivity is zero. The 'effective connectivity of the landscape' as a whole c_e is the root-mean-square value of effective connectivity across all cells in the landscape. The fragmented SAR can then be written as

$$\hat{S} = c_e^{\,2}\, \Psi\!\left(\frac{A_e}{c_e^{\,2}}, \nu\right), \tag{11.13}$$

where Ψ represents the Preston function, ν the per-capita speciation rate as before, A_e the effective area and c_e the effective connectivity. This result may appear similar to Equation (11.10), but we emphasize the key difference, which is that, in place of the straightforward parameter σ^2 representing the variance of the dispersal kernel, we now have the parameter c_e^2, which is dependent on both σ^2 and geometric properties of the fragmented landscape.

11.5 Fragmented Sampled SARs

Fragmented sampled SARs are defined on contiguous underlying landscapes, based on the spatially explicit patterns of sampling these landscapes. They are, in effect, mainland SARs where the sampling region is not a square or another simple shape. The general fragmented sampled SAR for a given set of parameter values can be computed via coalescence simulations, like the fragmented SAR, but limited analytical results are also available. Currently, formulas exist only for upper and lower bounds on the species richness and thus on the fragmented sampled SAR (Chisholm et al., 2018):

$$\sigma^2 \Psi\left(\frac{A_e}{\sigma^2}, \nu\right) \leq S \leq \sigma^2 \Psi\left(\frac{A_e}{\sigma^2}, 1 - (1 - \nu)^{\frac{A_{max}}{A_e}}\right), \qquad (11.14)$$

where S is the sampled species richness (vertical axis of the SAR), A_e is the effective area (horizontal axis of the SAR), Ψ is the Preston function, ν is the per-capita speciation rate, σ^2 is the variance of the dispersal kernel and $A_{max} \geq A_e$ is the largest possible effective area of the sample (i.e. the effective area of the focal landscape across which the sample is distributed). If $A_{max} = A_e$, this means that 100 per cent of the focal landscape is sampled and so there is no fragmentation and the original mainland SAR is recovered (the lower and upper bounds in Equation 11.14 reduce to the same value in this case). The upper bound for S (right-hand inequality in Equation 11.14) corresponds to the case where the fixed sample area is spread out as much as possible across the focal landscape, capturing as much as possible of the underlying beta diversity within the landscape (technically, individuals should be sampled in a spatially random manner). The lower bound for S corresponds to the case where the fixed sample area is aggregated into one contiguous clump, capturing as little as possible of the beta diversity across the landscape.

11.6 Applications

Neutral models make strong assumptions and so there will always be caveats associated with applying them. Nevertheless, we have seen that neutral models can qualitatively and quantitatively reproduce empirical SARs observed in nature, such as the power law, Preston's triphasic SAR (Figure 11.2A) and island SARs (Figure 11.2B). From this follows a number of exciting applications in fields where other quantitative mechanistic approaches are either non-existent, infeasible or make equally strong assumptions as neutral theory. In this section we will explore applications of neutral theory SARs in particular.

11.6.1 Predicting Diversity at One Scale based on Another

Often we have information on species diversity at one spatial scale and wish to use this to extrapolate richness at some other scale. For example, 'down-scaling' to estimate local species richness from a regional scale species list or 'up-scaling' to estimate regional species richness from smaller scale local surveys. In a recent down-scaling application, the spatially explicit neutral model with Gaussian dispersal was parameterized at a regional scale for a tropical forest tree community in Panama. It was then used to predict the local species diversity in a 50 ha plot. The result was underprediction by a factor of approximately two (O'Dwyer & Cornell, 2018). However, this study used a value of individual tree density of 0.04 per m^2, which is probably too high if we are considering only reproductively mature trees (Wright et al., 2005), as should be the case in applications of neutral models. Repeating the calculations of O'Dwyer and Cornell (2018) with more appropriate values of individual tree density is one priority for research in the near future and might resolve the problem. A version of the model incorporating distinct sapling and adult life stages might also provide a natural solution to the problem of reproductive adult density (Rosindell et al., 2012b). In such a model sampled individuals (usually those with a diameter at breast height over a certain arbitrary threshold) no longer need map directly to an exhaustive sample of reproductive adults.

Inconsistencies between the spatially explicit neutral model's SAR and empirical data may also arise because of discrepancies in the dispersal kernel rather than because of the neutrality assumption itself. Most dispersal kernels are not normal as was assumed by O'Dwyer and Cornell (2018) in their study. In particular, dispersal kernels may be fat tailed, which can have large effects on diversity, as discussed in Section 11.2.5. Developing analytical solutions for the SAR in the case of fat-tailed kernels should be another priority for future research. Of course, some discrepancies between neutral SARs and empirical data may also be due to the presence of niche mechanisms in the real world. In this sense, when a neutral model fails, it can do so informatively by pointing to situations in nature where non-neutral processes are important. For example, the failure of the neutral model to explain the small-island effect points to the role of niches in this phenomenon, as we discussed in Section 11.3.1 (see Figure 11.2B). This ability to fail informatively is a property of mechanistic models that is not shared by traditional phenomenological approaches to SARs, for example the power law.

11.6.2 Optimal Sampling Design

The fragmented sampling SAR formulas introduced in Section 11.5 (Equation 11.14) can be used to estimate regional richness from the collective richness of a series of small subplots spread throughout a region. This is essentially a mechanistic method of estimating sample completeness, that is, how much of regional diversity is captured by a certain empirical sample. A related application is to plan efficient surveys. If the goal of a survey is to estimate the true species richness of a region from multiple small survey plots, fragmented sampling SARs can in principle be used to estimate the expected value of sampling more smaller plots versus fewer larger plots and to balance this against the costs (in time and money) of the more complicated sampling designs. What is required for these applications is a suite of fragmented sampling SAR formulas. At present, such formulas do not exist for neutral models, nor for any models as far as we are aware, but the work described in Section 11.5 points towards possible avenues for developing such formulas by providing at least upper and lower bounds (Equation 11.14) that constrain the predictions that any such formulas would give.

11.6.3 Optimal Reserve Design

Another application of neutral SARs is reserve design. In this case, the goal is to predict the long term species richness of differently configured fragmented landscapes and to choose the configuration with the highest richness. This harks back to the classic SLOSS (single-large or several-small) debate (Diamond, 1975; Chapter 2), but brings more rigour to it. Because long term richness is the focus here, the fragmented SAR formulas described in Section 11.4 can be used to make predictions. A caveat here, however, is that these neutral formulas ignore edge effects and thus may identify a reserve configuration that is more fragmented than the truly optimal configuration. For these formulas to be translated into tools that can reliably be used for reserve design, versions that account for edge effects will need to be developed. In the meantime, a reasonable proxy would be to decrease the value of areas close to an edge in the calculations, thus reducing the effective area A_e by a certain amount in the calculations.

11.6.4 Predicting Extinction Debt in Response to Habitat Modification

A primary driver of species loss is habitat loss (Millennium Ecosystem Assessment, 2005). We need to understand the mechanisms of this species loss and make quantitative predictions about it – only then can we make informed decisions about the effects of habitat loss and habitat restoration. A key component of the question is time. When habitat is destroyed there are both immediate consequences and longer-term knock-on effects. Extinction debt refers to the species expected to be lost in the long term as a response to habitat loss that has already taken place (Tilman et al., 1994). For a long time, SARs have been the theoretical foundation for projections of future species richness following habitat loss (e.g. Thomas et al., 2004; Chapters 14, 16 and 17). The use of SARs for this purpose assumes that species richness is a function of one variable: habitat area. This approach has numerous well-documented problems (see Chapter 14). First, species richness is not a function of only habitat area; in reality it is also a function of spatial habitat structure. Second, as noted, species richness is a function of time with different responses in the short term compared to the long term. Third, although in principle a broad range of functional forms of the SAR could be used to predict species loss (see Chapter 7), in practice the function almost always used is, for convenience, a power law, even though we know the power law does not strictly apply across a broad range of spatial scales (Lewis, 2006; Rosindell & Cornell, 2007). Lastly, and related to the previous point, the widespread use of the power law SAR is unsatisfying because it is a purely statistical or correlative approach that does not shed light on the underlying ecological mechanisms behind species loss (see also Chapter 7).

Neutral theory, through the fragmented SAR (Section 11.4) and fragmented sampled SAR (Section 11.5) solutions, provides a means of estimating species loss and extinction debt in response to habitat loss. An advantage of neutral theory in this regard, over phenomenological approaches like the power law SAR is that neutral theory predictions can be attributed to ecological mechanisms. Neutral theory's strong assumptions can be seen as a disadvantage, but it is these assumptions that bring analytical tractability and make neutral models transparent. At least one study has used non-spatial versions of the neutral model to predict extinction debt in birds (Halley & Iwasa, 2011), though note this study's methods involve some circularity that guarantees its 'predicted' values are fairly close to its 'empirical' values. A spatially explicit neutral model has the potential to take this further by incorporating the effects of spatial habitat structure on extinction debt. Such a spatially explicit neutral model could of course be simulated directly on a given landscape in which there is a known spatially and temporally explicit pattern of habitat loss. Whilst this approach is certainly useful for solving some specific problems, a general analytical solution to the model would obviate the need for simulations and facilitate fast application to arbitrarily large landscapes. Indeed, one reason for the success of the classic power law SAR approach, despite its obvious limitations, is that it can be quickly and straightforwardly applied by plugging numbers into a simple equation.

Fortunately, analytical results for the spatially explicit neutral model are accumulating and the goal of a general analytical treatment of species loss and habitat loss in a neutral context is within reach. We presented earlier solutions for the spatially explicit neutral model's fragmented SAR (Equation 11.13) as well as upper and lower bounds for the fragmented sampled SAR (Equation 11.14). The fragmented sampled SAR can be used to describe the species remaining immediately after a habitat destruction event (Figure 11.6). This problem is equivalent to sampling all individuals from the habitat that remains after destruction, but none from outside the habitat. Complementarily, the fragmented SAR can be used to describe how many species remain in the long term after the system reaches a new equilibrium following a habitat destruction event. The difference between species richness immediately after destruction (from the fragmented sampled SAR) and the species richness in the longer term (from the fragmented SAR) gives the extinction debt (Thompson et al., 2019).

A number of satisfying broad inferences follow from these calculations. First, the classic power law SAR method overestimates short term extinction but underestimates long term extinction. Second, the two

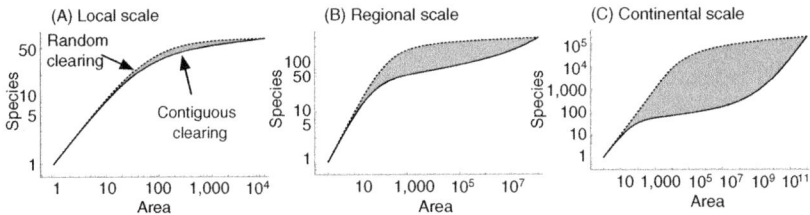

Figure 11.6 Species–area relationships (SAR) from a mainland SAR and a fragmented sampled SAR (with random sampling across space) give upper and lower bounds for the short term extinctions in a region (redrawn from Chisholm et al., 2018). The horizontal axis in each panel shows the area of habitat remaining after habitat destruction whilst the vertical axis in each panel shows the number of species remaining immediately after habitat destruction. The three panels show regions at different spatial scales (different total landscape areas). This is reflected in the maximal values of remaining area in the horizontal axis, which correspond to the original total landscape area and zero habitat loss.

key parameters that appear in the analytical solutions for long term species richness are effective area and effective connectivity, which directly mirror the conventional wisdom in conservation biology that area and connectivity are the two main elements required for supporting healthy biodiversity.

11.7 Conclusion and Future Directions

In this chapter we have discussed the SARs predicted by various neutral models: the non-spatial neutral model, the spatially implicit neutral model and spatially explicit neutral models. The resulting SARs find a variety of applications, from mainlands to islands and from vast intact landscapes to landscapes fragmented by human activities. It is clear that this class of models is versatile and able to make a broad range of predictions for very different systems. Overarching advantages of neutral models are the existence of analytical results in many cases and the ability to simulate the models efficiently in others.

There remains fertile ground for future research related to neutral SARs. For example, incorporating a sapling (or equivalently, juvenile) life stage may be necessary for comparing results to empirical data in cases where the density of reproductive adults is hard to quantify (Rosindell et al., 2012b). There is also much to be done in terms of further analytical and simulation studies of SARs for more general spatial structures and modes of speciation (Table 11.1). Most especially, the analytical results from spatially explicit neutral theory thus far relate only to a normally distributed dispersal kernel and should ideally be extended to fat-tailed dispersal kernels.

An important future research priority is to continue to build niches into the mathematical framework established by neutral theory. The long term goal is to have SAR models that encapsulate a more holistic array of ecological processes but retain the analytical tractability of neutral theory. One straightforward but crude way of adding niche structure is via non-overlapping niches (Allouche & Kadmon, 2009; Chisholm & Pacala, 2010). This was the basis of the niche-neutral model we briefly discussed in the context of island SARs and the small-island effect (see Section 11.3.1). Although this non-overlapping niche structure is overly simple when compared to real niches, there is reason to suspect that many of the predictions of the non-overlapping niche models may be robust and consistent with more sophisticated niche models (e.g. Bewick et al., 2015). The non-overlapping niche structure has the further advantage that it can be easily incorporated into spatial models and coalescence algorithms.

We have studied four different types of SAR in this chapter. Perhaps the most fruitful goal for future research will be to generalize and unify these apparently disparate concepts. For example, by developing more formulas in which both sample area and landscape area can be varied as independent parameters and thus generalizing both the island SAR and the mainland SAR (Figure 11.1). The results for species richness in fragmented habitats could also be generalized to incorporate endemism in isolated patches (or on islands), which the current formulas ignore (to achieve analytical tractability). This would generalize the fragmented SAR and island SAR; at present it can be done with simulations, but not analytically. Should a broad generalization be achieved in future, the result would be a single analytical and mechanistic model that can predict the transition between regions of parameter space where fragmentation is harmful for biodiversity and the regions where it promotes biodiversity in the long term through speciation, such as in some archipelagos. Such a model would find applications across a broad range of fields from evolution to conservation; this is the bright future we hope to see stemming from neutral theory and its predicted SARs.

Acknowledgements

J. R. was funded by an Individual Research Fellowship from the Natural Environment Research Council (NERC) (NE/L011611/1). Through J. R. this study is a contribution to Imperial College's Grand Challenges in Ecosystems and the Environment initiative. R. A. C. was supported by a grant from the James S. McDonnell Foundation (#220020470). We thank the editors and one anonymous referee for their helpful comments on the manuscript.

References

Abramowitz, M. & Stegun, I. A. (eds.) (1972) *Handbook of mathematical functions with formulas, graphs, and mathematical tables*, 10th Printing. National Bureau of Standards Applied Mathematics Series 55. Washington, DC: US Government Printing Office.

Allouche, O. & Kadmon, R. (2009) Demographic analysis of Hubbell's neutral theory of biodiversity. *Journal of Theoretical Biology*, **258**, 274–280.

Alzate, A., Janzen, T., Bonte, D., Rosindell, J. & Etienne, R. S. (2019) A simple spatially explicit neutral model explains the range size distribution of reef fishes. *Global Ecology & Biogeography*, **28**, 875–890.

Bell, G. (2001) Neutral macroecology. *Science*, **293**, 2413–2418.

Bewick, S., Chisholm, R. A., Akçay, E. & Godsoe, W. (2015) A stochastic bio-diversity model with overlapping niche structure. *Theoretical Ecology*, **8**, 81–109.

Chisholm, R. A. & Lichstein, J. W. (2009) Linking dispersal, immigration and scale in the neutral theory of biodiversity. *Ecology Letters*, **12**, 1385–1393.

Chisholm, R. A. & Pacala, S. W. (2010) Niche and neutral models predict asymp-totically equivalent species abundance distributions in high-diversity ecological communities. *Proceedings of the National Academy of Sciences USA*, **107**, 15821–15825.

Chisholm, R. A., Fung, T., Chimalakonda, D. & O'Dwyer, J. P. (2016) Mainten-ance of biodiversity on islands. *Proceedings of the Royal Society B: Biological Sciences*, **283**, 20160102.

Chisholm, R. A., Lim, F., Yeoh, Y. S., Seah, W. W., Condit, R. & Rosindell, J. (2018) Species–area relationships and biodiversity loss in fragmented land-scapes. *Ecology Letters*, **21**, 804–813.

Clark, J. S. (2009) Beyond neutral science. *Trends in Ecology & Evolution*, **24**, 8–15.

Clark, J. S. (2012) The coherence problem with the unified neutral theory of biodiversity. *Trends in Ecology & Evolution*, **27**, 198–202.

Clark, J. S., Silman, M., Kern, R., Macklin, E. & HilleRisLambers, J. (1999) Seed dispersal near and far: Patterns across temperate and tropical forests. *Ecology*, **80**, 1475–1494.

Diamond, J. M. (1975) The island dilemma: Lessons of modern biogeographic studies for the design of natural reserves. *Biological Conservation*, **7**, 129–146.

Durrett, R. & Levin, S. (1996) Spatial models for species–area curves. *Journal of Theoretical Biology*, **179**, 119–127.

Etienne, R. S. & Alonso, D. (2005) A dispersal-limited sampling theory for species and alleles. *Ecology Letters*, **8**, 1147–1156.

Haegeman, B. & Etienne, R. S. (2017) A general sampling formula for community structure data. *Methods in Ecology and Evolution*, **8**, 1506–1519.

Halley, J. M. & Iwasa, Y. (2011) Neutral theory as a predictor of avifaunal extinc-tions after habitat loss. *Proceedings of the National Academy of Sciences USA*, **108**, 2316–2321.

Hansen, M. C., Potapov, P. V., Moore, R., Hancher, M., Turubanova, S. A., Tyukavina, A., Thau, D., Stehman, S. V., Goetz, S. J., Loveland, T. R., Kommareddy, A., Egorov, A., Chini, L., Justice, C. O. & Townshend, J. R. G. (2013) High-resolution global maps of 21st-century forest cover change. *Science*, **342**, 850–853.

Hanski, I., Zurita, G. A., Bellocq, M. I. & Rybicki, J. (2013) Species–fragmented area relationship. *Proceedings of the National Academy of Sciences USA*, **110**, 12715–12720.

He, F. (2005) Deriving a neutral model of species abundance from fundamental mechanisms of population dynamics. *Functional Ecology*, **19**, 187–193.

He, F. & Legendre, P. (2002) Species diversity patterns derived from species–area models. *Ecology*, **83**, 1185–1198.

Heaney, L. R. (2000) Dynamic disequilibrium: A long-term, large-scale perspective on the equilibrium model of island biogeography. *Global Ecology & Biogeog-raphy*, **9**, 59–74.

Holley, R. A. & Liggett, T. M. (1975) Ergodic theorems for weakly interacting systems and the voter model. *The Annals of Probability*, **3**, 643–663.

Hubbell, S. P. (1997) A unified theory of biogeography and relative species abundance and its application to tropical rain forests and coral reefs. *Coral Reefs*, **16**, S9–S21.

Hubbell, S. P. (2001) *The unified neutral theory of biodiversity and biogeography*. Princeton, NJ: Princeton University Press.

Kingman, J. F. C. (1982) The coalescent. *Stochastic Processes and their Applications*, **13**, 235–248.

Lewis, O. T. (2006) Climate change, species–area curves and the extinction crisis. *Philosophical Transactions of the Royal Society B: Biological Sciences*, **361**, 163–171.

Lomolino, M. V. & Weiser, M. D. (2001) Towards a more general species–area relationship: Diversity on all islands, great and small. *Journal of Biogeography*, **28**, 431–445.

MacArthur, R. H. & Wilson, E. O. (1963) An equilibrium theory of insular zoogeography. *Evolution*, **17**, 373–387.

MacArthur, R. H. & Wilson, E. O. (1967) *The theory of island biogeography*. Princeton, NJ: Princeton University Press.

Millennium Ecosystem Assessment (2005) *Ecosystems and human bell-being*. Washington, DC: Island Press.

McGarigal, K. & Marks, B. J. (1995) *FRAGSTATS: Spatial pattern analysis program for quantifying landscape structure*. General Technical Report PNW-GTR-351. Portland, OR: Northwest Research Station, USDA-Forest Service.

McGill, B. J. (2003) A test of the unified neutral theory of biodiversity. *Nature*, **422**, 881–885.

Niering, W. A. (1963) Terrestrial ecology of Kapingamarangi Atoll, Caroline Islands. *Ecological Monographs*, **33**, 131–160.

O'Dwyer, J. P. & Cornell, S. J. (2018) Cross-scale neutral ecology and the maintenance of biodiversity. *Scientific Reports*, **8**, 10200.

O'Dwyer, J. P. & Green, J. L. (2010) Field theory for biogeography: A spatially explicit model for predicting patterns of biodiversity. *Ecology Letters*, **13**, 87–95.

Pigolotti, S. & Cencini, M. (2009) Speciation-rate dependence in species–area relationships. *Journal of Theoretical Biology*, **260**, 83–89.

Plotkin, J. B., Potts, M. D., Yu, D. W., Bunyavejchewin, S., Condit, R., Foster, R., Hubbell, S., LaFrankie, J., Manokaran, N., Seng, L. H., Sukumar, R., Nowak, M. A. & Ashton, P. S. (2000) Predicting species diversity in tropical forests. *Proceedings of the National Academy of Sciences USA*, **97**, 10850–10854.

Preston, F. W. (1960) Time and space variation of species. *Ecology*, **41**, 611–627.

Ricklefs, R. E. (2006) The unified neutral theory of biodiversity: Do the numbers add up? *Ecology*, **87**, 1424–1431.

Rosindell, J. & Cornell, S. J. (2007) Species–area relationships from a spatially explicit neutral model in an infinite landscape. *Ecology Letters*, **10**, 586–595.

Rosindell, J. & Cornell, S. J. (2009) Species–area curves, neutral models, and long-distance dispersal. *Ecology*, **90**, 1743–1750.

Rosindell, J. & Phillimore, A. B. (2011) A unified model of island biogeography sheds light on the zone of radiation. *Ecology Letters*, **14**, 552–560.

Rosindell, J., Cornell, S. J., Hubbell, S. P. & Etienne, R. S. (2010) Protracted speciation revitalizes the neutral theory of biodiversity. *Ecology Letters*, **13**, 716–727.

Rosindell, J., Hubbell, S. P. & Etienne, R. S. (2011) The unified neutral theory of biodiversity and biogeography at age ten. *Trends in Ecology & Evolution*, **26**, 340–348.

Rosindell, J., Hubbell, S. P., He, F., Harmon, L. J. & Etienne, R. S. (2012a) The case for ecological neutral theory. *Trends in Ecology & Evolution*, **27**, 203–208.

Rosindell, J., Jansen, P. A. & Etienne, R. S. (2012b) Age structure in neutral theory resolves inconsistencies related to reproductive-size threshold. *Journal of Plant Ecology*, **5**, 64–71.

Rosindell, J., Wong, Y. & Etienne, R. S. (2008) A coalescence approach to spatial neutral ecology. *Ecological Informatics*, **3**, 259–271.

Scheiner, S. M. (2003) Six types of species–area curves. *Global Ecology & Biogeography*, **12**, 441–447.

Storch, D., Keil, P. & Jetz, W. (2012) Universal species–area and endemics–area relationships at continental scales. *Nature*, **488**, 78–81.

Thomas, C. D., Cameron, A., Green, R. E., Bakkenes, M., Beaumont, L. J., Collingham, Y. C., Erasmus, B. F. N., Siqueira, M. F. D., Grainger, A., Hannah, L., Hughes, L., Huntley, B., Jaarsveld, A. S. V., Midgley, G. F., Miles, L., Ortega-Huerta, M. A., Peterson, A. T., Phillips, O. L. & Williams, S. E. (2004) Extinction risk from climate change. *Nature*, **427**, 145–148.

Thompson, S. E., Chisholm, R. A. & Rosindell, J. (2019) Characterising extinction debt following habitat fragmentation using neutral theory. *Ecology Letters*, **22**, 2087–2096.

Tilman, D., May, R. M., Lehman, C. L. & Nowak, M. A. (1994) Habitat destruction and the extinction debt. *Nature*, **371**, 65–66.

Vallade, M. & Houchmandzadeh, B. (2003) Analytical solution of a neutral model of biodiversity. *Physical Review E*, **68**, 0619021–0619025.

Volkov, I., Banavar, J. R., Hubbell, S. P. & Maritan, A. (2003) Neutral theory and relative species abundance in ecology. *Nature*, **424**, 1035–1037.

Volkov, I., Banavar, J. R., Hubbell, S. P. & Maritan, A. (2007) Patterns of relative species abundance in rainforests and coral reefs. *Nature*, **450**, 45–49.

Wright, S. J., Jaramillo, M. A., Pavon, J., Condit, R., Hubbell, S. P. & Foster, R. B. (2005) Reproductive size thresholds in tropical trees: Variation among individuals, species and forests. *Journal of Tropical Ecology*, **21**, 307–315.

Zillio, T., Volkov, I., Banavar, J., Hubbell, S. P. & Maritan, A. (2005) Spatial scaling in model plant communities. *Physical Review Letters*, **95**, 098101.

12 · On the Interface of Food Webs and Spatial Ecology: The Trophic Dimension of Species–Area Relationships

ROBERT D. HOLT, DOMINIQUE GRAVEL,
ADRIAN STIER AND JAMES ROSINDELL

12.1 Introduction

A hidden dimension of food web ecology almost since its inception has been a concern with space and spatial dynamics. One of the first food webs ever scribbled onto paper was a hand-drawn diagram Charles Elton sketched as an undergraduate, during an expedition to the Svalbard region north of Norway (Summerhayes & Elton, 1923). The food web he drew for Bear Island depicts spatial interactions among the marine realm, terrestrial habitats and bodies of freshwater, mediated by the movement of colonial seabirds from their foraging grounds to their nesting arenas (a process now called 'spatial subsidy'; Polis et al., 1997; Graham et al., 2018). Elton maintained an interest in the spatial aspects of communities through his life. One of his last publications, the rarely read tome *The pattern of animal communities* (Elton, 1966), contained two essential spatial insights about food webs. Elton noted that 'The pyramid of numbers, really a pyramid of consumer layers [trophic levels], is matched by . . . the inverse pyramid of habitat'. Namely, species at higher trophic levels often range further and so have to be examined at larger spatial scales. This means species at higher trophic ranks can spatially couple dynamics of communities at lower trophic ranks, an insight that resonates to the present day (e.g. McCann et al., 2005). Moreover, echoing his youthful excursion to the frigid Arctic, he states 'no habitat component with its animal community is a closed system . . . the structural boundaries . . . are constantly passed by population movement . . . every community unit is partly interlocked with others, not necessarily its nearest neighbours . . .'. This statement could be viewed as a harbinger of today's concern with metacommunity processes (Holt, 1997; Leibold et al., 2004; Leibold & Chase, 2018).

However, for most of its history, food web ecology has largely focussed on local interactions. Amarasekare (2008) notes 'scant empirical evidence of spatial effects in food webs'. Even the fine recent volume *Adaptive food webs* (2018) only marginally deals with food webs in a spatial context. Montoya and Galiana (2018), however, do provide a useful perspective on how to relate species interaction networks to biogeography. They note several distinct modalities of spatial processes at play in food web ecology. First, there now exist preliminary attempts to explicitly integrate island biogeography and food web theory via influences of food web interactions on the colonization and extinction processes generating species' occupancy patterns (Holt, 1996, 1997, 2010; Gravel et al., 2011; Massol et al., 2017). Second, there is growing recognition that taxa at different trophic levels can have distinct spatial strategies, for example the coupling of spatially separated habitats by mobile consumers (e.g. McCann et al., 2005; Rooney et al., 2008), related to the broader theme of spatial subsidies and spatial ecosystem ecology (Polis et al., 1996; Massol et al., 2011). Third, dispersal in metacommunities can permit the persistence of otherwise unfeasible configurations of species interactions (Amarasekare, 2008); area effects on species persistence can reflect within-island spatial dynamics (Holt, 1992), which can be particularly important in fostering the persistence of strong food web interactions (e.g. see Wilson et al., 1998 for a food chain example). Yet these three components of merging food web and spatial ecology have yet to be comprehensively integrated.

Montoya and Galiana (2018) note that one potential approach to integrating food web ecology with biogeography is to examine how food web properties vary along major environmental gradients. Preeminent among such gradients are those spatial attributes of habitats at the heart of island biogeography theory (MacArthur & Wilson, 1967; Whittaker & Fernández-Palacios, 2007) – the area of islands or isolated habitat patches and their distance to potential sources of colonists. Baiser et al. (2012) examined the food webs of pitcher plants (*Sarracenia purpurea*) and showed that food web attributes – food chain length, total species richness, linkage density and incidence of omnivores – increased with pitcher volume. Gravel et al. (2018) point to the rapidly growing evidence that the structure of ecological networks varies by habitat, through time and across gradients. A systematic understanding of how area influences network attributes – such as the number of species in each trophic rank – is clearly emerging as a key desideratum for current research. Species coexistence mechanisms can also be scale dependent (Holt, 1993; Chesson, 2018) and variation in species richness with area could reflect the breakdown of such mechanisms at small spatial scales. For example, Orrock and Fletcher (2005) showed that the respective roles of stochasticity and competitive ability in driving competitive outcomes can change with habitat size. The principal ingredient of quantitative, dynamical food webs – interaction strength – can itself vary with island area (e.g. Schoener et al., 2016). Martinson and Fagan (2014) carried out a meta-analysis of plant–insect herbivore interactions in fragmented landscapes and found that plants on small, isolated fragments enjoyed almost 50 per cent less herbivory than plants in large patches. Across a range of island and patch sizes, Connor et al. (2000) observed that animal densities are higher in larger areas. Ecological interactions are typically 'dosage-dependent', so for instance competition intensity increases with the densities of competitors, implying a likely systematic signal of area on interspecific interaction strength (as found in parasitoid–host interactions by Fenoglio et al., 2013) – with implications for the spatial scaling of species richness.

Island biogeography has dealt with trophic interactions and food web structure, but not as a central theme. As Holt (2010) noted, MacArthur and Wilson (1967) do not directly discuss food webs at all, but do obliquely hint at how local food web interactions can govern colonization and extinction and, thus, island community structure. The synthesis of island biogeography by Whittaker and Fernández-Palacios (2007) touches on predation and food chains several times. Two telling examples they cite from classic island studies are: i) that Thornton (1996) ascribed several extinctions of birds on Anak Krakatau to top avian predators (whose numbers were sustained, it appears, by the ability of raptors to move among several islands, an example of the spatial scaling of movement with trophic rank hypothesized by Elton in 1966) and ii) that Lomolino (1984) demonstrated that the carnivorous shrew *Blarina brevicauda* drove extinctions of a mammalian prey species, *Microtus pennsylvanicus*, on islands in the Thousand Island region of the St. Lawrence River. Harold Heatwole (Heatwole & Levins, 1972; Heatwole, 2018) examined islands in the Caribbean and off Australia and argued that insular trophic structure (the relative proportion of species in major trophic groups such as detritivores, herbivores and predators) can be relatively stable, even if species composition is in continual flux. Piechnik et al. (2008) re-examined the classic Simberloff-Wilson experimental defaunation of mangrove islets and reported a signature of trophic breadth in the order of colonization, with generalist consumers arriving first in community assembly (see also Cirtwill & Stouffer, 2016). These examples show that the presence or absence of predators and trophic attributes such as dietary specialization can influence the primary drivers of community assembly on islands – extinction and colonization – an issue we examine more closely in Sections 12.3 and 12.4.

A principal goal of the theory of island biogeography was to develop a dynamic explanation for the species–area and species–distance relationships on true islands. There are many kinds of species–area relationships (SARs), including for instance different kinds of species accumulation curves within contiguous land masses (Scheiner, 2003) and we touch on several below. In island biogeography and analyses of habitat fragmentation, the focus is typically on the Type IV curves of Scheiner (2003), which describe species richness across 'true' or 'habitat' islands or distinct geographical domains varying in area (also termed island species–area relationships or ISARs, see Chapter 1). Over some spatial scales, ISARs nicely fit the classic power law, $S = cA^{z}$, where c is a parameter representing a kind of carrying capacity per unit area and z is a parameter that indicates how area (A) boosts species richness (S) (Triantis et al., 2012; Matthews et al., 2016). Equivalently, we have $z = \frac{\partial \log(S)}{\partial \log(A)}$. This expression usefully characterizes the strength of the relationship between species richness and area, even if the power law does not hold (i.e. z can vary with A).

A null expectation might be that the parameters defining the strength of the species–area relationship are not influenced by trophic rank or interactions. Indeed, Drakare et al. (2006) reviewed species–area relationships (largely for nested or contiguous species–area curves) from a wide range of systems and concluded that there was no significant signal of trophic guild '. . . across autotrophs, herbivores, omnivores, carnivores, microbivores, parasites and decomposers'. So maybe this chapter could end right here!

But we won't do that; instead we will revisit these issues. We will build simple models of SARs across multiple trophic levels, beginning from the unlikely starting point of ecological neutral theory (Hubbell, 2001), which at its core shares the assumption that trophic position is an unimportant predictor of spatial variation in biodiversity. After presenting this novel bit of theory, we then examine some decidedly non-neutral models, including communities with tightly specialized food

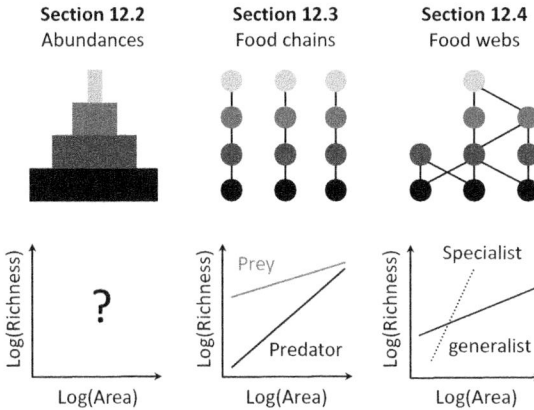

Figure 12.1 The three scenarios considered in this chapter. Left: a trophic abundance pyramid, with lower abundances at higher trophic ranks. Centre: a community comprised of 'stacked specialists', where each plant sustains a specialist consumer, which in turn supports a specialist hyper-consumer. Right: a community with a mixture of trophic specialists and generalists within trophic levels.

chains and mixtures of specialist and generalist predators, complementing several important reviews and theoretical advances which have recently appeared (e.g. Gravel et al., 2011; Massol et al., 2017; Galiana et al., 2018). Figure 12.1 shows schematically the three aspects of community structure we explore in the three main sections of this chapter: trophic pyramids (Section 12.2), food chains (Section 12.3) and more complex food webs (Section 12.4).

12.2 Trophic Pyramids and the SAR

Ecological neutral theory assumes an individual's chances of reproduction and death are independent of its species identity. This may seem an extreme simplification of a richly complex reality, especially in the context of food web interactions. However, the merits of neutral theory (like the theory of island biogeography) lie in its ability to act as a minimal starting point on which more complex ideas or inferences can be built, a 'yardstick' for assessing implications of additional biological assumptions. Furthermore, neutral theory actually does an excellent job of predicting SARs, both on islands and on contiguous mainland (see Chapter 11). In the simplest neutral community model there is a 'zero-sum' competitive game at play: the total size of the community is fixed so that the abundance of one species can only increase when another decreases. This fixed total abundance is set by unexamined factors, such as total resource supply or top-down effects of higher trophic levels acting to constrain total abundance at lower levels. In each time step, one individual is chosen at random to die and another is comparably chosen to reproduce, filling the 'gap' left in the community by the death. Relative abundances of species drift over time and species go extinct. Based on these rules alone, the community would drift eventually to monodominance of one species. To maintain diversity, an input of new species is required to counterbalance extinction, either by speciation or immigration from an external species pool or both. One can use neutral theory to construct species–area relationships in either case (see Rosindell & Cornell, 2007; Chapter 11). To do so over the full range of spatial scales requires a fully spatially explicit model in which individuals occupy a precise location in space. The ubiquitous power-law relationship emerges at intermediate sample areas from these models and is especially prominent when dispersal across space follows a fat-tailed distribution in which long-distance dispersal events are common (Rosindell & Cornell, 2009).

What could possibly be neutral about food webs? In some systems, trophic levels or guilds are cleanly delineated, so that one can discern distinct trophic levels. Interactions across trophic levels are obviously not neutral. But maybe interactions *within* trophic levels or 'guilds' could be treated as if governed by neutral dynamics, at least as a simplification (see Krishna et al., 2008 for a comparable approach to mutualistic networks; see Chave, 2004 and Adler et al., 2007 for more discussion on the equivalence assumption). This has previously been done to provide a neutral model of predator–prey interaction networks (Canard et al., 2012) and later in host–parasite systems (Canard et al., 2014). Here we consider the simpler question of SARs in linked neutral models across two or more trophic levels. One could imagine for the sake of argument that the total number of individuals (across all species) in a given trophic level is determined by trophic interactions, whilst dynamics among species within a trophic level are a competitive, zero-sum game. If all species are competitively equivalent on the same trophic level, their numbers should drift, comparable to Hubbell's neutral tree community model (Hubbell, 2001). In a food web diagram, all species (or alternatively, all individuals) on one trophic level would be equally connected to all species (or individuals) in levels below and above. In later sections, we explore the implications of more specialized interconnections across trophic levels.

We first explore the consequences of trophic levels for a non–spatial neutral model, which is suitable for capturing SARs on isolated islands or at very large continental scales. The species richness S in a community containing J individual organisms and having a per capita speciation rate of u can be described by (Etienne & Alonso, 2005):

$$S = \theta(\psi_0(\theta + J) - \psi_0(\theta)). \tag{12.1}$$

Here ψ_0 is the digamma function and the quantity θ is the 'fundamental biodiversity number' (Hubbell, 2001) given by $\theta = \frac{(J-1)u}{1-u}$ (see also Chapter 11).

Now consider a system with n trophic levels; trophic level i has a total community size of $J_i = A \cdot \rho_i$, where A is area and ρ_i is the density of individuals per unit area for trophic level i. Total community size is thus assumed to be proportional to area; more complex relationships between density and area are known (Connor et al., 2000) and could straightforwardly be accounted for. The SAR for species from trophic level i is given by an extension of Equation (12.1):

$$S_i(A) = \theta_i(A)(\psi_0(\theta_i(A) + \rho_i A) - \psi_0(\theta_i(A))), \tag{12.2}$$

where $\theta_i(A) = \frac{(\rho_i A - 1) u_i}{1 - u_i}$ (see also Chapter 11). The SAR incorporating species from all n trophic levels is a sum over Equation (12.2):

$$S(A) = \sum_{i=1}^{n} \theta_i(A)(\psi_0(\theta_i(A) + \rho_i A) - \psi_0(\theta_i(A))). \tag{12.3}$$

Two patterns emerge from these formulas. First, very small metacommunities will harbour only a single species in each trophic level. Second, once metacommunity size (area) becomes larger, the SAR becomes linear. This is consistent with an approximation of Equation (12.2) in which, provided species richness is not small, $\theta_i \gg 1$, we can write

$$S_i(A) \approx A \frac{\rho_i u}{1 - u} \log\left(\frac{1}{v}\right). \tag{12.4}$$

These findings essentially restate the results of Chisholm et al. (2016), who studied an extension of neutral theory, in the context of ISARs, where the community consists of multiple niches and where within each niche multiple species interact neutrally. Here, we simply replace 'niches' with 'trophic levels'.

Note that this SAR for large A is only consistent with the classic power law $S = cA^z$ when $z = 1$. This does not match typical observations for small to moderate-sized oceanic islands, but resembles the SAR of some isolated islands which receive very limited immigration, so that all species richness emerges from in situ diversification; every island is then an independent evolutionary arena (Losos & Schluter, 2000; Chisholm et al., 2016).

We now use the above machinery to examine SARs as a function of trophic rank. A general rule of thumb is that predators, collectively, are rarer than their prey. Charles Elton, in his *Animal ecology* (1927), referred to this as the 'pyramid of numbers', which often (not always, see figure 3.15 in Odum, 1971) describes how total abundance varies with trophic rank (when such ranks are cleanly delineated). Let us imagine that species in different trophic ranks are demographically equivalent and have the same speciation rates, but that trophic ranks differ in total community size. For illustration, we make the simplifying assumption that the number of individuals declines by a factor of 10 with each increase in trophic level. This assumption has no effect on our qualitative conclusions.

At sufficiently large areas, where the relationship is effectively linear, one expects the relative species richness of two trophic ranks to match their relative abundances, or

$$\frac{S_{i+1}}{S_i} \approx \frac{J_{i+1}}{J_i}. \tag{12.5}$$

In an earlier era of food web theory, it was suggested that the relative richness of top, intermediate and basal trophic species within communities are relatively invariant across communities (Cohen, 1977; Cohen & Briand, 1984; Cohen & Newman, 1991). Where this is the case, the null expectation might be that SARs (after accounting for differences in individual densities) would match, across trophic levels. Combined with our conclusions from the above simple neutral model, it might appear that Drakare et al. (2006) are correct and that trophic rank or guild are not informative relative to SARs.

But organisms at different trophic levels can differ in many important ways, for instance in body size and mobility. Genetic estimates suggest substantial variation in dispersal across trophic levels (Kinlan & Gaines, 2003). McCann et al. (2005) argued that, in oceanic food webs, there is an allometric relationship between body size and dispersal ability, so top-ranked predators roam widely and link distinct ecosystems (Figure 12.2). For example, in coral reef fish communities, predator dispersal scale increases with increasing body size (Stier et al., 2014a). We build on

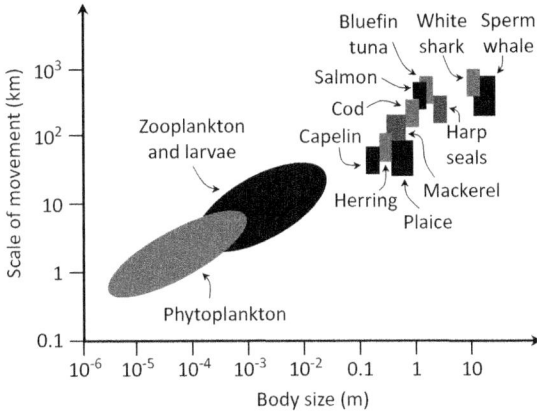

Figure 12.2 The spatial scale of movement increases with body size, which is correlated with trophic rank (redrawn using data from a figure in McCann et al., 2005; under license from John Wiley and Sons). The shades used are for graphic clarity only and are not intended to map to any property of the system.

results in Chapter 11 to provide a spatially explicit neutral SAR for mainlands (SAR types I, IIA; Scheiner, 2003) across multiple trophic levels. We assume, as a starting point, that the primary drivers are trophic level, abundance and dispersal ability, rather than the precise interaction structure of the food web.

First we state the density of individuals δ_i in trophic level i using the '10 per cent' rule of thumb given by

$$\delta_i = 0.1 \cdot \delta_{i-1}, \ \delta_1 = \delta. \tag{12.6}$$

Next, we determine the individual body mass m_i of an individual in trophic level i using a scaling law (Cohen et al., 2003)

$$\delta_i = N_0 \cdot m_i^{-\beta}, \tag{12.7}$$

which can be rearranged to produce

$$m_i = \left(N_0/\delta_i\right)^{1/\beta}, \tag{12.8}$$

where $\beta = \frac{3}{4}$ or $\beta = \frac{2}{3}$ are typical (Cohen et al., 2003); we assume $\beta = \frac{3}{4}$ in our results.

In many systems, dispersal range scales with body size, which increases with trophic rank (Figure 12.2). We use individual body mass in a further scaling law to predict total lifetime dispersal distance σ_i for each trophic level

$$\sigma_i = B_0 m_i^{\alpha}. \tag{12.9}$$

The allometric relationship itself is well supported (Brown et al., 2000), but the value of the scaling exponent α may be hard to determine in this case. For example, 0.21 was inferred for scaling of active dispersal velocity in terrestrial mammalian carnivores (Rizzuto et al., 2018) and 0.48 was used more generally for scaling of passive dispersal in the Madingley model (Harfoot et al., 2014). However, neither is quite the same as total lifetime dispersal (in general, the precise scaling of dispersal with body size and how it interacts with other factors such as fertility and lifespan is unknown and an important desideratum for future research). Dispersal distance σ_i can be multiplied by $\sqrt{\delta_i}$ to give dispersal measured in units of individual widths at their natural density, rather than in units of geometric distance.

We will use a spatially explicit neutral model for which the SAR on a contiguous mainland is closely approximated by an analytical formula (see Chapter 11):

$$S_i\left(A, v, \sigma_i^2\right) \sim \delta_i \sigma_i^2 \Psi\left(\frac{A}{\sigma_i^2}, v\right). \tag{12.10}$$

Here A is area (measured in the same units as σ_i^2), v is a per capita speciation rate and Ψ is the 'Preston Function' (Chisholm et al., 2018; Chapter 11), which can be approximated analytically (O'Dwyer and Cornell, 2018). This can be written with substitutions from Equations (12.6), (12.8) and (12.9) into Equation (12.10) to obtain

$$S_i(A, v, \sigma_i^2) \sim x \cdot \delta^{1 - 2\alpha/\beta} 0.1^{(i-1)(1-2\alpha/\beta)} \cdot \Psi(A \cdot x^{-1} \cdot \delta^{2\alpha/\beta} \cdot 0.1^{2\alpha(i-1)/\beta}, v), \tag{12.11}$$

where $x = B_0^2 \cdot N_0^{2\alpha/\beta}$. The total SAR across multiple trophic levels is given by the sum across all trophic levels of the individual SARs:

$$S(A) = \sum_{i=1}^{n} S_i\left(A, v, \sigma_i^2\right). \tag{12.12}$$

Figure 12.3 depicts what happens when dispersal rates are assumed equal across trophic levels ($\alpha = 0$). As area increases, food chain length increases (upper left; the minimal number of species cannot drop below one). However, the spatial dependence of richness on area (the value of z) is the *same* across trophic levels (upper right). Thus, in the simplest neutral model that is spatially explicit, where all individuals disperse in the same way, we conclude: trophic rank has no effect on the SAR. This may

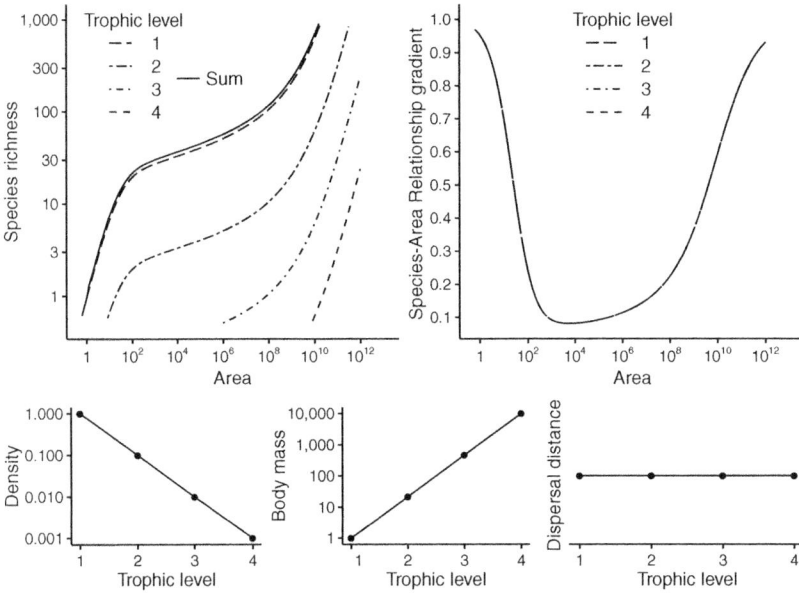

Figure 12.3 Top left: the species–area relationship on a contiguous mainland for multiple trophic levels. The data come from an independent spatially explicit neutral model for each trophic level and are based on an analytical solution (the Preston Function) for the mean total number of species in different sized (nested) areas. It is assumed that there are several trophic levels which differ systematically by a factor of 10 in total abundance, but with equal dispersal ability. Top right: the species–area relationship gradient, $z = \frac{\partial \log (S)}{\partial \log (A)}$ as a function of log area. The strength of the species–area relationship depends on the spatial scale in question – but does not differ by trophic level. Bottom panels: underlying allometric relationships between trophic level and density (bottom left), body mass (bottom centre) and dispersal (bottom right). The parameters used were $\nu = 10^{-9}$, $\delta = 10^{6}$, $N_0 = 10^{-3}$, $\beta = \frac{3}{4}$, $\alpha = 0$, $B_0 = 3$.

seem surprising, but it happens because dispersal limitation is the key driver of spatial distribution in a neutral model and in this first example this has been assumed fixed across all trophic levels.

Figure 12.4 shows a more interesting observation, where now we assume that individuals at higher trophic rank disperse further (with exponent $\alpha = 0.25$). The gradient of the SAR in logarithmic space is now dependent on spatial scale as a function of trophic level. At smaller spatial scales, higher trophic levels are predicted to have *steeper* SAR gradients. However, at larger spatial scales the opposite is predicted, with higher trophic levels having *shallower* SAR slopes. Thus, the effect of trophic level on the strength of the SAR exponent is predicted to be scale dependent.

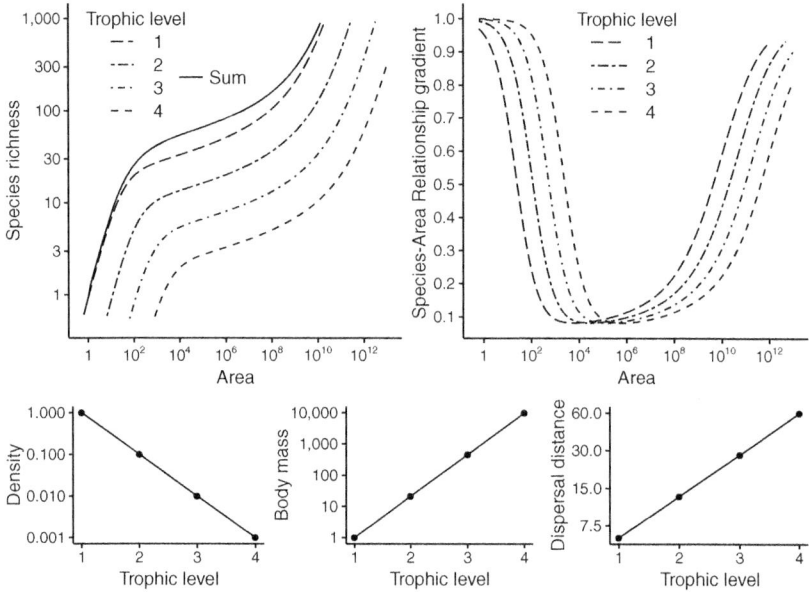

Figure 12.4 As in Figure 12.3, but now dispersal distances increase with trophic level $\alpha = 0.25$. This implies that species–area relationships can vary by trophic level, but in a manner that is scale dependent.

12.2.1 Beyond Neutrality

There are several ways to relate trophic interactions and the species–area relationship that go beyond these neutral expectations (Holt, 2010). First, trophic status might be correlated with individual or population-level attributes which influence extinction or colonization rates or evolutionary rates. Alonso et al. (2015) provided an excellent example from coral reef fish experiencing mass mortality events: extinction rates increase at higher trophic ranks. Jacquet et al. (2017) showed that including trophic position improved predictive accuracy of occupancies in tropical coral reefs, using a MacArthur–Wilson style model. Stier et al. (2014a) convincingly argued that colonization rates should be higher at higher trophic ranks in fish communities, because predators with relatively larger body sizes in their larval stages can disperse further through longer larval durations, a factor that we have taken into account in the analyses leading to Figure 12.4 because trophic levels are treated as having different dispersal abilities. Colonization makes rescue effects more likely, dampening extinction. This may help explain the observation that predator:prey richness ratios increase with distance from sources across the Pacific Ocean (Stier et al., 2014a). Top fish predators have highly generalized diets and so are not likely to be constrained by the occurrence of particular prey species; together with high mobility, this should flatten SARs in contrast to specialist enemies, as we will investigate later.

12.3 Food Chains and the SAR

Trophic interactions themselves could directly drive extinction and colonization, via 'bottom-up' and 'top-down' forces. Trophic interactions can be complex (nonlinear functional responses, predator–prey cycles, chaos and the like) and in the future it would be valuable to link the rich body of theory that exists on food web ecology more explicitly to the spatial processes that underlie SARs. But it is useful to start with simpler, more schematic models, a 'minimalist' community ecology that goes beyond neutral theory, but does so by utilizing the simpler abstraction of MacArthur and Wilson (1967), which focuses on the extinction and colonization of entire species. The simplest, non-interactive island biogeographic model (where the rate parameters for each species do not depend upon other community members) can be viewed as a limiting niche model, where each species potentially has its own separate niche on an island.

A fundamental descriptor in spatial ecology is 'occupancy' – the fraction of habitat patches or islands occupied by a species. Occupancies expressed as a function of area are called 'incidence functions'. Given incidence functions, one can construct SARs for entire communities (Ovaskainen & Hanski, 2003). Interactions among species, including food web interactions, can be built into incidence functions. It is an ecological truism that all species require resources and a food web at the very least describes asymmetrical resource dependencies among species. We start by assuming 'donor-control', so that predators need prey, to colonize and avoid extinction, but do not themselves alter prey colonization or extinction. This sequential trophic dependency, all by itself, has consequences for community structure, including species–area relationships. In empirical studies of extinction, specialist herbivores usually go extinct before their required host plants (e.g. Sang et al., 2010) and differences in establishment by basal host species constrain colonization by specialist trophic guilds dependent upon those hosts (Harvey & MacDougall, 2014; Cirtwill & Stouffer, 2016). We will first consider simple communities of multiple unbranched food chains – 'stacked specialists' (e.g. host plants, with specialist herbivores, sustaining specialist parasitoids; Figure 12.1). Take a single such food chain and consider a taxon at trophic rank i. A mainland community is assumed to have n such food chains.

Previous papers by one of us (Holt, 1993, 2010; Holt et al., 1999) developed simple models incorporating sequential dependencies among specialist consumers and their biotic resources, drawing out implications for the species–area relationship built on this 'bottom-up' effect. These models include static incidence function models and dynamic patch occupancy models; we here just sketch the former. The incidence function for a given species of rank i is the percentage of islands that are occupied, as a function of area and rank, $p(i)$ (we leave the functional relationship on area, A, implicit and for simplicity assume all species of a given rank have the same incidence function). The expected number of species of rank i on a given island is $S_i = np(i)$.

Given tight specialization up food chains, a species of rank i cannot persist on any island lacking its resource species of rank $i - 1$, but it might not persist even if that resource is present. Colling and Matthies (2004) provided an example of an incidence function for a specialist fungal pathogen, which is not sustained on small host populations. Small islands often contain the host, but not the pathogen. This sequential dependency leads to nested spatial distributions, which we can formalize as follows.

A resource species of rank $i - 1$, which a species of rank i requires, has its own incidence function, $p(i - 1)$. If we focus on one food chain, we expect nested distributions across islands, for populations that persist without recurrent immigration: if species $i - 1$ is absent, so too should be any dependent consumer, species i, $i + 1$, etc.

Define the conditional incidence function of species i as the probability it will be present, given that its required resource is present, $p(i | i - 1)$, as a function of (say) area. The unconditional incidence for species i can be written as a compounding of such conditional incidence functions

$$p(i) = p(i | i - 1)p(i - 1) \ldots = p(1) \prod_{j=2}^{i} p(j | j - 1) \qquad (12.13)$$

If we assume that there are n stacked specialist food chains on a mainland, on the islands the expected number of species of rank i is $S_i = np(i)$. On a log–log plot, the strength of the SAR across trophic ranks is as follows

$$z_i = \frac{\partial \log(S_i)}{\partial \log(A)} = \frac{\partial \log(p(i))}{\partial \log(A)} = \frac{\partial \log(p(i-1))}{\partial \log(A)} + \frac{\partial \log(p(i|i-1))}{\partial \log(A)}$$

$$= z_{i-1} + \frac{\partial \log(p(i|i-1))}{\partial \log(A)}, \tag{12.14}$$

implying $z_1 < z_2 < z_3$, etc., if the rightmost term is positive. Thus, if conditional incidence increases with area, the z-value increases with trophic rank. As a limiting case, conditional incidence might be independent of area, in which case z-values across trophic levels will match. Similar results emerge for dynamic occupancy models given donor control with sequential colonization and linked extinction dynamics. The basic idea is that a specialist consumer cannot colonize an island unless its required resource is present and if that resource goes extinct – so should it and, in turn, any species depending solely on it. Based on these simple models, Holt et al. (1999) and Holt (2010) argued that on islands relatively closed to immigration, specialist consumers (which attack just a single prey species) with weak top-down effects on their prey should often have stronger SARs (i.e. higher z-values) than do their prey. Because the same argument holds at each trophic level, this suggests that z-values should increase, as one marches up higher trophic ranks through specialist food chains in a community. Roslin et al. (2014) provided a fine empirical demonstration of this prediction, using a system where the natural history fits the model assumptions – herbivorous insects on an island archipelago in the Baltic, where those insects sustained primary parasitoids, which in turn supported secondary parasitoids. Each trophic level was comprised of relatively specialized consumers (herbivores on their plant hosts, parasitoids on their insect hosts, etc.). Consistent with theory, the slope of the log–log ISAR '... steepened from plants through herbivores and primary parasitoids, to secondary parasitoids'.

12.4 Food Webs and the SAR

The tightly specialized food chains explored in Section 12.3 are an abstraction; few, if any, natural communities will show such marked specialization across trophic levels. Holt et al. (1999) suggested circumstances in which trophic rank might not systematically affect the SAR, if for instance communities are open with rapid, recurrent immigration at high trophic levels, or if consumers are generalists, so their populations can be sustained by any of a number of lower-level prey species. If some consumers are generalists, but others specialists, there could be differences in the strength of the SAR associated with the degree of trophic specialization.

12.4.1 'Bottom-up' Effects

Holt (2010) developed a variant of the classic MacArthur-Wilson island model, to suggest that generalists could even have lower SARs (i.e. lower z-values) than their prey. There is growing evidence that trophic specialization versus generalization does influence colonization and persistence. Bagchi et al. (2018) reviewed evidence that specialist plant–herbivore interactions are differentially absent on the small, isolated patches that characterize fragmented landscapes, leading to steeper SARs for specialist, compared to generalist, herbivores. As they note, this reflects the fact that the abundance of dietary specialists depends on host plant availability, whereas generalists do not show such strong dependency. Cirtwill and Stouffer (2016) revisited the mangrove island experiment of Simberloff and Wilson and showed that knowledge of species-specific resource requirements increased predictability of extinctions. In butterfly and moth assemblages on islands, dietary specialists have higher z-values (Franzen et al., 2012). This matches earlier results by Steffan-Dewenter and Tscharntke (2000) for butterflies of meadows in central Europe: polyphages, strongly oligophagous and monophagous species, respectively, have z-values of 0.07, 0.16 and 0.22. In other words, small meadows are dominated by trophic generalists. This has been detected in oceanic island studies. For instance, Santos and Quicke (2011) showed that oceanic island parasitoid faunas are more typified by generalists than are continental parasitoid assemblages. Gravel et al. (2011) have extended the sequential dependency approach of Holt to more complex food webs and demonstrated that the model successfully predicts several patterns in an Adirondacks lake dataset: small areas are dominated by primary producers and generalist predators are relatively more prevalent on islands of intermediate size. The same held for coral reef fishes, but these patterns were not observed in the mangrove arthropod communities examined by Cirtwill and Stouffer (2016).

Bottom-up effects can percolate up multiple trophic levels. Fenoglio et al. (2012) reported how forest fragment area influences parasitism on leafminers by interlocked changes in plant, herbivore and parasitoid diversity, largely reflecting the differential loss of specialist parasitoids in small fragments. Even for generalists, trophic rank sometimes strongly influences the strength of SARs, for example Noordwijk et al. (2015) showed that the richness of zoophagous carabid beetles increased with the area of calcareous grassland, but phytophagous carabids did not. This contrasts with some of the examples cited in Holt et al. (1999), where an absence of a trophic rank effect on z was ascribed to generalization; these carabids are generalists, but nevertheless show the predicted pattern of stronger species–area relationships with trophic rank. Noordwijk et al. (2015) sensibly suggested that even generalist predators are likely scarcer than their prey, so are prone to higher extinction risks, and that generalization is no buffer against extinction, if prey numbers fluctuate synchronously (e.g. in response to major climatic events). Prairie dogs for instance have highly generalized diets, yet show a strong effect of plant species richness on local extinction rates (Ritchie, 1999). An ineffective generalist or one that draws different nutrients from different species might need a diverse array of prey types in order to gain a foothold and persist on an island.

Holt and Hoopes (2005) and Holt (2010) developed a model of bitrophic 'donor-controlled' island assembly, extending the classic immigration-extinction model of MacArthur and Wilson (1967). They assumed a prey trophic level colonizes from a mainland source pool, with non-interactive colonization and extinction rates dependent on island area. A predator trophic level likewise colonizes but, in addition to direct effects of area on its colonization and extinction rates, there are indirect effects, because increased prey species richness can facilitate predator colonization or reduce extinction. If predators are specialists, these effects should be strong (the sequential dependency effect discussed in Section 12.3). But if predators are generalists, they might well be weaker, depending on the details of trophic generalization. Generalist predators may be able to plastically use prey types that are not even part of their regular diet on the mainland, for instance, thereby promoting colonization and lowering extinction. The model made plausible that one might expect to see heterogeneous SARs by trophic rank. This theoretical prediction matched a finding in the Steffan-Dewenter and Tscharntke (2000) study: the z-value of the host plants was 0.14, even higher than that of the generalist butterflies, but lower than that of the specialists.

12.4.2 'Top-down' Impacts of Natural Enemies on Species–Area Relationships

The models summarized in Section 12.4.1 neglect the widespread observation that predators and other natural enemies alter the population sizes of their victims, with likely consequences for colonization, extinction and, thus, SARs. In the Bahamas, predatory lizards greatly lower the abundance of spider prey, boosting extinction risks (Toft & Schoener, 1983). Top-down effects in principle could have divergent impacts on SARs at lower trophic levels. Predators and parasites can at times facilitate coexistence among competing prey (e.g. Holt et al., 1994) or keep in check mesopredators that can wreak havoc at lower trophic levels. There are indications that elimination of top predators can often lead to extinction cascades at lower trophic ranks (Donohue et al., 2017). If these natural enemies are found mainly on large islands, prey diversity may collapse on smaller islands. Large islands may also harbour more refuges from predation or disturbance, reducing extinction (Schoener & Spiller, 2010). This should steepen SARs in the prey.

Conversely, strong, generalist predators can elevate extinctions in their prey, particularly if prey are not competing and some prey species are better able to withstand predation (Holt, 1977). Predators can prevent colonization, with systematic effects on SARs in the prey trophic guild (Ryberg & Chase, 2007; Holt, 2010). Strong predation can reverse positive impacts of island area on occupancy for vulnerable species (Grainger et al., 2017). In the island-like glades of the Ozarks, the eastern collared lizard is a top predator with a generalized insectivorous diet and is largely found within larger glades. Its prey had depressed z-values among glades harbouring the collared lizard, compared to glades without it. Ryberg and Chase (2007) developed a simple model where predators were assumed to increase extinction rates by a constant additive amount across islands and showed that this should depress z-values. This result does not hold in all systems. For instance, a test of the hypothesis that predators affect the SAR in coral reef systems found no evidence that predators affected the slope of the SAR in coral reef fish ecosystems in the South Pacific (Stier et al., 2014b). Moreover, at times, predators may exert stronger negative effects on their prey on larger islands. In the Bahamas, islands with lizards have fewer spider species than islands without, and this effect is strongest on large islands (Spiller & Schoener, 2009), leading to a lower value of z (0.16) for lizard islands than for lizard-free islands (0.4). This is consistent with a model presented in Holt (2010), where predators are assumed to have a multiplicative effect on prey extinction. Murakami and Hirao (2010) reported similar impacts of lizard predation upon z-values in insects, for a different suite of Bahamian islands. Such results could arise if predators are differentially more abundant, on larger islands. Ostman et al. (2007) suggest that heterogeneity of spatial distribution of predators and the strength of their impacts of predation could help explain variation in SARs among systems, particularly given that top predators often may be absent from small, isolated islands or habitat patches (Holt & Hoopes, 2005).

12.5 Conclusions and Future Directions

In an early foray toward linking food web ecology and island biogeography, Holt et al. (1999) presented simple theory suggesting that z-values should increase with trophic rank, particularly for specialist consumers, but sometimes also for generalists. They reported examples which fit this expectation, but also counter-examples. They suggested the latter might reflect several factors: '(1) ... strong top-down interactions leading to prey extinction; (2) communities are open, with recurrent immigration, particularly at higher trophic ranks; (3) consumers are facultative generalists, able to exist on a wide range of resource species, or (4) systems are far from equilibrium' (p. 1495).

In this chapter, we have presented new neutral theory that emphasizes how differences in dispersal rates among trophic ranks could influence species–area relationships in a scale dependent way, matching one idea broached in Holt et al. (1999). Leaving aside neutrality, we then summarized past theory and touched on illustrative empirical examples. When trophic interactions are relatively specialized, species–area relationships tend to be steeper at higher trophic ranks. Overall, specialist consumers have steeper species–area relationships than do generalists. Top-down effects of predation on their prey can have a diverse range of impacts on such relationships.

There are many directions one could envisage for further development at the interface of food web ecology and the spatial scaling of species richness. One realistic complication in multi-trophic systems is that cascading effects across multiple trophic levels can lead to non-monotonic relationships between predation pressure on a focal taxon and area. For example, if top predators are restricted to large islands, their prey can grow to larger numbers on smaller islands, imposing greater pressure on lower trophic levels (e.g. Genua et al., 2017). In principle, patch occupancy models built to include trophic influences on colonization and extinction can also include top-down and lateral competitive effects (Lafferty & Dunne, 2010; Pillai et al., 2010; Massol et al., 2017), although even seemingly simple models can rapidly become analytically intractable. This line of thought should be extended to incorporate a richer array of trophic phenomena, such as interference competition among consumers, apparent competition among prey, intraguild predation increasing extinction risk for consumers of intermediate rank, the influence of food web structure and predator behaviour on local population stability and thus extinction risk, and so on. The approach developed in Gravel et al. (2011) and Massol et al. (2017) could, in principle, provide a springboard to consider such effects.

One real-world complication is that, along the gradient from small to large islands, it may be insufficient to monitor occupancy on islands taken as a whole. As one of us observed long ago (Holt, 1992), organisms may have limited dispersal capacity *within* islands, so that one might view area as a proxy for, for example, lattice size, where cells in the lattice are domains of local interactions, connected by within-island dispersal. Large islands correspond to large lattices and small islands correspond to small lattices. Chase et al. (2019) make useful suggestions about how to use rarefaction at different scales within islands to discern possible mechanisms for area effects. This perspective is particularly important given strong natural enemy–victim interactions, which tend to self-destruct. Ever since the classic experiments of Huffaker (1958) it has been clear that predator–prey persistence can be enhanced by colonization among patches (Hastings, 1977) and other spatial effects such as refugia may also be important (Lampert & Hastings, 2016). Wilson et al. (1998) demonstrated that local instability could lead to strong area effects in the persistence of tritrophic (host–parasitoid–hyperparasitoid) food chains. On large islands, metacommunity processes operating within the island may buffer the many disparate ways species-rich and interconnected food webs can be unstable in their local dynamics (LeCraw et al., 2014; Wang & Loreau, 2014; Liao et al., 2017).

An important task for future work will be to articulate how the stabilizing attributes of space – against a backdrop of habitat heterogeneity and island ontogeny (Scherber et al., 2018) – in food web interactions help contribute to realized species–area relationships. One key factor affecting persistence is movement behaviour of species at different levels in the food web. Top predators can be highly mobile, coupling different local communities within a large island, modulating species coexistence at lower trophic ranks. Guzman et al. (2019) have recently argued that understanding spatial use properties is necessary to characterize spatially distributed food web dynamics, and outlined key ingredients in an emerging predictive framework. Articulating how dispersal varies across a food web will be crucial for a more precise characterization of how trophic rank influences SARs.

The importance of considering food webs and more broadly network development in the context of island biogeography is becoming increasingly recognized as an exciting direction for current research (Warren et al., 2015). Gravel et al. (2011; see also Massol et al., 2017) creatively extended the sequential dependency model proposed earlier by Holt to include generalist consumers, such that colonization required at least one suitable resource species (among an array of alternatives) to be present already and extinction was mandated if all potential resources went missing on an island. With these reasonable assumptions, they crafted predictions about food web structure and found that their predictions held in empirical datasets: consumers accumulate at larger areas, compared to primary producers, and small areas are dominated by generalist consumers (which experience only weak sequential dependency). Network relationships are implicit, not explicit, in the model of Holt (2010) mentioned above. It would be an instructive exercise to tie this minimalist model more directly to the generalized sequential dependency models developed by Gravel et al. (2011) and Massol et al. (2017).

Further integration of food webs and biogeography theories should also consider the role of trophic interactions in driving the turnover of species. Specifically, how do factors such as island size and isolation drive the turnover of predator and prey species? What is the role of top-down and bottom-up processes in governing the rate at which the species composition of prey communities turn over across a landscape? And does trophic specialization alter the effects of island characteristics on species turnover? Ryberg et al. (2012) developed a model that predicts how predators can alter the diversity of organisms within a patch as well as the rate of species turnover among patches. Yet empirical tests of this model remain absent and additional theory is required to understand how different types of predator foraging (i.e. generalists or specialists) alter the effects predators have on patterns of prey diversity within and among patches. An additional issue of great current importance is integrating humans as natural enemies into trophic biogeography. Humans act as top predators and also as agents mediating interjections of other natural enemies (e.g. rats, goats) onto even desolate, isolated oceanic islands.

There are likely to be substantial shifts in the form of SARs, concordant with the magnitude of such top-down anthropogenic influences.

Additionally, we need to explicitly consider the importance of spatial scale in the development and testing of theory linking island biogeography to food web ecology. The neutral model presented in Section 12.2 provides a first pass at this, since it shows that if dispersal rates vary systematically with trophic rank, the ranking of z-values with trophic rank shifts with increasing scale. Naturally, experiments testing new theory necessarily tend to operate at much smaller spatial scales than that for which theory is often developed, presenting challenges in rigorously testing theory. Moreover, the role of migration between patches, in addition to how predator–prey interactions alter patterns of biodiversity across patchy landscapes, remains poorly understood. There is a growing body of literature on predator–prey interactions in patchy landscapes (see e.g. Schmitz et al., 2017), yet this literature has yet to be integrated with the literature linking food web ecology and island biogeography, or SARs more broadly. We need theory linking island biogeography to food web ecology that better accounts for movements of animals among patches or sites at a range of spatial and temporal scales.

Another key direction for future research is to examine the joint influences of food web interactions, other kinds of ecological networks, non-trophic interactions and cross-system subsidies on species richness as a function of island area. There is increasing recognition that food web structure and dynamics cannot be fully understood without paying attention to interactions such as mutualism, ecological engineering and information flows (Olff et al., 2009). Specialists are differentially lost from mutualism networks on small fragments or islands, doubtless with consequences for the remaining food webs (Aizen et al., 2012). Spatial subsidies (Polis et al., 1997) are likely more important in small islands, because of their greater perimeter:area ratio, altering colonization or extinction rates for consumers directly or indirectly capable of exploiting those subsidies (Anderson & Wait, 2001), with consequences for the strength of top-down effects (Piovia-Scott et al., 2017). A particularly important avenue for future research will be tying trophic island biogeography more explicitly to modern coexistence theory, which increasingly recognizes the importance of food web interactions for maintaining coexistence in complex assemblages (Chesson, 2018). Some coexistence mechanisms may be ineffective on small islands (e.g. any that involve patch dynamics across a metacommunity) and this surely contributes to shaping the form of realized SARs. Understanding the interplay of coexistence mechanisms and space in maintaining diversity and in contributing to SARs is becoming ever more important in our world, so increasingly dominated by the relentless hammers of anthropogenic habitat destruction and fragmentation (among other drivers of global change) and the looming risk of mass extinctions.

Finally, all these issues should be given an evolutionary spin. Several authors have observed that trophic cascades play out very differently on isolated oceanic islands than in continental areas (Oksanen et al., 2010; Terborgh, 2010) and suggest that this reflects dramatically different evolutionary histories. On islands without top predators, herbivores can exert strong consistent selective pressure that leads to adaptations such as heterophylly and alters the allocation of plant resources to different anti-herbivore defences, depending on the suite of herbivores that are present. This evolutionary dimension at the interface of trophic interactions and species–area relationships has yet to be explored in any depth in the literature.

Acknowledgements

We thank the editors for their patience, Ryan Chisholm, Ben Baiser, Alyssa Cirtwill, Daniel Stouffer, Wade Ryberg, Mark Ritchie, Tomas Roslin and Jana Petermann for comments and references and Vitrell Sherif for assistance with manuscript preparation. R.D.H. was supported by the University of Florida Foundation. J.R. was funded by an Individual Research Fellowship from the Natural Environment Research Council (NERC) (NE/L011611/1). Through J.R. this study is a contribution to Imperial College's Grand Challenges in Ecosystems and the Environment initiative.

References

Adler, P. B., Hille Ris Lambers, J. & Levine, J. M. (2007) A niche for neutrality. *Ecology Letters*, **10**, 95–104.

Aizen, M. A., Sabatino, M. & Tylianakis, J. M. (2012) Specialization and rarity predict nonrandom loss of interactions from mutualist networks. *Science*, **335**, 1486–1489.

Alonso, D., Pinyol-Gallemi, A., Alcoverro, T. & Arthur, R. (2015) Fish community reassembly after a coral mass mortality: Higher trophic groups are subject to increased rates of extinction. *Ecology Letters*, **18**, 451–461.

Amarasekare, P. (2008) Spatial dynamics of foodwebs. *Annual Review of Ecology and Systematics*, **39**, 479–500.

Anderson, W. B. & Wait, D. A. (2001) Subsidized island biogeography: Another new twist on an old theory. *Ecology Letters*, **4**, 289–291.

Bagchi, R., Brown, L. M., Elphick, C. S., Wagner, D. L. & Singer, M. S. (2018) Anthropogenic fragmentation of landscapes: Mechanisms for eroding the specificity of plant-herbivore interactions. *Oecologia*, **187**, 521–533.

Baiser, B., Gotelli, N. J., Buckley, H. L., Miller, T. E. & Ellison, A. M. (2012) Geographic variation in network structure of a Nearctic aquatic food web. *Global Ecology & Biogeography*, **21**, 579–591.

Brown, J. H., West, G. B. & Enquist, B. J. (2000) Scaling in biology: Patterns and processes, causes and consequences. *Scaling in biology* (ed. by J. H. Brown and G. B. West), pp. 1–24. Oxford: Oxford University Press.

Canard, E., Mouillot, D., Mouquet, N. & Gravel, D. (2014) Empirical evaluation of neutral interactions in host-parasite networks. *The American Naturalist*, **183**, 468–479.

Canard, E., Mouquet, N., Marescot, L., Gaston, K. J., Gravel, D. & Mouillot, D. (2012) Emergence of structural patterns in neutral trophic networks. *PLoS One*, **7**, e38295.

Chase, J. M., Gooriah, L., May, F., Ryberg, W. A., Schuler, M. S., Craven, D. & Knight, T. M. (2019) A framework for disentangling ecological mechanisms underlying the island species–area relationship. *Frontiers of Biogeography*, **11**, e40844.

Chave, J. (2004) Neutral theory and community ecology. *Ecology Letters*, **7**, 241–253.

Chesson, P. (2018) Updates on mechanisms of maintenance of species diversity. *Journal of Ecology*, **107**, 1773–1794.

Chisholm, R. A., Fung, T., Chimalakonda, D. & O'Dwyer, J. P. (2016) Maintenance of biodiversity on islands. *Proceedings of the Royal Society B: Biological Sciences*, **283**, 20160102.

Chisholm, R. A., Lim, F., Yeoh, Y. S., Seah, W. W., Condit, R. & Rosindell, J. (2018) Species–area relationships and biodiversity loss in fragmented landscapes. *Ecology Letters*, **21**, 804–813.

Cirtwill, A. R. & Stouffer, D. B. (2016) Knowledge of predator–prey interactions improves predictions of immigration and extinction in island biogeography. *Global Ecology & Biogeography*, **25**, 900–911.

Cohen, J. E. (1977) Ratio of prey to predators in community food webs. *Nature*, **270**, 165–166.

Cohen, J. E. & Briand, F. (1984) Trophic links of community food webs. *Proceedings of the National Academy of Sciences USA*, **81**, 4105–4109.

Cohen, J. E. & Newman, C. M. (1991) Community area and food-chain length: Theoretical predictions. *The American Naturalist*, **138**, 1542–1554.

Cohen, J. E., Jonsson, T. & Carpenter, S. R. (2003) Ecological community description using the food web, species abundance, and body size. *Proceedings of the National Academy of Sciences USA*, **100**, 1781–1786.

Colling, G. & Matthies, D. (2004) The effects of plant population size on the interactions between the endangered plant *Scorzonera humilis* (Asteraceae), a specialized herbivore, and a phytopathogenic fungus. *Oikos*, **105**, 71–78.

Connor, E. F., Courtney, A. C. & Yoder, J. M. (2000) Individuals–area relationships: The relationship between animal population density and area. *Ecology*, **81**, 34–748.

Donohue, I., Petchey, O. L., Kefi, S., Genin, A., Jackson, A. L., Yang, Q. & O'Connor, N. E. (2017) Loss of predator species, not intermediate consumers, triggers rapid and dramatic extinction cascades. *Global Change Biology*, **23**, 2962–2972.

Drakare, S., Lennon, J. J. & Hillebrand, H. (2006) The imprint of the geographical, evolutionary and ecological context on species–area relationships. *Ecology Letters*, **9**, 215–227.

Elton, C. (1927; reprinted 2001) *Animal ecology*. Chicago, IL: University of Chicago Press.

Elton, C. (1966) *The pattern of animal communities*. New York: John Wiley.

Etienne, R. S. & Alonso, D. (2005) A dispersal-limited sampling theory for species and alleles. *Ecology Letters*, **8**, 1147–1156.

Fenoglio, M. S., Srivastava, D., Valladares, G., Cagnolo, L. & Salvo, A. (2012) Forest fragmentation reduces parasitism via species loss at multiple trophic levels. *Ecology*, **93**, 2407–2420.

Fenoglio, M. S., Videla, M. A., Salvo, M. A. & Valladares, G. (2013) Beneficial insects in urban environments: Parasitism rates increase in large and less isolated plant patches via enhanced parasitoid species richness. *Biological Conservation*, **164**, 82–89.

Franzen, M., Schweiger, O. & Betzholtz, P.-E. (2012) Species–area relationships are controlled by species traits. *PLoS One*, **7**, e37359.

Galiana, N., Lurgi, M., Claramunt-Lopez, B., Fortin, M.-J., Leroux, S., Cazelles, K., Gravel, D. & Montoya, J. M. (2018) The spatial scaling of species interaction networks. *Nature Ecology & Evolution*, **2**, 782–790.

Genua, L., Start, D. & Gilbert, B. (2017) Fragment size affects plant herbivory via predator loss. *Oikos*, **126**, 1357–1365.

Graham, N. A. J., Wilson, S. K., Carr, P., Hoey, A. S., Jennings, S. & MacNeil, M. A. (2018) Seabirds enhance coral reef productivity and functioning in the absence of invasive rats. *Nature*, **559**, 250–253.

Grainger, T. N., Germain, R. M., Jones, N. T. & Gilbert, B. (2017) Predators modify biogeographic constraints on species distributions in an insect meta-community. *Ecology*, **98**, 851–860.

Gravel, D., Baiser, B., Dunne, J. A., Kopelke, J.-P., Martinez, N. D., Nyman, T., Poisot, T., Stouffer, D. B., Tylianakis, J. M., Wood, S. A. & Roslin, T. (2018) Bringing Elton and Grinnell together: A quantitative framework to represent the biogeography of ecological interaction networks. *Ecography*, **41**, 1–15.

Gravel, D., Massol, F., Canard, E., Mouillot, D. & Mouquet, N. (2011) Trophic theory of island biogeography. *Ecology Letters*, **14**, 1010–1016.

Guzman, L. M., Germain, R. M., Forbes, C., Straus, S., O'Connor, M. I., Gravel, D., Srivastava, D. S. & Thompson, P. L. (2019) Towards a multi-trophic extension of metacommunity ecology. *Ecology Letters*, **22**, 19–33.

Harfoot, M. B., Newbold, T., Tittensor, D. P., Emmott, S., Hutton, J., Lyutsarev, V., Smith, M. J., Scharlemann, J. P. & Purves, D. W. (2014) Emergent global patterns of ecosystem structure and function from a mechanistic general ecosystem model. *PLoS Biology*, **12**, e1001841.

Harvey, E. & MacDougall, A. S. (2014) Trophic island biogeography drives spatial divergence of community establishment. *Ecology*, **95**, 2870–2878.

Hastings, A. (1977) Spatial heterogeneity and the stability of predator-prey systems. *Theoretical Population Biology*, **12**, 37–48.

Heatwole, H. (2018) Trophic structure stability in insular biotic communities. *The truth is the whole: Essays in honor of Richard Levins* (ed. by T. Awerbuch, M. S. Clark and P. J. Taylor), pp. 220–243. Arlington, MA: The Pumping Station.

Heatwole, H. & Levins, R. (1972) Trophic structure stability and faunal change during recolonization. *Ecology*, **53**, 531–534.

Holt, R. D. (1977) Predation, apparent competition, and the structure of prey communities. *Theoretical Population Biology*, **12**, 197–229.

Holt, R. D. (1992) A neglected facet of island biogeography: The role of internal spatial dynamics in area effects. *Theoretical Population Biology*, **41**, 354–371.

Holt, R. D. (1993) Ecology at the mesoscale: The influence of regional processes on local communities. *Species diversity in ecological communities* (ed. by R. Ricklefs and D. Schluter), pp. 77–88. Chicago, IL: University of Chicago Press.

Holt, R. D. (1996) Food webs in space: An island biogeographic perspective. *Food webs: Contemporary perspectives* (ed. by G. Polis and K. Winemiller), pp. 313–323. New York: Chapman and Hall.

Holt, R. D. (1997) From metapopulation dynamics to community structure: Some consequences of spatial heterogeneity. *Metapopulation biology* (ed. by I. Hanski and M. Gilpin), pp. 149–164. New York: Academic Press.

Holt, R. D. (2010) Towards a trophic island biogeography: Reflections on the interface of island biogeography and food web ecology. *The theory of island biogeography revisited* (ed. by J. B. Losos and R. E. Ricklefs), pp. 143–185. Princeton, NJ: Princeton University Press.

Holt, R. D. & Hoopes, M. F. (2005) Food web dynamics in a metacommunity context: Modules and beyond. *Metacommunities: Spatial dynamics and ecological communities* (ed. by M. Holyoak, M. A. Leibold and R. D. Holt), pp. 68–94. Chicago, IL: University of Chicago Press.

Holt, R. D., Grover, J. & Tilman, D. (1994) Simple rules for interspecific dominance in systems with exploitative and apparent competition. *The American Naturalist*, **144**, 741–777.

Holt, R. D., Lawton, J. H., Polis, G. A. & Martinez, N. (1999) Trophic rank and the species–area relation. *Ecology*, **80**, 1495–1504.

Hubbell, S. P. (2001) *The unified neutral theory of biodiversity and biogeography*. Princeton, NJ: Princeton University Press.

Huffaker, C. B. (1958) Experimental studies on predation: Dispersion factors and predator-prey oscillations. *Hilgardia*, **27**, 343–383.

Jacquet, C., Mouillot, D., Kulbicki, M. & Gravel, D. (2017) Extensions of island biogeography theory predict the scaling of functional trait composition with habitat area and isolation. *Ecology Letters*, **20**, 135–146.

Kinlan, B. P. & Gaines, S. D. (2003) Propagule dispersal in marine and terrestrial environments: A community perspective. *Ecology*, **84**, 2007–2020.

Krishna, A., Guimaraes Jr., P. R., Jordano, P. & Bascompte, J. (2008) A neutral-niche theory of nestedness in mutualistic networks. *Oikos*, **117**, 1609–1618.

Lafferty, K. D. & Dunne, J. A. (2010) Stochastic ecological network occupancy (SENO) models: A new tool for modeling ecological networks across spatial scales. *Theoretical Ecology*, **3**, 123–135.

Lampert, A. & Hastings, A. (2016) Stability and distribution of predator–prey systems: Regional mechanisms and patterns. *Ecology Letters*, **19**, 279–288.

LeCraw, R. M., Kratina, P. & Srivastava, D. S. (2014) Food web complexity and stability across habitat connectivity gradients. *Oecologia*, **176**, 903–915.

Leibold, M. A. & Chase, J. M. (2018) *Metacommunity ecology*. Princeton, NJ: Princeton University Press.

Leibold, M. A., Holyoak, M., Mouquet, N., Amarasekare, P., Chase, J. M., Hoopes, M. F., Holt, R. D., Shurin, J. B., Law, R, Tilman, D., Loreau, M. & Gonzalez, A. (2004) The metacommunity concept: A framework for multi-scale community ecology. *Ecology Letters*, **7**, 601–613.

Liao, J., Bearup, D. & Blasius, B. (2017) Food web persistence in fragmented landscapes. *Proceedings of the Royal Society B: Biological Sciences*, **284**, 20170350.

Lomolino, M. V. (1984) Immigrant selection, predation, and the distribution of *Microtus pennsylvanicus* and *Blarina brevicauda* on islands. *The American Naturalist*, **123**, 468–483.

Losos, J. B. & Schluter, D. (2000) Analysis of an evolutionary species–area relationship. *Nature*, **408**, 847–850.

MacArthur, R. H. & Wilson, E. O. (1967) *The theory of island biogeography*. Princeton, NJ: Princeton University Press.

Martinson, H. M. & Fagan, W. F. (2014) Trophic disruption: A meta-analysis of how habitat fragmentation affects resource consumption in terrestrial arthropod systems. *Ecology Letters*, **17**, 1178–1189.

Massol, F., Dubart, M., Calcagno, V., Cazelles, K., Jacquet, C., Kefi, S. & Gravel, D. (2017) Island biogeography of food webs. *Advances in Ecological Research*, **56**, 183–262.

Massol, F., Gravel, D., Mouquet, N., Cadotte, M. W., Fukami, T. & Leibold, M. A. (2011) Linking community and ecosystem dynamics through spatial ecology. *Ecology Letters*, **14**, 313–323.

Matthews, T. J., Guilhaumon, F., Triantis, K. A., Borregaard, M. K. & Whitaker, R. J. (2016) On the form of species–area relationships in habitat islands and true islands. *Global Ecology & Biogeography*, **25**, 847–858.

McCann, K. S., Rasmussen, J. B. & Umbanhowar, J. (2005) The dynamics of spatially coupled food webs. *Ecology Letters*, **8**, 513–523.

Montoya, J. M. & Galiana, N. (2018) Integrating species interaction networks and biogeography. *Adaptive food webs: Stability and transitions of real and model ecosystems* (ed. by J. C. Moore, P. C. de Ruiter, K. S. McCann and V. Wolters), pp. 289–304. Cambridge: Cambridge University Press.

Murakami, M. & Hirao, T. (2010) Lizard predation alters the effect of habitat area on the species richness of insect assemblages on Bahamian isles. *Diversity and Distributions*, **16**, 952–958.

van Noordwijk, C. G. E., Verberk, W. C. E. P., Turin, H., Heuerman, T., Alders, K., Dekoninck, W., Hannig, K., Regan, E., McCormack, S., Brown, M. J. F., Remke, E., Siepel, H., Berg, M. P. & Bonte, D. (2015) Species–area relationships are modulated by trophic rank, habitat affinity and dispersal ability. *Ecology*, **96**, 518–531.

O'Dwyer, J. P. & Cornell, S. J. (2018) Cross-scale neutral ecology and the maintenance of biodiversity. *Science Reports*, **8**, 10200.

Odum, E. P. (1971) *Fundamentals of ecology*. Philadelphia, PA: W.B. Saunders.

Oksanen, L., Oksanen, T., Dahlgren, J., Hambäck, P., Ekerholm, P., Lindgren, Å. & Olofsson, J. (2010) Islands as tests of the green world hypothesis. *Trophic cascades: Predators, prey, and the changing dynamics of nature* (ed. by J. Terborgh and J. A. Estes), pp. 163–178. Washington, DC: Island Press.

Olff, H., Alonso, D., Berg, M. P., Eriksson, B. P., Loreau, M., Piersma, T. & Rooney, N. (2009) Parallel ecological networks in ecosystems. *Philosophical Transactions of the Royal Society B: Biological Sciences*, **364**, 1755–1779.

Orrock, J. L. & Fletcher Jr., R. J. (2005) Changes in community size affect the outcome of competition. *The American Naturalist*, **166**, 107–111.

Ostman, O., Griffin, N. W., Strasburg, J. L., Brisson, J. A., Templeton, A. R., Knight, T. M. & Chase, J. M. (2007) Habitat area affects arthropod communities directly and indirectly through top predators. *Ecography*, **30**, 359–366.

Ovaskainen, O. & Hanski, I. (2003) The species–area relationship derived from species-specific incidence functions. *Ecology Letters*, **6**, 903–909.

Piechnik, D. A., Lawler, S. P. & Martinez, N. D. (2008) Food-web assembly during a classic biogeographic study: Species 'trophic breadth' corresponds to colonization order. *Oikos*, **117**, 665–674.

Pillai, P., Loreau, M. & Gonzalez, A. (2010) A patch-dynamic framework for food web metacommunities. *Theoretical Ecology*, **3**, 223–237.

Piovia-Scott, J., Yang, L. H., Wright, A. N., Spiller, D. A. & Schoener, T. W. (2017) The effect of lizards on spiders and wasps: Variation with island size and marine subsidy. *Ecosphere*, **8**, e01909.

Polis, G. A., Anderson, W. B. & Holt, R. D. (1997) Toward an integration of landscape ecology and food web ecology: The dynamics of spatially subsidized food webs. *Annual Review of Ecology and Systematics*, **28**, 289–316.

Polis, G. A., Holt, R. D., Menge, B. A. & Winemiller, K. O. (1996) Time, space, and life history: Influences on food webs. *Food webs: Integration of patterns and dynamics* (ed. by G. A. Polis and K. O. Winemiller), pp. 435–460. London: Chapman and Hall.

Ritchie, M. E. (1999) Biodiversity and reduced extinction risks spatially isolated rodent populations. *Ecology Letters*, **2**, 11–13.

Rizzuto, M., Carbone, C. & Pawar, S. (2018) Foraging constraints reverse the scaling of activity time in carnivores. *Nature Ecology & Evolution*, **2**, 247–253.

Rooney, N., McCann, K. S. & Moore, J. C. (2008) A landscape theory for food web architecture. *Ecology Letters*, **11**, 867–881.

Rosindell, J. & Cornell, S. J. (2007) Species–area relationships from a spatially explicit neutral model in an infinite landscape. *Ecology Letters*, **10**, 586–595.

Rosindell, J. & Cornell, S. J. (2009) Species–area curves, neutral models, and long-distance dispersal. *Ecology*, **90**, 1743–1750.

Roslin, T., Várkonyi, G., Koponen, M., Vikberg, V. & Nieminen, M. (2014) Species–area relationships across four trophic levels – decreasing island size truncates food chains. *Ecography*, **37**, 443–453.

Ryberg, R. A. & Chase, J. M. (2007) Predator-dependent species–area relationships. *The American Naturalist*, **170**, 636–642.

Ryberg, W. A., Smith, K. G. & Chase, J. M. (2012) Predators alter the scaling of diversity in prey metacommunities. *Oikos*, **121**, 1995–2000.

Sang, A., Teder, T., Helm, A. & Partel, M. (2010) Indirect evidence for an extinction debt of grassland butterflies half century after habitat loss. *Biological Conservation*, **143**, 1405–1413.

Santos, A. M. C. & Quicke, D. L. J. (2011) Large-scale diversity patterns of parasitoid insects. *Entomological Science*, **14**, 371–382.

Scheiner, S. M. (2003) Six types of species–area curves. *Global Ecology & Biogeography*, **12**, 441–447.

Scherber, C., Andert, H., Niedringhaus, R. & Tscharntke, T. (2018) A barrier island perspective on species–area relationships. *Ecology and Evolution*, **8**, 12879–12889.

Schmitz, O. J., Miller, J. R., Trainor, A. M. & Abrahms, B. (2017) Toward a community ecology of landscapes: Predicting multiple predator–prey interactions across geographic space. *Ecology*, **98**, 2281–2292.

Schoener, T. W. & Spiller, D. A. (2010) Trophic cascades on islands. *Trophic cascades: Predators, prey, and the changing dynamics of nature* (ed. by J. Terborgh and J. A. Estes), pp. 179–202. Washington, DC: Island Press.

Schoener, T. W., Spiller, D. A. & Piovia-Scott, J. (2016) Variation in ecological interaction strength with island area: Theory and data from the Bahamian archipelago. *Global Ecology & Biogeography*, **25**, 891–899.

Spiller, D. A. & Schoener, T. W. (2009) *Species–area relationship. Encyclopedia of islands* (ed. by R. G. Gillespie and D. A. Clague), pp. 857–861. Berkeley, CA: University of California Press.

Steffan-Dewenter, I. & Tscharntke, T. (2000) Butterfly community structure in fragmented habitats. *Ecology Letters*, **3**, 449–456.

Stier, A. C., Hanson, K. M., Holbrook, S. J., Schmitt, R. J. & Brooks, A. J. (2014b) Predation and landscape characteristics independently affect reef fish community organization. *Ecology*, **95**, 1294–1307.

Stier, A. C., Hein, A. M., Parravicini, V. & Kulbicki, M. (2014a) Larval dispersal drives trophic structure across Pacific coral reefs. *Nature Communications*, **5**, 5575.

Summerhayes, V. S. & Elton, C. S. (1923) Contributions to the ecology of Spitsbergen and Bear Island. *Journal of Ecology*, **11**, 214–286.

Terborgh, J. (2010) The trophic cascade on islands. *The theory of island biogeography revisited* (ed. by J. B. Losos and R. E. Ricklefs), pp. 116–142. Princeton, NJ: Princeton University Press.

Thornton, I. W. B. (1996) *Krakatau – The destruction and reassembly of an island ecosystem*. Cambridge, MA: Harvard University Press.

Toft, A. & Schoener, T. W. (1983) Abundance and diversity of orb spiders on 106 Bahamian Islands: Biogeography at an intermediate trophic level. *Oikos*, **41**, 411–426.

Triantis, K. A., Guilhaumon, F. & Whittaker, R. J. (2012) The island species–area relationship: Biology and statistics. *Journal of Biogeography*, **39**, 215–231.

Wang, S. & Loreau, M. (2014) Ecosystem stability in space: α, β and δ variability. *Ecology Letters*, **17**, 891–901.

Warren, B. H., Simberloff, D., Ricklefs, R. E., Aguilée, R., Condamine, F. L., Gravel, D., Morlon, H., Mouquet, N., Rosindel, J., Casquet, J., Conti, E., Cornuault, J., Fernández-Palacios, J. M., Hengl, T., Norder, S. J., Rijsdijk, K.

F., Sanmartín, I., Strasberg, D., Triantis, K. A., Valente, L. M., Whittaker, R. J., Gillespie, R. G., Emerson, B. C. & Thébaud, C. (2015) Islands as model systems in ecology and evolution: Prospects fifty years after MacArthur-Wilson. *Ecology Letters*, **18**, 200–217.

Whittaker, R. J. & Fernández-Palacios, J. M. (2007) *Island biogeography: Ecology, evolution, and conservation*, 2nd ed. Oxford: Oxford University Press.

Wilson, H. B., Holt, R. D. & Hassell, M. P. (1998) Persistence and area effects in a stochastic tritrophic model. *The American Naturalist*, **151**, 587–596.

Part IV

The Species–Area Relationship in Applied Ecology

13 · *The Identification of Biodiversity Hotspots Using the Species–Area Relationship*

SIMONE FATTORINI

13.1 The Concept of Biodiversity Hotspots

The concept of biodiversity hotspots was introduced by Myers (1988) to indicate areas with high levels of endemism (measured as the number of endemic vascular plants) and which are, at the same time, acutely threatened (measured by the percentage of habitat loss). Based on this qualitative definition, that is, without the adoption of threshold values, Myers (1988) initially identified ten tropical forests as global hotspots, to which a further eight areas (including four in Mediterranean regions) were included in a subsequent analysis (Myers, 1990). Myers' concept of hotspots was adopted by Conservation International (see Conservation International, 2019) as its institutional blueprint in 1989 and has become increasingly popular. To surpass the original subjectivity in hotspot definition and identification, Mittermeier et al. (1999) and Myers et al. (2000) introduced quantitative criteria based on a priori established threshold values. According to these criteria, to qualify as a biodiversity hotspot, a region must have more than 1,500 endemic vascular plants (i.e. about 0.5 per cent of global plant richness, currently estimated to be 308,312 species; Christenhusz & Byng, 2016) and have lost at least 70 per cent of its original natural vegetation. Using this logic, Myers et al. (2000) identified twenty-five biodiversity hotspots. Further updates and revisions were undertaken by Mittermeier et al. (2004, 2011) leading to the identification of thirty-five regions which, despite comprising only 17 per cent of the Earth's land surface (excluding Antarctica), sustain about 50 per cent of the world's plant diversity, 77 per cent of all mammal, bird, reptile and amphibian species and 43 per cent of all terrestrial vertebrate species as endemics, as well as 80 per cent of all threatened amphibians (Brooks et al., 2002; Mittermeier et al., 1998, 2004; Williams et al., 2011). Recently, a further hotspot has been proposed by Noss et al. (2015) and has been designated as the thirty-sixth hotspot by the Critical Ecosystem Partnership Fund (Noss, 2016).

The approach proposed by Myers and coworkers has, however, raised a number of critiques (Marchese, 2015). For example:

1) The concept of hotspots has been applied only to terrestrial ecosystems and thus excluding the marine realm, where, for example, coral reefs represent areas that host high biodiversity and are under severe threat (Roberts et al., 2002; Bellwood et al., 2004).
2) Only focusing on hotspots may lead us to lose large, natural and ecologically important areas that have few species (coldspots; Kareiva & Marvier, 2003), but which contribute to many different ecosystem services.
3) According to Smith et al. (2001), focusing on particular habitats means biodiversity hotspots ignore regions of ecological transition. However, this has been questioned by Araújo (2002), who highlighted that, in fact, biodiversity hotspots tend to be located in areas of ecological transition.
4) Plant endemism may not be a good proxy for overall diversity. In particular, although the proponents of the global hotspot strategy highlight that these areas also host large numbers of species of other taxa, such as birds and mammals, there is no indication that they can be considered a good proxy for other taxa, such as arthropods, which represent the largest animal group (e.g. Stork et al., 2014).
5) The identification of global hotspots may help to focus our attention, but their extent is too large for effective conservation actions (Brummitt & Nic Lughadha, 2003; Cañadas et al., 2014), which in many cases would involve multiple national governments. Actually, as observed by Myers and Mittermeier (2003), several of the multinational hotspots are in fact targeted by multigovernmental efforts. In fact, global hotspots can be viewed as large areas where conservationists should concentrate their attention and, for example, create one or more reserves. This process requires shifting from the macrogeographical to the local scale, which generates several problems. For example, at the local scale, hotspots may host large numbers of common and widespread species, instead of endemic, rare or imperilled species (Prendergast et al., 1993a, b).

Despite these limitations, the hotspot strategy has become the principal global conservation-prioritization approach (Mittermeier et al., 2011) and has attracted over \$1 billion in conservation investments from entities such as the Critical Ecosystem Partnership Fund, The MacArthur Foundation, The Global Conservation Fund and Conservation International (Sloan et al., 2014).

The idea of identifying priority areas for biological conservation has attracted much attention and stimulated researchers to investigate definitions and procedures different from those proposed by Mittermeier et al. (1999) and Myers et al. (2000); not only for the identification of global hotspots, but also to rank areas at smaller spatial scales. The original hotspots scheme had three primary characteristics, all of which can be expanded upon:

1) It is based on a coarse scale, that is, it is aimed at identifying large regions at the global level, but the idea of area prioritization can be applied at any scale;
2) It is based on endemics, but other diversity metrics could be used (for example, species richness and functional and phylogenetic diversity);
3) It is based on only one taxon (vascular plants), but any taxon could be considered.

Thus, using a more extensive concept of hotspots (see Reid, 1998), various authors have used different metrics and have applied them to different taxa at varying scales, from global to very local. Hotspots have been defined using species richness (e.g. Myers et al., 2000; Veech, 2000; Brummitt & Nic Lughadha, 2003; Maes et al., 2003; Ovadia, 2003; Fattorini, 2006a; Guilhaumon et al., 2008; Jenkins et al., 2013), species rarity or conservation status (Prendergast et al., 1993a; Dobson et al., 1997; Troumbis & Dimitrakopoulos, 1998; Griffin, 1999; Possingham & Wilson, 2005; Funk & Fa, 2010), a combination of richness and endemism (e.g. Kier & Barthlott, 2001; Hobohm, 2003), evolutionary distinctiveness (Cadotte & Davies, 2010; Jetz et al., 2014), functional diversity (Stuart-Smith et al., 2013) and a combination of species richness, phylogenetic diversity and functional diversity (Mazel et al., 2014).

When applied simultaneously to the same dataset, different metrics produce different prioritization schemes (Devictor et al., 2010). For example, using worldwide data on birds, Orme et al. (2005) found a lack of congruence between hotspots defined with the criterion of species endemism and those areas with high species richness or concentrated threat.

In regard to the taxonomic coverage, it has been observed that the use of different taxa tends to produce different prioritizations. In general, the correlation of species richness and coincidence of diversity hotspots between pairs of taxa may be dependent on the spatial scale considered. As a rule, studies at coarse resolutions (e.g. countries to continents) have typically found relatively high cross-taxon congruence in species richness and endemism patterns, because of common responses of different organisms to broad-scale variations in climate and geological history and as a statistical consequence of differences in range size between species (Fattorini et al., 2011). By contrast, studies of cross-taxon congruence at finer resolutions (local scales) have revealed moderate to low correlations, although there are examples of high correlations (Fattorini et al., 2011). The differences between coarse and fine resolution studies is related to the size of the observational units (grain) (Caro, 2010) and a low congruence at small scales is expected because of the small range size of many species, which reduces the probability of overlap when a fine resolution is used to map them (Grenyer et al., 2006). Recently, Jenkins et al. (2013), using a scale 100-times finer than previous assessments (Ceballos & Ehrlich, 2006) to map global priority areas for mammals, amphibians and birds, found a high congruence among taxa when total species richness was considered, but important differences emerged in the patterns of small ranged species, particularly the amphibians.

13.2 Species Richness and the Species–Area Relationship

When species richness is used to identify hotspots, the areas with an exceptionally large number of species are considered as hotspots. If the compared areas are of the same size, as in cases where a grid system is used, it is possible to simply rank them according to the number of species they contain and to designate as hotspots those that rank highest, that is, that contain the highest species richness (see Williams et al., 1996; Maes et al., 2003; Balletto et al., 2010). However, if the compared areas are of different size, their species richness values cannot be compared directly, because larger areas tend to have more species, a pattern known as the species–area relationship (SAR). For this reason, Mittermeier et al. (1998) and Myers et al. (2000) proposed the use of the species–area ratio. The species–area relationship is not necessarily linear, however, so one cannot divide the total species richness by the area and report the result as species per unit area (e.g. Veech, 2000; Brummitt & Nic Lughadha, 2003; Ovadia, 2003). To overcome this problem, various approaches have been proposed and are now discussed in-turn.

13.3 Use of the c-Parameter of the Power Function for Hotspot Identification

Although several mathematical functions have been proposed to model the SAR (Tjørve, 2003, 2009; Dengler, 2009; Williams et al., 2009; Chapter 7), comparative studies identify the Arrhenius power function as the model which, in general, provides the best fit to the most empirical datasets (Triantis et al., 2012; Matthews et al., 2016) and which is best supported by various ecological theories (e.g. Rosenzweig, 1995; Martín & Goldenfeld, 2006). The Arrhenius power function is:

$$S = cA^z, \tag{13.1}$$

where S represents species richness, A the area and c and z are fitting parameters. This function can be linearized with a double log-transformation:

$$\log(S) = \log(c) + z\log(A). \tag{13.2}$$

It should be noted that, since different quantities are minimized, Equations (13.1) and (13.2) are not statistically equivalent for least squares regression and thus they give different estimates of z and c. Therefore, Equations (13.1) and (13.2) should be considered two different models. Whereas Equation (13.2) can be modelled with ordinary least squares regressions, Equation (13.1) must be modelled using a non-linear fitting algorithm. Although Equation (13.2) is the most frequently applied approach, a non-linear fit may provide a better fit, depending on the dataset; thus, the best fitted model can only be evaluated by comparing the two models using empirical data (Fattorini, 2006a, 2007a).

Assuming the power function to be the best SAR model, Rosenzweig (1995), Ovadia (2003) and Brummitt and Nic Lughadha (2003) used the c-parameter as a measure of species richness standardized by area. Because c is the ratio of diversity (S) to A^z, the higher the c-value the faster the increase in species richness and, consequently, the higher the qualitative rank for the area. To obtain realistic scores of relative diversity, it is however important to use appropriate values of z. For example, Ovadia (2003) and Brummitt and Nic Lughadha (2003) used a priori z-values such as 0.14 (Brummitt & Nic Lughadha), 0.18 (Ovadia) and 0.25 (Brummitt & Nic Lughadha), because these values are among those most commonly found in mainland and island systems. By applying this approach to the twenty-five hotspots of Myers et al. (2000), the composition of leading hotspots (i.e. hotspots that appeared at least three times in the top ten listings for each of the following factors: number of endemic plants, number of endemic vertebrates, species–area ratio of endemic plants, species–area ratio of endemic vertebrates and habitat loss) changed considerably. Three hotspots were added (Tropical Andes, Mediterranean Basin and Mesoamerica) and two were removed (Western Ghats/Sri Lanka and Eastern Arc and Coastal Forests of Tanzania/Kenya) (Ovadia, 2003). A reanalysis of the hotspots of Mittermeier et al. (1999) with this approach also resulted in major changes to the hotspot rankings based on the number of vascular plants, number of endemic plants, number of vertebrates and number of endemic vertebrates (Brummitt & Nic Lughadha, 2003). As a general pattern, Brummitt and Nic Lughadha (2003) found that, whereas diversity increased with area for the power-transformed values, it appeared to decrease with area for untransformed species–area ratios. Of course, the hotspot rankings changed considerably. For example, the Tropical Andes, which were not ranked among the highest in Mittermeier et al. (1999), ranked highest in Brummitt and Nic Lughadha's (2003) analysis.

Although Brummitt and Nic Lughadha (2003) obtained similar results using $z = 0.25$ and $z = 0.14$, z varies according to the study system and appropriate choices should be made on the basis of empirical evaluations (Fattorini, 2006a). Of course, the simple S/A ratio would be appropriate only if $z = 1$, a value that seems to be quite rare in real-world systems (see Triantis et al., 2012).

Another popular function used to model the species–area relationship is the Gleason function:

$$S = c + z \log (A), \tag{13.3}$$

where c and z are again fitted parameters. Assuming the Gleason function is the best fit SAR model, various authors have adopted strategies to compare the species richness of areas of different size. For example, Evans et al. (1955) proposed the following equation:

$$S = [s/ \log (n + 1)] \times \log (N + 1), \tag{13.4}$$

where S is an estimate of the number of species expected in an area consisting of N units and s is the number of species encountered in n units selected at random. The transformation using the logarithm of $n + 1$ causes the species–area line to pass through the intersection of the coordinates and thus permits a better estimate of S when this is based on relatively few samples. This equation has been used to calculate relative species richness for areas of different size, where S is the species number sought in a standard unit of area (e.g. per 10,000 km^2), s is the given species number in a given unit of area (e.g. of a country), n is the size of the given unit of area and N is the size of the chosen standard unit of area (see, for example, Barthlott et al., 1996, and references therein). However, Equation (13.4) would be correct only if $c = 0$ or if c was identical in all systems. As c-values differ among systems, Equation (13.4) does not produce comparable estimates of S. As such, we do not recommend using it for hotspot identification.

13.4 Use of Residuals from SAR Model Fitting

Another relevant problem with using c-values from the power model as a measure of species richness corrected for area is that neither the c-value nor the z-value represents the magnitude of species diversity, because both parameters are responsible for the shape of the power function (in arithmetic space; see Chapter 4). Thus, some authors (e.g. Veech, 2000; Ulrich & Buszko, 2005; Fattorini, 2006a) have proposed that a SAR model be fitted to empirical data and then the areas located above the fitted curve to be considered as hotspots (i.e. those areas with a positive residual), because these areas have more species than predicted by the fitted function. This procedure has also been applied by Pomeroy (1993) and Ceballos and Brown (1995) to identify areas of exceptionally high richness, although in neither study do the authors refer to these areas as hotspots.

An application of this procedure is shown here for the vascular plants of the thirty-five global biodiversity hotspots (Table 13.1). Results of ordinary least squares regressions to fit the linearized power function indicate that the original hotspot area is a better predictor than the current extent of pristine vegetation (Table 13.2). Using species richness (Figure 13.1), the regions which ranked highest (top 10 per cent) in regards to their residuals were the Tropical Andes, Sundaland, the Caribbean Islands and the Atlantic Forest. On the other hand, if only endemics are considered (Figure 13.2), the regions that ranked highest were Sundaland, the Tropical Andes, Madagascar and the Cape Floristic Region. These results are in contrast with those obtained by Myers et al. (2000), who, using the endemics–area ratio, identified the Eastern Arc and Coastal Forests of Tanzania/Kenya, the Philippines, New Caledonia and Polynesia/Micronesia as the top ranked hotspots. However, the ranking based on the residuals for total richness is mostly in agreement with Myers et al.'s (2000) ranking based on the percentage of endemics, which placed the Tropical Andes, Sundaland, the Mediterranean Basin and Madagascar as the top ranked hotspots.

Table 13.1 *Values of the original and current extent (km^2) of natural vegetation, species richness of vascular plants and endemic plant species in thirty-five global biodiversity hotspots. Values of richness and number of endemics are taken from Mittermeier et al. (2011). Data on the extent of natural vegetation are from Conservation International (2005)*

Hotspot	Original extent of natural vegetation	Current extent of natural vegetation	Species richness	Endemics
Tropical Andes	1,542,644	385,661	30,000	15,000
Tumbes-Chocó-Magdalena	274,597	65,903	11,000	2,750
Atlantic Forest	1,233,875	99,944	20,000	8,000
Cerrado	2,031,990	438,910	10,000	4,400
Chilean Winter Rainfall and Valdivian	397,142	119,143	3,892	1,957
Mesoamerica	1,130,019	226,004	17,000	2,941
Madrean Pine-Oak Woodlands	461,265	92,253	5,300	3,975
Caribbean Islands	229,549	22,955	13,000	6,550
California Floristic Province	293,804	73,451	3,488	2,124
Guinean Forests of West Africa	620,314	93,047	9,000	1,800
Cape Floristic Region	78,555	15,711	9,000	6,210
Succulent Karoo	102,691	29,780	6,356	2,439
Maputaland-Pondoland-Albany	274,136	67,163	8,100	1,900
Coastal Forest of Eastern Africa	291,250	29,125	4,000	1,750
Eastern Afromontane	1,017,806	106,870	7,598	2,356
Horn of Africa	1,659,363	82,968	5,000	2,750
Madagascar and the Indian Ocean	600,461	60,046	13,000	11,600
Mediterranean Basin	2,085,292	98,009	22,500	11,700
Caucasus	532,658	143,818	6,400	1,600
Irano-Anatolian	899,773	134,966	6,000	2,500
Mountains of Central Asia	863,362	172,672	5,500	1,500
Western Ghats and Sri Lanka	189,611	43,611	5,916	3,049
Himalaya	741,706	185,427	10,000	3,160
Mountains of Southwest China	262446	20,996	12,000	3,500
Indo-Burma	2,373,057	118,653	13,500	7,000
Sundaland	1,501,063	100,571	25,000	15,000
Wallacea	338,494	50,774	10,000	1,500
Philippines	297,179	20,803	9,253	6,091

Table 13.1 (*cont.*)

Hotspot	Original extent of natural vegetation	Current extent of natural vegetation	Species richness	Endemics
Japan	373,490	74,698	5,600	1,950
Southwest Australia	356,717	107,015	5,571	2,948
East Melanesian Islands	99,384	29,815	8,000	3,000
New Zealand	270,197	59,443	2,300	1,865
New Caledonia	18,972	5,122	3,270	2,432
Polynesia-Micronesia	47,239	10,015	5,330	3,074
Forests of East Australia	253,200	58,900	8,257	2,144

Table 13.2 *Results of regression analyses (linearized power function) for the species–area relationship and endemics–area relationship for vascular plants in thirty-five global hotspots of biodiversity*

Model	$\log(c)$	t	P	z	t	P	$F_{1,33}$	R^2
Species richness versus original extent of natural vegetation	2.370 ±0.453	5.234	<0.0001	0.274 ±0.008	3.413	0.002	11.649	0.261
Endemic species versus original extent of natural vegetation	2.358 ±0.553	4.262	<0.0001	0.209 ±0.098	2.127	0.041	4.526	0.121
Species richness versus current extent of natural vegetation	2.926 ±0.491	5.953	<0.0001	0.204 ±0.102	2.012	0.052	4.305	0.109
Endemic species versus current extent of natural vegetation	3.195 ±0.580	5.507	<0.0001	0.070 ±0.120	0.581	0.565	0.337	0.101

To identify hotposts on the basis of the SAR, Hobohm (2003) proposed a metric, called the α-index, defined as:

$$\alpha = \log S - [z \log (A) + \log (c)], \tag{13.5}$$

or, where the analysis is undertaken using endemics (E), as:

$$\alpha = \log E - [z \log (A) + \log (c)]. \tag{13.6}$$

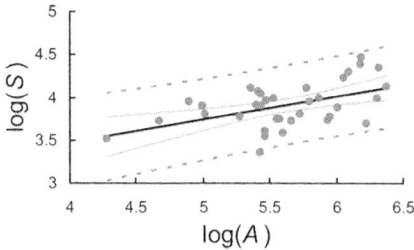

Figure 13.1 Relationship (linearized power function) between species richness (S) and area (A) for vascular plants in thirty-five global hotspots of biodiversity, using data on the original extent of natural vegetation (Table 13.1). Solid black line: ordinary least squares regression fit; solid grey lines: 95 per cent confidence intervals; dotted grey lines: 95 per cent prediction intervals.

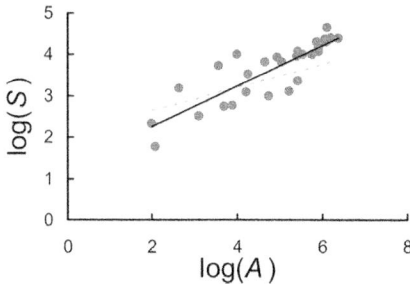

Figure 13.2 Relationship (linearized power function) between species richness (S) and area (A) for vascular plants from twenty-eight worldwide regions as given by Hobohm (2003). Solid black line: ordinary least squares regression fit; dotted grey line: expected relationship on the basis of the c- and z-values calculated by Hobohm (2003) for the world flora.

In fact, this index value is exactly the residual of an area from the linearized power function regression line. Thus, this method is identical to the standard approach using residuals. The advantage of Hobohm's (2003) equations is that they can be used to calculate residuals from a hypothetical regression line, even when there are no data points to perform a regression analysis; if c- and z-values from other systems can be used with any degree of confidence. For example, Hobohm (2003), using plant richness values from 121 worldwide areas (including previously recognized hotspots), obtained the following SAR:

$$\log(S) = 0.28 \log (A) + 2.08, \tag{13.7}$$

with an R^2 of 0.59.

These values for c and z would make it possible to calculate the hypothetical residual from this line for the flora of any given region. While this approach may be appealing to generate an estimate of the species (or endemic species) richness corrected for the size of an area, it is questionable because c- and z-values are fitted parameters and are thus dependent on the particular dataset for which they are calculated. For example, for the Tropical Andes, a hotspot with an area of 1,258,000 km^2 and a richness of 45,000 species, using z- and c-values from Equation (13.7), we obtain:

$$\alpha = \log(45,000) - [0.28 \log(1,258,000) + 2.08] = 0.865. \qquad (13.8)$$

This hypothetical residual is, however, much larger than can be obtained from a true regression line. Using values of area and plant species richness for twenty-eight worldwide regions considered as hotspots by Hobohm (2003) and which are a subset of his 121 areas, the equation resulting from an ordinary least squares regression is:

$$\log(S) = (1.266 \pm 0.286) + (0.491 \pm 0.058) \log(A), \qquad (13.9)$$

with $R^2 = 0.736$, $F_{1,26} = 72.385$, and $P < 0.0001$ (Figure 13.2). From this equation, it is possible to calculate the residuals and they can be compared with the values of the α-index obtained using $z = 0.28$ and $c = 2.08$, as in Hobohm's study. In the case of the Tropical Andes, the residual from this regression is 0.392, less than half that calculated using Equation (13.8). Perhaps more importantly, although residuals from the fitted equation and values of the α-index in this exercise were correlated (Spearman rank correlation coefficient, $r_s = 0.777$, $P < 0.0001$), agreement with regards to prioritization was low. If the areas corresponding to the top 25 per cent of ranks are selected as hotspots (i.e. the seven areas with highest values of residuals and α-index), the percentage agreement was 43 per cent (i.e. only three areas were prioritized by both approaches). In addition, New Caledonia, which was the area with the lowest positive α-index value, was included amongst the top-ranked hotspots when true residuals for the selected twenty-eight biodiversity hotspots were used. This exercise indicates that the α-index approach, when used in combination with c- and z-values from other systems, gives different results to the α-index approach when using values from the true regression line.

13.5 Use of Multiple Models

Although the power function has been applied extremely widely, several alternative functions have been proposed to model the SAR

(Chapter 7) and it seems that the best model, if any, can only be established empirically. In fact, it is not only that the SAR of different taxa may be best modelled by different functions, but there is often substantial uncertainty as to which function provides the best fit in cases where different functions are applied to the same dataset. In addition, the power function may produce very different results when applied in its original form (with an algorithm for non-linear fitting) and in the linearized form (using ordinary least squares regression) (Fattorini, 2007a).

For this reason, Veech (2000) proposed that various fitting functions should be applied and then the areas with higher positive residuals in the most models be considered as hotspots. This approach has been followed by Fattorini (2006a), who applied fourteen fitting functions (a linear model; five convex models: exponential, power, linearized power and negative exponential with and without location parameter; and eight sigmoidal models: logistic, Gompertz, Lomolino, He and Legendre, cumulative Weibull distribution and cumulative Weibull distribution with location parameter and the latter two calculated using both a fitted and a pre-established asymptote) to the same dataset (the tenebrionid beetles of thirty-two Aegean Islands) and calculated the percentage similarity of identified hotspots selected by the different models. To achieve this, he ranked the displacement of each island from each curve, based on the value of its residual. Then, for the residuals of each curve, he selected as hotspots the highest 10 per cent and 25 per cent of ranks (i.e. the first three and eight islands, respectively) from all fourteen functions. Then, following Veech (2000), to evaluate the congruence of different functions in recovering the same areas as hotspots, percentage similarity among the fourteen functions was determined as the number of hotspots identified by all fourteen functions divided by the total number of hotspots. In this exercise, the percentage similarity of identified hotspots was never below 72 per cent for the first 25 per cent ranked islands and in several cases the group of hotspots identified by the fourteen functions had 90–100 per cent of the members common to all fourteen groups.

The use of residuals from different functions, however, may produce some paradoxical results. If a function fits the data extremely well, the residuals will be very small and no area may be considered as a hotspot. Moreover, with the possible exception of areas that lie exactly on the regression line, all areas will have a positive or a negative residual. This means that in all cases there will be some areas with positive, albeit extremely small, residuals. In other words, the use of residuals may lead to areas being identified as hotspots if they have higher species richness than predicted by the best fit model for any value of richness and for any size of the residuals, thus including areas with few species and/or with small residuals. To overcome this problem, one could decide that a residual must exceed some predetermined value in order for the region to be considered a hotspot. To have a threshold criterion, one might use the 95 per cent confidence intervals of the regression line (see Thomasson, 1999). However, this procedure is problematic at very large or very small areas due to the expanding of the confidence limits (see Ulrich & Buszko, 2005). For this reason, Ulrich and Buszko (2005) used the 95 per cent confidence limits of the log-transformed intercept value of the power function for constructing boundaries and only areas with residuals above the upper limit were identified as hotspots. This approach, however, may still be problematic. First, it is possible that there are no areas with a residual exceeding this threshold value. This is the case, for example, when the power model is applied to Hobohm's global hotspot data discussed in Section 13.4, where no areas were outside of the 95 per cent confidence limits of the log-transformed intercept value (Figures 13.1 and 13.2). A second problem is that, when multiple models are compared, the use of residuals may lead to the unsettling conclusion that the models that fit the data the poorest are also the best at identifying hotspots, in that they are the ones that return the largest residuals (Veech, 2000). Veech (2000) justified the procedure of using several models because different functions may give different results. Actually, he found that the functions he used failed to agree on hotspots of species richness; a result that was, however, mostly due to the particular behaviour of the extreme value function. In fact, the analysis of the Aegean tenebrionids (Fattorini, 2006a) revealed that different models tended to produce similar results in hotspot identification. Based on these results it seems to be epistemologically and practically preferable to identify a best fit model and then to restrict the analysis of residuals to this model.

However, when the data provide support for several models, relying on only the best model is inadequate and multimodel inference is recommended as a way to construct a robust final inference (Guilhaumon et al., 2008). For example, to take into account uncertainty in the identification of the best fit model(s), Guilhaumon et al. (2008) proposed the use of multimodel SARs and applied this approach to the identification of global hotspots using species richness of vascular plants and vertebrates across the world's terrestrial ecoregions. They considered four convex models (power, exponential, negative exponential and Monod) and four sigmoidal models (rational function, logistic, Lomolino and cumulative Weibull) and evaluated the relative likelihood of each model using Akaike weights derived from the Akaike Information Criterion and normalized across the set of candidate models to sum to one. Then, to construct a robust final inference, they used model averaging and considered the weighted average of model predictions with respect to model weights. Finally, they ranked ecoregions with respect to their positions in the confidence interval of the model-averaged SAR (see Guilhaumon et al., 2008, for technical details). They found that: 1) although in most cases individual models provided good fits, in several circumstances no single model adequately described the SAR; 2) where one or more of the models tested did provide satisfactory fits to a dataset, the model which provided the best fit was extremely variable and 3) where more than one of the models tested provided a satisfactory fit to a dataset, there was often substantial uncertainty regarding which of these provided the best fit. For these reasons, they recommended the use of multimodel SARs. To investigate the effect of accounting for uncertainty when comparing the species richness of places with varying size, Guilhaumon et al. (2008) assessed the percentage similarity between the rankings obtained using their multimodel inference approach and from the residuals of the linearized power function. They found variable results. Using the cut-off of distinguishing the richest 2.5 per cent of ecoregions as hotspots, between 30 per cent (birds) and 78 per cent (amphibians) of the hotspots identified by the two approaches were the same. These findings stress the need for testing multiple hotspot identification approaches in combination with a range of taxa when undertaking assessments of conservation priorities.

13.6 Endemics–Area Relationships (EARs) and Hotspot Identification

Incorporating the concept of geographical rarity in area prioritization is of key importance in biological conservation because species with small ranges may also have small population sizes (e.g. Borregaard & Rahbek, 2010), which increases their extinction risk (Fattorini, 2017). Thus, prioritizing areas that host high levels of endemism has long been used when setting targets for conservation actions. The use of endemic species, instead of species richness, in the definition of hotspots given by Mittermeier et al. (1999) and Myers et al. (2000), as well as in other conservation approaches (such as, for example, 'endemic bird areas'; Stattersfield et al., 1998), also guarantees that there is no overlap in the diversity of selected areas. Since endemics are, by definition, exclusive to single areas, one can sum the number of endemics occurring in the selected areas and have the total number of endemics that are included using this set of areas. In contrast, summing the values of species richness of a set of areas does not give us the value of the total number of included species across all areas because of the presence of shared species. For this reason, some authors have concentrated on the number of endemics and compared regions of different size by using the density of endemics, calculated as the number of endemics (E) over area (A). Examples of the use of E/A ratios in hotspot identification can be found in Mittermeier et al. (1999) and Myers et al. (2000). The power function (Equation 13.1) can be applied to data on endemics (E) to model the endemics–area relationship (EAR) as:

$$E = cA^z, \tag{13.10}$$

which can be linearized as:

$$\log(E) = \log(c) + z \log(A). \tag{13.11}$$

However, the use of E/A ratios in cases where the endemics–area relationship is not linear suffers from the same aforementioned issue as when using species/area ratios. Again, the E/A ratio would be appropriate only if $z = 1$.

In a nested plot design, where every plot (except the largest) is included in a larger one, the z-value of the EAR modelled by the power function should be as high as, or higher than, one, because the number of endemics should increase at the same rate as, or more rapidly than, the total number of species. This was shown empirically by Storch et al. (2012; see also Chapter 8), who found that the EARs for amphibians, birds and mammals at a global scale can be assimilated into one universal power law with a z-value close to one (after the area is rescaled by using the range sizes of the taxa). If empirical EARs do really have z-values close to one, the use of the E/A ratio to locate hotspots would be correct. However, the situation is very different for isolated and independent areas, such as islands or regions, where the largest areas are not the sum of the smaller ones. In these cases, the z-values are much more variable and often lower than one (cf. Hobohm, 2000; Ulrich & Buszko, 2005; Triantis et al., 2008; Georghiou & Delipetrou, 2010; Chapter 14), which indicates that the EAR is not linear.

Of course, the impact of EAR non-linearity is larger when the compared areas have very different sizes, whereas it may be negligible for areas of similar size as, for small differences in area size, the curvilinear shape of the EAR can be well approximated by a straight line. Thus, the E/A ratio may be an appropriate measure of endemism density for areas of similar size. However, if the compared areas are approximatively of the same size, it would be easier to compare absolute numbers of endemics (E), with no need to standardize for A. Similarly, it is possible to directly compare areas that have different sizes but very similar numbers of endemics. In this case, the smaller area obviously has the highest density of endemic species (Hobohm & Tucker, 2014).

As previously observed for the SAR, one could use c-values from Equations (13.10) or (13.11) to express the number of endemics corrected for the size of an area (Brummitt & Nic Lughadha, 2003; Ovadia, 2003). However, as in the case of SARs, if z-values differ between taxa and regions, as is frequently observed for datasets comprising independent areas, using EAR c-values may also be problematic, as c-values should only be calculated after the fit of the model to each dataset in order to obtain appropriate estimates of z. As discussed in the context of the SAR, it is also possible to rank areas using residuals from the EAR (Hobohm, 2000; Ulrich & Buszko, 2005; Fattorini, 2006a). This approach is, however, even more problematic than for the SAR, because it can produce misleading results when areas with zero endemics are included in the calculations (Fattorini, 2006a). For these reasons, Fattorini (2006a) stressed the need for using different methods, such as the percentage of endemics occurring in the compared areas, to take into account different levels of endemism and different kinds of endemic species (e.g. endemic to single islands versus endemic to the archipelago).

Recently, Hobohm et al. (2019) used a nested circle design, with grain sizes of 10^4, 10^5, 10^6, 10^7 and 10^8 km^2 randomly located across sixty points on the Earth's surface, to construct vascular plant EARs for 300 datasets (five grain sizes across sixty points). Species richness values for each dataset were estimated by a panel of experts. Using this method, Hobohm et al. (2019) provided reference values of endemic vascular plants for circular areas of different sizes across the Earth and noted some interesting observations. For example, while no endemic is expected, on average, to be found in areas below 13,000 km^2, a random area of 10^7 km^2 is expected to host, on average, between 400 and 928 endemic vascular plants. These findings may potentially be used as a reference to identify areas with particularly large numbers of endemic species.

13.7 Proportion of Endemics (E/S)

The percentage of taxa endemic to an area (i.e. the proportion of endemic species, E, relative to the total number of species, S, living there) is the most frequently used measure to characterize the relative endemism of an area's biota (see Anderson, 1994) and to compare the number of species in areas of different size (see Fattorini, 2007b) for hotspot identification. The use of E/S ratios (or percentages) has some advantages. Larger areas tend to have more species and hence more endemics. Thus, a certain region may appear poorer in endemics only because of its smaller size in comparison with other regions. However, the use of percentages is problematic for two important reasons. First, percentages obscure differences in raw numbers that can be important in conservation biology (Hobohm, 2000; Ungricht, 2004; Magurran & McGill, 2011). The same percentage value can be obtained using completely different values of E and S. For example, an area with a total of ten species, five of which are endemic, will have the same percentage of endemics as an area with a hundred species, fifty of which are endemic, but it is obvious that the second area should be considered much more important if our aim is to preserve as many endemics as possible. Second, if the number of endemic species is correlated with the number of non-endemic species, then the percentage of endemic species is not an unbiased measure of the level of endemism. This problem has been stressed by Vilenkin and Chikatunov (1998) with regards to regression approaches. If E is the number of endemics (i.e. the species occurring only within the limits of a well-defined area) and NE is the number of non-endemics (i.e. all species with ranges extending outside the focal area), the percentage of endemism is:

$$PE = 100 \times E/(E + NE), \qquad (13.12)$$

and hence:

$$E = NE \times PE/(100 - PE). \qquad (13.13)$$

Vilenkin and Chikatunov (1998) noted that when the separate faunas in a series of areas are the focus of interest, the term $PE/(100 - PE) = b$ can be considered as the regression coefficient of E on NE. If the regression of E on NE is not significant, E and NE are statistically independent and PE, the endemism percentage, may be an appropriate measure of the endemism level in each particular area. By contrast, a significant regression indicates a certain trend in endemism values and, according to this trend, the mean endemism percentage is expressed by the equation:

$$PE = 100 \times b/(b + 1). \tag{13.14}$$

PE will also depend on the total number of species if the intercept in the regression of E on NE (or vice versa) is not zero. In all such cases, according to Vilenkin and Chikatunov (1998), the endemism percentage is not an adequate measurement of the comparative distinctiveness of the local fauna under consideration and residuals from a regression model should be used as a more appropriate measure.

Vilenkin and Chikatunov (1998) developed several distinct procedures for analysing the relationships between endemic species and species belonging to other biogeographical ranks. To calculate the values of endemic species expected on the basis of possible relationships with non-endemics, Vilenkin and Chikatunov (1998) used the following model:

$$\ln(E + 1) = a + bNE, \tag{13.15}$$

where a and b are fitted parameters (see also Vilenkin & Chikatunov, 2000; Vilenkin et al., 2009). This equation allows the investigator to assess whether the number of endemic species is influenced by the number of non-endemics. The expected values according to calculations can be compared with the observed values and the residuals can be used as an estimation of the level of endemism. Areas are then ranked according to both the magnitude of residuals and the simple percentage of endemic species. If the two methods agree, the number of endemic species is not influenced by the number of non-endemics. If the two methods give different arrangements of areas, the number of endemic species is influenced by the number of non-endemics. An empirical analysis of the tenebrionids of the Aegean Islands using this method (Fattorini, 2007b) revealed that, as hypothesized by Vilenkin and Chikatunov (1998), the number of endemic taxa was influenced by both the number of non-endemics and by the sizes of the islands. However, the reasons behind the observed positive relationship between the number of endemics and non-endemics are not yet fully understood (Fattorini, 2007b). In addition, this relationship did not necessarily influence the level of endemism, which, at least in this case study, could be expressed adequately by percentages.

13.8 Conclusions

Despite the various aforementioned problems, hotspot identification remains a crucial strategy for setting conservation priorities and ensuring a large proportion of the world's biological diversity is maintained (Marchese, 2015). A major issue with hotspot identification is that the areas under study often differ in size. Since both the total number of species and the number of endemic species tend to increase with area, prioritizing sites according to their (endemic) species richness can produce rankings that simply mirror the sizes of the sites. Thus, it is important to control for the dependence of species richness on the size of the sites. For this reason, some authors have proposed that the (endemic) species–area relationship should be modelled and then the areas located above the fitted curve(s) (i.e. those having a positive residual) designated as hotspots. However, the use of residuals may lead to areas being identified as hotspots when they have, in fact, few species. Thus, it is important to evaluate the ability of the hotspots designated by these procedures to really conserve total and endemic species diversity. For this, the best strategy might be to apply multiple approaches, combining the use of SARs and EARs with other approaches, including the presence of imperilled species (e.g. Fattorini, 2006b, 2009, 2010a), cross-taxon comparisons (e.g. Fattorini, 2010b; Fattorini et al., 2011), complementarity (Balletto et al., 2010; Kullberg et al., 2015), endemicity analysis, phylogenetic diversity, phylogenetic endemism and evolutionary distinctiveness (e.g. Daru et al., 2015; Fattorini, 2017).

Acknowledgements

I would like to thank T. J. Matthews for his comments on previous versions of this chapter.

References

Anderson, S. (1994) Area and endemism. *Quarterly Review of Biology*, **69**, 451–471.

Araújo, M. B. (2002) Biodiversity hotspots and zones of ecological transition. *Conservation Biology*, **16**, 1662–1663.

Balletto, E., Bonelli, S., Borghesio, L., Casale, A., Brandmayr P. & Vigna Taglianti, A. (2010) Hotspots of biodiversity and conservation priorities: A methodological approach. *Italian Journal of Zoology*, **77**, 2–13.

Barthlott, W., Lauer, W. & Placke, A. (1996) Global distribution of species diversity in vascular plants: Towards a world map of phytodiversity. *Erdkunde*, **50**, 317–327.

Bellwood, D. R., Hughes, T. P., Folke, C. & Nystrom, M. (2004) Confronting the coral reef crisis. *Nature*, **429**, 827–833.

Borregaard, M. K. & Rahbek, C. (2010) Causality of the relationship between geographic distribution and species abundance. *The Quarterly Review of Biology*, **85**, 3–25.

Brooks, T. M., Mittermeier, R. A., Mittermeier, C. G., Da Fonseca, G. A. B., Rylands, A. B., Konstant, W. R., Flick, P., Pilgrim, J., Oldfield, S., Magin, G. & Hilton-Taylor, C. (2002) Habitat loss and extinction in the hotspots of biodiversity. *Conservation Biology*, **16**, 909–923.

Brummitt, N. & Nic Lughadha, E. (2003) Biodiversity: Where's hot and where's not. *Conservation Biology*, **17**, 1442–1448.

Cadotte, M. W. & Davies, T. J. (2010) Rarest of the rare: Advances in combining evolutionary distinctiveness and scarcity to inform conservation at biogeographical scales. *Diversity and Distributions*, **16**, 376–385.

Cañadas, E. M., Fenu, G., Peñas, J., Lorite, J., Mattana, E. & Bacchetta, G. (2014) Hotspots within hotspots: Endemic plant richness, environmental drivers, and implications for conservation. *Biological Conservation*, **170**, 282–291.

Caro, T. M. (2010) *Conservation by proxy: Indicator, umbrella, keystone, flagship, and other surrogate species.* Washington, DC: Island Press.

Ceballos, G. & Brown, J. H. (1995) Global patterns of mammalian diversity, endemism, and endangerment. *Conservation Biology*, **9**, 559–568.

Ceballos, G. & Ehrlich, P. (2006) Global mammal distributions, biodiversity hotspots, and conservation. *Proceedings of the National Academy of Sciences USA*, **103**, 19374–19379.

Christenhusz, M. J. M. & Byng, J. W. (2016) The number of known plants species in the world and its annual increase. *Phytotaxa*, **261**, 201–217.

Conservation International (2005) *Biodiversity hotspots: Hotspots by region.* http://www.biodiversityhotspots.org/xp/Hotspots/hotspots_by_region.

Conservation International (2019) *Biodiversity hotspots.* www.conservation.org/How/Pages/Hotspots.aspx.

Daru, B. H., Bank, M. & Davies, T. J. (2015) Spatial incongruence among hotspots and complementary areas of tree diversity in southern Africa. *Diversity and Distributions*, **21**, 769–780.

Dengler, J. (2009) Which function describes the species–area relationship the best? A review and empirical evaluation. *Journal of Biogeography*, **36**, 728–744.

Devictor, V., Mouillot, D., Meynard, C., Jiguet, F., Thuiller, W. & Mouquet, N. (2010) Spatial mismatch and congruence between taxonomic, phylogenetic and functional diversity: The need for integrative conservation strategies in a changing world. *Ecology Letters*, **13**, 1030–1040.

Dobson, A. P., Rodriguez, J. P., Roberts, W. M. & Wilcove, S. S. (1997) Geographic distribution of endangered species in the United States. *Science*, **275**, 550–553.

Evans, E. C., Clark, P. J. & Brandt, R. H. (1955) Estimation of the number of species present in a given area. *Ecology*, **36**, 342–343.

Fattorini, S. (2006a) Detecting biodiversity hotspots by species–area relationships: A case study of Mediterranean beetles. *Conservation Biology*, **20**, 1169–1180.

Fattorini, S. (2006b) A new method to identify important conservation areas applied to the butterflies of the Aegean Islands (Greece). *Animal Conservation*, **9**, 75–83.

Fattorini, S. (2007a) To fit or not to fit? A poorly fitting procedure produces inconsistent results when the species–area relationship is used to locate hotspots. *Biodiversity and Conservation*, **16**, 2531–2538.

Fattorini, S. (2007b) Levels of endemism are not necessarily biased by the co-presence of species with different size ranges: A case study of Vilenkin and Chikatunov's models. *Journal of Biogeography*, **34**, 994–1007.

Fattorini, S. (2009) Assessing priority areas by imperilled species: Insights from the European butterflies. *Animal Conservation*, **12**, 313–320.

Fattorini, S. (2010a) Use of insect rarity for biotope prioritisation: The tenebrionid beetles of the Central Apennines (Italy). *Journal of Insect Conservation*, **14**, 367–378.

Fattorini, S. (2010b) Biotope prioritisation in the Central Apennines (Italy): Species rarity and cross-taxon congruence. *Biodiversity and Conservation*, **19**, 3413–3429.

Fattorini, S. (2017) Endemism in historical biogeography and conservation biology: Concepts and implications. *Biogeographia – The Journal of Integrative Biogeography*, **32**, 47–75.

Fattorini, S., Dennis, R. L. H. & Cook, L. M. (2011) Conserving organisms over large regions requires multi-taxa indicators: One taxon's diversity-vacant area is another taxon's diversity zone. *Biological Conservation*, **144**, 1690–1701.

Funk, S. M. & Fa, J. E. (2010) Ecoregion prioritization suggests an armoury not a silver bullet for conservation planning. *PLoS One*, **5**, e8923.

Georghiou, K. & Delipetrou, P. (2010) Patterns and traits of the endemic plants of Greece. *Botanical Journal of the Linnean Society*, **162**, 130–422.

Grenyer, R., Orme, C. D., Jackson, S. F., Thomas, G. H., Davies, R. G., Davies, T. J., Jones, K. E., Olson, V. A., Ridgely, R. S., Rasmussen, P. C., Ding, T. S., Bennett, P. M., Blackburn, T. M., Gaston, K. J., Gittleman, J. L. & Owens, I. P. (2006) Global distribution and conservation of rare and threatened vertebrates. *Nature*, **444**, 93–96.

Griffin, P. C. (1999) Endangered species diversity 'hot spots' in Russia and centers of endemism. *Biodiversity and Conservation*, **8**, 497–511.

Guilhaumon, F., Gimenez, O., Gaston, K. J. & Mouillot, D. (2008) Taxonomic and regional uncertainty in species–area relationships and the identification of richness hotspots. *Proceedings of the National Academy of Sciences USA*, **105**, 15458–15463.

Hobohm, C. (2000) Plant species diversity and endemism on islands and archipelagos, with special reference to the Macaronesian Islands. *Flora*, **195**, 9–24.

Hobohm, C. (2003) Characterization and ranking of biodiversity hotspots: Centres of species richness and endemism. *Biodiversity and Conservation*, **12**, 279–287.

Hobohm, C. & Tucker, C. M. (2014) How to quantify endemism. *Endemism in vascular plants, plant and vegetation*, vol. 9 (ed. by C. Hobohm), pp. 11–48. Heidelberg: Springer.

Hobohm, C., Janišová, M., Steinbauer, M., Landi, S., Field, R., Vanderplank, S., Beierkuhnlein, C., Grytnes, J.-A., Vetaas, O. R., Fidelis, A., de Nascimento, L., Clark, V. R., Fernández-Palacios, J. M., Franklin, S., Guarino, R., Huang, J., Krestov, P., Ma, K., Onipchenko, V., Palmer, M. W., Simon, M. F., Stolz, C. & Chiarucci, A. (2019) Global endemics–area relationships of vascular plants. *Perspectives in Ecology and Conservation*, **17**, 41–49.

Jenkins, C. N., Pimm, S. L. & Joppa, L. N. (2013) Global patterns of terrestrial vertebrate diversity and conservation. *Proceedings of the National Academy of Sciences USA*, **110**, E2602–E2610.

Jetz, W., Thomas, G. H., Joy, J. B., Redding, D., Hartmann, K. & Moore, A. O. (2014) Global distribution and conservation of evolutionary distinctness in birds. *Current Biology*, **24**, 919–930.

Kareiva, P. & Marvier, M. (2003) Conserving biodiversity coldspots. *American Scientist*, **91**, 344–351.

Kier, G. & Barthlott, W. (2001) Measuring and mapping endemism and species richness: A new methodological approach and its application on the flora of Africa. *Biodiversity and Conservation*, **10**, 1513–1529.

Kullberg, P., Toivonen, T., Montesino Pouzols, F., Lehtomäki, J., Di Minin, E. & Moilanen, A. (2015) Complementarity and area-efficiency in the prioritization of the global protected area network. *PLoS One*, **10**, e0145231.

Maes, D., Gilbert, M., Titeux, N., Goffart, P. & Dennis, R. L. H. (2003) Prediction of butterfly diversity hotspots in Belgium: A comparison of statistically focused and land use-focused models. *Journal of Biogeography*, **30**, 1907–1920.

Magurran, A. E. & McGill, B. J. (eds.) (2011) *Biological diversity: Frontiers in measurement and assessment*. Oxford: Oxford University Press.

Marchese, C. (2015) Biodiversity hotspots: A shortcut for a more complicated concept. *Global Ecology & Conservation*, **3**, 297–309.

Martín, H. G. & Goldenfeld, N. (2006) On the origin and robustness of power-law species–area relationships in ecology. *Proceedings of the National Academy of Sciences USA*, **103**, 10310–10315.

Matthews, T. J., Guilhaumon, F., Triantis, K. A., Borregaard, M. K. & Whittaker, R. J. (2016) On the form of species–area relationships in habitat islands and true islands. *Global Ecology & Biogeography*, **25**, 847–858.

Mazel, F., Guilhaumon, F., Mouquet, N., Devictor, V., Gravel, D., Renaud, J., Cianciaruso, M. V., Loyola, R., Diniz-Filho, J. A., Mouillot, D. & Thuiller, W. (2014) Global hotspots of multifaceted mammal diversity. *Global Ecology & Biogeography*, **23**, 836–847.

Mittermeier, R. A., Gil, P. R., Hoffman, M., Pilgrim, J., Brooks, T., Mittermeier, C. G., Lamoreux, J. & da Fonseconda, G. A. B. (2004) *Hotspots revisited: Earth's biologically richest and most endangered terrestrial ecoregions*. Mexico City: Cemex.

Mittermeier, R. A., Myers, N., Robles-Gil, P. & Mittermeier, C. G. (eds.) (1999) *Hotspots: Earth's biologically richest and most endangered terrestrial ecoregions*. Mexico City: CEMEX and Agrupación Sierra Madre.

Mittermeier, R. A., Myers, N., Thomsen, J. B., da Fonseca, G. A. B. & Olivieri, S. (1998) Biodiversity hotspots and major tropical wilderness areas: Approaches to setting conservation priorities. *Conservation Biology*, **12**, 516–520.

Mittermeier, R. A., Turner, W. R., Larsen, F. W., Brooks, T. M. & Gascon, C. (2011) Global biodiversity conservation: The critical role of hotspots. *Biodiversity hotspots* (ed. by F. E. Zachos and J. C. Habel), pp. 3–22. London: Springer Publishers.

Myers, N. (1988) Threatened biotas: 'Hotspots' in tropical forests. *Environmentalist*, **8**, 1–20.

Myers, N. (1990) The biodiversity challenge: Expanded hot-spots analysis. *Environmentalist*, **10**, 243–256.

Myers, N. & Mittermeier, R. A. (2003) Impact and acceptance of the hotspots strategy: Response to Ovadia and to Brummit and Lughadha. *Conservation Biology*, **17**, 1449–1450.

Myers, N., Mittermeier, R. A., Mittermeier, C. G., da Fonseca, G. A. B. & Kent, J. (2000) Biodiversity hotspots for conservation priorities. *Nature*, **403**, 853–858.

Noss, R. (2016) *Announcing the World's 36th biodiversity hotspot: The North American Coastal Plain*. www.cepf.net/stories/announcing-worlds-36th-biodiversity-hotspot-north-american-coastal-plain.

Noss, R. F., Platt, W. J., Sorrie, B. A., Weakley, A. S., Means, D. B., Costanza, J., Peet, R. K. & Richardson, D. (2015) How global biodiversity hotspots may go unrecognized: Lessons from the North American Coastal Plain. *Diversity and Distributions*, **21**, 236–244.

Orme, C. L., Davies, R. G., Burgess, M. H., Eigenbrod, F., Pickup, N. J., Olson, V. A., Webster, A. J., Ding, T., Rasmussen, P. C., Ridgely, R. S., Stattersfield, A. J., Bennett, P. M., Blackburn, T. M., Gaston, K. J. & Owens, I. P. (2005) Global hotspots of species richness are not congruent with endemism or threat. *Nature*, **436**, 1016–1019.

Ovadia, O. (2003) Ranking hotspots of varying sizes: A lesson from the nonlinearity of the species–area relationship. *Conservation Biology*, **17**, 1440–1441.

Pomeroy, D. (1993) Centres of high biodiversity in Africa. *Conservation Biology*, **7**, 901–907.

Possingham, H. P. & Wilson, K. A. (2005) Turning up the heat on hotspots. *Nature*, **436**, 919–920.

Prendergast, J. R., Quinn, R. M., Lawton, J. H., Eversham, B. C. & Gibbons, D. W. (1993a) Rare species, the coincidence of diversity hotspots and conservation strategies. *Nature*, **365**, 335–337.

Prendergast, J., Wood, S., Lawton, J. & Eversham, B. (1993b) Correcting for variation in recording effort in analyses of diversity hotspots. *Biodiversity Letters*, **1**, 39–53.

Reid, W. V. (1998) Biodiversity hotspots. *Trends in Ecology & Evolution*, **13**, 275–280.

Roberts, C. M., McClean, C. J., Veron, J. E. N., Hawkins, J. P., Allen, G. R., McAllister, D. E., Mittermeier, C. G., Schueler, F. W., Spalding, M., Wells, F., Vynne, C. & Werner, T. B. (2002) Marine biodiversity hotspots and conservation priorities for tropical reefs. *Science*, **295**, 1280–1284.

Rosenzweig, M. L. (1995) *Species diversity in space and time*. Cambridge: Cambridge University Press.

Sloan, S., Jenkins, C. N., Joppa, L. N., Gaveau, D. L. A. & Laurance, W. F. (2014) Remaining natural vegetation in the global biodiversity hotspots. *Biological Conservation*, **177**, 12–24.

Smith, T. B., Kark, S., Schneider, C. J., Wayne, R. K. & Moritz, C. (2001) Biodiversity hotspots and beyond: The need for preserving environmental transitions. *Trends in Ecology & Evolution*, **16**, 431.

Stattersfield, A. J., Crosby, M. J., Long, A. J. & Wege, D. C. (1998) *Endemic bird areas of the world: Priorities for biodiversity conservation*. BirdLife Conservation Series 7. Cambridge, UK: BirdLife International.

Storch, D., Keil, P. & Jetz, W. (2012) Universal species–area and endemics–area relationships at continental scales. *Nature*, **488**, 78–81.

Stork, N. E., Habel, J. C. & Ladle, R. (2014) Can biodiversity hotspots protect more than tropical forest plants and vertebrates? *Journal of Biogeography*, **41**, 421–428.

Stuart-Smith, R. D., Bates, A. E., Lefcheck, J. S., Duffy, J. E., Baker, S. C., Thomson, R. J., Stuart-Smith, J. F., Hill, N. A., Kininmonth, S. J., Airoldi, L., Becerro, M. A., Campbell, S. J., Dawson, T. P., Navarrete, S. A., Soler, G. A., Strain, E. M. A., Willis, T. J. & Edgar, G. J. (2013) Integrating abundance and functional traits reveals new global hotspots of fish diversity. *Nature*, **501**, 539–542.

Thomasson, M. (1999) Réflexions sur la biodiversité: Richesse, originalité et endémicité floristiques. *Acta Botanica Gallica*, **146**, 403–419.

Tjørve, E. (2003) Shapes and functions of species–area curves: A review of possible models. *Journal of Biogeography*, **30**, 827–835.

Tjørve, E. (2009) Shapes and functions of species–area curves (II): A review of new models and parameterizations. *Journal of Biogeography*, **36**, 1435–1445.

Triantis, K. A., Guilhaumon, F. & Whittaker, R. J. (2012) The island species–area relationship: Biology and statistics. *Journal of Biogeography*, **39**, 215–231.

Triantis, K. A., Mylonas, M. & Whittaker, R. J. (2008) Evolutionary species–area curves as revealed by single-island endemics: Insights for the inter-provincial species–area relationship. *Ecography*, **31**, 401–407.

Troumbis, A. Y. & Dimitrakopoulos, P. G. (1998) Geographic coincidence of diversity threatspots for three taxa and conservation planning in Greece. *Biological Conservation*, **84**, 1–6.

Ulrich, W. & Buszko, J. (2005) Detecting biodiversity hotspots using species–area and endemics–area relationships: The case of butterflies. *Biodiversity and Conservation*, **14**, 1977–1988.

Ungricht, S. (2004) How many plant species are there? And how many are threatened with extinction? Endemic species in global biodiversity and conservation assessments. *Taxon*, **53**, 481–484.

Veech, J. A. (2000) Choice of species–area function affects identification of hotspots. *Conservation Biology*, **14**, 140–147.

Vilenkin, B. Y. & Chikatunov, V. I. (1998) Co-occurrence of species with various geographical ranges, and correlation between area size and number of species in geographical scale. *Journal of Biogeography*, **25**, 275–284.

Vilenkin, B. Y. & Chikatunov, V. I. (2000) Participation of species with different zoogeographical ranks in the formation of local fauna: A case study. *Journal of Biogeography*, **27**, 1201–1208.

Vilenkin, B. Y., Chikatunov, V. I., Coad, B. W. & Schileyko, A. A. (2009) A random process may control the number of endemic species. *Biologia*, **64**, 107–112.

Williams, K. J., Ford, A., Rosauer, D. F., De Silva, N., Mittermeier, R., Bruce, C., Larsen, F. W. & Margules, C. (2011) Forests of East Australia: The 35th biodiversity hotspot. *Biodiversity hotspots* (ed. by J. C. Habel and F. Zachos), pp. 295–310. Heidelberg: Springer.

Williams, M. R., Lamont, B. B. & Henstridge, J. D. (2009) Species–area functions revisited. *Journal of Biogeography*, **36**, 1994–2004.

Williams, P., Gibbons, D., Margules, C., Rebelo, A., Humphries, C. & Pressey, R. (1996) A comparison of richness hotspots, rarity hotspots, and complementary areas for conserving diversity of British birds. *Conservation Biology*, **10**, 155–174.

14 · Using the Species–Area Relationship to Predict Extinctions Resulting from Habitat Loss

SIMONE FATTORINI, WERNER ULRICH
AND THOMAS J. MATTHEWS

14.1 Introduction

Species–area relationships (SARs) have a number of major applications, including the estimation of the species richness of large areas by extrapolating beyond the richness of observed areas (Kunin et al., 2018; Matthews & Aspin, 2019) and the assessment of whether areas within the observed range of data have an excess (hotspots) of species or are depauperate (coldspots) (Ulrich & Buszko, 2005; Chapter 13). However, perhaps the SAR's most prevalent application is in the prediction of extinctions resulting from habitat loss (He & Hubbell, 2011; Halley et al., 2013; Chapter 16). It is widely acknowledged that we are in the midst of an extinction crisis; however, providing accurate estimates of extinction rates has proven to be problematic (Stork, 2010; He & Hubbell, 2011; Costello et al., 2013).

The purpose of the present chapter is to provide a review of the use of the SAR to predict the number of extinctions resulting from habitat loss. By doing so, we highlight the pitfalls of using the SAR in such a way and discuss the myriad ways in which studies have extended and built on standard SAR models and approaches to better model and predict extinctions resulting from habitat loss, a process that is widely considered to be the biggest contemporary threat to biodiversity (Haddad et al., 2015).

The common procedure in SAR extinction studies is to apply the power function backwards, that is, to calculate the number of species lost due to a reduction in habitat area (e.g. Reid & Miller, 1989; Wilson, 1992; May et al., 1995; Rosenzweig, 1995; Brooks & Balmford, 1996; Ney-Nifle & Mangel, 2000; Brook et al., 2003; Wilsey et al., 2005; Koh & Ghazoul, 2010; Triantis et al., 2010; Whittaker & Matthews, 2014) (Figure 14.1).

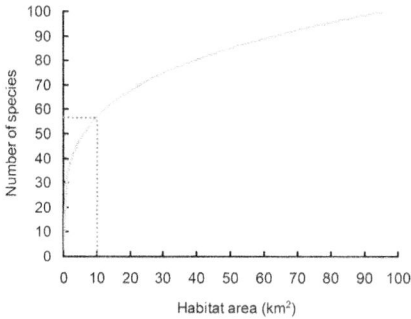

Figure 14.1 A hypothetical species–area curve following a power function, where S_0 = approximately thirty-two species and z = 0.25. Here, habitat loss of 90 per cent (from 100 km² to 10 km²) reduces the number of species by about 40 per cent (dashed line).

Applied in a naïve way, the power function describing the species–area relationship is:

$$S = S_0 A^z, \qquad (14.1)$$

where S represents species richness at area A and S_0 is the expected average species richness at unit area. The exponent z indicates the increase in species richness with area and is an often-applied metric of beta diversity (see Chapter 8), when species accumulation curves are the focus of study.

If we suppose a reduction of area A_1 to A_2, so that the average number of species S_1 is expected to decline to S_2, we obtain, after rearrangements:

$$S_2 = S_1 \times (A_2/A_1)^z. \qquad (14.2)$$

The term $(A_2/A_1)^z$ thus represents the expected average proportion of remaining species after a reduction in area. Equation (14.2) also predicts that the proportion of species lost after a certain loss of area is:

$$f_{SAR} = 1 - (1 - A_2/A_1)^z. \qquad (14.3)$$

Although this calculation is intuitively appealing, the use of a reversed SAR to calculate how many species are lost after a reduction in habitat area may produce flawed results. This approach relies on a number of theoretical assumptions that cannot be met in some contexts or that are arguably intrinsically wrong, including:

1) The z-value can be calculated from empirical data or a suitable value chosen when such data do not exist. Different z-values can generate

widely different extinction estimates (see Figure 14.2) and it is not always clear which value is appropriate for a given system and taxon in the absence of empirical data.

2) The power function provides the best fit to the data. While the power model has been found to provide the best fit to many datasets, this is not always the case (Triantis et al., 2012; Matthews et al., 2016). Thus, its use a priori is not necessarily justified.

3) The use of a nested SAR is appropriate to model habitat loss. Nested means that smaller areas are always perfect subsamples of the next larger area. Mainland SARs in continuous habitats are generally constructed using a nested design (May, 1975; Durrett & Levin, 1996; Harte et al., 1999a, b; Ney-Nifle & Mangel, 1999; Hubbell, 2001; Allen & White, 2003). However, habitat loss does not proceed in a nested manner. For example, using various existing stages of habitat fragmentation to infer patterns of species extinction, Seabloom et al. (2002) found that aggregated patterns of habitat loss might lead to higher extinction rates in California vascular plant communities than those predicted by SARs.

4) Another point related to nested designs is that they tend to produce power function SARs, whereas SARs constructed in other ways may deviate from power functions (Palmer & White, 1994; Rosenzweig, 1995). Nested designs tend to equalize the effects of environmental variation as this variation enters the model at subsequent stages of the nested sampling. For fractal landscapes, Harte et al. (1999a) demonstrated that this effect leads to power function SARs. This mechanism does not apply to non-nested designs and the variability in richness between independent areas, for instance in the case of island SARs, caused by environmental differences might force such SARs to deviate from the power function in certain cases (Matthews et al. 2014a).

5) The reverse application of a power function model implies that the exponent (z) remains constant. However, Ney-Nifle and Mangel (2000) and Ulrich and Buszko (2004) showed that habitat reduction causes changes in SAR exponents. Harte and Kinzig (1997) and Kinzig and Harte (2000) assumed that, to construct a SAR for species loss forecasting, an area should be subsequently subdivided into smaller parts and that in each part the original allometric species–area relation (Equation 14.1) should be applied. This assumption refers in particular to a situation where species have similar scale invariant spatial distribution patterns (Harte et al., 1999a). Ulrich and Buszko (2004) noted that nested SARs constructed for Polish

Power model: 60 x (20/100)^z

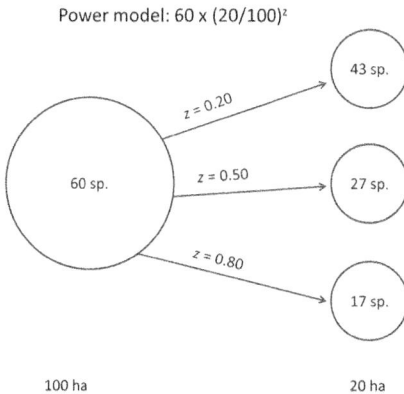

Figure 14.2 The effect on extinction predictions of using different z-values in the power SAR function. The large circle represents a habitat patch of 100 ha, containing sixty species. The power SAR predicts that, when this patch is reduced in size to 20 ha, the number of species is also reduced, but the exact number depends on the choice of z-value.

butterflies are well approximated by power functions, indicating a self-similar process of species accumulation, but species loss deviated from self-similarity and could not be described by a single power function. Such changes in the z-value indicate breaking points in the process of species loss, beyond which loss either accelerates or possibly even decelerates. There is still no general (ecological) model that predicts and accounts for such breakpoints. A study of island SARs (ISARs) in fragmented landscapes also found evidence of breakpoints in the ISAR in several cases (Matthews et al. 2014a).

14.2 Species–Area Relationships and Endemics–Area Relationships

A further problem with the (nested) SAR in the context of extinction predictions is that by definition it incorporates all species in an area. However, the only species which can possibly be driven to extinction in the area due to habitat loss are those species endemic to that focal area.

If only endemics are considered, a nested endemics–area relationship (EAR) can be constructed. In this context, the term endemics has a slightly different meaning than is commonly used and refers to species that occur only in the focal subset of area and not in any other subset of equal or larger size within the studied region (see Chapter 16). Kinzig and Harte (2000) argued EARs provide more reliable estimates of species loss during a process of habitat destruction and a number of studies have recommended the use of the EAR rather than the standard SAR when predicting extinctions (Smith, 2010; He & Hubbell, 2011, 2013), although with the caveat that the data necessary to construct an EAR are rarely available. According to Harte and Kinzig (1997), the EAR should take the form:

$$E(A) = S_0(A/A_0)^{z'}, \tag{14.4}$$

where $E(A)$ is the number of species found *only* in an area A and nowhere else in a larger area A_0, S_0 is the total number of species found in A_0, and z' is a fitted parameter, related to the SAR exponent z by the formula:

$$z' = -\ln(1 - 1/2^z)/\ln(2). \tag{14.5}$$

If applied to endemics only, Equation (14.2) implies that the fraction of species lost after habitat destruction should be:

$$f_{EAR} = 1 - (A_1/A_0)^{z'}. \tag{14.6}$$

In the case of nested SARs, Equations (14.3) and (14.6) predict an identical number of lost species, however Equation (14.6) cannot be applied to non-nested situations, such as a set of distinct habitat fragments.

He and Hubbell (2011) pointed to another pitfall in the naïve application of (nested) SARs when predicting extinctions, showing that, when species are aggregated in their spatial distribution, which is generally considered to be the case (Condit et al., 2000), the power model SAR and EAR are not mirror images and one cannot use the SAR to infer the EAR. In particular, He and Hubbell (2011) observed that, contrary to common thought (e.g. Harte & Kinzig, 1997; Harte, 2000; Kinzig & Harte, 2000; Ulrich & Buszko, 2003a), (nested) EARs do not have higher, but lower (typically half the size) z-values than those of the respective (nested) SARs, meaning that the use of z-values derived from standard SAR studies (e.g. 0.25) leads to overestimation of species loss. However, Storch et al. (2012) noted that non-nested EARs of amphibian, bird and mammal species at the global scale have z-values close to unity, indicating scale invariant loss of endemic species after habitat loss. More generally, Ulrich (2005) and Storch et al. (2012) have found that the z-values of various types of EARs are only weakly correlated with those of their respective SARs; again suggesting that

EARs are relations in their own right and cannot be derived from the underlying SARs (He & Hubbell, 2011). Of more practical relevance, due to these differences between EARs and SARs, estimates of extinctions based only on endemics (i.e. if EARs are applied) differ substantially from those achieved using all species.

Ulrich and Buszko (2003a) argued that if widespread and endemic species are differently affected by habitat loss, this will produce different changes in the S_0 and z parameters of the power function EAR as habitat loss progresses. Namely, they identified four possible scenarios:

1) S_0 values remain similar or decrease slightly, and z-values decrease. If locally restricted (endemic) species are less abundant and hence more prone to extinction than widespread species, global extinction of locally restricted species is more probable than extinction of widespread species. As a result, the mean local species richness (expressed by the parameter S_0 in Equation 14.1) decreases slightly and global species richness decreases substantially, leading to a decrease in the z-value.

2) S_0 decreases, but z increases. This happens if there are few locally restricted (endemic) species, which become globally extinct and many widespread species, which become locally extinct in some areas. In this case, the mean local species richness is strongly affected, but global species richness remains similar. As a result, S_0 decreases and z increases.

3) S_0 decreases, but z remains constant. This may occur if restricted and widespread species have approximately similar extinction probabilities.

4) S_0 increases, but z decreases. This means that habitat destruction would result in increases in local diversity, which could occur, for example, if a fragmentation process allows some species to fill open niches in the remaining habitat patches.

Ulrich and Buszko (2003a) used butterfly species distributions in European countries to explore these four scenarios and the influence of species extinction on SAR parameters through a simulation process that implemented stepwise density dependent random elimination of species. They conducted separate analyses for Mediterranean ($S = 0.54A^{0.49}$) and northern and eastern European countries ($S = 2.5A^{0.10}$), because these two groups of areas have different SARs. In the Mediterranean countries, 43 per cent of species were endemic (i.e. restricted to only one country). The higher probability of eliminating these species during the simulation process resulted in decreasing z-values (and irregular changes in S_0-values), because regional species numbers were more affected than local ones, as expected in the scenario 1. The northern and eastern countries had a high number of widespread species (with 29 per cent of endemics) and thus elimination affected local species numbers more than regional species numbers. The effect was an initial rise in z (followed by a rapid decline) and a decline in S_0, with a pattern similar to that of scenario 2. Ulrich and Buszko (2003a) also found that in most cases the use of the SAR overestimated simulated species loss in the Mediterranean countries. In contrast, the opposite occurred in the northern and eastern countries, where the SAR approach always underestimated simulated species loss when 75 per cent of species had been eliminated.

14.3 Are SAR-based Estimates of Species Loss Intrinsically Flawed? The Importance of How Area Is Lost

He and Hubbell's (2011) conclusions that SARs overestimate species loss and that EARs should be used were controversial and have been criticized on the basis that they are dependent on the geometry of habitat loss (Pereira et al., 2012) and that their analytical approach was flawed (Axelsen et al., 2013; but see He & Hubbell, 2013). It should also be noted that their analyses relate to the species accumulation curve (i.e. nested SAR) rather than the island SAR (see Chapter 1). In particular, the geometry of habitat loss (e.g. whether habitat loss occurs from the outside of a block of habitat towards the centre or vice versa; see Figure 14.3) has been argued to be a key, but often overlooked, consideration in SAR- and EAR-based extinction prediction studies (Keil et al., 2015; discussed in Chapter 16).

In an evaluation of habitat loss geometry, Ulrich and Buszko (2004) used four different models of habitat loss to simulate species extinction in Polish butterflies: 1) a random elimination of cells in the species distribution matrix (Figure 14.3); 2) an aggregated area loss, in which a small number of randomly chosen cells that are eliminated are used as the nucleus for further aggregated elimination (see also Figure 14.3), with the cell elimination probability declining exponentially with distance; 3) elimination of the more elevated cells in simulated fractal landscapes; and 4) a nested habitat reduction design. They found that, using the aggregated, fractal or random models, the process of species extinction was severe only when large fractions of habitat are lost, with 50 per cent of species lost when habitat area was reduced to 99 per cent. The nested design produced higher rates of extinction (the absolute differences between the nested and the non-nested designs were around fifteen to twenty-five species). In Ulrich and Buszko's (2004) simulations, non-random patterns of area loss (fractal or aggregated) did not differ significantly from each other or from a random pattern with respect to species loss. This result is in contrast with the finding of Seabloom et al. (2002) who, in a study of Californian vascular plant communities, reported that non-random patterns of area loss caused higher species loss than that resulting from a random pattern. However, Seabloom et al. (2002) used standard power function models with fixed z-values, without controlling for systematic deviations at different scales of resolution. In Ulrich and Buszko's (2004) study, both the power function and the logarithmic models overestimated species loss, in line with the argument of He and Hubbell (2011).

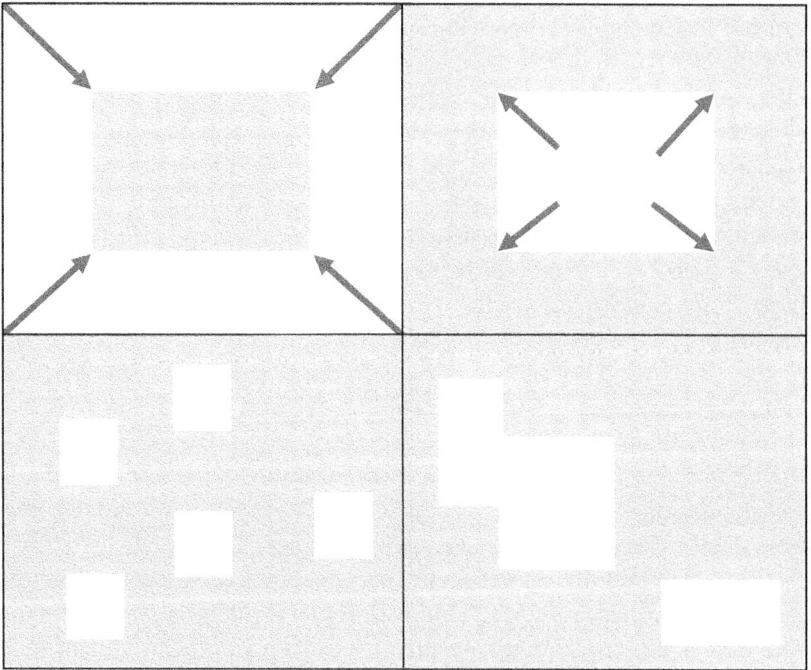

Figure 14.3 The different ways in which habitat loss can progress (focal habitat = grey; non-focal habitat = white). For a given area of habitat (top row), habitat loss can either proceed from the outside inwards (left side) or from the inside outwards (right side). For a given landscape (bottom row), habitat loss can occur in a random fashion (left side) or in an aggregated manner (right side); the latter could occur, for example, if the destruction of an area of habitat provides opportunities (e.g. through the creation of roads and infrastructure) for habitat loss in adjacent areas.

Studies have identified other factors related to how habitat is lost that can affect SAR predictions. For example, Ney-Nifle and Mangel (2000), using a patch occupancy approach, found that range sizes, fragmentation patterns and edge effects influence SARs, leading either to higher or lower species loss than predicted by the initial SAR pattern. In particular, smaller edge effects during area reduction (introduced by more aggregated patterns of area loss) reduced species loss.

Ulrich (2005) proposed a patch occupancy approach for studying EARs and SARs with simulated communities. This approach allows the generation and analysis of multiple replicable spatial distribution patterns, with sample areas that have the same spatial properties as those that were assumed for the theoretical derivations of SARs and EARs by Harte and Kinzig (1997) and Harte et al. (1999a). The procedure involved generating virtual communities with random species abundance distribution patterns. Then, the individuals of these model communities were placed at random into the cells of a grid. For each species, a different grid was used and the carrying capacity of the grid cells was variable. After placement, groups of a fixed number of cells were sampled using the nested design used by Harte and Kinzig (1997) and Harte et al. (1999a) to construct nested SARs and EARs. Finally, species occurrences in the cell groups of the nested design were used to determine the number of species lost after area reduction. Based on this approach, Ulrich (2005) found that the use of nested EARs did not provide better results than when using classic nested SARs. Predictions varied widely (especially below 10 per cent habitat loss), because of 1) the small number of species that went extinct, 2) large differences in local species numbers and 3) aggregated species spatial distributions. Predictions of species loss using regression lines were therefore strongly influenced by which patches were lost.

In conclusion, the higher number of extinctions, relative to observed extinctions, predicted using the SAR in several studies may be due to the fact that the pattern of habitat loss does not match patterns of species spatial distributions. When this is the case, the parameters of the model change, while the extinction predictions are based on the initial parameter values (Smith, 2010); a problem that might also affect the use of the EAR.

14.4 Additional Reasons Why SARs Might Overestimate Species Loss

The above discussion indicates that a number of authors are of the view that the SAR generally overestimates species loss resulting from habitat loss.

That is, it is not simply that the SAR provides erroneous extinction predictions, but that these predictions are systematically biased. Putting to one side the fact that this view may be incorrect (discussed in Section 14.5), we review here four additional reasons (i.e. additional to those provided in the preceding sections) why SARs might overestimate species loss.

14.4.1 Matrix Effects and Countryside Biogeography

The SAR (and indeed the EAR) assume that all species go extinct when an area of native habitat is nearly completely lost; more precisely, when only an area smaller than the minimal spatial niche (i.e. minimum area requirement) for any of the species remains. However, it is known that at least some species can survive in anthropogenic habitats, such as plantation forestry and agricultural land (Daily et al., 2003; Mendenhall et al., 2014). Therefore, the average minimal area, given by $A_{min} = (1/S_0)^{1/z}$ (i.e. the habitat area where, on average, exactly one species occurs), does not mark the lower boundary of species survival because, in some fragments smaller than this, species will still be able to survive due to utilization of the matrix habitat. Standard SAR extinction studies in fragmented landscapes are often affected by this issue as they tend to consider matrix habitat simply as 'non-habitat'. In such situations, the SAR will always overestimate species loss.

A growing realization that today most native habitat is embedded within a matrix of human habitats and that certain species can utilize these human habitats, has given rise to fields such as countryside biogeography (Daily et al., 2003; Mendenhall et al., 2014). These areas of research are, at least partly, focused on calibrating the SAR for the application in anthropogenically fragmented landscapes; often incorporating additional parameters specific to such systems, such as matrix permeability and landscape connectivity. For instance, Koh and Ghazoul (2010) have shown that by using different z-values based on the type of matrix, the predictive power of the power SAR model (i.e. the ability to predict extinctions resulting from habitat loss) can be improved by up to 60 per cent. Other studies have undertaken variations on this approach to include, for example, edge effects and the varying tolerances of different species groups to different habitat types (Pereira & Daily, 2006; Koh et al., 2010; Mendenhall et al., 2014). Tests of these types of models using simulated data and data on observed extinctions (or species close to extinction based on the classifications of the International Union for the Conservation of Nature (IUCN)) have generally found them to provide more accurate extinction estimates (e.g. Koh & Ghazoul, 2010; Martins & Pereira, 2017). However, there is a need for more research in this area, and further broad-scale tests of these previously published 'calibrated' and 'countryside biogeographic' SAR models.

14.4.2 The Extinction Debt

It is possible that, while SAR and EAR extinction predictions may seem like overestimates, they are actually correct. One often-discussed argument in support of this view is the potential existence of an extinction debt, that is, the number of species predicted to become extinct as a community relaxes to a new equilibrium following a disturbance, such as habitat loss (Tilman et al., 1994). For example, according to Brooks et al. (2011), He and Hubbell's (2011) assumptions in the use of the SAR are wrong because species that seem to persist in the remaining fragments are likely condemned to extinction, if fragments are too small.

Thus, perhaps SAR extinction predictions are accurate, it is just that we have not yet seen enough time passed for the debt to be paid in full. The presence of an extinction debt means it is also possible that in many systems the conservation actions of humans have prevented a number of species from becoming extinct that otherwise would have based on the reduction in habitat amount. Due to the possibility of an extinction debt and the positive impact of conservation actions, a number of studies have used the number of species classified as endangered and critically endangered by the IUCN as the variable of interest rather than the number of observed extinctions; and have subsequently observed closer matches between predicted and observed 'extinctions' (see Halley et al., 2013). However, IUCN classifications are not necessarily always without bias and/or error as they are based on subjective expert views often relying on limited data. Further, the IUCN categories, particularly the endangered category, are highly dependent on the spatial scale used for assessment.

14.4.3 Ecology Plasticity and Evolutionary Change

Ecological process dynamics, particularly niche plasticity leading to changes in habitat demands, and evolutionary dynamics have often been overlooked in the discussion around SARs and their application in biodiversity conservation (Chevin et al., 2010). Biodiversity forecasting treats species as being static, without the ability to adapt to new habitats, habitat conditions or resources. However, the role of ecological plasticity has become increasingly appreciated in ecology and an increasing number of studies has demonstrated high ecological plasticity in vertebrates (Canale & Henry, 2010) and arthropods (Moczek, 2010). In the course of habitat loss, this plasticity (i.e. the ability to adapt to the new conditions) might lead to much lower extinction rates than predicted by static SARs. These types of adaptations may occur at very fast rates (Schilthuizen, 2018) due to the high selective pressures resulting from reduced population sizes and novel environmental conditions.

More speculatively, ecological plasticity might also act at the community level, where reduced habitat availability might lead to increased species densities and possibly to changes in species interactions. This is clearly a neglected aspect of community ecology that has mainly been studied theoretically as part of the discussion on tipping points (Boettiger & Hastings, 2013; Rohr et al., 2014), while respective field studies are largely missing (e.g. Goulson et al., 2015). In theory, habitat loss should cause higher local species densities before extinction loss sets in. During this transition phase species interactions, particularly those involved in trophic networks and competitive hierarchies, might change. These changes in species interactions might then change the fate of species in the course of ongoing habitat loss. More research is needed to provide more definitive answers to these ideas.

14.4.4 The Use of Total Species Richness Masks Unequal Sensitivity to Habitat Loss across Taxa

Habitat (or more generally niche) specialist and generalist species react differently to habitat loss. For example, Habel et al. (2019) have recently demonstrated that specialist butterfly species are much more negatively affected by land use change than generalist species. Changes in the proportion of generalist to specialist species in smaller areas would necessarily affect SAR exponents (see Matthews et al., 2014b), which would in turn affect SAR extinction projections. Therefore, it may be better, for diversity forecasting, to split taxa into sufficiently homogeneous ecological guilds and to use separate SARs for different guilds (Matthews et al., 2014b). A similar argument holds for species with different dispersal abilities. Clearly, weak dispersers are more affected by habitat loss than species with high colonization ability (e.g. Bommarco et al., 2010). Within a meta-population framework, the latter type of species will be better able to temporarily colonize small habitat patches not suited for long term survival. Occurrence data, often stemming from long term repeated observations, will count these species as being present, leading to a false impression about the real process of species loss. As such, SAR extrapolations would thus benefit from accounting for differential dispersal ability between species. Respective projection models have already been developed within the meta-population framework (Hanski & Ovaskainen, 2003; MacKenzie et al., 2006).

14.5 SARs Do Not Always Overestimate Extinction Rates

Readers of the above sections may be forgiven for thinking that the SAR universally overestimates extinctions. However, the situation is unfortunately not that simple (if it was, finding solutions would be more straightforward) and He and Hubbell's (2011) conclusion that SARs always overestimate extinction rates from habitat loss is mostly based on theoretical reasoning, with a few examples from the real world (Brooks et al., 2011; Evans et al., 2011). Surprisingly, few papers have actually investigated the effectiveness of SARs (May et al., 1995; Pimm & Askins, 1995; Pimm, 1998; Ulrich & Buszko, 2003a, b) for predicting species loss. However, empirical evidence indicates that SARs do not necessarily always overestimate, but often actually underestimate extinction rates (Fattorini & Borges, 2012).

First, it is important to stress that there are enormous uncertainties regarding documented extinctions. For example, early SAR extrapolations of species extinctions provided global extinction estimates of around 5–15 per cent of rainforest species to be lost by the year 2020, 2–8 per cent of closed forest species between 1990 and 2015, and 15–20 per cent of all species between 1985 and 2015 (e.g. World Conservation Monitoring Centre, 1992); numbers which seem wildly excessive. Figure 14.4 illustrates the number of extinct and endangered species for a range of taxa and it can be seen that these numbers are far below these early SAR predictions. However, the number of documented extinct species is likely less than the number of true extinctions, because of the unknown number of extinctions of undiscovered species (so-called Linnaean extinction; Riddle et al., 2011).

In certain regions, information on species extinction is more accurate and in north Europe and North America extinction trends are well documented, at least for certain charismatic taxa. For example, distribution and incidence data for butterflies in north Belgium (Flanders) are available from 1830 onwards and indicate that about 30 per cent of butterflies have gone locally extinct because of large scale land use change

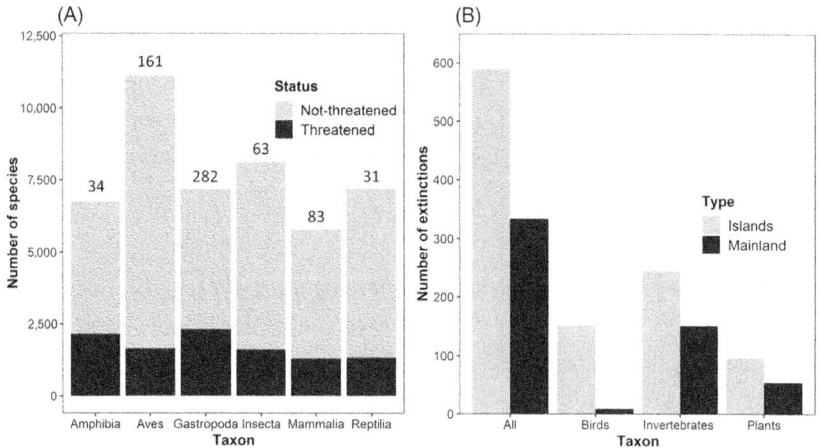

Figure 14.4 Threat status and observed extinction numbers for various taxa. (A) The number of threatened and not-threatened species within the various taxa; data are from IUCN (2019). Species classified as 'Extinct', 'Extinct in the Wild', 'Critically Endangered', 'Endangered' and 'Vulnerable' by the IUCN are grouped as 'Threatened' here. The numbers at the top of the bars are the number of documented extinctions (number of 'Extinct' and 'Extinct in the Wild' classification). The total number of species (i.e. height of the bar) in each taxon is the number of species classified by the IUCN, not the total number of described species. (B) The number of documented extinctions on islands and the mainland according to the IUCN for various taxa; the figure is from Whittaker et al. (2017). Reprinted with permission from AAAS

over the twentieth century (Maes & Van Dyck, 2001). During this period, natural and semi-natural habitats in Flanders declined by 61 per cent. Using this observed reduction in habitat availability, a SAR with $z = 0.25$ would predict only about 12 per cent of species going extinct and even fewer extinctions would be predicted using lower z-values as commonly found in mainland SARs (Fattorini & Borges, 2012). However, contrasting findings have been reported for the same taxon and in a similar system. Habel et al. (2019) studied long-term data on butterfly monitoring in southern Germany, ranging from the end of the eighteenth century to the present. During this period, land use change and agricultural intensification has led to a massive decline in suitable butterfly habitat. Despite local declines, these authors did not find a regional decline, but in fact an increase in regional richness. Nor did they observe any regional extinctions. However, Habel et al. (2019) did report negative trends in abundances. In particular, the abundance of specialist species was negatively affected by habitat changes. Clearly, the discussion around extinction forecasting needs more focus on the spatial and the temporal scales involved.

To take another example, in parts of eastern North America, out of the twenty-eight species of birds restricted to forest habitat, three or four have gone extinct since the arrival of Europeans (Pimm & Askins, 1995); the extinction of the ivory-billed woodpecker, *Campephilus principalis*, is disputed (Donahue, 2017), although the species is probably extinct. This equates to a 14 per cent loss of species. Forest clearing in North America reached a peak in 1872, when only half of the area covered by the eastern forest at the time of European arrival (1620) was still forested. With a 50 per cent loss of habitat, a SAR with a value of $z = 0.15$ (which is in the range typically found for continental sample areas: 0.1–0.2; Rosenzweig, 1995) predicts a 10 per cent loss of species, slightly less than the observed number (Fattorini & Borges, 2012). However, as pointed out by Pimm and Askins (1995), the observed extinctions (four species) refer to species that have gone globally extinct, whereas SAR predictions refer to species that become locally extinct. Local extinction is obviously (much) higher than global extinction, because only species globally endemic to the area of lost habitat go globally as well as locally extinct.

As an example of a tropical study, Fattorini and Borges (2012) compared observed and SAR-predicted extinctions for Helictopleurini dung beetles in Madagascar, a global biodiversity hotspot that has lost about half of its forest cover since 1953. Hanski et al. (2007) report that twenty-two out of fifty-one Helictopleurini species are considered extinct (43 per cent), whereas the use of SARs with z-values ranging from 0.15 to 0.25 predicts that around 10–16 per cent of species should have been lost. Finally, Seabloom et al. (2002), who examined the impact of habitat fragmentation on Californian vascular plant communities, reported that, given the observed amount of habitat loss, species loss had proceeded much faster than predicted by the SAR.

Several factors might explain why SARs can underestimate extinction rates:

1) The use of SARs (and EARs) to forecast species extinctions assumes that habitat reduction is the main cause of species extinction. Although habitat reduction is recognized as a major driver of extinction, many other factors may be involved, such as *inter alia* pollution, global climatic change, overexploitation, introduction of exotics and matrix

and edge effects (e.g. Pullin, 2002; Primack, 2014). These additional factors, which may increase with habitat loss in certain cases, might lead to more extinctions than predicted by the SAR alone. Put another way, habitat loss rarely occurs in isolation and the co-variation between habitat loss and other extinction drivers may render simple SAR calculations based only on habitat loss problematic.

2) Linked to number 1), habitat fragmentation, resulting in greater habitat isolation, reduction in habitat quality, increased edge effects and changes in the relative abundances of species within different trophic groups, almost always accompanies habitat loss (e.g. Didham et al., 1998; Borges et al., 2006; Fischer & Lindenmayer, 2007; Laurance, 2008; Lomolino et al., 2010). As SARs (and EARs) consider only habitat amount, they do not take into account these additional processes that drive species loss.

3) SAR (and EAR) extrapolations of species loss based on the power function assume that conversion of habitat occurs at random with respect to the distribution of species. However, habitat loss does not generally proceed randomly, but often follows aggregated patterns (see Figure 14.3). If individuals of each species are also aggregated in their distribution, then aggregated patterns of habitat loss could lead to more extinctions than predicted by standard SAR approaches (Seabloom et al., 2002).

As a final aside, SARs (and EARs) consider only species numbers, not numbers of individuals. As He and Hubbell (2011) note, extinction is generally regarded to have occurred only after the last individual of the species has been lost. However, arguably, if a species has only a few individuals remaining, it can be considered extinct. The precise number of remaining individuals that should be used to classify practical extinction (i.e. the minimum viable population size; Babu, 2011) is debatable and will vary between species. Regardless, if a large proportion of remaining species in, for example, a deforested landscape is represented by one or two individuals, then our use of known extinctions (i.e. extirpation of the last individual) to validate SAR predictions will underestimate the (arguably) true extinction rate in this system.

14.6 Conclusions and Future Directions

The arguments and examples provided may be interpreted as evidence that the SAR is an ineffective tool for predicting the number of extinctions resulting from habitat loss. Indeed, the fact that the SAR has been found to overestimate and underestimate extinctions in different studies, often by quite substantial margins, is worrying. However, despite these criticisms, a strong position can be taken that the SAR is a valuable conservation and policy 'tool' as it is simple to use and has the *potential* to make semi-reasonable predictions without the need for large amounts of species-specific data; data that for many taxa are unlikely to be collected in the near-future. Perhaps then, rather than abandoning the approach completely, an effective way forward is to build on recent studies that have focused on calibrating and adapting the SAR to incorporate factors relevant to fragmented systems (e.g. connectivity) so that it can be more effectively applied in such systems. The aforementioned discussion of standard SAR models being adapted to include factors such as fragmentation and matrix effects (e.g. Koh et al., 2010; Hanski et al., 2013) and the resultant increased prediction and forecasting accuracy provides examples of the benefits of such a research agenda.

To take another example, one issue with using the SAR to predict extinctions (discussed briefly in Section 14.5) is that the approach implicitly assumes that a species is present in the landscape if one individual remains; a fact that is obviously true in the literal sense, but which is evidently problematic for conservation purposes (Kitzes & Harte, 2014). To address these shortcomings, Kitzes and Harte (2014) have recently introduced two new concepts, termed the extinction–area relationship (the probability of a given species going extinct following habitat loss) and the probabilistic species–area relationship (the probability a species survives in the landscape following habitat loss). These concepts are based on empirical spatially explicit species abundance distributions and have been found to provide more accurate predictions than the standard SAR approach.

Perhaps it is not that surprising that the SAR and EAR cannot generate perfect extinction estimates, given the fact that area never explains all of the variance in species richness. For example, in a study involving 449 significant power function SARs for island ecosystems, Triantis et al. (2012) found that the explained variance as measured by the coefficient of determination R^2 ranged from 0.060 to 0.993, with a mean value of 0.640 ± 0.204. Thus, on average more than 30 per cent of the variation in species richness is explained by other factors, such as habitat quality or isolation. Given the complexity of the factors influencing species richness, the fact that area alone is able to explain, on average, more than 50 per cent of the variance in species richness indicates a highly prominent role of this environmental variable. However, this level of explained variance may be too low for highly accurate predictive purposes. For example, Ulrich (2005) noted that, because of the often-observed low goodness of fit values of SARs, even moderate extrapolations beyond the range of measurements (and hence estimates of extinctions) are doubtful (see Matthews & Aspin, 2019, for a further example focused on island species richness extrapolation). As such, perhaps we need to decide what degree of precision in extinction estimation is sufficient to make reasonable inference about the ecological consequences of habitat loss. Exaggerated demands with regard to precision might actually cause more harm than good with respect to diversity management and political decisions.

In sum, moving forward, there is a need to recognize that any approach based on a single variable (i.e. area) is unlikely to provide a perfect extinction prediction, regardless of the specific details. As such, we would argue that the SAR or EAR should be used to provide a rough first guess of the effects of habitat loss rather than a detailed specific estimate of the number of extinctions. At the same time, we should progress with the recent research agenda focused on how the SAR (and EAR) approach can be adapted and improved such that this rough first guess is as accurate as possible.

References

Allen, A. P. & White, E. P. (2003) Effects of range size on species–area relationships. *Evolutionary Ecology Research*, **5**, 493–499.

Axelsen, J. B., Roll, U., Stone, L. & Solow, A. (2013) Species–area relationships always overestimate extinction rates from habitat loss: Comment. *Ecology*, **94**, 761–763.

Babu, S. (2011) Online comment on 'species–area relationships always overestimate extinction rates from habitat loss'. www.nature.com/nature/journal/v473/n7347/full/nature09985.html#/comments.

Boettiger, C. & Hastings, A. (2013) Tipping points: From patterns to predictions. *Nature*, **493**, 157–158.

Bommarco, R., Biesmeijer, J. C., Meyer, B., Potts, S. G., Pöyry, J., Roberts, S. P., Steffan-Dewenter, I. & Öckinger, E. (2010) Dispersal capacity and diet breadth modify the response of wild bees to habitat loss. *Proceedings of the Royal Society B: Biological Sciences*, **277**, 2075–2082.

Borges, P. A. V., Lobo, J. M., Azevedo, E. B., Gaspar, C., Melo, C. & Nunes, L. V. (2006) Invasibility and species richness of island endemic arthropods: A general model of endemic vs. exotic species. *Journal of Biogeography*, **33**, 169–187.

Brook, B. W., Sodhi, N. S. & Ng, P. K. L. (2003) Catastrophic extinctions follow deforestation in Singapore. *Nature*, **424**, 420–426.

Brooks, M. T. Brook, B. W., Koh, L. P., Pereira, H. M., Pimm, S. L., Rosenzweig, M. L. & Sodhi, N. S. (2011) Extinctions: Consider all species. *Nature*, **474**, 284.

Brooks, T. & Balmford, A. (1996) Atlantic forest extinctions. *Nature*, **380**, 115.

Canale, C. I. & Henry, P.-Y. (2010) Adaptive phenotypic plasticity and resilience of vertebrates to increasing climatic unpredictability. *Climate Research*, **43**, 135–147.

Chevin, L.-M., Lande, R. & Mace, G. M. (2010) Adaptation, plasticity, and extinction in a changing environment: Towards a predictive theory. *PLoS One*, **8**, e1000357.

Condit, R., Ashton, P. S., Baker, P. Bunyavejchewin, S., Gunatilleke, S., Gunatilleke, N., Hubbell, S. P., Foster, R. B., Itoh, A., LaFrankie, J. V., Lee, H. S., Losos, E., Manokaran, N., Sukumar, R. & Yamakura, T. (2000) Spatial patterns in the distribution of tropical tree species. *Science*, **288**, 1414–1418.

Costello, M. J., May, R. M. & Stork, N. E. (2013) Can we name Earth's species before they go extinct? *Science*, **339**, 413–416.

Daily, G. C., Ceballos, G., Pacheco, J., Suzán, G. & Sánchez-Azofeifa, A. (2003) Countryside biogeography of Neotropical mammals: Conservation opportunities in agricultural landscapes of Costa Rica. *Conservation Biology*, **17**, 1814–1826.

Didham, R. K., Lawton, J. H., Hammond, P. M. & Eggleton, P. (1998) Trophic structure stability and extinction dynamics of beetles (Coleoptera) in tropical forest fragments. *Philosophical Transactions of the Royal Society B: Biological Sciences*, **353**, 437–451.

Donahue, M. (2017) Possible ivory-billed woodpecker footage breathes life into extinction debate. www.audubon.org/news/possible-ivory-billed-woodpecker-footage-breathes-life-extinction-debate.

Durrett, R. & Levin, S. A. (1996) Spatial models for species–area curves. *Journal of Theoretical Biology*, **179**, 119–127.

Evans, M., Possingham, H. & Wilson, K. (2011) Extinctions: Conserve not collate. *Nature*, **474**, 284.

Fattorini, S. &. Borges, P. V. A. (2012) Species–area relationships underestimate extinction rates. *Acta Oecologica*, **40**, 27–30.

Fischer, J. & Lindenmayer, D. B. (2007) Landscape modification and habitat fragmentation: A synthesis. *Global Ecology & Biogeography*, **16**, 265–280.

Goulson, D., Nicholls, E., Botias, C. & Rotheray, E. L. (2015) Bee declines driven by combined stress from parasites, pesticides, and lack of flowers. *Science*, **347**, 1255957.

Habel, J. C., Trusch, R., Schmitt, T., Ochse, M. & Ulrich, W. (2019) Long-term large-scale decline in relative abundances of butterfly and burnet moth species across south-western Germany. *Scientific Reports*, **9**, 14921.

Haddad, N. M., Brudvig, L. A., Clobert, J., Davies, K. F., Gonzalez, A., Holt, R. D., Lovejoy, T. E., Sexton, J. O., Austin, M. P., Collins, C. D., Cook, W. M., Damschen, E. I., Ewers, R. M., Foster, B. L., Jenkins, C. N., King, A. J., Laurance, W. F., Levey, D. J., Margules, C. R., Melbourne, B. A., Nicholls, A. O., Orrock, J. L., Song, D.-X. & Townshend, J. R. (2015) Habitat fragmentation and its lasting impact on Earth's ecosystems. *Science Advances*, **1**, e1500052.

Halley, J. M., Sgardeli, V. & Monokrousos, N. (2013) Species–area relationships and extinction forecasts. *Annals of the New York Academy of Sciences*, **1286**, 50–61.

Hanski, I. & Ovaskainen, O. (2003) Metapopulation theory for fragmented landscapes. *Theoretical Population Biology*, **64**, 119–127.

Hanski, I., Koivulehto, H., Cameron, A. & Rahagalala, P. (2007) Deforestation and apparent extinctions of endemic forest beetles in Madagascar. *Biology Letters*, **3**, 344–347.

Hanski, I., Zurita, G. A., Bellocq, M. I. & Rybicki, J. (2013) Species–fragmented area relationship. *Proceedings of the National Academy of Sciences USA*, **110**, 12715–12720.

Harte, J. (2000) Scaling and self-similarity in species distributions: Implications for extinction, species richness, abundance, and range. *Scaling in biology: Patterns and processes, causes and consequences* (ed. by J. H. Brown, J. H. West and B. J. Enquist), pp. 325–342. Oxford: Oxford University Press.

Harte, J. & Kinzig, A. P. (1997) On the implications of species–area relationships for endemism, spatial turnover, and food web patterns. *Oikos*, **80**, 417–427.

Harte, J., Kinzig, A. P. & Green, J. (1999a) Self-similarity in the distribution and abundance of species. *Science*, **284**, 334–336.

Harte, J., McCarthy, S., Kinzig, A. P. & Fischer, M. L. (1999b) Estimating species–area relationships from plot to landscape scale using species spatial-turnover data. *Oikos*, **86**, 45–54.

He, F. & Hubbell, S. P. (2011) Species–area relationships always overestimate extinction rates from habitat loss. *Nature*, **473**, 368–371.

He, F. & Hubbell, S. P. (2013) Estimating extinction from species–area relationships: Why the numbers do not add up. *Ecology*, **94**, 1905–1912.

Hubbell, S. P. (2001) *The unified neutral theory of biodiversity and biogeography*. Princeton, NJ: Princeton University Press.

IUCN (2019) *Summary statistics*. www.iucnredlist.org/resources/summary-statistics.

Keil, P., Storch, D. & Jetz, W. (2015) On the decline of biodiversity due to area loss. *Nature Communications*, **6**, 8837.

Kinzig, A. P. & Harte, J. (2000) Implications of endemics–area relationships for estimates of species extinction. *Ecology*, **81**, 3305–3311.

Kitzes, J. & Harte, J. (2014) Beyond the species–area relationship: Improving macroecological extinction estimates. *Methods in Ecology and Evolution*, **5**, 1–8.

Koh, L. P. & Ghazoul, J. (2010) A matrix-calibrated species–area model for predicting biodiversity losses due to land-use change. *Conservation Biology*, **24**, 994–1001.

Koh, L. P., Lee, T. M., Sodhi, N. S. & Ghazoul, J. (2010) An overhaul of the species–area approach for predicting biodiversity loss: Incorporating matrix and edge effects. *Journal of Applied Ecology*, **47**, 1063–1070.

Kunin, W. E. Harte, J., He, F., Hui C., Jobe, R. T., Ostling, A., Polce, C., Šizling, A., Smith, A. B., Smith, K., Smart, S., Storch, D., Tjørve E., Ugland, K.-E., Ulrich, W. & Varma V. (2018) Upscaling biodiversity: Estimating the species–area relationship from small samples. *Ecological Monographs*, **88**, 170–187.

Laurance, W. F. (2008) Theory meets reality: How habitat fragmentation research has transcended island biogeographic theory. *Biological Conservation*, **141**, 1731–1744.

Lomolino, M. V., Riddle, B. R., Whittaker, R. J. & Brown, J. H. (2010) *Biogeography*, 4th ed. Sunderland, MA: Sinauer Associates.

MacKenzie, D. I., Nichols, J. D., Royle, J. A., Pollock, K. H., Bailey, L. L. & Hines, J. E. (2006) *Occupancy estimation and modeling: Inferring patterns and dynamics of species occurrence*. Burlington, MA: Academic Press.

Maes, D. & Van Dyck, H. (2001) Butterfly diversity loss in Flanders (north Belgium): Europe's worst case scenario? *Biological Conservation*, **99**, 263–276.

Martins, I. S. & Pereira, H. M. (2017) Improving extinction projections across scales and habitats using the countryside species–area relationship. *Scientific Reports*, **7**, 12899.

Matthews, T. J. & Aspin, T. (2019) Model averaging fails to improve the extrapolation capability of the island species–area relationship. *Journal of Biogeography*, **46**, 1558–1568.

Matthews, T. J., Cottee-Jones, H. E. & Whittaker, R. J. (2014b) Habitat fragmentation and the species–area relationship: A focus on total species richness obscures the impact of habitat loss on habitat specialists. *Diversity and Distributions*, **20**, 1136–1146.

Matthews, T. J., Guilhaumon, F., Triantis, K. A., Borregaard, M. K. & Whittaker, R. J. (2016) On the form of species–area relationships in habitat islands and true islands. *Global Ecology & Biogeography*, **25**, 847–858.

Matthews, T. J., Steinbauer, M. J., Tzirkalli, E., Triantis, K. A. & Whittaker, R. J. (2014a) Thresholds and the species–area relationship: A synthetic analysis of habitat island datasets. *Journal of Biogeography*, **41**, 1018–1028.

May, R. M. (1975) Patterns of species abundance and diversity. *Ecology and evolution of communities* (ed. by M. L. Cody and J. M. Diamond), pp. 81–120. Cambridge MA: Harvard University Press.

May, R. M., Lawton, J. H. & Stork, N. E. (1995) Assessing extinction rates. *Extinction rates* (ed. by J. H. Lawton and R. M. May), pp. 1–24. Oxford: Oxford University Press.

Mendenhall, C. D., Karp, D. S., Meyer, C. F. J., Hadly, E. A. & Daily, G. C. (2014) Predicting biodiversity change and averting collapse in agricultural landscapes. *Nature*, **509**, 213–217.

Moczek, A. P. (2010) Phenotypic plasticity and diversity in insects. *Philosophical Transactions of the Royal Society B: Biological Sciences*, **365**, 593–603.

Ney-Nifle, M. & Mangel, M. (1999) Species–area curves based on geographic range and occupancy. *Journal of Theoretical Biology*, **196**, 327–342.

Ney-Nifle, M. & Mangel, M. (2000) Habitat loss and changes in the species–area relationship. *Conservation Biology*, **14**, 893–898.

Palmer, M. W. & White, P. S. (1994) Scale dependence and the species–area relationship. *The American Naturalist*, **144**, 717–740.

Pereira, H. M. & Daily, G. C. (2006) Modeling biodiversity dynamics in countryside landscapes. *Ecology*, **87**, 1877–1885.

Pereira, H. M., Borda-de-Água, L. & Martins, I. S. (2012) Geometry and scale in species–area relationships. *Nature*, **482**, E3–E4.

Pimm, S. L. (1998) Extinction. *Conservation science and action* (ed. by W. J. Sutherland), pp. 28–38. Oxford: Blackwell.

Pimm, S. L. & Askins, R. A. (1995) Forest losses predict bird extinction in eastern North America. *Proceedings of the National Academy of Sciences USA*, **92**, 9343–9347.

Primack, R. B. (2014) *Essentials of conservation biology*, 6th ed. Oxford: Oxford University Press.

Pullin, A. S. (2002) *Conservation biology*. Cambridge: Cambridge University Press.

Reid, W. V. & Miller, K. R. (1989) *Keeping options alive: The scientific basis for conserving biodiversity*. Washington, DC: World Resources Institute.

Riddle, B. R., Ladle, R. J., Lourie, S. A. & Whittaker, R. J. (2011) Basic biogeography: Estimating biodiversity and mapping nature. *Conservation biogeography* (ed. by R. J. Ladle and R. J. Whittaker), pp. 47–92. Chichester: Wiley-Blackwell.

Rohr, R. P., Saavedra, S. & Bascompte, J. (2014) On the structural stability of mutualistic systems. *Science*, **345**, 1253497.

Rosenzweig, M.L. (1995) *Species diversity in space and time*. Cambridge: Cambridge University Press.

Schilthuizen, M. (2018) *Darwin comes to town: How the urban jungle drives evolution*. New York: Picador.

Seabloom, E. W., Dobson, A. P. & Stoms, D. M. (2002) Extinction rates under non-random patterns of habitat loss. *Proceedings of the National Academy of Sciences USA*, **99**, 11229–11234.

Smith, A. B. (2010) Caution with curves: Caveats for using the species–area relationship in conservation. *Biological Conservation*, **143**, 555–564.

Storch, D., Keil, P. & Jetz, W. (2012) Universal species–area and endemics–area relationships at continental scales. *Nature*, **488**, 78–81.

Stork, N. (2010) Re-assessing current extinction rates. *Biodiversity and Conservation*, **19**, 357–371.

Tilman, D., May, R. M., Lehman, C. L. & Nowak, M. A. (1994) Habitat destruction and the extinction debt. *Nature*, **371**, 65–66.

Triantis, K. A., Borges, P. A. V., Ladle, R. J., Hortal, J., Cardoso, P., Gaspar, C., Dinis, F., Mendonça, E., Silveira, L. M. A., Gabriel, R., Melo, C., Santos, A. M. C., Amorim, I. R., Ribeiro, S. P., Scrrano, A. R. M., Quaitau, J. A. & Whittaker, R. J. (2010) Extinction debt on oceanic islands. *Ecography*, **33**, 285–294.

Triantis, K. A., Guilhaumon, F. & Whittaker, R. J. (2012) The island species–area relationship: Biology and statistics. *Journal of Biogeography*, **39**, 215–231.

Ulrich, W. (2005) Predicting species numbers using species–area and endemics–area relations. *Biodiversity and Conservation*, **14**, 3351–3362.

Ulrich, W. & Buszko, J. (2003a) Species–area relationship of butterflies in Europe: The simulation of extinction processes reveals different patterns between Northern and Southern Europe. *Ecography*, **26**, 365–374.

Ulrich, W. & Buszko, J. (2003b) Self-similarity and the species–area relation of Polish butterflies. *Basic and Applied Ecology*, **4**, 263–270.

Ulrich, W. & Buszko, J. (2004) Habitat reduction and patterns of species loss. *Basic and Applied Ecology*, **5**, 231–240.

Ulrich, W. & Buszko, J. (2005) Detecting biodiversity hotspots using species–area and endemics–area relationships: The case of butterflies. *Biodiversity and Conservation*, **14**, 1977–1988.

Whittaker, R. J. & Matthews, T. J. (2014) The varied form of species–area relationships. *Journal of Biogeography*, **41**, 209–210.

Whittaker, R. J., Fernández-Palacios, J. M., Matthews, T. J., Borregaard, M. K. & Triantis, K. A. (2017) Island biogeography: Taking the long view of nature's laboratories. *Science*, **357**, eaam8326.

Wilsey, B. J., Martin, L. M. & Polley, H. W. (2005) Predicting plant extinction based on species–area curves in prairie fragments with high beta richness. *Conservation Biology*, **19**, 1835–1841.

Wilson, E. O. (1992) *The diversity of life*. Cambridge, MA: Belknap Press.

World Conservation Monitoring Centre (1992) *Global biodiversity: Status of the Earth's living resources*. London: Chapman and Hall.

15 · *Using Network Analysis to Explore the Role of Dispersal in Producing and Maintaining Island Species–Area Relationships*

JOSEPH A. VEECH AND
GIOVANNI STRONA

15.1 Introduction

The species–area relationship is essentially an explanation of how geography affects species diversity. Since first described by Arrhenius (1921; at least the first mathematical description; see Chapter 2) it has proven to be a robust and near-universal principle in ecology (Rosenzweig, 1995; Lomolino, 2000; Drakare et al., 2006; McGill, 2010; O'Dwyer & Green, 2010). However, understanding the determinants of species diversity patterns continues to be a primary challenge for ecologists. Moreover, ecologists are increasingly recognizing the fundamental role of dispersal in maintaining patterns of species diversity (Kneitel & Miller, 2003; Mouquet & Loreau, 2003; Cadotte, 2006; Seymour et al., 2015; Gómez-Rodríguez et al., 2019). To some extent, individuals of all species disperse from their natal habitat over various spatial distances to arrive at new locations. Movements can involve the progressive colonization of close locations through short steps or longer 'jumps' to more distant locations. In both cases, establishment and persistence of a species is determined by how well the species can survive and reproduce in the new locality given the local abiotic and biotic conditions, including the compatibility between the species' niche and the physical environment as well as presence of other species (Case, 1983; Janzen, 1985; Ackerly, 2003; MacDougall et al., 2009; Skóra et al., 2015). The ecological mechanisms determining whether a species can successfully colonize and establish a population in a new area can be extremely complex and highly dependent on the specific situation. Similarly, it is very difficult to predict (or even interpret) the broad effects that a colonizer can have on a local community and environment. In some cases, such effects can be very detrimental, as we are observing with increasing frequency in the broad context of biological invasions (Mooney & Cleland, 2001; Simberloff et al., 2013).

Thus, the composition of a community emerges (and changes through time) as a consequence of the spatial arrangement of localities, mechanisms of species dispersal, environmental features at the various localities and composition of other communities that receive and send out species. This depicts a very dynamic scenario. Evolutionary and ecological processes at play in the various localities as well as geography (such as the presence of barriers) may affect both the ability of local species to move and colonize other localities and in turn affect the ability of potential colonizing species to establish themselves in the local community (Thomas et al., 2001; Lomolino et al., 2006; Altieri et al., 2010). Colonization can have very strong effects on the ecological and the evolutionary processes occurring at a given locality (Kokko & López-Sepulcre, 2006; Ricciardi et al., 2013). Further, events occurring at a given locality may have broader effects at the regional level and on the greater meta-community (Ricklefs, 1987; Cornell & Lawton, 1992; Leibold et al., 2004). Despite this complexity, simplified models have permitted ecologists to identify some seemingly universal patterns that have fundamental implications for the study and conservation of biodiversity. Among these, the species–area relationship (SAR), a near universal law describing the increase in species richness with increase in the area surveyed or sampled, occupies a primary position (Connor & McCoy, 1979; Triantis et al., 2012). Although there are several different varieties of SAR (including those based on nested sampling areas; see Chapters 1 and 8), we focus on SARs derived from true islands and pseudo-islands (habitat islands) such as distinct patches of habitat embedded in a non-habitat landscape matrix. Importantly, island SARs (ISARs) are constructed from data that come from non-overlapping areas that are physically separated.

Ecologists have typically viewed the SAR as a temporally static pattern. Often, various functions (e.g. the power function) are fitted to data with the implicit assumption that species composition on islands, habitat fragments or other sampling units is invariant or at least species richness is stable over time, even though species might be replacing one another (e.g. species turnover). Such studies then test and seek explanations for the SAR primarily by examining factors that are directly related to area. For instance, these explanations could be larger islands 1) have greater variety of habitat and hence greater species richness, 2) support larger population sizes and therefore species are less likely to become locally extinct and 3) are simply larger 'targets' for randomly dispersing species and so at some point in the past they collected more species (Connor & McCoy, 1979; Drakare et al., 2006). These three are among many other explanations invoking area as the ultimate driver of the pattern (see Chapters 3 and 4). This makes sense, after all 'area' is the independent variable (placed on the x-axis) when visualizing the pattern in a scatter-plot, quantifying the relationship with any of various mathematical functions (Chapter 7) and testing for statistical significance with methods as straightforward as least squares regression.

The irony in this perspective based on temporal stasis is that dynamic ecological processes are responsible for producing and maintaining SARs. ISARs emerge from colonization–extinction dynamics and, in some situations, speciation also plays a role (Whittaker & Fernández-Palacios, 2007). Colonization depends on mainland-island or inter-island dispersal and extinction is simply the inverse of species population persistence on an island. Thus, colonization–extinction dynamics are determined locally by the capacity for a species population to grow and spread on the island and regionally by metacommunity dispersal/colonization processes (Brown & Kodric-Brown, 1977; He et al., 2005). Regardless of island size, a population must increase (spread) after the initial colonization event if it is to persist and be more likely to send out colonists (dispersing individuals) to other islands. As such, the spatial arrangement of islands relative to the mainland and each other should influence the temporal development and maintenance of the ISAR (MacArthur & Wilson, 1967).

The arrangement of islands is the stage upon which the colonization–extinction dynamics play out. Details of the dispersal process are likely to be very important in determining the strength of an ISAR. For example, whether mainland-island and inter-island dispersal occurs at a low, inter-mediate or high rate should in part determine the slope of an ISAR curve, in addition to having an effect on the overall number of species on each island. The rate of local extinction should also influence the slope of the ISAR curve. Again, both are temporally dynamic processes and thus changes in the rates over time should lead to a dynamic ISAR. To be exact, dispersal does not necessarily equal colonization. The latter is defined as a dispersing individual successfully reaching an island and being able to settle. Thus, colonization depends greatly on mainland-island and inter-island distances as well as island size (i.e. the target effect).

Island size (area) does not solely determine the form of an ISAR. Islands in an archipelago compose a dynamic system in which each island may experience different rates of colonization and extinction depending on its location relative to other islands (Matthews et al., 2019; see also Chapter 3). For example, we might expect a relatively small island to have more species (than its size alone would predict) if substantially larger islands are nearby. The smaller island will receive more colonists from the nearby larger islands than it would if it were isolated far off at one end or perimeter of the archipelago. Continuing this logic, the larger islands are expected to have greater species richness (than the small island) because they are large and, hence, the colonists from the larger islands will often represent species new to the smaller island. In this way, the species richness of the smaller island might be continually reinforced even if the extinction rate is relatively high on the smaller island. Of course, this hypothetical scenario can be expanded to consider that all the islands in an archipelago are potentially connected and mutually influence each other's species composition. This conceptual perspective on ISARs is an ideal match for using connectivity networks to model and analyse the colonization–extinction dynamics that determine the quantitative characteristics (exact mathematical form) of the ISAR.

15.2 Archipelago Geometry, Species Dispersal and Connectivity Networks

Intuition suggests that the spatial arrangement of islands, that is, the geometry of an archipelago, may play an important role in the evolution and maintenance of ISARs (see also MacArthur & Wilson, 1967; Chapter 3). In particular, the archipelago geometry is expected to strongly affect colonization patterns through different scenarios of connectivity between islands. For example, an elongated archipelago (e.g. the Hawaiian Islands) is expected to promote sequential patterns of colonization, with species being progressively transmitted through subsequent steps of dispersal and colonization, starting from the oldest islands or perhaps those closest to the mainland (MacArthur & Wilson, 1967; Wagner & Funk, 1995; Shaw, 1996; Hormiga et al., 2003; Medeiros & Gillespie, 2011; Santamaria et al., 2013). Conversely, colonization in archipelagos lacking a major dimensional axis (e.g. the Aegean Islands) might happen in less predictable ways.

However, the colonization process does not depend solely on the relative positioning of islands in the geographic space. When a species disperses naturally between islands (i.e. in the absence of human translocation), the dispersal ability of the species determines which islands can be colonized from other islands. Thus, the theoretical connectivity matrix depicting the probability of species movements from one island to any other island is a product of island geometry and species dispersal ability. This also implies that the connectivity matrix will vary within the same archipelago for species having different dispersal ability. Various additional factors might generate further complexity, by affecting both the 'intrinsic' connectivity between different islands and/or species-specific differences. For example, physical factors such as ocean currents might increase a species' chances of dispersing through certain routes and preclude other routes for those species dispersing by oceanic drift or by swimming. Species dispersing through the air would be unaffected by ocean currents but perhaps affected by prevailing wind currents. Then, of course, the ecological requirements of a species might combine with environmental factors in multiple ways so that, depending on the degree of environmental heterogeneity, the actual set of islands capable of hosting a given species could be a small subset of the archipelago. Although the colonization and establishment processes can be very species-specific, in this chapter we focus on idealized scenarios in which the only factors affecting colonization patterns – and the emergence of ISARs – are archipelago geometry and species dispersal ability. Our aim is to develop an ideal, synthetic and general theoretical framework rather than develop an approach that would simulate the exact details of every possible species dispersal scenario.

By assuming island geometry and species dispersal ability as the only elements controlling colonization, we can represent any island-species system (i.e. a given assemblage of species dispersed among a set of islands) as a spatially explicit network. The network is based on inter-island connectivity, with nodes corresponding to islands and with edges linking any pair of nodes (i.e. islands) whose geographical separation (inter-island distance) is less than the maximum dispersal ability of the focal species/group. Previously, Economo and Keitt (2008, 2010) used spatially explicit connectivity networks to analyse species diversity patterns of metacommunities, but not in the context of analysing ISARs.

For islands placed within an ideal and limited space (simulated as a square) and for species with *low dispersal ability*, links exist only between nodes that are very close, hence leading to a network with a very strong spatial signature. Such a network is characterized by a large diameter; that is, a large number of intermediate steps is needed to move across the network from one node to the most distant one through the shortest path. Furthermore, the network has a relatively uniform degree distribution; that is, all nodes will tend to have a similar number of connected neighbouring nodes.

Conversely, for species with *high dispersal ability*, the spatial character of the network is weaker. If species can disperse across distances that are comparable to the maximum distance between any two islands, the network is close to being fully connected (that is, each node is linked to every other node in the network). For *intermediate dispersal ability* the network is mostly randomly connected. However, different geometrical arrangements of islands might significantly modulate the differences between these scenarios of connectivity based on species dispersal ability.

The network of connectivity might have various effects on the assembly of communities on different islands. Connectivity might determine the timing and frequency of arrival of the different species and hence have a strong effect on the persistence of local populations, as well as on the emergence of structural patterns at different ecological scales (from a local population to the entire archipelago's metacommunity). Thus, there is an expectation that different network structures will generate differences in both the dynamic patterns of species diversity across different islands, as well as in the final diversity pattern at equilibrium (the ISAR).

Here we use a simulation model to explore the potential of such an approach to better understand the role of dispersal and colonization in producing and maintaining the ISAR. We simulated various archipelago arrangements and examined the dispersal–colonization process within a spatially explicit modelling environment. We tracked the accumulation of species on islands through time until the metacommunity and its ISAR reached an equilibrium (see Section 15.3). We explored variation in species dispersal abilities that, when combined with the different geometric arrangement of islands, permitted us to evaluate the effect of various network properties on the resulting ISARs.

15.3 The Model

We developed a spatially-explicit model in a two-dimensional (non-toroidal) plane. A grid is super-imposed on the plane with all cells being classified either as island or ocean. Islands are modelled as sets (of variable size) of contiguous cells in the grid with each set/island forming an approximate circle with a varying radius. Different rules for island placement permitted us to simulate archipelagos with specific geometries. The model operates at the population level, wherein only one population (of any species) occupies a given grid cell; that is, one grid cell = one complete population regardless of species identity. Not all cells need be occupied by a population. A population established in a cell can produce a 'propagule', meant as a minimum set of progeny capable of potentially colonizing another island. Note that the minimum set does not need to be defined. In all cases, colonization success is granted to propagules landing in an unoccupied grid cell, while different rules were set to identify the outcome of a colonization attempt in an occupied cell (see text later in this section).

In the model, after having generated the spatially explicit archipelago, we start with a global pool of 1,000 species. This global pool represents a set of species that could develop over evolutionary time (on the mainland) and serve as the ultimate source of species that might enter the archipelago at any time. Species enter the archipelago by dispersing at random from a hypothetical continent, with all islands being uninhabited at the beginning of the simulation. The continent is not modelled explicitly. Instead, the spatial source of species is a 'window' on one edge of the grid. Propagules of species are sampled at random from the species pool and enter the system at each simulation step from that window, dispersing in a random direction (0 to 180°) to generate a gradient of colonization. During each time step, a random number of species between one and five enters the system.

Because all species are modelled as equivalent in their requirements for space, each cell of an island can host no more than one population (of any species). Consequently, the total number of cells composing an island is also the maximum number of populations that can inhabit the island given the overall constraint set by the size of the species pool. In turn, this determines an upper boundary for the species richness of the island, with maximum richness occurring when each cell of the island is occupied by a population of a species not occurring in any other cell. The lower theoretical boundary for species richness (for inhabited islands) is one. Note that, although unlikely, even a fully occupied, large island might have diversity equal to one (when all of its cells are occupied by populations of the same species).

Throughout a simulation, species can disperse from the mainland in addition to dispersing from already occupied grid cells. At each step of a

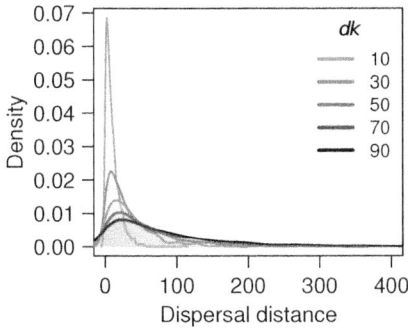

Figure 15.1 Dispersal kernels modelled using the exponential distribution with mean *dk*. Dispersal distance is measured in grid units (see text for further details). The parameter *dk* is used in the model as a synthetic measure of species dispersal ability.

simulation, with a given probability '*b*', each population produces a dispersal 'propagule'. The propagule disperses in a random direction, and moves along a linear trajectory for a distance defined by a dispersal kernel drawn from an exponential distribution with mean '*dk*' (Figure 15.1). Note that the dispersal distance is sampled at random from the same exponential distribution for each dispersal event, regardless of species identity (i.e. the dispersal kernel is assumed to be identical for each species in the simulation).

Whenever the trajectory intersects a cell belonging to an island different from the source island, the propagule may either continue along the trajectory with probability *s* (without attempting colonization of the intersected island unless the trajectory terminates on the island) or it may attempt colonization of the cell with probability $1 - s$. If the propagule attempts colonization then it is always successful if the intersected cell is empty while, if the cell is occupied, colonization succeeds with probability *p*. If the propagule continues across its dispersal trajectory beyond the intersected island the propagule might intersect another island and attempt colonization there or eventually end up in the sea. In any case, the propagule will stop at the end of the trajectory. If this happens when the propagule is still on the source island, it will attempt intra-island range expansion ('colonization' of an empty cell on the same island), with the same rules as above. Hence the model can simulate both inter- and intra-island dispersal. With a probability *e*, at each step, populations go extinct. Whenever this happens, the cell previously occupied by the extinct population is emptied. For example, if $e = 0.01$ then 1 per cent of all cells occupied by the species will be emptied regardless of their location. Because extinction is random, we refer to this as the neutral version of the model.

In addition to this model (now referred to as 'neutral'), we also developed a slightly modified non-neutral version in which we assigned species-specific differences in both competitive strength and dispersal ability. To model differences in competitive strength, for each simulation we defined an adjacency species competition matrix $[C]$, with each C_{ij} entry indicating the probability of species i outcompeting species j. For any $i < j$, we attributed a random value sampled from $U(0, 1)$ to each C_{ij} cell; then we set each C_{ji} entry to $1 - C_{ij}$, while we set each C_{ii} entry to 0 (i.e. assuming no intraspecific competition).

To model variability in species dispersal abilities, we attributed to each species a unique random dispersal kernel. We did this by assigning each species a kernel randomly selected from an exponential distribution with a mean dk itself sampled from $U(1, DK\star1.5)$, where DK was a specified upper bound. By intentionally specifying DK as a high (or low) value in each simulation, we were able to generate scenarios with different variability in species dispersal abilities. A small DK value produced a scenario (simulation) where all species had reduced dispersal ability, whereas large DK values represented scenarios in which there was a wide array of dispersal abilities among the species, possibly including species with very high dispersal ability.

15.4 Geographical Setting (Simulated Archipelagos)

We considered an arbitrary grid of size $10,000 \times 10,000$ units (where one unit = one cell). We generated sets of archipelagos (each having fifty islands) with four different idealized spatial patterns:

Random: Islands are randomly placed within the grid; each island is a set of contiguous cells in the grid approximating a circle with the radius varying at random between one and twenty units. A minimum distance of ten units is required between an island and another (Figure 15.2A).

Star: A circular island with a radius of forty units is placed at the centre of a $1,000 \times 1,000$ square that itself is at the centre of the overall grid. Then the other islands, circles with radii varying at random between 1 and 18.6, are placed at random within the $1,000 \times 1,000$ area. The maximum size is set to 18.6 to compensate for the large size of the central island, so as to obtain, on average, the same total insular area obtained for the random and the elongated archipelago. Again, placement is specified so that islands are separated by at least ten units (excluding the central, big

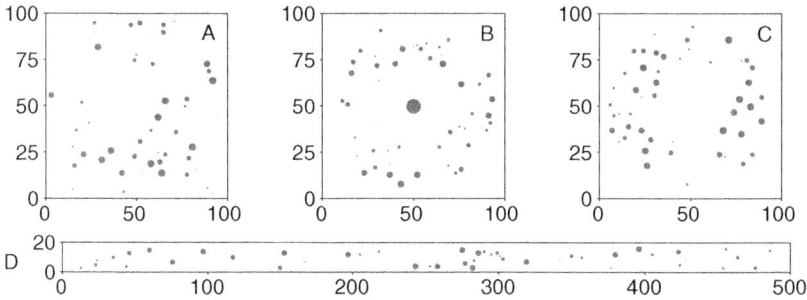

Figure 15.2 Examples of simulated island archipelagos: (A) random, (B) star, (C) ring and (D) elongated. Axes represent the two arbitrary spatial dimensions of the underlying regular grid. See text for descriptions of archipelago geometries.

island). Also, islands are positioned so that they are 200–500 units from the central island (Figure 15.2B).

Ring: Same as star but with the large central island replaced by a smaller island positioned in the same way as the other islands (Figure 15.2C).

Elongated: The archipelago is generated with the same procedure as the random one, but the grid is modified to a rectangle of size 5,000 × 200 (Figure 15.2D).

15.5 Identifying the Archipelago Connectivity Network

To identify the network representing the potential movement of species among islands in the archipelago, we used a straightforward procedure that attributed a link between any pair of islands separated by a distance smaller than the average dispersal ability of a species in the neutral simulations (i.e. *dk*). Thus, the structure of each connectivity network depends both on the geometry of the archipelago and on the specific *dk* of the corresponding simulation.

We emphasize that this type of connectivity network represents *potential* colonization pathways. Alternatively, by keeping track of the actual frequency of dispersal events from one island to another throughout a complete simulation, we could obtain the actual colonization pathways (realized connectivity) among islands. However, we chose to use a simplified analytical criterion for connectivity, since our goal is to provide a theoretical framework applicable to real datasets. For most species–area datasets, detailed information about historical dispersal/colonization patterns (as would be needed for modelling realized connectivity) is not available.

After constructing the connectivity network, we then measured basic network properties potentially relevant for the overall dispersal patterns across the archipelago. In particular, we used the following metrics to measure network structure (Newman et al., 2006):

Connectance – the ratio between the number of observed and the maximum theoretical number of links, with the latter defined as links existing between any and all pairs of nodes (islands in the archipelago). Connectance can range from near zero to one. Larger values indicate greater connectivity among the nodes of a network.

Diameter – the length of the shortest path connecting the two most distant nodes in the network (with a 'path' being a set of steps across network links, with length identical to the number of steps). Diameter can range from one to the number of nodes minus one. Lower values indicate greater connectivity among the nodes of the network.

Mean and standard deviation of the node degree distribution – the degree of a given node is simply the number of its neighbours (that is, connected islands or islands closer than dk in the geometric space). A large mean paired with a small SD indicates greater connectivity among the nodes of the network.

In our explorative analyses, we also considered several other potentially informative network-level properties (such as network 'clustering'). However, the metrics for these properties were strongly correlated with the three basic measures described above and hence we only used those three. At the level of individual nodes, we used two metrics of connectedness: node degree (as described above) and PageRank centrality. The latter is one of many centrality metrics that characterize the extent that a given node is connected to other nodes that themselves are well connected (Allesina & Pascual, 2006; Bodin & Saura, 2010; de Arruda et al., 2014).

15.6 Model Simulations and Analysis

We conducted 800 simulations of the model, as described in Section 15.3; 100 simulations for each of the four archipelago geometries in the neutral and non-neutral versions of the model. Each simulation was run for 15,000 time steps. During each step the computer code tallied the number of species on each island and fit a species–area curve to the data as the power function, $S = cA^z$, where S = species richness on an island of area A, and c and z are function parameters that control the shape of the curve (see Chapters 3 and 4 for a thorough discussion on the mathematical and ecological meaning of c and z). The first 10,000 steps were more than enough to permit the metacommunity and the ISARs to reach an equilibrium. We defined equilibrium as the parameters of the ISAR (namely c and z) becoming stable and not changing by more than 1 per cent from one time step to the next. In the last 5,000 steps we intentionally reduced the frequency of inter-island dispersal events, while we promoted intra-island expansion. We did this by allowing only 25 per cent of the dispersal events to be attempted off-island according to the rules described in Section 15.3, while 75 per cent of the dispersal events happened between a source cell and a randomly sampled cell adjacent to the source cell. In such short-distance dispersal events, colonization succeeded under the same rules applied in the first 10,000 simulation steps. That is, the probability of colonization was zero if the random target cell was ocean, one if the cell was empty and either p or C_{ij} otherwise. Our intention in reducing inter-island dispersal was to further examine its role in maintaining the equilibrium ISAR. In particular, we were interested in verifying if continual dispersal is a necessary condition for the maintenance of ISARs, by testing to what extent a reduction or cessation of dispersal and hence inter-island colonization would modify – most likely weaken and deteriorate – the relationships observed at equilibrium.

After the simulations we performed several analyses intended to explore and demonstrate the theoretical framework detailed in Section 15.1. First, we examined how dispersal ability combines with archipelago geometry to generate different kinds of connectivity networks. For each archipelago, we compared different network properties with the corresponding values of dk, both considering the different archipelago geometries separately and together. Note that this comparison did not require running the simulations; it was solely based on the different model settings and constructing the connectivity network (see Section 15.7). Second, we examined the relationship between network properties and the parameters of the ISAR. For this, we calculated Spearman's rank-correlation coefficient between each network property and each ISAR parameter separately for each archipelago geometry and each version (neutral and non-neutral) of the model (see Sections 15.8 and 15.9). Third, we compared the relative effects of island area, degree and PageRank centrality on island species richness; this was accomplished by conducting multiple regression separately for each ISAR dataset resulting from a given simulation. For each dataset, we applied the full regression model (area + degree + PageRank as predictor variables), two partial models (area + degree, area + PageRank) and each of the three single-variable models. We then compared the adjusted R^2 values of the partial models to the full model to determine the additional amount of variance in species richness explained by either of the two metrics (degree and PageRank) assessing an island's connectedness to other islands. This analysis was particularly intriguing because it explored the possibility of using hybrid ISARs combining an island's topological position and area to better explain species richness (see Section 15.10).

15.7 Relationship between *dk* and Network Properties

As expected, there were some major intuitive differences in the connectivity networks for the different types of island archipelagoes. In particular, the average pairwise distance between any two islands in the archipelago was much larger in the elongated geometry (163.3 grid units) than in the random (47.3), ring (42.7) or star (42.3) geometries (with averages calculated across all the simulated archipelagos). This translated into some clear effects on the connectivity networks. Not surprisingly, increasing *dk* led to networks with greater connectance, larger average node degree and smaller diameter. However, for the same value of *dk*, networks of elongated archipelagos had a much lower connectance and average node degree and a much longer diameter. This is simply a consequence of the fact that, in the elongated setting, islands have a much larger average pairwise distance than in the other geometrical settings (random, ring, star). However, when *dk* is standardized to mean pairwise distance between islands then it becomes evident that structure of the connectivity networks is mainly driven by the distances separating islands, not by archipelago geometry. For all archipelago geometries, as species dispersal ability increases (increase in *dk*) connectance increases and tends to reach the maximum value of one when *dk* is twice the mean inter-island distance (Figure 15.3). The standard deviation of node degree reaches a maximum value when *dk* is slightly greater than mean inter-island distance and then rapidly approaches zero as dispersal distance becomes much greater than mean inter-island distance (Figure 15.3). Also, as expected, network diameter decreases with increasing dispersal distance and reaches the minimum possible value of one when dispersal distance is about twice the mean inter-island distance (Figure 15.3). To summarize, regardless of archipelago geometry, when dispersal distance is great enough relative to distances separating islands, then all of the

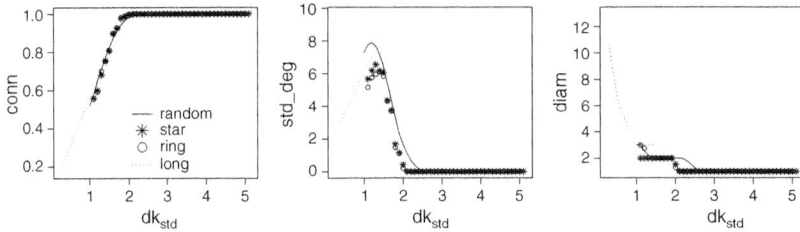

Figure 15.3 Basic structural properties of connectivity networks obtained for increasing hypothetical values of species dispersal ability for the four simulated archipelago geometries (random, star, ring and elongated). Dispersal distance (*dk*) is standardized to the mean pairwise distance between islands (*dk*$_{std}$). See text for description of the network properties and archipelago geometries. Note that the same patterns are obtained when *dk* is standardized to maximum inter-island distance. Abbreviations: conn, network connectance; st_deg, standard deviation of the network node degree distribution; diam, network diameter.

islands will have all fifty species such that mean degree = 49, SD of mean degree = 0, connectance = 1 and diameter = 1.

15.8 Emergence of ISARs

In all simulations, ISARs emerged quite rapidly (typically within a few hundred time steps) and achieved structural properties (*z* and *c*) that were maintained throughout the simulation prior to the intentional reduction in inter-island dispersal after step 10,000 (Figure 15.4). With regard to the initial source pool of 1,000 species, most model runs tended to finish (i.e. arrive at an equilibrium ISAR) with 20–136 species in the system. The *z* and *c* parameters of most simulated ISARs were in the range typically observed for real-world ISARs, $0.1 < z < 0.4$ and $0.3 < c < 1.5$ (Connor & McCoy, 1979; Drakare et al., 2006; Sólymos & Lele, 2012; Triantis et al., 2012; Patiño et al., 2014; Fattorini et al., 2017; Chapter 3) (Figure 15.5). The trajectories of ISAR development were very consistent among the random, star and ring archipelagos, but slightly different for the elongated archipelagos (Figure 15.4). In addition, ISARs tended to emerge and reach equilibrium sooner in the non-neutral version of the model (Figure 15.4, bottom panels). When dispersal was reduced, all the metacommunities departed quite rapidly from the structure shown at the equilibrium and reached a new equilibrium characterized by a shallower ISAR (lower *z*-value) that did not fit the data as well (lower R^2) (Table 15.1). Interestingly, in both versions of the model (i.e. the

Table 15.1 *Summary of ISAR parameters (mean ± SD) for both the fully neutral model and the model accounting for species-specific competitive and dispersal abilities, at equilibrium before and after the reduction in inter-island dispersal. Values are the mean and standard deviation of simulations combined among different archipelago geometries*

	Fully neutral model		Model with species-specific competition and dispersal	
	Before reduction	After reduction	Before reduction	After reduction
z	0.28 ± 0.09	0.16 ± 0.04	0.45 ± 0.05	0.19 ± 0.03
c	1.07 ± 0.32	0.17 ± 0.05	1.94 ± 0.26	0.21 ± 0.05
R^2	0.69 ± 0.16	0.55 ± 0.09	0.91 ± 0.04	0.57 ± 0.06

Table 15.2 *Summary of ISAR parameters (mean ± SD) for the fully neutral model and the model accounting for species-specific competitive and dispersal abilities, at equilibrium before and after the reduction in inter-island dispersal, for each archipelago geometry*

Model version	Dispersal reduction	Geometry	z	c	R^2
Neutral	Before	Random	0.29 ± 0.10	1.03 ± 0.36	0.65 ± 0.18
–	–	Star	0.28 ± 0.09	1.10 ± 0.38	0.70 ± 0.17
–	–	Ring	0.28 ± 0.09	1.18 ± 0.31	0.66 ± 0.18
–	–	Long	0.29 ± 0.06	0.96 ± 0.17	0.74 ± 0.08
–	After	Random	0.16 ± 0.04	0.17 ± 0.04	0.54 ± 0.10
–	–	Star	0.15 ± 0.04	0.15 ± 0.04	0.54 ± 0.11
–	–	Ring	0.16 ± 0.04	0.17 ± 0.05	0.55 ± 0.09
–	–	Long	0.16 ± 0.02	0.19 ± 0.05	0.56 ± 0.06
Non-neutral	Before	Random	0.47 ± 0.04	2.02 ± 0.13	0.92 ± 0.02
–	–	Star	0.45 ± 0.05	2.09 ± 0.10	0.92 ± 0.02
–	–	Ring	0.46 ± 0.04	2.09 ± 0.10	0.93 ± 0.03
–	–	Long	0.42 ± 0.05	1.54 ± 0.11	0.86 ± 0.05
–	After	Random	0.19 ± 0.03	0.22 ± 0.05	0.56 ± 0.07
–	–	Star	0.20 ± 0.02	0.18 ± 0.04	0.57 ± 0.05
–	–	Ring	0.19 ± 0.03	0.22 ± 0.06	0.58 ± 0.06
–	–	Long	0.19 ± 0.03	0.21 ± 0.05	0.58 ± 0.05

fully neutral version and the version simulating species-specific variability in competitive and dispersal abilities), when inter-island dispersal was reduced, the ISARs at the new equilibrium were less variable among the four archipelago geometries than they were at the previous equilibrium. At the new equilibrium, z-values tended to be 0.15–0.20, c-values were 0.15–0.22 and R^2 values were 0.54–0.58, regardless of archipelago geometry (also compare the SD values for 'before' and 'after' in Table 15.2). These results make sense; with a substantial reduction in species dispersal among the islands, the spatial arrangement (geometry) of the islands will not matter as much to the ISAR and the ISAR will be weaker than when there is greater dispersal among the islands.

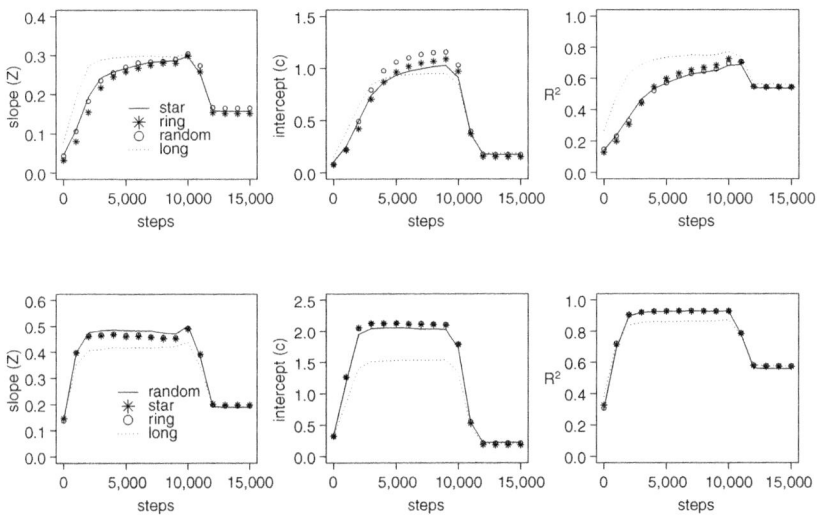

Figure 15.4 Emergence of ISARs through time (i.e. model simulation steps) in both the neutral version of the model (top panels) and the non-neutral version (bottom panels) for the different archipelago geometries. Results are shown (from left to right in each panel) for slope (z), intercept (c) and R^2 of the ISAR for a single simulation. After 10,000 steps, inter-island dispersal was intentionally reduced (see text for details) to verify if continual dispersal is a necessary condition for the maintenance of ISARs and to test to what extent a reduction or cessation of dispersal and hence inter-island colonization would modify – most likely weaken and deteriorate – the relationships observed at equilibrium.

15.9 Effect of Network Properties on the ISAR

In the previous sections we showed that species dispersal ability combines with archipelago geometry to produce connectivity networks that lead to ISARs that reach a stable equilibrium. Perhaps the simplest measure of connectivity among the islands is the mean degree of the islands. Depending on archipelago geometry, mean degree ranged from nine to fifty. Not surprisingly, mean degree was lowest for the elongated geometry (Figure 15.5). However, the c and z parameters of the ISAR did not appear to be tightly correlated with mean degree except possibly for the elongated geometry in the neutral model (Figure 15.5). Thus, perhaps mean degree is not the most thorough measure of the overall connectivity among nodes (islands) of a network.

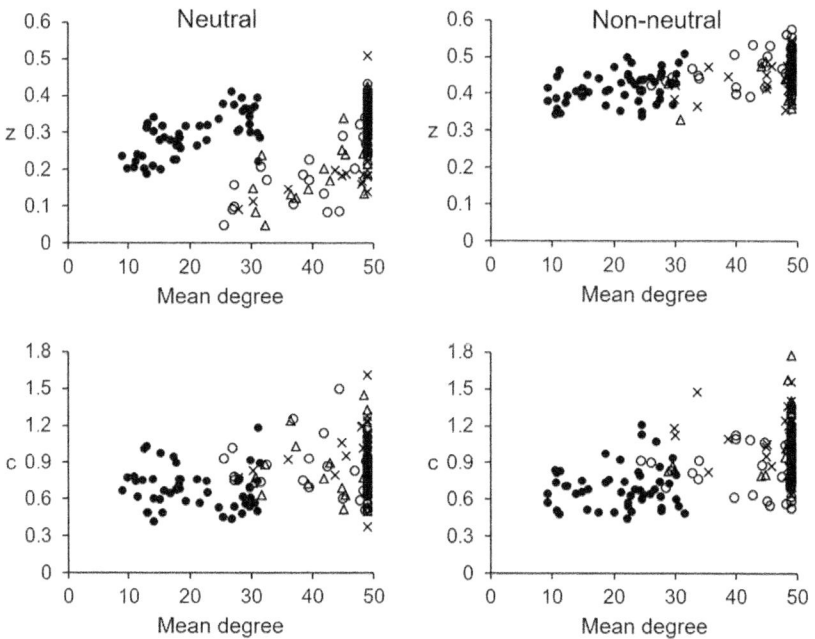

Figure 15.5 ISAR parameters (c- and z-values) plotted against mean island degree for each of the 200 simulations of the neutral and non-neutral versions of the model. Archipelago geometries are random (open circles), star (open triangles), ring (X) and elongated (filled circles).

We used Spearman's rank-correlation coefficients (r_s) to further examine whether characteristics of the ISAR correlate with other properties of the connectivity network. Of the 144 correlations analysed, only 15 (10 per cent) were non-significant ($P > 0.05$). In general, c, z and R^2 were correlated with each of the network properties (connectance, degree SD and diameter) in the expected way. That is, for connectance, r_s values were always positive, whereas for degree SD and diameter r_s values were typically negative. Such correlations represent greater network connectivity giving rise to stronger ISARs (i.e. steeper curves and better least squares fit to the data). For the *neutral* version of the model and ISARs prior to dispersal reduction, thirty-three of thirty-six (92 per cent) correlations had either $r_s > 0.6$ or < -0.6, indicating very strong and statistically significant correlation ($P < 1 \times 10^{-10}$) of the given ISAR parameter with the given network property (Table 15.3). When dispersal was reduced at simulation step 10,000, all of the correlations became weaker (i.e. all r_s values shifted toward zero). However, thirty-two of the thirty-six correlations maintained statistical significance ($r_s > 0.2$ or < -0.2, $P < 0.05$) and many were still highly significant ($r_s > 0.6$ or < -0.6, $P < 1 \times 10^{-10}$) (Table 15.3).

For the *non-neutral* version of the model, parameters of the ISAR were not as strongly correlated with network properties. Prior to dispersal reduction, only five of thirty-six (14 per cent) correlations had either $r_s > 0.6$ or < -0.6, although thirty-three of thirty-six correlations had either $r_s > 0.2$ or < -0.2 (Table 15.3). When dispersal was reduced at simulation step 10,000, thirty-one of thirty-six correlations became weaker, although all thirty-one of these were still significant ($P < 0.05$).

There were only slight differences among the three network properties in the magnitude of their correlation with the ISAR parameters (Tables 15.3 and 15.4). In particular, network connectance and diameter were more strongly correlated with c, z and R^2 than was degree SD (Table 15.4). There were also no huge differences in r_s values among the different archipelago geometries except for the r_s values for degree SD versus c, z and R^2 all being close to zero for the elongated island geometry (the values in italic in Table 15.4). The greatest differences in r_s values could be attributed to neutral versus non-neutral models and before versus after dispersal reduction (see also Table 15.4).

Table 15.3 *Spearman's rank correlation coefficients[1] (r_s) depicting relationships between properties of the connectivity network (con, sd deg and dia) and ISAR parameters (c, z and R^2) for the fully neutral model and the non-neutral model allowing for species-specific competitive and dispersal abilities. Results are shown for ISARs at equilibrium before and after reduction in inter-island dispersal, for each archipelago geometry (Geo). Coefficients correspond to P-values in the following way: $r_s > 0.2$ or < -0.2 ($P < 0.05$), $r_s > 0.4$ or < -0.4 ($P < 0.0001$) and $r_s > 0.6$ or < -0.6 ($P < 1 \times 10^{-10}$). Abbreviations: con, connectance; sd deg, standard deviation of node degree; dia, diameter. See text for descriptions of these metrics. Each coefficient is based on $N = 100$ simulations.*

Model version	Geo.	Disp. reduc.	z			c			R^2		
			con	sd deg	dia	con	sd deg	dia	con	sd deg	dia
Neutral	Rand	Before	0.630	−0.628	−0.629	0.632	−0.629	−0.631	0.628	−0.626	−0.627
—	—	After	0.488	−0.481	−0.487	0.589	−0.584	−0.586	0.259	−0.243	−0.255
—	Star	Before	0.647	−0.645	−0.648	0.653	−0.641	−0.654	0.648	−0.646	−0.648
—	—	After	0.643	−0.640	−0.644	0.373	−0.323	−0.378	0.621	−0.614	−0.621
—	Ring	Before	0.654	−0.643	−0.654	0.653	−0.642	−0.653	0.652	−0.641	−0.652
—	—	After	0.648	−0.635	−0.647	0.593	−0.573	−0.592	0.624	−0.607	−0.624
—	Long	Before	0.749	0.160	−0.754	0.942	−0.061	−0.859	0.708	0.104	−0.727
—	—	After	0.589	−0.076	−0.445	0.632	0.082	−0.602	0.301	0.019	−0.182
Non-neutral	Rand	Before	0.592	−0.542	−0.587	0.760	−0.728	−0.759	0.600	−0.558	−0.587
—	—	After	0.529	−0.487	−0.527	0.545	−0.490	−0.537	0.235	−0.184	−0.234
—	Star	Before	0.463	−0.448	−0.463	0.593	−0.583	−0.594	0.361	−0.355	−0.363
—	—	After	0.362	−0.336	−0.359	0.476	−0.463	−0.475	0.272	−0.250	−0.268
—	Ring	Before	0.380	−0.358	−0.379	0.579	−0.574	−0.579	0.510	−0.495	−0.510
—	—	After	0.474	−0.463	−0.474	0.469	−0.458	−0.467	0.355	−0.348	−0.352
—	Long	Before	0.553	0.108	−0.503	0.642	0.148	−0.615	0.525	0.156	−0.536
—	—	After	0.357	0.120	−0.336	0.419	0.209	−0.381	0.108	0.123	−0.149

[1] Spearman's rank correlation coefficient (a non-parametric descriptor) was used because the variables were generally non-normally distributed.

Table 15.4 *Mean values for Spearman's rank correlation coefficients (r_s) calculated for neutral (N = 8) versus non-neutral (N = 8) versions of the model, before and after dispersal reduction (N = 8 for each group) and for the four different island geometries (N = 4 for each group). Means are based on results given in Table 15.3. The italic values represent those close to zero (see text).*

Group	Connectance			Degree SD			Diameter		
	z	c	R^2	z	c	R^2	z	c	R^2
Neutral	0.63	0.63	0.56	−0.45	−0.42	−0.41	−0.61	−0.62	−0.54
Non-neutral	0.46	0.56	0.37	−0.30	−0.37	−0.24	−0.45	−0.55	−0.37
Before Disp. Reduc.	0.58	0.68	0.58	−0.37	−0.46	−0.38	−0.58	−0.67	−0.58
After Disp. Reduc.	0.51	0.51	0.35	−0.37	−0.33	−0.26	−0.49	−0.50	−0.34
Random	0.56	0.63	0.43	−0.53	−0.61	−0.40	−0.56	−0.63	−0.43
Star	0.53	0.52	0.48	−0.52	−0.50	−0.47	−0.53	−0.53	−0.48
Ring	0.54	0.57	0.54	−0.52	−0.56	−0.52	−0.54	−0.57	−0.53
Long	0.56	0.66	0.41	*0.08*	*0.09*	*0.10*	−0.51	−0.61	−0.40
Overall mean (N = 16)	0.55	0.60	0.46	−0.38	−0.39	−0.32	−0.53	−0.59	−0.46

15.10 Including Island Connectivity in the ISAR

The analyses presented in Section 15.9 examined the effect of network-level properties on the ISAR. However, it is also possible and worthwhile to examine whether node-level properties (such as degree and PageRank centrality) can be added to an ISAR as a means of further improving it as a tool for predicting the species richness of an island. That is, we addressed the question of whether area, degree and PageRank centrality together in a multiple regression equation perform better in predicting species richness than does area alone. Degree and PageRank centrality measure the extent to which a given node is connected to other nodes. In our application, an island with a high degree and PageRank is highly accessible by many species via dispersal from other islands. As such, we predicted that degree and PageRank should be meaningful contributors in explaining the overall variance in species richness among a set of islands. That is, adjusted R^2 values should increase as each of these predictors is added to a multiple regression equation that also includes island area. This prediction was confirmed. In particular, the neutral model applied to the random, star and ring geometries produced ISARs in which area tended to explain only about 15–27 per cent of the variance in island species richness on average ('A only' model in Table 15.5). However, by including island degree and PageRank centrality in the regression equation an additional

13–15 per cent of richness variance can be explained and for some datasets the additional explained variance was as great as 64 per cent (Table 15.5). Some datasets (e.g. ring geometry in the non–neutral model) had very little additional variance explained by adding degree and PageRank centrality to the ISAR. This was likely because those ISARs fit the richness values very well to begin with; that is, for most of the non–neutral model simulations the area-only regression models (i.e. the ISARs) had relatively high R^2 values (Table 15.5). Hence, there was not much 'room for improvement' by adding degree and PageRank centrality as predictors.

Comparison of the regression models also indicated that area is clearly the best single predictor of island species richness regardless of archipelago geometry and whether the species are completely neutral or different in competitive and dispersal abilities (non-neutral). Of the 652 regression models that included area, 598 models had a positive and significant coefficient ($P < 0.05$) for area; 458 of these coefficients were highly significant ($P < 0.0001$). Of the 489 models that included degree, 121 and 6 had a significant ($P < 0.05$) or a highly significant ($P < 0.0001$) coefficient for degree. Likewise, 117 and 5 out of 489 models had a significant ($P < 0.05$) or highly significant ($P < 0.0001$) coefficient for PageRank centrality. Further, these two measures of island connectivity had approximately equal effects on species richness, as shown by comparing the adjusted R^2 values of the 'A+D' and 'A+PR' models (Table 15.5). In general, the 'D-only' and 'PR-only' regression models (not shown in Table 15.5) were similar to one another and very weak with $R^2 < 0.02$. Thus, for our simulated datasets, island connectivity *alone* (measured as either degree or PageRank centrality, without area) is a very poor predictor of island species richness.

Although island connectivity alone is a poor predictor of species richness, measures of connectivity can improve the predictive performance (fit to the data) of an equation that also includes island area (i.e. the ISAR), although the improvement is very slight, such as only 1–4 per cent of additional richness variance explained by degree or PageRank (Table 15.5). The effect of degree and PageRank on species richness was more likely to be positive than negative. The regression coefficient for degree was statistically significant ($P < 0.05$) and positive for 105 of the 489 regression models and significant and negative for only 16. Similarly, the regression coefficient for PageRank was significant and positive for 92, and significant and negative for 25 out of 489 regression models. Herein lies an irony of island connectivity and species richness. An increase in connectivity at both the level of individual islands and the

Table 15.5 *Mean adjusted R^2 values from the analysis of island species richness regressed against area (A), degree (D) and PageRank centrality (PR) of the islands for simulations of the neutral and non-neutral model versions and the four archipelago geometries (Geo). The table also shows the number of datasets (N) that were analysed and the number of those in which area was the best single predictor of species richness as indicated by the R^2 value for the area-only model being greater than those for D-only and PR-only (the latter are not shown). The two right-most columns give the percentages of additional variance in species richness explained by adding degree and PageRank to the area-only model. Note that the area-only model is the ISAR in the linearized form of the power function: ln(S) = ln(c) + z × ln(A).*

Model version	Geo	N[1]	Area best predictor	Mean adjusted R^2 of regression models				Additional variance explained (%)	
				A only	A+D	A+PR	A+D+PR	Mean	Max
Neutral	Rand	14	11	0.18	0.24	0.23	0.34	15.3	63.9
–	Star	12	10	0.27	0.31	0.30	0.40	13.5	52.7
–	Ring	6	4	0.15	0.17	0.17	0.29	14.5	48.2
–	Long	50	49	0.53	0.56	0.56	0.57	4.3	26.9
Non-neutral	Rand	16	16	0.83	0.84	0.84	0.86	3.3	20.1
–	Star	6	6	0.86	0.87	0.87	0.89	2.5	10.2
–	Ring	9	9	0.83	0.84	0.84	0.84	0.8	2.9
–	Long	50	49	0.72	0.77	0.76	0.78	6.1	31.1

[1] The regression analyses were limited to those simulated ISAR datasets that had a connectivity network with connectance < 0.95 (163 out of 400 simulated datasets). The islands of the excluded datasets were so well-connected that there was very little variation among them in degree or PageRank centrality and hence little scope to test the effects of those predictors on island species richness.

entire archipelago network tends to increase island species richness (particularly for neutral models). This is because an island must be accessible from other islands in order to gain and maintain species richness in a 'cyclical' scenario of dispersal–colonization–extinction, particularly if colonization from a mainland is just a trickle. However, if islands are too well-connected, then dispersal from any one island to any other is relatively easy and each island will tend to have an equilibrium number of species that is determined simply by island area (available space) and non–neutral processes such as interspecific competition occurring on the island. Moreover, this effect transcends archipelago geometry. The importance of dispersal and island connectivity was also revealed when we intentionally reduced dispersal once a simulation had achieved a stable (equilibrium) ISAR. The ISARs then tended to weaken.

15.11 The Dynamic Nature of the ISAR

In this chapter we have presented a spatial model simulating the process of mainland-to-island and inter-island dispersal and colonization to examine how the structure of the underlying connectivity network contributes to the emergence and maintenance of ISARs. The connectivity network itself is determined by both the dispersal ability of species and the distances separating islands. Historically, the ISAR has been analyzed as a static pattern, particularly when fitting a species–area function (curve) to data, although ecologists also seem to recognize that it is based on dynamic processes. The well-known and widely-studied theory of island biogeography (MacArthur & Wilson, 1963, 1967; Losos & Ricklefs, 2010) explains species richness of an island as a function not only of area but also of the isolation (distance to the mainland or other source of potential colonists) of the island. Isolation affects colonization rate and island area affects extinction rate; thus, area and isolation are simply proxies for the two dynamic processes of colonization and extinction. Notably, in the classic formulation of the equilibrium theory of island biogeography (sometimes referred to as just 'island biogeography theory') greater emphasis is placed on the equilibrium number of species on an island rather than on equilibrium of the entire ISAR and its parameters through time, although the latter is discussed to some extent (MacArthur & Wilson, 1967). Nonetheless, if every island in a given archipelago has reached an equilibrium number of species then any ISAR fit to the data would also represent an equilibrium and one that presumably persists through time as colonization and extinction continue to balance one another.

The original exposition of the neutral theory of biodiversity also recognized that dynamic processes were behind the SAR, namely colonization, extinction and speciation (Hubbell, 2001). Indeed, Hubbell (2001) suggested that the often-observed change in the form of the SAR across spatial scales from small plots to regions was due to these processes having varying influence at different spatial scales. Hubbell (2001) also referred to SARs as being in a 'steady-state', which was due to immigration rate and his fundamental biodiversity number (θ) (see Chapter 11). These two parameters also determine the equilibrium of the species abundance distribution. Although neutral theory and island biogeography theory invoke dynamic processes to explain the form of the SAR, neither one makes detailed predictions about how (or if) SARs should change through time. MacArthur and Wilson (1967) did suggest that speciation and adaptive radiation within an archipelago could lead to an overall increase in species and hence an increase in the intercept of an ISAR. However, they did not explicitly model this process. Moreover, island biogeography theory and neutral theory are generally tested or explained simply by fitting a mathematical function (species–area curve) to static data.

The dynamic nature of the SAR is also recognized in the species–time relationship (White, 2004; Adler et al., 2005; Chapter 19) and the general dynamic model of island biogeography (Whittaker et al., 2008; Fattorini, 2009). Formally, these perspectives incorporate area, time, and sometimes time^2 and a time × area interaction as explanatory variables in equations that predict species richness. Time is measured by island age. As such, time is representing the occurrence of past processes such as dispersal/colonization, extinction and speciation in producing the current SAR for a set of islands. However, again, the SAR is treated as a static outcome of dynamic processes.

Our simulation model arrived at an equilibrium ISAR after a sufficient number of time steps (species dispersal and colonization events) as evidenced by temporal stability in c, z and R^2. However, although stable, the ISAR was not static in that ongoing dispersal and colonization were needed to maintain it. This was partly due to our model also including the process of extinction. Without a background system-wide rate of extinction, all species would eventually accumulate on every sufficiently-large island that is accessible through dispersal from the mainland and other islands. Although smaller islands would lack some species (particularly those that are dispersal-restricted) due to space constraints, even relatively small islands would have many species in the absence of extinction – and the ISAR would be relatively flat.

An ISAR can be at equilibrium, defined as relatively little change in c, z and R^2 over time and yet still be very dynamic. The dynamism exists due to species turnover (cf. MacArthur & Wilson, 1967), that is, the replacement of one species in a community (or on an island) by another. Our simulation model incorporated species turnover in two ways: 1) through extinction – species disappeared from grid cells and possibly entire islands, leaving behind unoccupied grid cells available for colonization by a different species; and 2) in the non-neutral version of the model species could competitively displace one another from a grid cell and presumably on some occasions from an entire island. Although we did not keep track of it, some amount of turnover was likely continually happening at some rate on most islands during most simulations.

Importantly, in real-world contexts, turnover might be relatively high and yet the species richness of islands remains relatively constant over time such that the ISAR is at an equilibrium.

In our model, each step in a simulation is characterized by a number of dispersal/colonization events which depends on the total number of populations in the system. Therefore, colonization is infrequent at the beginning and increases through time as more populations establish themselves in the system. The increase, however, declines as the system approaches the 'carrying capacity' (i.e. when most cells of every island are occupied). Even in the hypothetical situation where all cells have been occupied, however, new successful colonization events are still possible, because the dispersing propagules can sometimes outcompete established populations (in both the neutral and non-neutral versions of the model). Thus, even in a relatively saturated system, dynamic processes can continue to occur such as species displacement of one another. When the system is far from saturation then these processes are likely even more dynamic and influential in affecting the form of the ISAR. Perhaps, given a long enough time period (e.g. decades) most ISARs are not at a *stable* equilibrium. Schoener (2010) thoroughly discussed how an ISAR equilibrium can take on several different forms (including cyclical change and slow unidirectional change) other than a steady state. Further, ecosystem disturbance of some magnitude and rate may be paramount in determining if or how often a species–area system is at equilibrium – the greater the rate, the less likely any kind of equilibrium is attained (Schoener, 2010). Further, species that are good dispersers will reach an equilibrium sooner and remain at it more consistently over time (Schoener, 2010).

Our ability to understand the dynamic nature and equilibrium of the ISAR was facilitated by constructing and using the island connectivity network. At the very heart of network analysis and network thinking is the idea that networks represent the flow or sharing of something such as information in the broadest sense and species in the context of our simulation model. Species are constantly flowing among islands. Moreover, networks allow one to identify the nodes that may be more conducive to the flow of information and/or have more of the information (Proulx et al., 2005). Node-level measurements of centrality and connectivity are used for this purpose, and these metrics are influenced by a node's position in the overall network. This approach is not completely novel in ecology. Various studies, especially in the context of conservation biology, have used spatially explicit networks to investigate the importance of habitat patches to population connectivity in fragmented landscapes (e.g. Bodin & Norberg, 2007; Baranyi et al., 2011). Yet, to the best of our knowledge, only a few studies have used spatially explicit networks mapping the connections between different isolated localities/islands to explore biogeographical patterns (Economo & Keitt, 2008, 2010) and none in the context of ISARs.

Our model clearly reveals that the pathways of dispersal of species towards a given island will be affected by the position of that island in the connectivity network. This concept is analogous to several ideas which have received much interest in the context of network theory. For example, various studies focusing on modelling the spread of epidemics (or information) across networks whose nodes represent potential hosts and whose links represent potential transmission routes for the infection have tried to identify which properties make a node more susceptible to be infected or more active in promoting the epidemic spread (Kitsak et al., 2010; Radicchi & Castellano, 2017). Intuitively, some of the structural features (e.g. number of links, position in the network) that might affect the probability of a node to be reached by an infection might also affect the probability of an island to receive dispersing species and therefore play an important role in determining the species richness of the island. Indeed, the multiple regression models applied to our simulated ISARs revealed that the connectivity of an island to other islands (as measured by mean degree and PageRank centrality) often affected the species richness of the island, although island area was always the best single predictor of richness.

15.12 Conclusion

To investigate the role of dispersal in producing and maintaining the ISAR, we developed a model that simulated the dispersal–colonization–extinction process. Further, the model was combined with an island connectivity network as a means of directly exploring the potential for dispersal to affect island species richness in the context of an island's position within the network. In the connectivity network, links between nodes (islands) were based on inter-island distances as well as the potential dispersal distances of the species being simulated. This overall investigative approach revealed that relatively strong ISARs emerge rather quickly regardless of the geometry of an archipelago and remain at an equilibrium as long as dispersal and colonization continue. An increase in network- and node-level connectivity tended to produce a stronger ISAR as well as greater species richness for a given island, except for highly connected networks, in which case island area alone determined the form of the ISAR. Our approach has much more heuristic potential that could be explored. For example, the step when we intentionally reduced dispersal could be treated much more intricately. If dispersal were restored to initial levels would the same equilibrial ISAR return? Is the effect of the dispersal reduction on the ISAR manifested as a continuous effect or is there a threshold? In our simulations, the ISAR eroded (weakened) fairly quickly when dispersal was reduced; however, if dispersal were reduced much more gradually would the parameters of the ISAR also decline gradually or would a threshold level of reduction be needed to induce any decrease of the parameters and would the decrease then be rapid?

We anticipate that these and other questions could be addressed with further simulations to learn more about the role of dispersal/colonization in maintaining the ISAR and other patterns that are based on species diversity. For example, species–area data are amenable to diversity partitioning, a way of analysing and comparing the mean species richness within islands (alpha diversity) to the richness found among islands (beta diversity) (Crist & Veech, 2006). We encourage other researchers to use the basic framework of our simulation model in conjunction with connectivity networks to further explore how processes such as dispersal, colonization, extinction and competition affect the distribution of species among true islands as well as habitat fragments embedded in a non-habitat matrix. In addition, major modifications of the simulation model and connectivity network could be undertaken to enhance the realism of each. In the connectivity networks that we constructed, links between islands were put in if the inter-island distance was less than the mean dispersal distance of all the species; that is, the links were not weighted. Alternatively, links could be weighted and based on a probability that decreases as inter-island distance increases. The probability could be derived from a function that corresponds with the dispersal kernel or even some property or trait of the species. Finally, our modelling approach could have practical application for understanding the structure of real metacommunities wherein species are dispersed among islands or habitat fragments of varying sizes and spatial arrangement. A key challenge in such an application would be obtaining information on dispersal distances (capacities) so as to construct a realistic connectivity network. But once that task is accomplished then the model and connectivity network could be used in various ways to better understand how the spatial distribution of species depends on the ongoing processes of dispersal and colonization.

References

Ackerly, D. D. (2003) Community assembly, niche conservatism, and adaptive evolution in changing environments. *International Journal of Plant Sciences*, **164**, S165–S184.

Adler, P. B., White, E. P., Laurenroth, W. K., Kaufman, D. M., Rassweiler, A. & Rusak, J. A. (2005) Evidence for a general species–time–area relationship. *Ecology*, **86**, 2032–2039.

Allesina, S. & Pascual, M. (2006) Googling food webs: Can an eigenvector measure species' importance for coextinctions? *PLoS Computational Biology*, **9**, e1000494.

Altieri, A. H., van Wesenbeeck, B. K., Bertness, M. D. & Silliman, B. R. (2010) Facilitation cascade drives positive relationship between native biodiversity and invasion success. *Ecology*, **91**, 1269–1275.

Arrhenius, O. (1921) Species and area. *Journal of Ecology*, **9**, 95–99.

Baranyi, G., Saura, S., Podani, J. & Jordán, F. (2011) Contribution of habitat patches to network connectivity: Redundancy and uniqueness of topological indices. *Ecological Indicators*, **11**, 1301–1310.

Bodin, Ö. & Norberg, J. (2007) A network approach for analyzing spatially structured populations in fragmented landscape. *Landscape Ecology*, **22**, 31–44.

Bodin, Ö. & Saura, S. (2010) Ranking individual habitat patches as connectivity providers: Integrating network analysis and patch removal experiments. *Ecological Modelling*, **221**, 2393–2405.

Brown, J. H. & Kodric-Brown, A. (1977) Turnover rates in insular biogeography: Effect of immigration on extinction. *Ecology*, **58**, 445–449.

Cadotte, M. W. (2006) Dispersal and species diversity: A meta-analysis. *The American Naturalist*, **167**, 913–924.

Case, T. J. (1983) Niche overlap and the assembly of island lizard communities. *Oikos*, **41**, 427–433.

Connor, E. F. & McCoy, E. D. (1979) The statistics and biology of the species–area relationship. *The American Naturalist*, **113**, 791–833.

Cornell, H. V. & Lawton, J. H. (1992) Species interactions, local and regional processes, and limits to the richness of ecological communities: A theoretical perspective. *Journal of Animal Ecology*, **61**, 1–12.

Crist, T. O. & Veech, J. A. (2006) Additive partitioning of rarefaction curves and species–area relationships: Unifying α-, β-, and γ-diversity with sample size and habitat area. *Ecology Letters*, **9**, 923–932.

de Arruda, G. F., Barbieri, A. L., Rodríguez, P. M., Rodrigues, F. A., Moreno, Y. & Costa, L. F. (2014) Role of centrality for the identification of influential spreaders in complex networks. *Physical Review E*, **90**, 032812.

Drakare, S., Lennon, J. J. & Hillebrand, H. (2006) The imprint of the geographical, evolutionary, and ecological context on species–area relationships. *Ecology Letters*, **9**, 215–227.

Economo, E. P. & Keitt, T. H. (2008) Species diversity in neutral metacommunities: A network approach. *Ecology Letters*, **11**, 52–62.

Economo, E. P. & Keitt, T. H. (2010) Network isolation and local diversity in neutral metacommunities. *Oikos*, **119**, 1355–1363.

Fattorini, S. (2009) On the general dynamic model of oceanic island biogeography. *Journal of Biogeography*, **36**, 1100–1110.

Fattorini, S., Borges, P. A. V., Dapporto, L. & Strona, G. (2017) What can the parameters of the species–area relationship (SAR) tell us? Insights from Mediterranean islands. *Journal of Biogeography*, **44**, 1018–1028.

Gómez-Rodríguez, C., Miller, K. E., Castillejo, J., Iglesias-Piñeiro, J. & Baselga, A. (2019) Understanding dispersal limitation through the assessment of diversity patterns across phylogenetic scales below the species level. *Global Ecology & Biogeography*, **28**, 353–364.

He, F., Gaston, K. J., Connor, E. F. & Srivastava, D. S. (2005) The local–regional relationship: Immigration, extinction, and scale. *Ecology*, **86**, 360–365.

Hormiga, G., Arnedo, M. & Gillespie, R. G. (2003) Speciation on a conveyor belt: Sequential colonization of the Hawaiian Islands by *Orsonwelles* spiders (Araneae, Linyphiidae). *Systematic Biology*, **52**, 70–88.

Hubbell, S. P. (2001) *The unified neutral theory of biodiversity and biogeography*. Princeton, NJ: Princeton University Press.

Janzen, D. H. (1985) On ecological fitting. *Oikos*, **45**, 308–310.

Kitsak, M., Gallos, L. K., Havlin, S., Liljeros, F., Muchnik, L., Stanley, H. E. & Makse, H. A. (2010) Identification of influential spreaders in complex networks. *Nature Physics*, **6**, 888–893.

Kneitel, J. M. & Miller, T. E. (2003) Dispersal rates affect species composition in metacommunities of *Sarraenia purpurea* inquilines. *The American Naturalist*, **162**, 165–171.

Kokko, H. & López-Sepulcre, A. (2006) From individual dispersal to species ranges: Perspectives for a changing world. *Science*, **313**, 789–791.

Leibold, M. A., Holyoak, M., Mouquet, N., Amarasekare, P., Chase, J. M., Hoopes, M. F., Holt, R. D., Shurin, J. B., Law, R., Tilman, D., Loreau, M. & Gonzalez, A. (2004) The metacommunity concept: A framework for multiscale community ecology. *Ecology Letters*, **7**, 601–613.

Lomolino, M. V. (2000) Ecology's most general, yet protean pattern: The species–area relationship. *Journal of Biogeography*, **27**, 17–26.

Lomolino, M. V., Riddle, B. R. & Brown, J. H. (2006) *Biogeography*, 3rd ed. Sunderland, MA: Sinauer Associates.

Losos, J. B. & Ricklefs, R. E. (2010) *The theory of island biogeography revisited*. Princeton, NJ: Princeton University Press.

MacArthur, R. H. & Wilson, E. O. (1963) An equilibrium theory of insular zoogeography. *Evolution*, **17**, 373–387.

MacArthur, R. H. & Wilson, E. O. (1967) *The theory of island biogeography*. Princeton, NJ: Princeton University Press.

MacDougall, A. S., Gilbert, B. & Levine, J. M. (2009) Plant invasions and the niche. *Journal of Ecology*, **97**, 609–615.

Matthews, T. J., Rigal, F., Triantis, K. A. & Whittaker, R. J. (2019) A global model of island species–area relationships. *Proceedings of the National Academy of Sciences USA*, **116**, 12337–12342.

McGill, B. J. (2010) Towards a unification of unified theories of biodiversity. *Ecology Letters*, **13**, 627–642.

Medeiros, M. J. & Gillespie, R. G. (2011) Biogeography and the evolution of flightlessness in a radiation of Hawaiian moths (Xyloryctidae: *Thyrocopa*). *Journal of Biogeography*, **38**, 101–111.

Mooney, H. A. & Cleland, E. E. (2001) The evolutionary impact of invasive species. *Proceedings of the National Academy of Sciences USA*, **98**, 5446–5451.

Mouquet, N. & Loreau, M. (2003) Community patterns in source-sink metacommunities. *The American Naturalist*, **162**, 544–557.

Newman, M. E., Barabási, A. L. E. & Watts, D. J. (2006) *The structure and dynamics of networks.* Princeton, NJ: Princeton University Press.

O'Dwyer, J. P. & Green, J. L. (2010) Field theory for biogeography: A spatially explicit model for predicting patterns of biodiversity. *Ecology Letters*, **13**, 87–95.

Patiño, J., Weigelt, P., Guilhaumon, F., Kreft, H., Triantis, K. A., Naranjo-Cigala, A., Sólymos, P. & Vanderpoorten, A. (2014) Differences in species–area relationships among the major lineages of land plants: A macroecological perspective. *Global Ecology & Biogeography*, **23**, 1275–1283.

Proulx, S. R., Promislow, D. E. L. & Phillips, P. C. (2005) Network thinking in ecology and evolution. *Trends in Ecology & Evolution*, **20**, 345–353.

Radicchi, F. & Castellano, C. (2017) Fundamental difference between superblockers and superspreaders in networks. *Physical Review E*, **95**, 012318.

Ricciardi, A., Hoopes, M. F., Marchetti, M. P. & Lockwood, J. A. (2013) Progress toward understanding the ecological impacts of nonnative species. *Ecology*, **83**, 263–282.

Ricklefs, R. E. (1987) Community diversity: Relative roles of local and regional processes. *Science*, **235**, 167–171.

Rosenzweig, M. L. (1995) *Species diversity in space and time.* Cambridge: Cambridge University Press.

Santamaria, C. A., Mateos, M., Taiti, S., DeWitt, T. J. & Hurtado, L. A. (2013) A complex evolutionary history in a remote archipelago: Phylogeography and morphometrics of the Hawaiian endemic *Ligia* isopods. *PLoS One*, **12**, e85199.

Schoener, T. W. (2010) The MacArthur-Wilson equilibrium model: A chronicle of what it said and how it was tested. *The theory of island biogeography revisited* (ed. by J. B. Losos and R. E. Ricklefs), pp. 52–87. Princeton, NJ: Princeton University Press.

Seymour, M., Fronhofer, E. A. & Altermatt, F. (2015) Dendritic network structure and dispersal affect temporal dynamics of diversity and species persistence. *Oikos*, **124**, 908–916.

Shaw, K. L. (1996) Sequential radiations and patterns of speciation in the Hawaiian cricket genus *Laupala* inferred from DNA sequences. *Evolution*, **50**, 237–255.

Simberloff, D., Martin, J. L., Genovesi, P., Maris, V., Wardle, D. A., Aronson, J., Courchamp, F., Galil, B., García-Berthou, E., Pascal, M., Pyšek, P., Sousa, R., Tabacchi, E. & Vilà, M. (2013) Impacts of biological invasions: What's what and the way forward. *Trends in Ecology & Evolution*, **28**, 58–66.

Skóra, F., Abilhoa, V., Padial, A. A. & Vitule, J. R. S. (2015) Darwin's hypotheses to explain colonization trends: Evidence from a quasi-natural experiment and a new conceptual model. *Diversity and Distributions*, **21**, 583–594.

Sólymos, P. & Lele, S. R. (2012) Global pattern and local variation in species–area relationships. *Global Ecology & Biogeography*, **21**, 109–120.

Thomas, C. D., Bodsworth, E. J., Wilson, R. J., Simmons, A. D., Davies, Z. G., Musche, M. & Conradt, L. (2001) Ecological and evolutionary processes at expanding range margins. *Nature*, **411**, 577–581.

Triantis, K. A., Guilhaumon, F. & Whittaker, R. J. (2012) The island species–area relationship: Biology and statistics. *Journal of Biogeography*, **39**, 215–231.

Wagner, W. L. & Funk, V. A. (1995) *Hawaiian biogeography: Evolution on a hot spot archipelago.* Washington, DC: Smithsonian Institution Press.

White, E. P. (2004) Two-phase species–time relationships in North American land birds. *Ecology Letters*, **7**, 329–336.

Whittaker, R. J. & Fernández-Palacios, J. M. (2007) *Island biogeography: Ecology, evolution, and conservation*, 2nd ed. Oxford: Oxford University Press.

Whittaker, R. J., Triantis, K. A. & Ladle, R. J. (2008) A general dynamic theory of oceanic island biogeography. *Journal of Biogeography*, **35**, 977–994.

16 · Does Geometry Dominate Extinction due to Habitat Loss?

ATHANASIOS S. KALLIMANIS AND
JOHN M. HALLEY

16.1 Introduction

The species–area relationship (SAR) has received its fair share of research focus in ecology over the past two centuries. The mathematical formulation of the SAR and its implications have been extensively investigated. The characteristics of sampling such as focus, extent and grain size (Scheiner et al., 2011) affect the shape of the SAR and may influence the conclusions about the spatial variation of species richness and the comparability of studies done at different scales (Rahbek, 2005; Dengler & Oldeland, 2010). One aspect that remains less well explained, yet continues to generate discussion, is how the geometry of sampling (i.e. the spatial pattern of the sample plots) affects the shape of the SAR and the inferences drawn (He & Hubbell, 2011; Šizling et al., 2011; Pereira et al., 2012; Chapter 8).

SARs are usually constructed from sparse ecological samples due to the difficulty and cost of obtaining complete species lists for an area or performing exhaustive biodiversity surveys. This may be a source of bias when constructing SARs, because, as is well established in the literature, the spatial distribution of biodiversity displays autocorrelation (Legendre, 1993; Diniz-Filho et al., 2003; Rahbek, 2005). This property means that sampling plots that are located close by are occupied by communities with more similar species composition, a phenomenon described as distance decay in community similarity (Soininen et al., 2007; Steinbauer et al., 2012). Therefore, for a given sampling area, we accumulate more species when the samples are dispersed in space than if they are contiguous (Lazarina et al., 2014). Thus, when applying SARs in ecological biogeography and conservation biology (e.g. projecting biodiversity responses to habitat reduction; Chapter 14), ignoring the geometry of sampling may lead to erroneous conclusions (e.g. He & Hubbell, 2011; Šizling et al., 2011; Pereira et al., 2012).

Habitat loss is arguably the single greatest threat to biodiversity (Brooks et al., 2006; Pimm et al., 2014). SARs have been widely used to predict biodiversity loss resulting from habitat destruction (Fattorini & Borges, 2012; Halley et al., 2014; Keil et al., 2015; Chapter 14). Habitat loss is expected to cause extinctions either directly and immediately or indirectly in the long term via the process of 'relaxation' (Diamond, 1972) and extinction debt. These extinction processes may be referred to as *imminent* or *delayed*, respectively (He & Hubbell, 2011, Halley et al., 2014). There is now a considerable literature on how to estimate delayed extinctions (Kuussaari et al., 2009; Halley & Iwasa, 2011). Imminent extinctions are considered easier to predict since they are associated directly with area loss, although the associated extinction–area relationship may have a complex nonlinear shape (Keil et al., 2018). Nevertheless, for predicting imminent extinctions the endemics–area relationship (EAR) is usually applied (Kinzig & Harte, 2000; Rybicki & Hanski, 2013). The EAR reflects the number of species whose entire range lies within a given area and, thus, if we lose this area, we are bound to lose the corresponding endemic species. This application of the EAR is of immense importance for conservation biology and thus the EAR has received a lot of research interest for estimating extinction risk, for identifying biodiversity hotspots and for a deeper understanding of the species–area relationship and the factors affecting it (Green & Ostling, 2003; Ulrich & Buszko, 2005; Rahbek & Colwell, 2011; Chapters 13 and 14).

Although the geometry of area loss has not received a lot of attention, Keil et al. (2015), following earlier studies (Pereira et al., 2012; Pan, 2013; Matias et al., 2014), highlighted the importance of the spatial pattern of habitat loss on the shape of the EAR and thus on extinction (see also Keil et al., 2018). They argued that the inward loss of habitats (i.e. losing contiguous habitat from the edge of a region towards its core) leads to a much more pronounced decline of species richness than when habitat is lost outwards (i.e. from the centre of the study region towards the edge) (see also Figure 14.3). In their study, extinction values given inward losses were up to one *order of magnitude higher* than for outward loss. The importance of such a finding is apparent not only for ecological theory, but also for conservation biology and the efforts to preserve biodiversity globally, where species–area relationships are widely used (Halley et al., 2013).

In this study, we are focused primarily on how the geometry of sampling affects the shape of nested SARs/SACs and the EAR, with an emphasis on the implications for species extinctions after habitat loss. To understand better the importance of geometry, we consider a series of calculations on how these relationships are affected by the spatial pattern of habitat loss. First, does it matter if area is lost in a random pattern or in large clusters? We examine how species accumulate with area, for continuous and non-continuous area loss. Another question is orientation – if habitat is lost inwards (from the edge towards the centre) does it always lead to greater rates of species loss than if habitat loss begins in the centre. This in turn raises the question of the definition of endemism. In many studies of the EAR, endemics are defined in regard to the study area (a species is considered 'endemic' to part of the study area if it does not occur anywhere else in the study area) even though these species may persist somewhere outside the study region. Hence, we examined the 'global' EAR, which identifies the true endemics of the study region, that is, species whose entire range lies within the study region. We find that recognizing this distinction resolves some of the apparent conflicts of this method.

16.2 Methods

16.2.1 Datasets

In large-scale analyses of biodiversity patterns two types of data are typically used: species-range data and atlases. The two types of data present different patterns of biodiversity (Tsianou et al., 2016), so we analysed an example of both datasets. Atlas data consist of grids of presence–absence data and we assume that the species is present only in the cells where it has been observed. For atlas data, we analysed the presences of the 245 species of breeding birds in New York State (Andrle & Carroll, 1988). The atlas data cover New York State in a grid format with a grain size of 5×5 km^2 (www.dec.ny.gov). All analyses of this dataset will refer to cells rather than km^2 as units of area. Species-range data consist of polygons defining each range and we assume that the species is present in every location within the range. For species-range data, we analysed the species ranges of amphibians, reptiles and terrestrial mammals in three continents (Africa, South America and North America) that were drawn by experts on behalf of IUCN (2019) for the estimation of species extinction risk. While a number of problems arise with species-range data as compared to atlas data, such as artificially low levels of fragmentation (Tsianou et al., 2016), these data have the added advantage of containing global endemics and are available for a wider range of taxonomic groups.

16.2.2 Species Accumulation Curves and the Endemics–Area Relationship

The species accumulation curve (SAC) is the number of species accumulated as sampled area a increases, $S(a)$. In order to estimate the impact of the spatial arrangement of habitat loss we used the data of the atlas of the breeding birds of New York State. To account for the role of sampling geometry we assumed that sampling was performed in different sample designs. The sampling designs differed in the grain of area analysed (i.e. a single atlas cell at a time or blocks of cells of different sizes up to 9×9) and in the spatial arrangement of habitat loss (i.e. from totally random, to more in the centre of the study area or more in the periphery of the study area, to a contiguous area). In each case, we accumulated samples across the extent of the data until all the extent was covered. At each step, we calculated how many species were present within the accumulated sample area to obtain the species accumulation curve (SAC; see Chapter 1). We obtained twenty replicates of each curve by simulation, finding the average and the extremes.

Endemic species are defined as the species that are present in a specific location and nowhere else. While this definition makes intuitive sense, its practical application is not always straightforward since it is a scale dependent concept. A species may occupy a restricted range and not be considered endemic if this range covers the boundary between different regions. Or a species may be widespread but its range occurs entirely within a single nation and thus the species is considered a national endemic. A similar issue arises in the study of the endemics–area relationship, because endemics in this case are defined in relation to the study area extent and not any further. A species is considered as a 'local endemic' if its distribution within the study area is restricted to the area lost, even if it is widespread outside the study area. To counterbalance this distortion we also use the 'global endemic' concept, when the entire species range is within the area lost.

Extinction curves, the expected biodiversity decline following loss of habitat, can be produced from the endemics–area relationship. The EAR tells us the number of endemic species found within an area. If this area of habitat is destroyed, unless a species whose range lies entirely within this area can migrate upon area loss, this species will be globally extinct. Such extinction curves can be generated by 'reversing' the SAC. This is done by noting that, given an area subset a of a total area A, the sum of the species accumulated in a, which is $S(a)$, plus $E(A-a)$, those *only* found in the remaining area, $A-a$, will be the total number of species in A. Thus:

$$S(a) = S(A) - E(A - a). \tag{16.1}$$

16.2.3 Comparing Different Geometries of Habitat Loss

We produced SACs and EARs for the different sampling schemes using the bird species of New York. Samples could be single cells (i.e. one cell of 5×5 km^2, at the finest scale) and grow to larger samples (4×4 cells or 9×9 cells). This compares the extinction process when habitat is removed either randomly or in blocks (see Keil et al., 2018, for a similar use of this method). We also compared what happens when habitat is removed starting either at the centre or at the edge. This was done by choosing a point close to the centroid of the area of New York (NY) State, but displaced slightly from it. We calculated the distances of all cells from the centre and began removing cells one-by-one starting either with those furthest from the centre (inwards) or with those closest to the centre and moving outwards. We then constructed the SAC and then the EAR from this. This compares the inward versus the outward EAR.

The NY dataset is limited in a number of ways. Primarily because, as no breeding bird is endemic to the state of NY, we cannot explore the importance of global versus local endemicity. Thus, we also used the second dataset of species ranges for amphibians, reptiles and mammals. For the species-range data, we explored two scenarios of habitat area loss: inwards (from the edge to the centre) and outwards (from the centre to the edge). We defined a series of concentric circles in different continents. For each continent, the centre of the circles was chosen so as to maximize the area of the terrestrial disk around it. The circle radius started from 40 km (inner circle) and reached up to 1,600 km, in steps of 40 km. The maximum radius in each case was set by the requirement that the total disk only covered terrestrial areas. Thus, the total study region covered different areas in different continents. To standardize for this difference in the graphs, the area is always presented as a proportion of the study region. We overlaid those circles on the IUCN expertly drawn species ranges and for each ring we estimated the area of each species range that was included. To estimate the extinction curves or the SACs we started either from the central circle (for the outwards scenario) or from the outer ring (for the inwards scenario) and started accumulating the area each species occupied.

We built the extinction curve (or endemics–area relationship) in two ways reflecting local extinctions and global extinctions. An extinction was classified as *global* if the species' entire worldwide range was included in the area lost. Conversely, this extinction was classified as *local* if the species had a proportion of its range outside the study area, so it would not go extinct even if the entire study area was lost.

16.3 Results

16.3.1 The Effects of the Scale of Units of Loss

Figure 16.1A, using the NY dataset for birds, compares the SAC obtained if an area is sampled in large blocks (9 × 9 cells) to that obtained if it is sampled randomly (one cell at a time). We see that fewer species are counted if the same sampling areas are larger blocks. More species are observed if the sampling area is scattered at random over the entire area. This can be understood through spatial autocorrelation. In the case of block sampling of area, species accumulate more slowly, since neighbouring cells have more similar species composition, so we need to cover a large proportion of the extent to observe all species, compared to the case of the non-continuous sampling where the SAC is steeper and reaches the point of saturation faster. Cells lying close to each other will contain similar communities, whereas if the cells are widely dispersed, fewer common species will occur. If one's purpose is to sample as many species as possible then it is best to disperse the sampling over as wide an area as possible.

The corresponding EAR or extinction curve (Figure 16.1B) shows how loss occurring in blocks of the same or different sizes will affect the extinction process. We note that the two curves begin and end at the same points. We also note that the extinction curve for the loss of area in blocks is above that for random removal. This means that losing area in large blocks causes more extirpations than if the same area is lost by removing individual cells at random, as has been observed in other scenarios (see Kallimanis et al., 2005). Another interesting feature of Figure 16.1B is that the natural scale to use is seemingly species number versus log(Area).

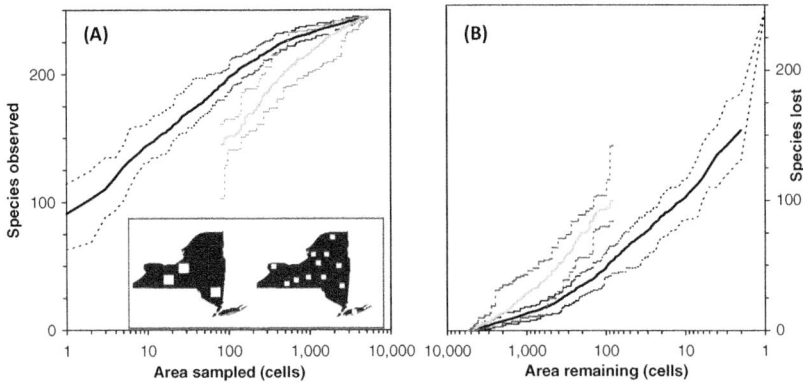

Figure 16.1 (A) Species accumulation curves (SAC) for two types of sampling: either a random distribution of single cells (black) or a random distribution of blocks of 9 × 9 cells (grey). Heavy continuous lines are the average values of species richness from twenty replicates. The broken lines are the corresponding minimum and maximum values. The average and associated extremes for the 9 × 9 blocks are not defined for areas less than eighty-one cells. (Inset) Patterns of sampling for NY state using different block sizes with the same total area. Such patterns can be interpreted either as sampling areas or as areas left behind after major habitat loss. (B) The corresponding endemics–area curves constructed by reversing the SACs according to Equation (16.1). The x-axis in both (A) and (B) is log(Area).

16.3.2 Expected Role of Extinction Debt

The preceding analysis suggests that, in relation to immediate extinctions, losing large blocks of habitat is more problematic than losing the same area through more incremental losses. However, the aggregation (or lack thereof) of individual losses plays a contrary role if we consider delayed extinctions (Halley et al., 2014). If a habitat area quickly shrinks or is isolated, the remaining fragment may contain more species than can be supported in the long term. Before equilibrium is re-established, a proportion of the species must be lost: this represents the extinction debt (Kuussaari et al., 2009; Halley et al., 2016). In situations where habitat loss continues block-by-block, there is a point where most or all habitat remaining is in the form of isolated fragments with a size typically in the same order as the blocks themselves. Each isolated fragment suffers delayed extinctions with the characteristic time (half-life) of biodiversity loss for which the clearest predictor is the area of the fragment (Halley et al., 2016).

We used different sizes of sampler on the NY birds dataset to construct a continental SAR. For each sampler size we took twenty different randomly placed samples and found the average number of species observed. The average values of species number for a single block with the 1 × 1, 4 × 4 and 9 × 9 samplers were 91.3, 125.0 and 145.0, respectively. We then fitted a continental Arrhenius SAR to these values for which $z = 1.10$. We are interested in habitat loss that occurs in blocks of characteristic size and with large levels of habitat loss (over 90 per cent) so that only isolated blocks remain behind, with most fragment sizes the same as the block size. After loss of all area apart from a single block sized 9 × 9 = 81 cells, the block contains on average 145 species. If the same area is lost but what remains is the 81 cells arranged in series of fragments of size 4 × 4, on average 169.9 species remain, while 191.2 remain if the blocks remaining are single cells (1 × 1). Using the results from Halley et al. (2016), we find the half-life for fragments consisting of one block at each of these three sizes. We then use equation (S4) parameterized from table 1 (both in Halley et al., 2016) to give results for predicted losses at times up to 1,000 years into the future (Figure 16.2).

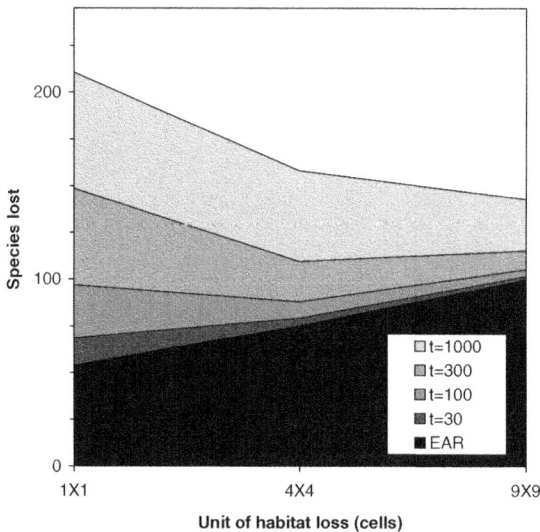

Figure 16.2 Number of extinctions expected at different times after habitat loss, as a function of the unit of habitat removal for the New York birds dataset. In each case, the total area remaining at t = 0 is eighty-one cells. The vertical extent of each shaded region represents the number of extinctions accumulated within a specific time interval. Black represents values given by the EAR, the darkest shade of grey represents extinctions that happened in the first thirty years after habit loss and so on, while the white represents the species richness remaining at t = 1,000.

Figure 16.2 shows how the passage of time changes the relative extinction rate. The number of extinctions expected if habitat loss occurs through the loss of individual cells, following the predictions using the EAR, is initially much smaller than that occurring when blocks are removed. Later, however, species are lost from fragments as they pay off their extinction debt. This effect is faster in the smaller segments to such an extent that the number of extinctions rises faster; so that eventually the number of extinctions is greater in small fragments than in the larger fragments created by removing larger blocks.

16.3.3 Inward versus Outward Habitat Loss

Using the gridded data for habitat loss yielded the two extinction curves in Figure 16.3, which show that there is a difference between inwards-out and outwards-in habitat loss. The idiosyncratic shape of these curves is due in part to the shape of Long Island, which resides furthest from the centre of the state (see Figure 16.1). Hence, the inwards curve increases much more rapidly than the outwards curve.

In order to draw more general conclusions, results provided by analysis of the species-range data prove more fruitful since they

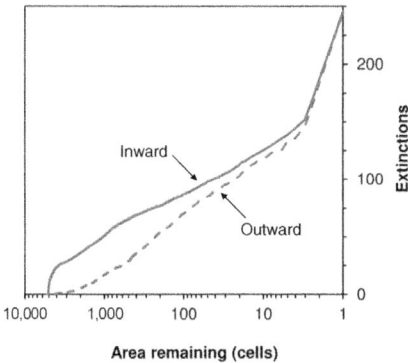

Figure 16.3 Number of extinctions expected as a function of habitat area loss (log-transformed) for two different scenarios of continuous habitat removal using the New York birds dataset. The continuous line represents inwards area accumulation (i.e. from the edge to the centre) and the dashed line represents the outwards area accumulation (i.e. from the centre to the edge).

comprise three taxonomic groups on three continents. These are shown in Figures 16.4–16.6. Using the species-range data for the two scenarios of habitat area loss, it is clear that the SAC of the study area appears distinctly different if we start accumulating species from the centre of the study area towards the edge or from the edge towards the centre (Figure 16.4). In both cases the end point is the same (the total number of species present), but the starting point is very different, with the centre to edge curve starting from a low value of species richness and the edge to centre curve starting from a considerably higher value of species richness (Figure 16.4). In many of the cases examined, the 5 per cent of the area distributed all over the edge of the study region includes approximately 80 or even 90 per cent of the regional species pool (see for example the case of North America; Figure 16.4C, F and G), while the corresponding 5 per cent of the area in the centre of the study region includes less than 40 per cent of the regional species pool.

Our results show that the spatial pattern of area loss does indeed affect the local extinction curve, with the inwards local extinction curve systematically located above the outwards local extinction curve (Figure 16.5). In some cases the two curves are located close to each other (for example, see the case of mammals in Africa; Figure 16.5G), but systematically the inwards curve lies well above the outwards one. Indeed, in most cases, for some intermediate value of area loss the inwards estimate of local extinctions is several times higher than the outward estimate of local extinctions. Note that for small areas the difference (ratio) becomes very great. It is always possible to find marginal intersections but not marginal true endemics. Also, this is not a scale related effect, since the different extent of the different areas analysed led to a similar pattern. This finding is strongly associated with the SAR.

16.3.4 Local versus Global Endemics

The most interesting results are the global extinction curves (Figure 16.6). First, in all cases, the number of global extinctions is only a small proportion of the local extinctions. In none of the study regions do the regional endemics exceed 50 per cent of the species pool (in most cases it is lower than 20 per cent) and we selected study areas that include considerable numbers of endemics. This is responsible for the more jagged appearance of some curves. Other regions in the world, such as those at higher latitudes or with adverse/extreme environmental conditions (e.g. the Sahara desert), have virtually no endemics. In this case the two scenarios of area loss produce more similar curves than in Figure 16.5.

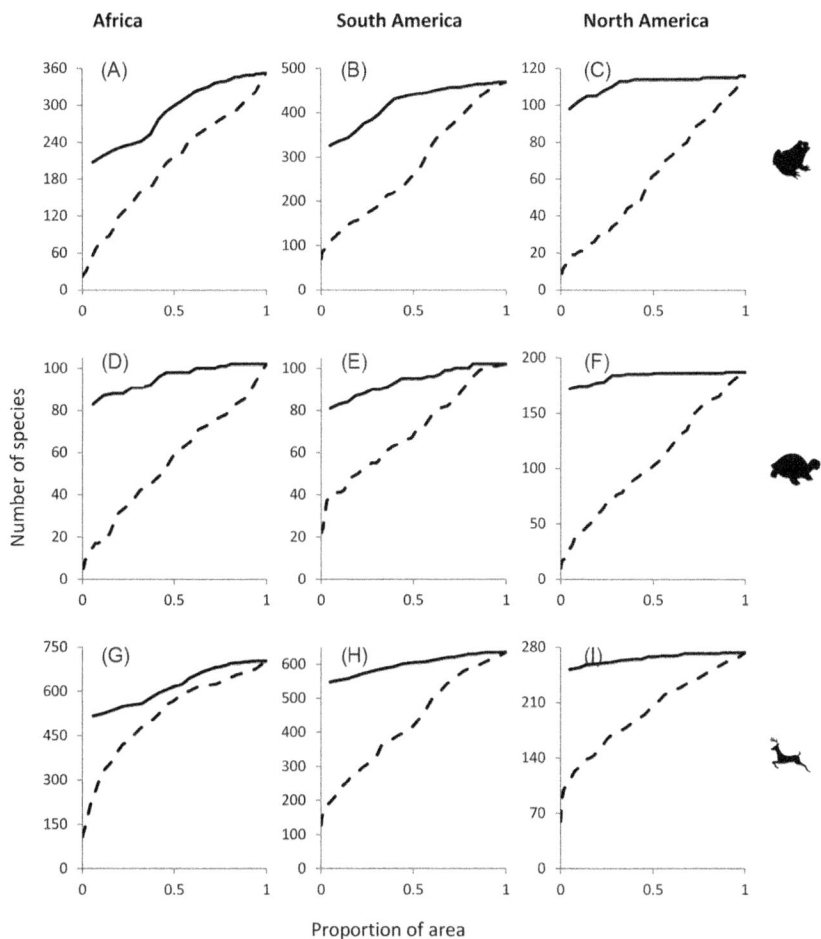

Figure 16.4 Species accumulation curves. Continuous lines represent inwards area accumulation (i.e. from the edge to the centre) and dashed lines represent outwards area accumulation (i.e. from the centre to the edge). (A, B and C) Amphibians, (D, E and F) Reptiles and (G, H and I) Mammals. (A, D and G) Africa, (B, E and H) South America and (C, F and I) North America.

There are still some cases where the inward curve is above the outward curve (e.g. the amphibians of North America), but we also see the opposite (all mammals). In most cases, the results are characterized by overlap of the curves.

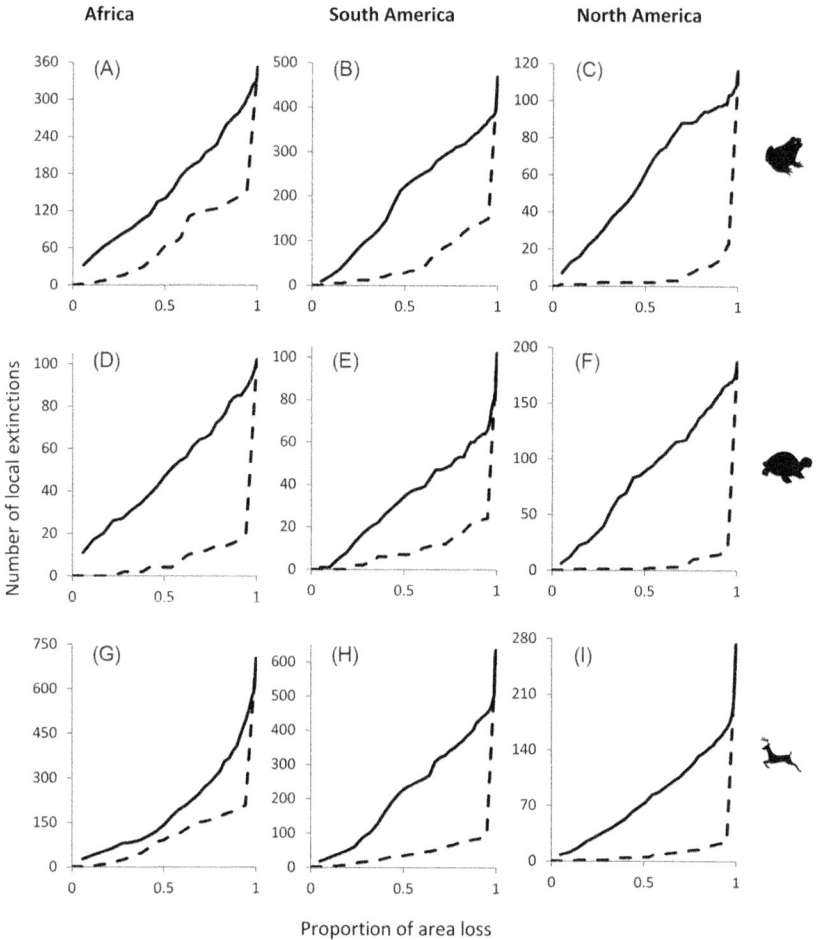

Figure 16.5 The relationship between the number of local extinctions and the proportion of area loss and how this relationship is affected by the spatial pattern of area loss. Continuous lines represent inwards area loss (i.e. from the edge to the centre) and dashed lines represent the outwards area loss (i.e. from the centre to the edge). (A, B and C) Amphibians, (D, E and F) Reptiles and (G, H and I) Mammals. (A, D and G) Africa, (B, E and H) South America and (C, F and I) North America.

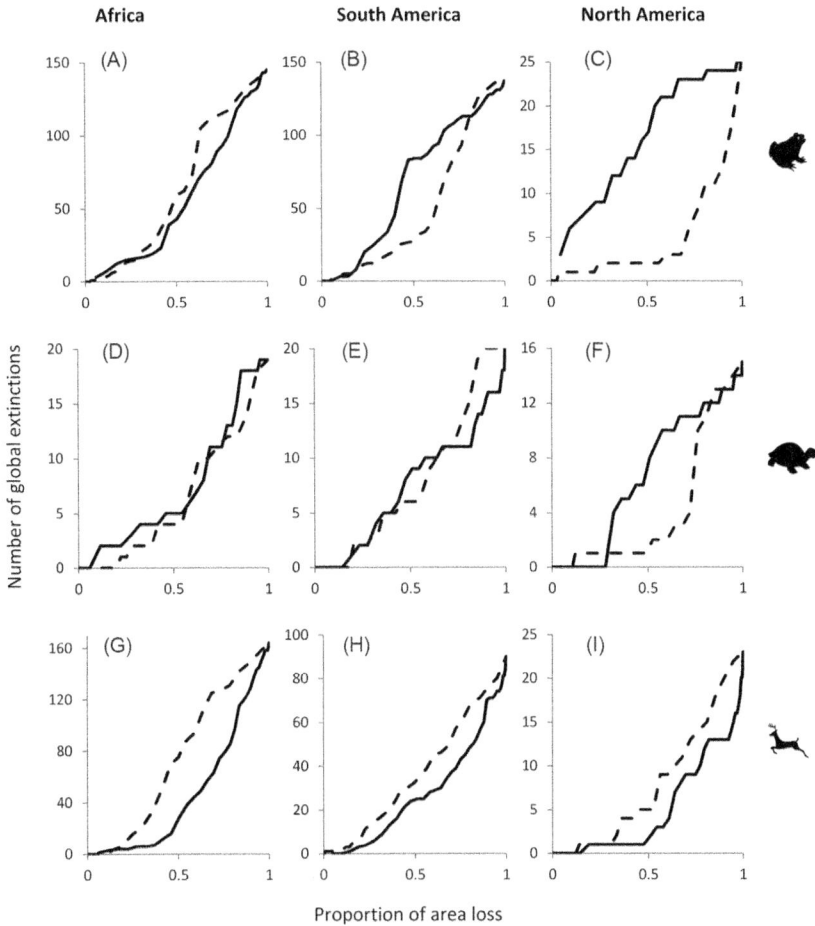

Figure 16.6 The relationship between the number of global extinctions (i.e. when the global range of the species lies within the area lost) and the proportion of the area loss and how this relationship is affected by the spatial pattern of area loss. Continuous lines represent inwards area loss (i.e. from the edge to the centre) and dashed lines represent the outwards area loss (i.e. from the centre to the edge). (A, B and C) Amphibians, (D, E and F) Reptiles and (G, H and I) Mammals. (A, D and G) Africa, (B, E and H) South America and (C, F and I) North America.

16.4 Discussion

If the removed area of habitat is discontinuous and random, an important aspect of geometry is the characteristic scale (block size) of habitat loss. As seen from the EARs in Figure 16.1B, the number of species lost is smaller when habitat destruction occurs through a random distribution of small events than when it is the result of removal in large blocks. This reflects spatial autocorrelation – random losses spread uniformly across an area will leave a pattern of wide-ranging fragments and hence will contain more species than if the same area loss is concentrated in a small number of large blocks. However, on average the number of individuals per surviving species must be smaller. Thus, the fragments will have a large extinction debt and, as a result, the number of delayed extinctions will be larger. Also, large-scale habitat loss in blocks will tend to leave behind larger remnants which are more robust against delayed extinction than many small fragments of the same total area. On the other hand, small fragments might also act in a complementary way so that a large number of small fragments may support species that cannot coexist on an island. Thus, geometry plays an important role in the size of imminent extinction rates when habitat is lost. As pointed out elsewhere, the total number of extinctions predicted by the SAR is the sum of the imminent extinctions and the delayed extinctions (Halley et al., 2014). Both of these depend on geometry. Our simulations show that if the loss pattern leads to fewer larger fragments, this will increase the number of immediate extinctions and reduce the number of delayed extinctions. By contrast, archipelagos of small fragments are likely to contain more species due to wider coverage but are subject to larger extinction debts and, when these are paid off, may have fewer species remaining.

If the removed area is continuous, the shape of the area will still affect the shape of the endemics–area curve. If we are losing area as a ring inwards from the edge to the centre then, as observed by Keil et al. (2015), there is a much greater extinction of endemics than if the same amount of habitat was being lost as a disc expanding outwards from the centre. This is because a given amount of continuous area that covers the edge of the study region, having a large geographic extent, will have considerably more species than the same amount concentrated in a single location. This again is a manifestation of the spatial autocorrelation of species distributions (Lazarina et al., 2014). The same total area includes more species if it is spread over a large extent, than if it is concentrated. Lazarina et al. (2014) defined an 'effective area' as that which is equal to the area lost in the outwards case but, in the inwards case, is higher than the area lost because of its huge extent. For example, for a circular geometry, in the outwards case we begin with a compact expanding disk but in the inwards case we have a vast but thin ring that is filling inwards that will span more bioclimatic zones. This finding is also highlighted by the SACs, where the inward curve starts from a point very close to the regional species pool and the outward curve starts at a much lower level. Thus, the use of SARs in predicting extinction should keep in mind the *effective* (*sensu* Lazarina et al., 2014) area of habitat loss.

However, the very large differences in extinctions observed by Keil et al. (2015) and our analysis (Figures 16.4 and 16.5) do not point to anything arbitrary in the SAR method. Rather, it is essential to take into consideration the definition of endemism. Many of the losses, especially in the outer sections, are not global extinctions. In reality, plenty of these species are characterized by extensive presence outside the study region and the loss of the edge of the study region means very little for the long-term persistence of these species (Figure 16.7). On the other hand, the majority of the species that are observed in the centre of the study region and not in the edge of it are true endemics of the region, in the sense that their entire species range is located in the study area and, if lost here, they will be lost everywhere. The number of these true endemics is only a fraction of the total species richness, even in endemism hotspots like the tropics. If we focus on these true endemics the picture totally changes. The spatial pattern of how habitat is lost plays only a minor role in the estimation of extinction risk and the amount of habitat lost is again the principal factor. In the case of global extinctions, the extinction curve is not tightly coupled to the SAR, since the number of global extinctions might be considerably less than the number of species (indeed there will be zero global extinctions in regions lacking endemic species). Hence, for

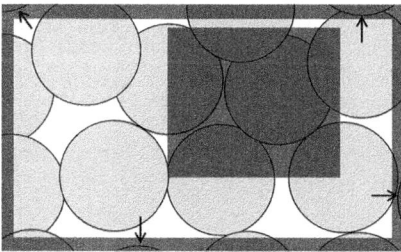

Figure 16.7 A simplified picture depicting the loss of species (ranges are circles) due to the destruction of habitat (dark areas). Losing an outer 'ring' of habitat from the study area (dark frame) causes the extirpation of four species (arrows). Losing a similar-sized area from the centre of the study area (dark rectangle) causes the loss of only one, since any species lost from the centre by removing the inner rectangle must be lost in its entirety. Also, any continuous range extending outside the study area that intersects an inner ring must also intersect the outer rings.

conservation of species diversity we cannot suggest that the pattern of habitat loss can mitigate the effect of habitat loss itself and preservation of as much habitat as possible should be the main goal for conservation efforts. Emphasis should also be placed on preserving habitat types characterized by high endemicity, like tropical rainforests.

16.5 Conclusions

The geometry of how we accumulate area to construct the SAC has direct impacts on the shape of the curve. This impact is also apparent on the accumulation of endemics along the EAR, which is used for extinction forecasting. Thus, both the SAC and the EAR should be constructed with care and must bear in mind the following qualifications:

- Due to the spatial autocorrelation of community composition, habitat loss leaving behind small fragments over a large area extent retains more species than a smaller number of large blocks with the same total area. However, archipelagos of small fragments will have larger extinction debts, to the extent that the delayed extinctions may reverse any advantage over the long term. Thus, an archipelago of small fragments could save more species only if habitat can be restored before delayed extinctions begin or if species are able to adapt to the new conditions.
- Losing area from the perimeter of the study area can lead to far more extirpations than losing habitat from the centre. Again, spatial autocorrelation plays a role in this but most of the difference reflects the local loss of species whose range centre falls outside the study area. This effect is much reduced if we limit our attention to global endemics.

Geometry cannot be ignored when applying the EAR in extinction forecasts, but it does not dominate species loss to such an extent as to nullify the central role of habitat area, which remains the principal determinant of biodiversity capacity. Taking into account the necessary qualifications, the SAR method remains an important tool for estimating extinction due to habitat loss.

References

Andrle, R. F. & Carroll, J. R. (eds.) (1988) *The atlas of breeding birds in New York State*. New York: Cornell University Press.

Brooks, T. M., Mittermeier, R. A., da Fonseca, G. A., Gerlach, J., Hoffmann, M., Lamoreux, J. F., Mittermeier, C. G., Pilgrim, J. D. & Rodrigues, A. S. (2006) Global biodiversity conservation priorities. *Science*, **313**, 58–61.

Dengler, J. & Oldeland, J. (2010) Effects of sampling protocol on the shapes of species richness curves. *Journal of Biogeography*, **37**, 1698–1705.

Diamond, J. M. (1972) Biogeographic kinetics: Estimation of relaxation times for avifaunas of southwest Pacific islands. *Proceedings of the National Academy of Sciences USA*, **69**, 3199–3203.

Diniz-Filho, J. A. F., Bini, L. M. & Hawkins, B. A. (2003) Spatial autocorrelation and red herrings in geographical ecology. *Global Ecology & Biogeography*, **12**, 53–64.

Fattorini, S. & Borges, P. A. (2012) Species–area relationships underestimate extinction rates. *Acta Oecologica*, **40**, 27–30.

Green, J. L. & Ostling, A. (2003) Endemics–area relationships: The influence of species dominance and spatial aggregation. *Ecology*, **84**, 3090–3097.

Halley, J. M. & Iwasa, Y. (2011) Neutral theory as a predictor of avifaunal extinctions after habitat loss. *Proceedings of the National Academy of Sciences USA*, **108**, 2316–2321.

Halley, J. M., Monokrousos, N., Mazaris, A. D., Newmark, W. D. & Vokou, D. (2016) Dynamics of extinction debt across five taxonomic groups. *Nature Communications*, **7**, 12283.

Halley, J. M., Sgardeli, V. & Monokrousos, N. (2013) Species–area relationships and extinction forecasts. *Annals of the New York Academy of Sciences*, **1286**, 50–61.

Halley, J. M., Sgardeli, V. & Triantis, K. A. (2014) Extinction debt and the species–area relationship: A neutral perspective. *Global Ecology & Biogeography*, **23**, 113–123.

He, F. & Hubbell, S. P. (2011) Species–area relationships always overestimate extinction rates from habitat loss. *Nature*, **473**, 368.

IUCN (2019) *The IUCN Red List of Threatened Species.* www.iucnredlist.org.

Kallimanis, A. S., Kunin, W. E., Halley, J. M. & Sgardelis, S. P. (2005) Metapopulation extinction risk under spatially autocorrelated disturbance. *Conservation Biology*, **19**, 534–546.

Keil, P., Pereira, H. M., Cabral, J. S., Chase, J. M., May, F., Martins, I. S. & Winter, M. (2018) Spatial scaling of extinction rates: Theory and data reveal nonlinearity and a major upscaling and downscaling challenge. *Global Ecology & Biogeography*, **27**, 2–13.

Keil, P., Storch, D. & Jetz, W. (2015) On the decline of biodiversity due to area loss. *Nature Communications*, **6**, 8837.

Kinzig, A. P. & Harte, J. (2000) Implications of endemics–area relationships for estimates of species extinctions. *Ecology*, **81**, 3305–3311.

Kuussaari, M., Bommarco, R., Heikkinen, R. K., Helm, A., Krauss, J., Lindborg, R., Öckinger, E., Pärtel, M., Pino, J., Rodà, F. & Stefanescu, C. (2009) Extinction debt: A challenge for biodiversity conservation. *Trends in Ecology & Evolution*, **24**, 564–571.

Lazarina, M., Kallimanis, A. S., Pantis, J. D. & Sgardelis, S. P. (2014) Linking species richness curves from non-contiguous sampling to contiguous-nested SAR: An empirical study. *Acta Oecologica*, **61**, 24–31.

Legendre, P. (1993) Spatial autocorrelation: Trouble or new paradigm? *Ecology*, **74**, 1659–1673.

Matias, M. G., Gravel, D., Guilhaumon, F., Desjardins-Proulx, P., Loreau, M., Münkemüller, T. & Mouquet, N. (2014) Estimates of species extinctions from species–area relationships strongly depend on ecological context. *Ecography*, **37**, 431–442.

Pan, X. (2013) Fundamental equations for species–area theory. *Scientific Reports*, **3**, 1334.

Pereira, H. M., Borda-de-Água, L. & Martins, I. S. (2012) Geometry and scale in species–area relationships. *Nature*, **482**, E3.

Pimm, S. L., Jenkins, C. N., Abell, R., Brooks, T. M., Gittleman, J. L., Joppa, L. N., Raven, P. H., Roberts, C. M. & Sexton, J. O. (2014) The biodiversity of species and their rates of extinction, distribution, and protection. *Science*, **344**, 1246752.

Rahbek, C. (2005) The role of spatial scale and the perception of large-scale species-richness patterns. *Ecology Letters*, **8**, 224–239.

Rahbek, C. & Colwell, R. K. (2011) Biodiversity: Species loss revisited. *Nature*, **473**, 288.

Rybicki, J. & Hanski, I. (2013) Species–area relationships and extinctions caused by habitat loss and fragmentation. *Ecology Letters*, **16**, 27–38.

Scheiner, S. M., Chiarucci, A., Fox, G. A., Helmus, M. R., McGlinn, D. J. & Willig, M. R. (2011) The underpinnings of the relationship of species richness with space and time. *Ecological Monographs*, **81**, 195–213.

Šizling, A. L., Kunin, W. E., Šizlingová, E., Reif, J. & Storch, D. (2011) Between geometry and biology: The problem of universality of the species–area relationship. *The American Naturalist*, **178**, 602–611.

Soininen, J., McDonald, R. & Hillebrand, H. (2007) The distance decay of similarity in ecological communities. *Ecography*, **30**, 3–12.

Steinbauer, M. J., Dolos, K., Reineking, B. & Beierkuhnlein, C. (2012) Current measures for distance decay in similarity of species composition are influenced by study extent and grain size. *Global Ecology & Biogeography*, **21**, 1203–1212.

Tsianou, M. A., Koutsias, N., Mazaris, A. D. & Kallimanis, A. S. (2016) Climate and landscape explain richness patterns depending on the type of species' distribution data. *Acta Oecologica*, **74**, 19–27.

Ulrich, W. & Buszko, J. (2005) Detecting biodiversity hotspots using species–area and endemics–area relationships: The case of butterflies. *Biodiversity and Conservation*, **14**, 1977–1988.

17 · Using Relict Species–Area Relationships to Estimate the Conservation Value of Reservoir Islands to Improve Environmental Impact Assessments of Dams

ISABEL L. JONES, ANDERSON SALDANHA
BUENO, MAÍRA BENCHIMOL, ANA FILIPA
PALMEIRIM, DANIELLE STORCK-TONON
AND CARLOS A. PERES

17.1 Introduction

The Amazon, Mekong and Congo are our most biodiverse river basin systems globally. Dams, both constructed and planned for construction within these basins, are emerging drivers of landscape- and basin-scale habitat loss and fragmentation. This is because as reservoirs are filled, terrestrial habitat is flooded and split apart to form smaller patches of remnant habitat 'islands' within a water matrix (Zarfl et al., 2015; Winemiller et al., 2016; Moran et al., 2018). The far-reaching ecological impacts of river impoundment are of particular concern in highly biodiverse habitats such as lowland tropical forests (Gibson et al., 2017). For example, the Amazon River basin is threatened by prolific dam development, with more than 280 dams in operation or planned for construction (Lees et al., 2016; Latrubesse et al., 2017).

Dam construction in moderately undulating lowland tropical regions results in the inundation of vast areas of tropical forest and wholesale changes to highly connected river systems, due to disruption of fisheries and flood pulses (Lees et al., 2016). Alongside the direct loss of terrestrial habitats during reservoir filling, archipelagos of forest islands are created from former hilltops. For example, the construction of the Tucuruí

Hydroelectric Reservoir (Brazil), the first major Amazonian dam (>15 m dam height; ICOLD, 2018) which began operation in 1984, led to the inundation of *ca.* 250,000 ha of forest and the creation of some 2,200 reservoir islands (Fearnside, 2016). In 1987, the Uatumã River, a first-order tributary of the Amazon River, was impounded by the Balbina Dam, which flooded *ca.* 300,000 ha of continuous old-growth forest (Fearnside, 2016) and created >3,500 islands (Benchimol & Peres, 2015a).

Dam-induced island archipelagos are true land-bridge islands, created simultaneously from previously continuous habitat. Reservoir islands are surrounded by a uniform and inhospitable water matrix, and are not buffered against the impacts of habitat fragmentation (i.e. edge-effects, such as wind damage) that even a low-quality terrestrial habitat matrix may provide (Ewers & Didham, 2006). Thus, species losses from habitat fragments embedded within a water matrix have been shown to be of a greater magnitude than species losses from fragments within a terrestrial matrix (Watling & Donnelly, 2006). For instance, when compared to continuous forest, forest islands in the Balbina archipelago have greatly altered tree assemblage composition due to edge-related fires and wind-throws (Benchimol & Peres, 2015a; Jones et al., 2017, 2019) and severely disrupted animal–plant mutualistic networks that have broken down due to landscape-scale habitat loss and fragmentation (Emer et al., 2013). Similar patterns of higher degrees of degradation of biological communities within a water matrix have been demonstrated for bats (Mendenhall et al., 2014) and birds (Wolfe et al., 2015).

Biological communities isolated on reservoir islands are largely comprised of relict species; that is, those species remaining from the previously continuous habitat (Watson, 2002). These relict communities undergo a process of disassembly (whereby species are lost from islands) and experience ecological 'relaxation' until new equilibrium communities are reached (Diamond, 1972; Gonzalez, 2000). Island area – and hence the island species–area relationship (SAR) – is a strong determinant of the rate of species loss from reservoir islands through time, with larger islands able to retain more species and for longer because conditions remain more closely aligned to those in continuous habitats (Jones et al., 2016). Conversely, smaller islands are at risk of rapid local extinctions because environmental conditions become much more degraded compared to continuous habitat, alongside the fact that fewer species can be sustained in a small area (Jones et al., 2016). Thus, the 'conservation value' (CV) of larger islands – in terms of the proportion of relict species retained from the once continuous forest – is higher than that of smaller islands.

There is widespread evidence of severe and long-term ecological impacts for relict biological communities isolated on reservoir islands, particularly in highly biodiverse tropical regions (Jones et al., 2016). For instance, islands within the Chiew Larn reservoir (Thailand) have experienced near total extinction of small mammals within twenty-six years of insularization, with 50 per cent of species becoming locally extinct in less than fourteen years, precipitated by the appearance of an invasive rat species (Gibson et al., 2013). In Lago Guri (Venezuela) local extinctions of top predators on islands caused trophic cascades, severely impacting tree regeneration due to hyper-herbivory (Terborgh et al., 2001, 2006). Moreover, in Brazilian Amazonia, wholesale avifaunal erosion was induced from hundreds of local extinctions on islands within the Tucuruí Hydroelectric Reservoir in twenty-two to twenty-three years of isolation history (Bueno et al., 2018). Thus, the local extinction of species from reservoir islands is an additional ecological impact associated with dam-induced habitat loss and fragmentation, which is not yet explicitly included in environmental impact assessments (EIAs) as a precondition of dam licensing (Gibson et al., 2017; Ritter et al., 2017).

The World Commission on Dams (WCD, 2000) outlined a comprehensive framework for assessing the social and environmental impacts of dams, including recommendations to assess ecological impacts at the basin-scale. Yet there is no binding signatory agreement for dam contractors, developers, financers or governments to forecast and mitigate dam-induced environmental impacts. Moreover, the lack of any systematic and long-term monitoring of the efficacy of environmental impact mitigation measures implemented precludes the advancement of policies aimed towards increasing the ecological sustainability of dams (Moran et al., 2018; Jones & Bull, 2020). In Brazil, a priori EIA has been required since 1986 to license any new major infrastructure project (Ritter et al., 2017). However, current EIAs are inadequate in terms of reporting the detrimental effects of hydroelectric dams on both natural and societal environments, including severe underestimates of forest and biodiversity loss, greenhouse gas emissions and displacement of indigenous communities (Fearnside, 2016; Ritter et al., 2017; Timpe & Kaplan, 2017).

In cases where environmental impact mitigation measures are proposed following EIAs of dams, they may include 'offsetting' the area of land flooded (i.e. the reservoir water surface area) through, for example, strictly protecting an equivalent or greater area of comparable mainland habitat (Bull et al., 2013). For instance, the Balbina Hydroelectric Reservoir (BHR; Brazilian Amazonia) triggered the creation of the *ca.* 940,000 ha

Uatumã Biological Reserve, which covers a portion of the reservoir and adjoining mainland continuous forest habitat. However, the majority of Amazonian dams do not have an appropriate offset associated with them, and the long-term ecological impacts of tropical dam construction remain poorly accounted for (Latrubesse et al., 2017; Sonter et al., 2018).

Tropical forest habitats are of global importance for biodiversity – due to the enormous number of species present – and ecosystem service provision including significant uptake and storage of atmospheric CO_2 (Pan et al., 2011). Yet these forests are under increasing threat from hydropower expansion, causing forest loss and fragmentation through reservoir creation (Latrubesse et al., 2017). Therefore, understanding the fate of terrestrial taxa isolated on tropical reservoir islands is an important component of the long-term environmental impact of tropical dams (Ritter et al., 2017). Despite the clear evidence of local species extinctions on reservoir islands, dam proponents have suggested that reservoir islands can act as an effective means for biodiversity conservation as part of environmental impact mitigation strategies (Trussart et al., 2002). Moreover, if the area of impacted terrestrial habitat is simply taken as the area of habitat flooded (i.e. the reservoir water surface area) this may severely underestimate the long-term environmental impact of dams due to species on reservoir islands being subject to local extinctions and a lasting extinction debt (Jones et al., 2016).

According to the SAR, islands sustain different numbers of relict species due to their different areas. Thus, SARs of relict species on islands (i.e. the proportion of remnant species from formally continuous habitat) present a powerful and accessible means for assessing the ecological impact of insularization on terrestrial species. Depending on landscape topography and reservoir levels, and hence whether many small or few large islands are created from the same amount of flooding, the outcomes for relict species retention on islands may be contrasting and require different impact mitigation strategies (Ewers & Didham, 2006; Jones & Bull, 2020). Applying relict species–area relationships ($_R$SARs) to quantify the proportion of relict species retained on reservoir islands compared to continuous habitat therefore indicates the 'conservation value' of islands. By including the area of reservoir islands that has reduced conservation value (i.e. the area of insular habitat that has been impacted by river impoundment) into the calculations of impacted habitat, EIAs can be made more accurate (Figure 17.1). Additionally, SARs comprise one of the fundamental patterns in biogeography (MacArthur & Wilson, 1967; Lomolino, 2000), making them highly applicable in conservation

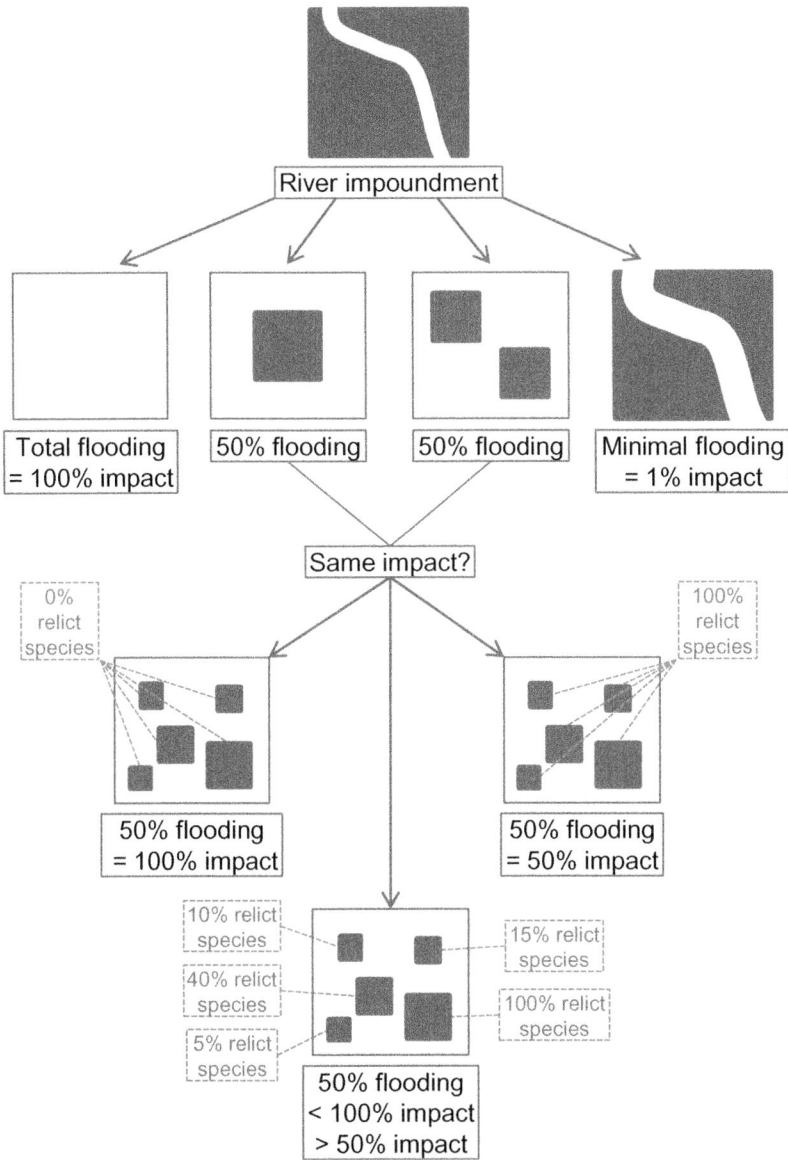

Figure 17.1 Schematic of a hypothetical scenario outlining our approach of using relict species–area relationships (ʀSARs) to determine the impact of insularization on biological communities. ʀSARs are used to estimate the conservation value (CV) of islands, which is expressed as the proportion of continuous forest species present on islands. Islands that harbour zero relict continuous forest species would therefore have a CV of zero. Conversely, if the island retains all species found in continuous forest, the CV would be one. In intermediate scenarios where a proportion of relict species are retained, the CV would be between zero and one.

measures, including the design of protected areas and estimation of local biological diversity (Matthews et al., 2016).

In this study, we use the Balbina Hydroelectric Reservoir (BHR) as a case study to 1) demonstrate how relict species–area relationships ($_R$SARs) can be used to assess the conservation value of islands; 2) present a novel area-of-impact correction tool which incorporates the conservation value of islands to estimate the *minimum* additional area of reservoir island habitat that must be considered in EIAs to account for impacted insular habitat; and, based on combined data for eight vertebrate, invertebrate and plant biological groups (as a proxy for the whole biological community); and 3) determine which individual groups can act as an indicator for the whole-community $_R$SAR. In doing so, we aim to assist with applying our conservation value and area-of-impact correction tool to other dam-induced archipelagos.

17.2 Methods

17.2.1 Study Area

We used the Balbina Hydroelectric Reservoir (BHR) system, central Brazilian Amazonia (01°01–01°55 S; 59°28–60°29 W) as our study area. The *ca.* 300,000 ha BHR was formed following the impoundment of the Uatumã River in 1987 (Fearnside, 2016). As the BHR filled, continuous old-growth lowland tropical forest was inundated, transforming the moderately undulating landscape into an archipelago of 3,546 islands, ranging in size from 0.3 to 4,878 ha (mean ± SD = 33.4 ± 156.3 ha). Island area, isolation and environmental disturbance at island edges shape the structure and composition of these insular forests (Benchimol & Peres, 2015a; Jones et al., 2017, 2019).

Within the BHR system, our survey sites comprised of a network of seventy-two focal islands (1.2–1,815 ha; mean ± SD = 132.5 ± 318.0 ha) and four mainland continuous forest sites positioned across a comparable elevation gradient (Figure 17.2). Focal islands were selected based on cloudless georeferenced Landsat ETM+ scenes from 2009 (230/061 and 231/061), were normally 1 km apart to ensure spatial independence, and were located at varying distances from the mainland. Using ArcGIS (ESRI, 2012) we calculated island areas for all 3,546 islands within the BHR using Rapid-Eye high-resolution (5 m pixel) imagery, covering *ca.* 700,000 ha of the BHR system (Benchimol & Peres, 2015b).

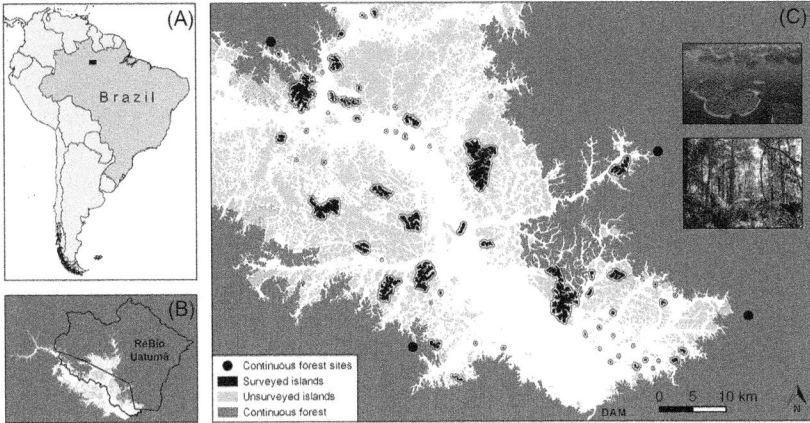

Figure 17.2 Geography of the Balbina Hydroelectric Reservoir (BHR) system.
(A) Location of the study area in central Brazilian Amazonia, indicated by a solid
rectangle containing (B) the BHR system, and showing the boundaries of the
Uatumã Biological Reserve, a strictly-protected area safeguarding a portion of the
BHR archipelago and mainland continuous forest; (C) a detailed map showing the
spatial distribution of the seventy-two survey islands and four continuous forest sites.
Photographs represent the BHR landscape (top; credit: Eduardo M. Venticinque)
and the forest interior of a surveyed island (bottom; credit: ASB).

17.2.2 Data Collection

Using original published datasets, we compiled presence/absence data for
eight biological groups across focal islands and mainland continuous
forest sites within the BHR system, surveyed within a 5-year period
(2011–2016; Table 17.1). Brief descriptions of the datasets follow and the
accompanying references should be consulted for full details regarding
species surveys.

Medium and large vertebrates: Medium and large-sized diurnal and noc-
turnal vertebrate species (>100 g and amenable to line transect censuses,
indirect sign surveys, armadillo surveys and camera trapping) were
surveyed between June 2011 and December 2012. Primate, carnivore,
xenarthran, ungulate, rodent, large bird and tortoise species were
recorded (Table 17.1; Benchimol & Peres, 2015b).

Small mammals: Surveys were conducted using transects with live traps
set at ground, understorey and sub-canopy heights, and pitfall units
connected by a drift fence. A total of 65,520 trap-nights across seventy-
nine transects were completed from April to November in 2014 and
2015 (Table 17.1; Palmeirim et al., 2018).

Understorey birds: Understorey birds were surveyed using mist nets deployed over 21,888 net-hours between July and December in 2015 and 2016 (Table 17.1; Bueno & Peres, 2020).

Orchid bees: Orchid (euglossine) bees were surveyed using scent trap-arrays baited with cineole, methyl salicylate, methyl cinnamate and vanillin in September 2012 and April 2013, with trap-arrays exposed at each sampling site for three consecutive days (Table 17.1; Storck-Tonon & Peres, 2017).

Lizards: Terrestrial diurnal lizard assemblages were sampled using pitfall traps connected by a drift fence. Sampling was undertaken between April and November in 2014 and 2015, totalling 5,447 trap-days across seventy-one trapping plots that were each sampled for sixteen consecutive days (Table 17.1; Palmeirim et al., 2017).

Table 17.1 *Summary of the datasets used in the analyses, comprising of eight biological groups surveyed on islands and in mainland continuous forest within the Balbina Hydroelectric Reservoir system (Brazilian Amazonia)*

Biological group	Number of species included in analyses	Number of sites surveyed within the BHR	
		Islands (size range in ha)	Mainland continuous forest
Medium and large vertebrates (Benchimol & Peres, 2015b)	35	37 (1.2–1,815)	3
Small mammals (Palmeirim et al., 2018)	19	25 (1.2–1,466)	4
Understorey birds (Bueno & Peres, 2020)	130	33 (1.4–1,815)	4
Orchid bees (Storck-Tonon & Peres, 2017)	25	34 (1.2–1,815)	3
Lizards (Palmeirim et al., 2017)	17	25 (1.2–1,466)	4
Frogs (Bueno et al., 2020)	37	72 (1.2–1,815)	4
Adult trees (Benchimol & Peres, 2015a)	360	34 (1.2–1,815)	3
Sapling trees (Jones et al., 2019)	391	36 (1.2–1,815)	3

Frogs: Frog calls were recorded in riparian and non-riparian habitats using autonomous recording units between July and December 2015. Frog calls were identified to species within a subset of 9,362 one-minute recordings totalling 156 hours (Table 17.1; Bueno et al., 2020).

Adult and sapling trees: Trees and arborescent palms ≥ 10 cm diameter at breast height (DBH; hereafter referred to as adult trees) were surveyed in 2012 (Table 17.1; Benchimol & Peres, 2015a). Saplings of trees and arborescent palms that had the potential to reach ≥ 10 cm DBH (hereafter, sapling trees) were surveyed in the same sites in 2014 (Table 17.1; Jones et al., 2019).

17.2.3 Data Analysis

17.2.3.1 Using $_R$SARs to Predict the Conservation Value of Islands

We compiled an 'island by species' and a 'continuous forest by species' presence/absence matrix for each biological group and for all groups combined as a proxy for the 'whole community'. We pooled species from all continuous forest sites to generate a representative 'reference' continuous forest community. We took this conservative approach to avoid possible undersampling of highly biodiverse habitats and to avoid bias towards very low island 'conservation values' (CVs): by using the maximum surveyed area of continuous forest available as a reference community, there was the best chance of detecting shared species between islands and continuous forest. We excluded species from our analyses that were present on islands but not in continuous forest because the focus of our study was on the capacity of islands to retain relict species from formerly continuous habitats, rather than their ability to support newly-immigrated disturbance-adapted species (Ewers & Didham, 2006). To estimate the CV of each island, we calculated the proportion of continuous forest species present on the island (i.e. relict species). Accordingly, an island harbouring all species found in continuous forest sites would have a maximum CV of one, whereas an island harbouring none of those species would have the minimum CV of zero (Figure 17.1).

For each biological group and the whole community, we generated species–area curves for relict species ($_R$SARs) for focal islands by modelling CV with $\log_{10}$ island area (ISLAND$_{AREA}$; ha) using a generalized linear model (GLM) with a quasibinomial error structure, and used these GLMs to predict the CV of unsurveyed islands across the entire BHR archipelago.

17.2.3.2 Estimating the Area of Islands with Reduced Conservation Value

We estimated the area of each of the 3,546 BHR islands with reduced CV (i.e. the area of island habitat impacted by insularization; $\text{IMPACTED}_{\text{ISLAND}}$; ha) using the following equation:

$$\text{IMPACTED}_{\text{ISLAND}} = \text{ISLAND}_{\text{AREA}} - (\text{ISLAND}_{\text{AREA}} * \text{CV}). \quad (17.1)$$

For each biological group and the whole community, we summed $\text{IMPACTED}_{\text{ISLAND}}$ for all 3,546 BHR islands to give $\text{IMPACTED}_{\text{TOTAL}}$. Taking the reservoir area of 300,000 ha (Fearnside, 2016), we then used the $\text{IMPACTED}_{\text{TOTAL}}$ estimates to ascertain the *minimum* percentage of additional reservoir water surface area – on top of reservoir water surface area alone – that should be included in EIAs to account for impacted island habitat. All analyses were performed using R (version 3.5.1; R Core Team, 2018).

17.2.3.3 Assessing Which Biological Groups Can Act as a Proxy for the Whole Community

We evaluated each biological group considering 1) our expert opinion, 2) survey feasibility, 3) alignment of individual biological group $_R$SARs with that of the proxy 'whole community' and 4) the percentage additional reservoir water surface area required to account for reduced CV on islands. This evaluation allowed us to assess which biological groups can act as an indicator for relict species retention patterns, in order to aid the application of our method to other landscapes affected by dam-induced habitat fragmentation (Gardner et al., 2008).

17.3 Results

17.3.1 Using $_R$SARs to Predict the Conservation Value of Islands

All biological groups demonstrated positive $_R$SARs, with the proportion of continuous forest species retained on islands – that is, the conservation value (CV) of islands – increasing with island area (Figure 17.3). Considering the proxy 'whole community', the CV of islands ranged from 0.06 (in a 1.4-ha island) to 0.66 (in a 4,878-ha island; mean CV $\pm$ SD = 0.22 $\pm$ 0.07; Table 17.2) and rapidly increased as a function of island area (slope = 0.738; Figure 17.4). Across the entire range of island sizes within the BHR (0.3–4,878 ha; mean $\pm$ SD = 33.4 $\pm$ 156.3 ha) the maximum CV of an island was 0.96 (4,878 ha; mid- to large-sized vertebrates; Table 17.2). Conversely, five small islands (1.2–11.5 ha) had zero observed CV for either mid- to large-bodied vertebrates or frogs (Table 17.2).

Table 17.2 *Summary of the proportion of relict species from continuous forest sites, that is, the conservation value (CV) of islands, aggregate area of island habitat with reduced CV across 3,546 islands in the Balbina Hydroelectric Reservoir archipelago and the percentage habitat area – in addition to the reservoir water surface area (300,000 ha) – to be included in EIAs*

Group	Conservation value (CV) of islands, that is, the proportion of mainland species retained on islands (CV; range, mean ± SD)	Aggregate island area with reduced CV in ha (% aggregate area impacted by reduced CV in brackets)	Per cent additional reservoir water surface area to be included in EIAs
Medium and large vertebrates	0–0.96 (0.24 ± 0.17)	47,031 (39.77)	15.68
Small mammals	0.05–0.89 (0.21 ± 0.09)	68,118 (57.60)	22.71
Understorey birds	0.01–0.45 (0.08 ± 0.04)	94,123 (79.58)	31.37
Orchid bees	0.11–0.72 (0.32 ± 0.05)	68,993 (58.34)	23.00
Lizards	0.05–0.81 (0.23 ± 0.11)	62,103 (52.51)	20.70
Frogs	0–0.63 (0.16 ± 0.07)	79,519 (67.24)	26.51
Adult trees	0.06–0.63 (0.26 ± 0.07)	69,965 (59.16)	23.32
Sapling trees	0.03–0.64 (0.19 ± 0.07)	75,814 (64.10)	25.27
Whole community	0.06–0.66 (0.22 ± 0.07)	71,940 (60.83)	23.98

17.3.2 Estimating the Area of Islands with Reduced Conservation Value

The aggregate area of all 3,546 islands within the BHR is 118,268 ha. Of that total area of insular habitat, the total area exhibiting reduced CV ranged from 47,031 ha (40 per cent) for medium and large vertebrates to 94,123 ha (80 per cent) for understorey birds. When the 'whole community' was considered, *ca.* 72,000 ha (*ca.* 60 per cent) of island habitat had reduced CV. If the overall reservoir water surface area (*ca.* 300,000 ha for the BHR; Fearnside, 2016) is used in area

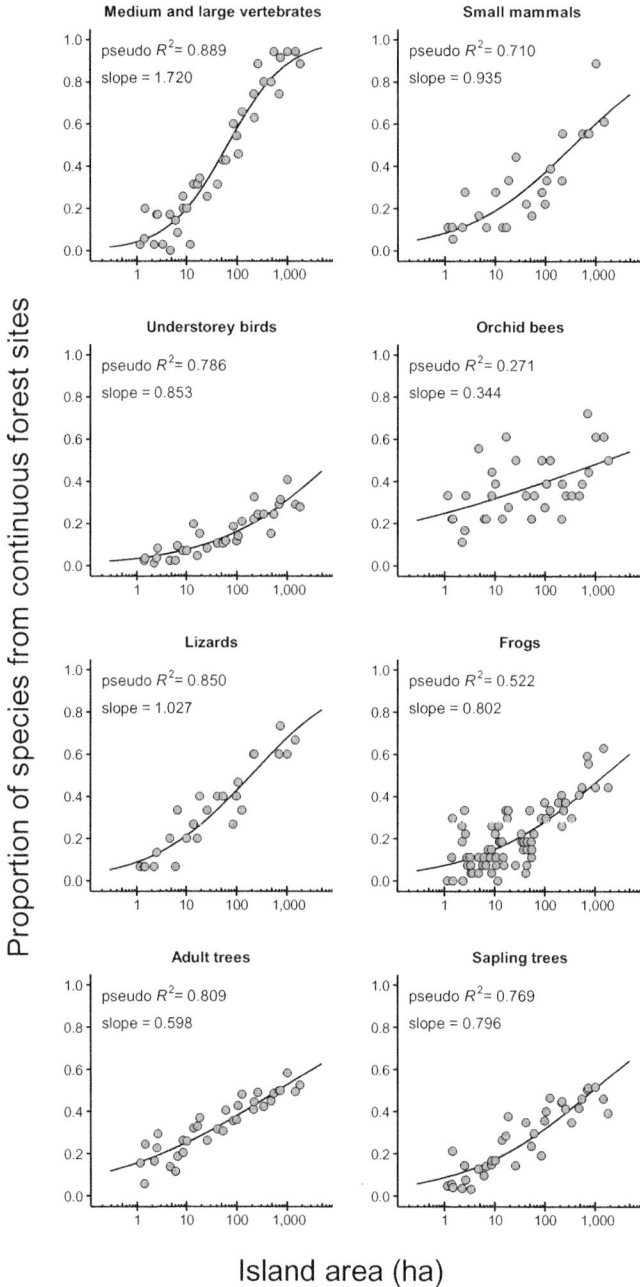

Figure 17.3 $_R$SARs for the eight biological groups surveyed in the Balbina Hydroelectric Reservoir system (BHR; Brazilian Amazonia). Grey points indicate the observed conservation value (CV) of surveyed islands and the black solid line represents predicted CVs across all 3,546 islands within the BHR. The x-axes are shown on a $\log_{10}$ scale.

calculations of impacted terrestrial habitat, then this *ca.* 72,000 ha of impacted island habitat represents an additional 24 per cent of the reservoir water surface area that should be included in EIAs (Table 17.2). In other words, the total impacted area of terrestrial habitat according to our method would be 372,000 ha (300,000 + 72,000 ha). However, certain biological groups – namely understorey birds, frogs and sapling trees – were impacted over an additional 25–31 per cent of the *ca.* 300,000 ha of reservoir area (Table 17.2).

17.3.3 Assessing Which Biological Groups Can Act as a Proxy for the Whole Community

The magnitude of impact (i.e. $_R$SAR slope) for all biological groups combined – the proxy 'whole community' – was 0.738 (Figure 17.4) and an additional 24 per cent of the reservoir water surface area would need to be included in EIAs to account for the reduced CV of reservoir islands at the community-level. Based on 1) our expert opinion, 2) survey

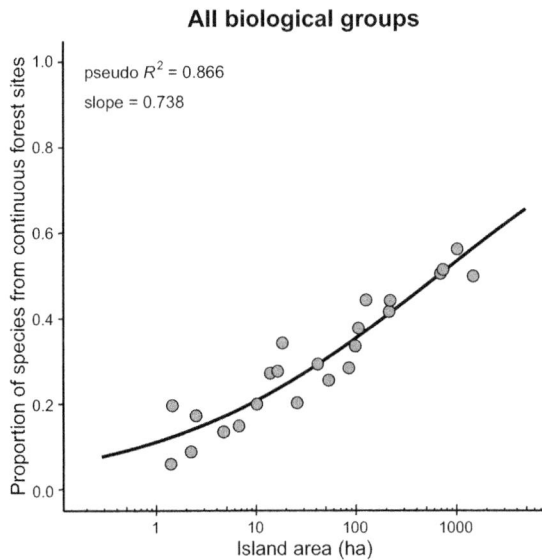

All biological groups

pseudo R^2 = 0.866

slope = 0.738

Figure 17.4 $_R$SAR for the proxy 'whole community' comprising of combined data for eight biological groups surveyed in the Balbina Hydroelectric Reservoir system (BHR; Brazilian Amazonia). Grey points indicate the observed conservation value (CV) of islands where all eight biological groups were surveyed (n = 23) and the black solid line represents predicted CVs across all 3,546 islands within the BHR. The x-axis is shown on a $\log_{10}$ scale.

feasibility, 3) alignment of individual biological group $_R$SARs with that of the proxy 'whole community' and 4) the percentage of additional reservoir water surface area required to account for reduced CV on islands, our results indicate that plants (both adult and sapling trees: slope = 0.598, 23.3 per cent; and 0.796, 25.3 per cent, respectively) and understorey birds (slope = 0.853, 31.4 per cent) can be used as appropriate indicator groups for the response of the proxy 'whole community' (Figures 17.3 and 17.4; Table 17.2; Watson, 2002; Gardner et al., 2008).

17.4 Discussion

We demonstrate that relict species–area relationships ($_R$SARs) can be effectively used to estimate the conservation value of reservoir islands to improve the accuracy of environmental impact assessments (EIAs) of major hydroelectric dams using a simple area-of-impact correction tool. Across eight biological groups within the Balbina Hydroelectric Reservoir (BHR) system, the conservation value of reservoir islands co-varied tightly with island area, with conservation value sharply declining with decreasing island area. Incorporating the conservation value of islands into assessments of impacted terrestrial habitat indicated that an additional 24 per cent of the reservoir water surface area should be included in EIAs to account for impacted insular habitats. Based on the degree to which $_R$SARs of any given biological group tracked that of the proxy 'whole community', we suggest that adult and sapling trees, as well as understorey birds, can serve as appropriate indicator groups for community-level responses to insularization induced by river impoundment.

17.4.1 Reservoir Islands Have Reduced Conservation Value

All islands within the BHR had reduced conservation value compared to mainland continuous forest, with smaller islands exhibiting the greatest reduction in conservation value (i.e. the lowest proportion of continuous forest species retained). Previous studies of biological communities isolated on reservoir islands have consistently shown that those isolated on small islands (<10 ha) are the most vulnerable to local extinctions (e.g. Terborgh et al., 2001; Benchimol & Peres, 2015b). However, disturbance to forest structure, including wind damage, fire and desiccation at fragment edges, can penetrate into remnant forests to varying extents and have been shown to affect forest biomass >1.5 km from fragment edges (Chaplin-Kramer et al., 2015). Thus, species isolated on much larger islands are still likely to be impacted by insularization. Indeed, our modelling exercise indicated that even the largest island within the BHR (4,878 ha) did not support a full complement of species found in mainland continuous forest.

17.4.2 Estimating Additional Island Habitat Area to Be Included in Environmental Impact Assessments

Incorporating the conservation value of islands into our area-of-impact correction tool revealed that of the total aggregate island area (118,268 ha, n = 3,546), *ca.* 60 per cent retained reduced CV when the entire biological community was considered. Thus, in our case study, the equivalent of at least 24 per cent of the BHR reservoir water surface area would need to be added on top of the reservoir water surface area alone, to account for the area of insular habitat with reduced conservation value. The surface area of the BHR is *ca.* 300,000 ha (Fearnside, 2016) and therefore an additional *ca.* 72,000 ha (i.e. 372,000 ha in total) would need to be incorporated into assessments of terrestrial habitat impacted by dam development.

Using $_R$SARs to estimate the conservation value of reservoir islands allows the impact of insularization on relict continuous habitat species to be assessed and accounted for in EIAs. $_R$SARs avoid artificial inflation of island conservation value by focussing on relict continuous forest species and excluding newly-immigrated disturbance-adapted species. Our $_R$SAR and island conservation value approach drives the EIA process forward beyond simply assuming that the reservoir water surface area equates to the total area of terrestrial habitat impacted (WCD, 2000; Ritter et al., 2017). EIAs do not explicitly consider reservoir island habitat in assessments of terrestrial land impacted by dam construction, neglect biodiversity losses from islands and any other impacted mainland habitat, and fall short of assessing many of the other direct and indirect ecological impacts of dam construction in mega-diverse tropical regions (Fearnside, 2016; Lees et al., 2016; Timpe & Kaplan, 2017; Jones & Bull, 2020). We show that species loss from islands induced by flooding moderately undulating terrain, such as in the Amazon Basin, is a significant additional environmental impact to be considered in the EIA process (Ritter et al., 2017; Jones & Bull, 2020).

However, we caution that the BHR archipelago likely represents a best-case-scenario for remnant insular habitat and relict species retention, due to the fact that almost half of BHR islands and mainland continuous forest to the east of the former Uatumã River bank are under strict protection from the Uatumã Biological Reserve. Thus, there is minimal impact from other drivers of species loss, such as subsistence farming, logging and hunting, which many other dams may be subject to and where species loss from islands may be far greater (Peres, 2001). Related to this, we also highlight that, depending on the condition of the 'baseline' habitat from which $_R$SARs are generated, the outcomes for island conservation values may be very different and may shift over time (Maron et al., 2018; Bueno & Peres, 2020). Moreover, our approach provides a snapshot of the habitat area correction needed to account for reduced conservation value of islands with *ca.* thirty years of post-isolation history. We therefore caution that the full effects of insularization on relict species is yet to be realized because our method does not include a means of incorporating ongoing extinction debts on islands, which have been shown to continue beyond ninety years of island isolation (Jones et al., 2016).

We focus on the role of island area in determining the degree of relict species retention, as this metric is likely the most accessible for dam developers and decision-makers during the EIA process, because the shape of the reservoir can be predicted using GIS and terrain maps with different water level scenarios. Yet other landscape metrics, such as isolation distance and the degree of habitat connectivity, may also play important roles in shaping $_R$SARs on islands following insularization (Storck-Tonon & Peres, 2017; Palmeirim et al., 2018). When focussing on individual biological groups, a more nuanced relationship may emerge regarding the ability of habitat fragments to retain species found in continuous habitats, as demonstrated by the variation in $_R$SARs among the eight biological groups in our study.

17.4.3 Trees and Understorey Birds Are Appropriate Indicators for Whole Community Responses to Insularization

The response of trees and birds to insularization is relatively well documented (e.g. Terborgh et al., 1997; Benchimol & Peres, 2015a; Aurélio-Silva et al., 2016; Bueno et al., 2018; Bueno & Peres, 2019; Jones et al., 2019). We found that the $_R$SARs for both adult and sapling trees and understorey birds aligned well with the community-level $_R$SAR. In addition, tree and bird communities play vital roles in the maintenance of critical ecological processes and services of forest ecosystems, such as carbon storage, pollination, seed dispersal and pest control

(Bregman et al., 2014; Anderson-Teixeira et al., 2016), are key groups in long-term monitoring programmes (Laurance et al., 2011), and species identities can be double-checked a posteriori using photographs and vegetative specimens, which reduces the potential for observer bias. Thus, we recommend that at a minimum trees or understorey birds should be surveyed across insular habitats when attempting to assess their conservation value.

Importantly, the conservation value of islands for understorey birds was lower than for the proxy 'whole community'. Therefore, using a general whole-community $_R$SAR to estimate conservation value leads to underestimating the impacts for certain biological groups and over-estimating for others. There may also be differences in $_R$SARs within biological groups depending on functional traits. For example, in another study of bird communities in an Amazonian dam-induced archipelago, forest specialist birds showed a higher rate of species loss as a function of island area reduction than habitat generalist birds (Bueno et al., 2018), a finding echoed in other groups in the BHR archipelago including lizards and small mammals (Palmeirim et al., 2017, 2018). By focusing on relict species from continuous mainland habitat, our approach avoids the confounding effect of replacing habitat specialist species with generalists: including newly-immigrated disturbance-adapted species would likely overestimate the conservation value of forest islands.

17.5 Conclusion

Using relict species–area relationships ($_R$SARs) and a simple area-of-impact correction tool, which incorporates the conservation value of reservoir islands, enables more reliable estimation of the total area of terrestrial habitat impacted by river impoundment. Thus, employing our method improves the accuracy of environmental impact assessments of dam development on terrestrial biodiversity.

References

Anderson-Teixeira, K. J., Wang, M. M. H., McGarvey, J. C. & LeBauer, D. S. (2016) Carbon dynamics of mature and regrowth tropical forests derived from a pantropical database (TropForC-db). *Global Change Biology*, **22**, 1690–1709.

Aurélio-Silva, M., Anciães, M., Henriques, L. M. P., Benchimol, M. & Peres, C. A. (2016) Patterns of local extinction in an Amazonian archipelagic avifauna following 25 years of insularization. *Biological Conservation*, **199**, 101–109.

Benchimol, M. & Peres, C. A. (2015a) Edge-mediated compositional and functional decay of tree assemblages in Amazonian forest islands after 26 years of isolation. *Journal of Ecology*, **103**, 408–420.

Benchimol, M. & Peres, C. A. (2015b) Predicting local extinctions of Amazonian vertebrates in forest islands created by a mega dam. *Biological Conservation*, **187**, 61–72.

Bregman, T. P., Sekercioglu, C. H. & Tobias, J. A. (2014) Global patterns and predictors of bird species responses to forest fragmentation: Implications for ecosystem function and conservation. *Biological Conservation*, **169**, 372–383.

Bueno, A. S. & Peres, C. A. (2019) Patch-scale biodiversity retention in fragmented landscapes: Reconciling the habitat amount hypothesis with the island bio-geography theory. *Journal of Biogeography*, **46**, 621–632.

Bueno, A. S. & Peres, C. A. (2020) The role of baseline suitability in assessing the impacts of land-use change on biodiversity. *Biological Conservation*, **243**, 108396.

Bueno, A. S., Dantas, S. M., Henriques, L. M. P. & Peres, C. A. (2018) Ecological traits modulate bird species responses to forest fragmentation in an Amazonian anthropogenic archipelago. *Diversity and Distributions*, **24**, 387–402.

Bueno, A. S., Masseli, G. S., Kaefer, I. L. & Peres, C. A. (2020) Sampling design may obscure species–area relationships in landscape-scale field studies. *Ecography*, **43**, 107–118.

Bull, J. W., Suttle, K. B., Gordon, A., Singh, N. J. & Milner-Gulland, E. J. (2013) Biodiversity offsets in theory and practice. *Oryx*, **47**, 369–380.

Chaplin-Kramer, R., Ramler, I., Sharp, R., Haddad, N. M., Gerber, J. S., West, P. C., Mandle, L., Engstrom, P., Baccini, A., Sim, S. & Mueller, C. (2015) Degradation in carbon stocks near tropical forest edges. *Nature Communications*, **6**, 10158.

Diamond, J. M. (1972) Biogeographic kinetics: Estimation of relaxation times for avifaunas of southwest pacific islands. *Proceedings of the National Academy of Sciences USA*, **69**, 3199–3203.

Emer, C., Venticinque, E. M. & Fonseca, C. R. (2013) Effects of dam-induced landscape fragmentation on Amazonian ant-plant mutualistic networks. *Conservation Biology*, **27**, 763–773.

ESRI (2012) *ArcGIS Desktop*, version 10.1. Redlands, CA: ESRI.

Ewers, R. M. & Didham, R. K. (2006) Confounding factors in the detection of species responses to habitat fragmentation. *Biological Reviews*, **81**, 117–142.

Fearnside, P. M. (2016) Environmental and social impacts of hydroelectric dams in Brazilian Amazonia: Implications for the aluminium industry. *World Development*, **77**, 48–65.

Gardner, T. A., Barlow, J., Araujo, I. S., Ávila-Pires, T. C., Bonaldo, A. B., Costa, J. E., Esposito, M. C., Ferreira, L. V., Hawes, J., Hernandez, M. I. & Hoogmoed, M. S. (2008) The cost-effectiveness of biodiversity surveys in tropical forests. *Ecology Letters*, **11**, 139–150.

Gibson, L., Lynam, A. J., Bradshaw, C. J., He, F., Bickford, D. P., Woodruff, D. S., Bumrungsri, S. & Laurance, W. F. (2013) Near-complete extinction of native small mammal fauna 25 years after forest fragmentation. *Science*, **341**, 1508–1510.

Gibson, L., Wilman, E. N. & Laurance, W. F. (2017) How green is green? *Trends in Ecology & Evolution*, **32**, 922–935.

Gonzalez, A. (2000) Community relaxation in fragmented landscapes: The relation between species richness, area and age. *Ecology Letters*, **3**, 441–448.

ICOLD (2018) *International Commission on Large Dams*. www.icold-cigb.org/GB/ World_register/general_synthesis.asp.

Jones, I. L. & Bull, J. W. (2020) Major dams and the challenge of achieving "No Net Loss" of biodiversity in the tropics. *Sustainable Development*, **28**, 435–443.

Jones, I. L., Bunnefeld, N., Jump, A. S., Peres, C. A. & Dent, D. H. (2016) Extinction debt on reservoir land-bridge islands. *Biological Conservation*, **199**, 75–83.

Jones, I. L., Peres, C. A., Benchimol, M., Bunnefeld, L. & Dent, D. H. (2017) Woody lianas increase in dominance and maintain compositional integrity across an Amazonian dam-induced fragmented landscape. *PLoS ONE*, **12**, 1–19.

Jones, I. L., Peres, C. A., Benchimol, M., Bunnefeld, L. & Dent, D. H. (2019) Instability of insular tree communities in an Amazonian mega-dam is driven by impaired recruitment and altered species composition. *Journal of Applied Ecology*, **56**, 779–791.

Latrubesse, E. M., Arima, E. Y., Dunne, T., Park, E., Baker, V. R., d'Horta, F. M., Wight, C., Wittmann, F., Zuanon, J., Baker, P. A. & Ribas, C. C. (2017) Damming the rivers of the Amazon basin. *Nature*, **546**, 363–369.

Laurance, W. F., Camargo, J. L., Luizão, R. C., Laurance, S. G., Pimm, S. L., Bruna, E. M., Stouffer, P. C., Williamson, G. B., Benítez-Malvido, J., Vasconcelos, H. L. & Van Houtan, K. S. (2011) The fate of Amazonian forest fragments: A 32-year investigation. *Biological Conservation*, **144**, 56–67.

Lees, A. C., Peres, C. A., Fearnside, P. M., Schneider, M. & Zuanon, J. A. (2016) Hydropower and the future of Amazonian biodiversity. *Biodiversity and Conservation*, **25**, 451–466.

Lomolino, M. V. (2000) Ecology's most general, yet protean pattern: The species–area relationship. *Journal of Biogeography*, **27**, 17–26.

MacArthur, R. H. & Wilson, E. O. (1967) *The theory of island biogeography*. Princeton, NJ: Princeton University Press.

Maron, M., Brownlie, S., Bull, J. W., Evans, M. C., von Hase, A., Quétier, F., Watson, J. E. & Gordon, A. (2018) The many meanings of no net loss in environmental policy. *Nature Sustainability*, **1**, 19–27.

Matthews, T. J., Guilhaumon, F., Triantis, K. A., Borregaard, M. K. & Whittaker, R. J. (2016) On the form of species–area relationships in habitat islands and true islands. *Global Ecology & Biogeography*, **25**, 847–858.

Mendenhall, C. D., Karp, D. S., Meyer, C. F., Hadly, E. A. & Daily, G. C. (2014) Predicting biodiversity change and averting collapse in agricultural landscapes. *Nature*, **509**, 213–217.

Moran, E. F., Lopez, M. C., Moore, N., Müller, N. & Hyndman, D. W. (2018) Sustainable hydropower in the 21st century. *Proceedings of the National Academy of Sciences USA*, **115**, 11891–11898.

Palmeirim, A. F., Benchimol, M., Vieira, M. V. & Peres, C. A. (2018) Small mammal responses to Amazonian forest islands are modulated by their forest dependence. *Oecologia*, **187**, 191–204.

Palmeirim, A. F., Vieira, M. V. & Peres, C. A. (2017) Non-random lizard extinctions in land-bridge Amazonian forest islands after 28 years of isolation. *Biological Conservation*, **214**, 55–65.

Pan, Y., Birdsey, R. A., Fang, J., Houghton, R., Kauppi, P. E., Kurz, W. A., Phillips, O. L., Shvidenko, A., Lewis, S. L., Canadell, J. G. & Ciais, P. (2011) A large and persistent carbon sink in the world's forests. *Science*, **333**, 988–993.

Peres, C. A. (2001) Synergistic effects of subsistence hunting and habitat fragmentation on Amazonian forest vertebrates. *Conservation Biology*, **15**, 1490–1505.

R Core Team (2018) *R: A language and environment for statistical computing*. Vienna: R Foundation for Statistical Computing. www.R-project.org

Ritter, C. D., McCrate, G., Nilsson, R. H., Fearnside, P. M., Palme, U. & Antonelli, A. (2017) Environmental impact assessment in Brazilian Amazonia: Challenges and prospects to assess biodiversity. *Biological Conservation*, **206**, 161–168.

Sonter, L. J., Gourevitch, J., Koh, I., Nicholson, C. C., Richardson, L. L., Schwartz, A. J., Singh, N. K., Watson, K. B., Maron, M. & Ricketts, T. H. (2018) Biodiversity offsets may miss opportunities to mitigate impacts on ecosystem services. *Frontiers in Ecology and the Environment*, **16**, 143–148.

Storck-Tonon, D. & Peres, C. A. (2017) Forest patch isolation drives local extinctions of Amazonian orchid bees in a 26 years old archipelago. *Biological Conservation*, **214**, 270–277.

Terborgh, J., Feeley, K., Silman, M., Nuñez, P. & Balukjian, B. (2006) Vegetation dynamics of predator-free land-bridge islands. *Journal of Ecology*, **94**, 253–263.

Terborgh, J., Lopez, L., Nuñez, P., Rao, M., Shahabuddin, G., Orihuela, G., Riveros, M., Ascanio, R., Adler, G. H., Lambert, T. D. & Balbas, L. (2001) Ecological meltdown in predator-free forest fragments. *Science*, **294**, 1923–1926.

Terborgh, J., Lopez, L. & Tello, J. S. (1997) Bird communities in transition: The Lago Guri islands. *Ecology*, **78**, 1494–1501.

Timpe, K. & Kaplan, D. (2017) The changing hydrology of a dammed Amazon. *Science Advances*, **3**, 1–14.

Trussart, S., Messier, D., Roquet, V. & Aki, S. (2002) Hydropower projects: A review of most effective mitigation measures. *Energy Policy*, **30**, 1251–1259.

Watling, J. I. & Donnelly, M. A. (2006) Fragments as islands: A synthesis of faunal responses to habitat patchiness. *Conservation Biology*, **20**, 1016–1025.

Watson, D. M. (2002) A conceptual framework for studying species composition in fragments, islands and other patchy ecosystems. *Journal of Biogeography*, **29**, 823–834.

WCD (2000) *Dams and development: A new framework for decision-making*. London: Earthscan Publications.

Winemiller, K. O., McIntyre, P. B., Castello, L., Fluet-Chouinard, E., Giarrizzo, T., Nam, S., Baird, I. G., Darwall, W., Lujan, N. K., Harrison, I. & Stiassny, M. L. J. (2016) Balancing hydropower and biodiversity in the Amazon, Congo, and Mekong. *Science*, **351**, 128–129.

Wolfe, J. D., Stouffer, P. C., Mokross, K., Powell, L. L. & Anciães, M. M. (2015) Island vs. countryside biogeography: An examination of how Amazonian birds respond to forest clearing and fragmentation. *Ecosphere*, **6**, 1–14.

Zarfl, C., Lumsdon, A. E. & Tockner, K. (2015) A global boom in hydropower dam construction. *Aquatic Sciences*, **77**, 161–170.

18 · *An Investigation of Species–Area Relationships in Marine Systems at Large Spatial Scales*

KARL INNE UGLAND AND ALEXANDRA KRABERG

18.1 Introduction

Species–area relationships (SARs) are one of the most studied ecological phenomena in terrestrial systems. It was Preston (1948, 1962) and MacArthur and Wilson (1967) who revolutionized biogeography by observing that, as area increases, the population sizes of many species also increase and consequently the probability of extinction declines. Typically, it is also expected that larger areas offer a greater number of habitats (habitat heterogeneity), thereby allowing a greater number of species to co-exist.

The above authors showed that in a wide range of communities the species abundance distribution is log-normal and this characteristic long-tailed distribution implies an Arrhenius power law (Arrhenius, 1921) SAR (but see Chapters 4 and 8). In MacArthur and Wilson's (1967) *The theory of island biogeography*, the biota of any island is governed by a dynamic equilibrium between immigration of new species into the island and extinction of species already present. The convincing theoretical connection between the power law SAR and the underlying equilibrium theory rapidly achieved the status of a paradigm: an adequate fit of the power law SAR was viewed as support for the equilibrium hypothesis. Since the 1960s, the SAR has often been referred to as 'the basic law of biodiversity' (Rosenzweig, 1999). While most SAR theory has been developed in the context of terrestrial ecosystems, SARs are just as relevant to marine systems, but have been much less frequently applied (Drakare et al., 2006).

When considering marine systems, we have first to define our terms of reference (Gray, 2000). Marine systems are often regarded as fundamentally different from terrestrial systems but this view is increasingly being challenged (Webb, 2012). In its broadest form the term 'marine system' includes both open water (and permanently submerged) environments, but also coastal intertidal systems on both rocky and sandy shores. However, these are structured in fundamentally different ways, which should lead to different expectations in terms of the accumulation of species in these systems (and in comparison to terrestrial systems). Coastal habitats, such as rocky shores, are highly fragmented and heterogeneous and characterized by understory and canopy plants (akin to forest systems on land?). Rocky shore habitats are found in all climatic zones (from pole to pole) but with strikingly different assemblages. Such differences are found not only horizontally but also with depth (shore level) and these habitats can exhibit very clear SARs at very small spatial scales (Hawkins & Hartnoll, 1980).

One fundamental difference between marine and terrestrial systems that is sometimes cited is the greater connectedness within marine systems (McCallum et al., 2004). While this might be the case for marine pelagic areas (McManus & Woodson, 2012), for coastal and shallow water habitats this is much less the case and these might in fact be expected to be more similar to terrestrial systems. The considerable dispersal ability of many marine organisms, particularly invertebrates, might potentially be expected to lead to the homogenization of marine communities across large spatial scales and, therefore, the slower accumulation of species with increasing area (and thus a flatter species–area curve; e.g. Leibold et al., 2004). However, this does not always seem to be the case (Drakare et al., 2006).

Open ocean systems have depths of hundreds to thousands of meters and in a study in the deep North East Atlantic, comprising several habitat types in two depth horizons, Foster et al. (2013) have shown that SARs of benthic communities can vary with depth. An overriding factor here was a known depth gradient (akin to a latitudinal gradient?) in biodiversity, with maximum diversity often observed at approximately 2,000 m (Howell et al., 2002; McClain & Barry, 2010). The depth structure of marine areas, with shallow sunlit shelf seas transitioning into the deep sea, is probably one of the clearest differences between the marine and terrestrial realms. In the ocean, the deep sea still offers a range of environments, inhabited by a diverse array of organisms to a depth of several kilometres.

Given the above, it is of considerable importance to properly define the habitats/realms as well as the spatial scale for which species–area curves (i.e. curves of the relationship of the cumulative number of species with increasing area of a habitat) are to be investigated. Otherwise, any results are difficult to interpret. Sandy, muddy and rocky shore habitats for instance are all 'benthos', but lumping them into one analysis of 'benthic species–area relationships' would clearly lead to misinterpretations. To take another example, Drakare et al. (2006) showed clear effects of latitude on the shape of the species–area curve, with a decrease in z (the slope of the power model in log–log space) with increasing latitude (i.e. towards the poles). Lomolino (2001) argued that at these very large (latitudinal) scales, as opposed to local and regional scales, speciation becomes an increasingly important factor in the accumulation of species with area (see also Section 18.3).

In this chapter, we provide some examples of the relationship between species number and area in selected marine environments in which SARs have been examined over very large spatial (latitudinal) scales and discuss why, at these large scales (and along latitudinal gradients), SARs in the marine realm might not follow terrestrial expectations. In the marine realm, SARs are increasingly used for conservation purposes, for example to design and determine the optimum size of new Marine Protected Areas (MPAs) or other protected sites, and this requires investigations of SARs at very large (biogeographic) scales. As such, we will also provide some examples of applied uses of the SAR in marine environments.

18.2 Example 1: Prosobranchs on the Continental Shelf Off North America

One marine habitat type in which area effects might be expected are continental shelf areas. These are continuous areas, albeit of very different spatial extent in different regions. There is a clear negative gradient in continental shelf area covered from the tropics towards the Arctic; an exception being the North American Shelf, which has its largest extent in the Arctic. An extensive study by Roy et al. (1998) analysed the geographic ranges of 3,916 species of marine prosobranch gastropods living on the western Atlantic and eastern Pacific shelves of the Americas (data collected from primary literature and museum collections). While steep gradients in species richness were found from the tropics to the Arctic (Figure 18.1), there was no evidence that the geographic area of the shelf had any appreciable influence on the prosobranch assemblages studied (Roy et al., 1998).

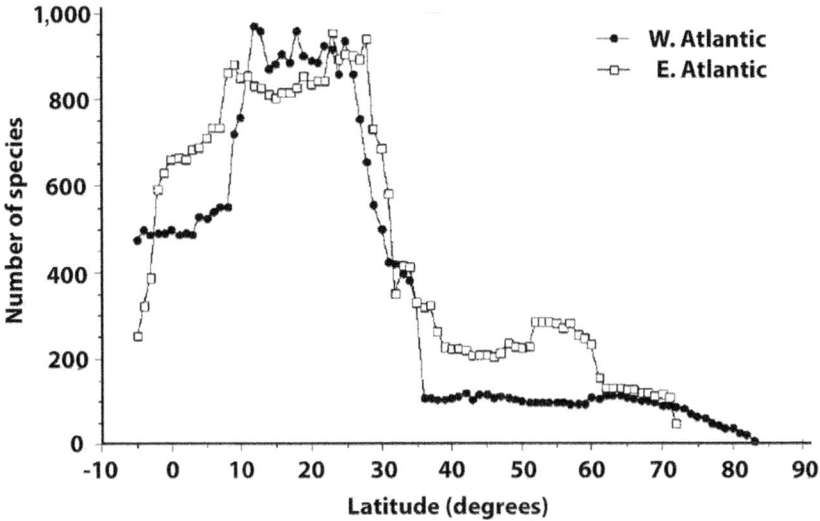

Figure 18.1 Latitudinal diversity gradients of western Atlantic and eastern Pacific marine prosobranch gastropods on continental shelves, binned per degree of latitude. After Roy et al. (1998). Copyright (1998) National Academy of Sciences, USA.

For example, for the continental shelves along the Americas, species richness was lowest and shelf area highest in high latitudes. In the eastern Pacific, the shelf area increases from 169,145 km^2 between 5° S and 20° N to 855,896 km^2 between 55° N and 70° N, but the species number of prosobranchs is about three-times larger in the small tropical area. Subdividing the huge coastline from the southern margin of the Caribbean to the Arctic Ocean into natural provinces revealed no significant relationship between species number and area on either coast. This finding again indicates the difficulty of extrapolating SARs over very large biogeographic scales, especially with very heterogeneous datasets (Figure 18.2).

18.3 Example 2: Worldwide Pattern of Fish Diversity in Estuaries

Vasconcelos et al. (2015) provided the first analysis of the worldwide pattern of fish biodiversity in estuaries. They compiled a comprehensive global database of community composition and environmental characteristics of estuaries (786 samples from 430 estuaries) and used this for

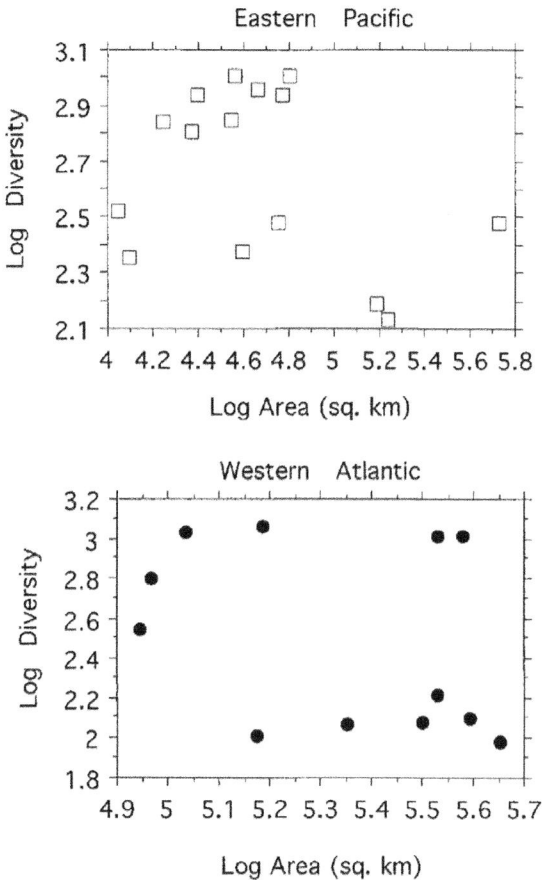

Figure 18.2 Relationships between shelf area and diversity for eastern Pacific (Top) and western Atlantic marine prosobranchs (Bottom). Diversity data were based on latitudinal range records assembled from primary literature and museum collections and assigned to 5-degree latitudinal bins (the number of records/samples contributing to the diversity estimates have not been provided). After Roy et al. (1998). Copyright (1998) National Academy of Sciences, USA.

analyses of species richness patterns at different spatial scales: the global scale, biogeographic realm/continent and individual estuaries.

Vasconcelos et al. (2015) highlighted the importance of considering spatial extent and sampling effects and also of combining geological history and contemporary environmental characteristics when exploring biodiversity. With species richness as the response variable, generalized linear models (GLMs) were used to quantify how variation in species richness among estuaries is related to historical events, energy dynamics and ecosystem characteristics. To evaluate the effects of spatial extent, GLMs were fitted hierarchically: (1) models were fitted to the whole database and (2) the database was partitioned by biogeographic realm or continent and models were fitted to each of the realms or continents. Realms and continents were defined by Vasconcelos et al. as follows:

Marine biogeographic realm: The largest spatial unit of the Marine Ecoregions of the World system (according to Spalding et al., 2007): Arctic, Temperate Southern Africa, Temperate Northern Atlantic, Temperate Northern Pacific, Temperate South America, Temperate Australasia, Tropical Atlantic, Tropical Eastern Pacific, Western Indo-Pacific, Central Indo-Pacific.
Continent: Africa, Europe, North America, South America, Asia, Oceania.

Fish species richness varied notably among estuaries (mean = 29, maximum = 214). A GLM fitted at the global extent had high explanatory power (63 per cent). In this global model, the variable realm explained 13 per cent of the variance, while area only made a very small contribution (<1 per cent). The low explanatory power of area, in terms of both estuary area and continental shelf width, is clearly seen in Figure 18.3A and B, respectively. The marine biogeographic realm was thus one of the most important predictors of species richness globally. Realms are large regions containing biota with a shared history (Spalding et al., 2007). At the global extent, species richness differed among marine biogeographic realms and increased with mean sea surface temperature. At smaller extents (within a marine biogeographic realm or continent), other characteristics were also important for explaining species richness. Only within a biogeographic realm or continent did the co-variability between species richness and estuary area become evident.

Vasconcelos et al. (2015) summarized their statistical analysis by stating that species richness in an estuary seems to be regulated by factors that are spatially hierarchical. First, the global species richness pattern is affected by history (marine biogeographic realm, continent) and energy dynamics (temperature and primary production). Second, the regional species richness patterns are affected by system connectivity (open/closed estuary), species–area relationships (estuary area) and habitat suitability (river flow, salinity gradient). Briefly, the global species richness is governed by the species pool that can potentially colonize an area and the regional species richness is governed by ecological factors that influence the

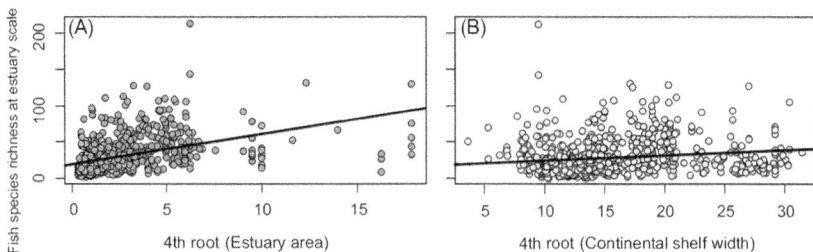

Figure 18.3 Results of an analysis using a generalized linear model (GLM) fitted to a global dataset of estuarine fish diversity. (A) Relationship between observed species richness of fish and estuary area (measured as the fourth root of km^2). (B) Relationship between species richness and continental shelf width. Each data point represents total species richness per study in an estuary. As in some cases more than one study per estuary was included in the database, this resulted in a total of 786 samples from 430 estuaries. The relationships were all statistically significant. After Vasconcelos et al. (2015). Re-used under license from John Wiley and Sons.

colonization of the estuary. In summary, area alone cannot explain much variation in estuarine fish diversity at these scales; thus, the utility of the SAR when applied to such systems may be limited.

18.4 Example 3: The Paradox of Polar Continental Shelf Diversity

For two centuries the continental shelves of the polar regions were regarded as areas of low marine diversity because of the perceived harshness of the environment. When the accumulated sampled species number started to increase considerably in the 1970s, this view was revised (see description for Antarctic benthos in Dell, 1972). By 1990 the observed number of macrobenthic species had risen to approximately 1,300 in the Arctic and 3,000 in the Antarctic. The numbers represent the diversity within eight major invertebrate groups (Grebmeier & Barry, 1991). Clarke (2008) also stated that the diversity of the Antarctic continental shelf exceeds that of the Arctic, and is comparable with temperate and even some non-reef tropical shelves. So, rather than explaining latitudinal gradients, the more interesting problem became to answer the question of why the diversity on the continental shelf in the Arctic Sea was so much less than in Antarctica? In the 1990s it became clear that an explanation would require a combination of geological history, oceanographic processes, glaciology, biogeography and community ecology. Briefly, the dominant opinion was that the greater richness of the Southern Ocean is probably due to 1) a greater geological age, 2) a much larger area and 3) a higher structural heterogeneity formed by living organisms.

However, more than twenty years ago, Kendall (1996) presented data showing that, contrary to the accepted opinion, the macrobenthic communities in the Arctic were not necessarily always impoverished, because muddy sand communities around Spitsbergen were no less diverse than similar assemblages in the North Sea. This is only one example of course and cannot be directly compared to biodiversity assessments across large areas in the Antarctic, but it again shows that faunas in both the Arctic and Antarctic are still under-investigated. In a review on Arctic benthos, Piepenburg (2005) argued that the common idea of a consistently poor Arctic benthos, in stark contrast to the rich Antarctic bottom fauna, is an unwarranted over-generalization. For example, increased research efforts in the Arctic, such as the ten-year Russian–German Laptev Sea study (Sirenko & Piepenburg, 1994), have brought into question the idea that the Arctic benthos is species poor.

Taxonomic and sampling effort are probably considerable confounding factors in the comparative assessment of Arctic and Antarctic diversity, as a significant number of species likely still remain to be discovered. Consequently, assessments of SARs are difficult. In the Antarctic, for instance, the number of described bivalve species increased from 0 to 160 between the early 1800s and the year 2000, while the number of described gastropod species increased from 0 to 500, with the biggest increases in species numbers occurring in the early twentieth century and between 1990 and 2000 (Clarke & Johnston, 2003; Clarke, 2008). This indicates, again, that polar species inventories are still far from comprehensive, with the most recent estimates for the total number of species now standing at approximately 2,900 for the Arctic (Sirenko, 2001; see also Bluhm et al., 2011, and Sirenko et al., 2020 for continually updated records), and over 4,100 for the Antarctic (Arntz et al., 1997; Clarke & Johnston, 2003).

Since the area of the continental shelf of the Arctic has been calculated to be about half that of the Antarctic (Dayton, 1990), it would be expected that species richness should be greater in the Antarctic than in the Arctic. But the application of the SAR in this context is rather difficult. This is partly due to the very complex and very heterogeneous muddy, sandy and rocky habitats present in both polar regions, not only along shorelines but also with increasing depth (e.g. intertidal versus subtidal habitats). This means that, for any given habitat mosaic (and species group), the available area of that habitat has to be carefully considered to facilitate comparable analyses of species richness. This can lead to rather surprising results. For instance, for shallow habitats, and based on data in Gutt (2001), we calculated the continental shelf areas to be 3.0 million km^2 in the Arctic and 2.2 million km^2 in the Antarctic. Thus, in contrast to the widely used measurements, the largest continental shelf is found in the Arctic and not the Antarctic! With reference to the SAR based on such enormous areas, at first sight it appears to be a paradox that the Antarctic shelf contains 70 per cent more species despite being only approximately 73 per cent of the size of the Arctic continental shelf. Obviously, as also indicated for the other examples in this chapter, several factors other than area have a dominant impact on polar diversity, possibly restricting the utility of species–area curves to accurately model richness in these systems.

We will now take a closer look at some special features of the two polar areas and how these features may lower the explanatory power of SARs (Dell, 1972; Dunton, 1992; Arntz & Gallardo, 1994, Dayton et al., 1994; Clarke, 2008; Griffiths, 2010; Gutt et al., 2013).

18.4.1 The Arctic

The Arctic has broad, shallow continental shelves (on average 100 metres deep) with strong seasonally fluctuating physical conditions, including a massive fresh water impact from large rivers in the coastal zones.

During the Quaternary Period, which began about 2,588 million years ago, the Earth's climate has alternated between numerous lengthy glacial periods and shorter warmer intervals of approximately equal duration of 10,000–20,000 years. The last major glaciation reached its maximum extent approximately 18,000–20,000 years ago. Significant deglaciation did not begin until 14,000 years ago and ended 6,000 years ago (Lowe & Walker, 1997; Denton et al., 2010). The glacial periods included sea level fluctuations of about 100 m. During the last glacial maximum for instance, sea level has been estimated to have been approximately 134 m below present sea levels (Lambeck et al., 2014). This lower sea level will have exposed vast areas of the large and shallow Arctic continental shelf, resulting in the near eradication of the shelf benthic assemblages. Re-invasion of the continental shelf was accomplished by species in the North Atlantic and North Pacific.

Thus, the marine benthic communities in the Arctic are composed of a young fauna comprising species of Atlantic and Pacific affinity and a small number of endemics. The variable predominance of species with Atlantic or Pacific affinities, and the low number of endemics indicate that the geographic distribution of the Arctic biota is in a highly dynamic state, far from equilibrium.

18.4.2 The Antarctic

The Southern Ocean is defined as the water masses extending south of the Antarctic Convergence to the coasts of the Antarctic continent. A characteristic feature of Antarctica is the narrow continental shelf, with a large average depth of 450 m.

When the Antarctic continent migrated poleward and the circum-Antarctic current was established, the continent started to become isolated. As the convergence, divergence and upwelling systems developed, the marine isolation became more prevalent and added a great deal of oceanographic complexity to the Southern Ocean system. For the past 20 million years, the physical patterns have been relatively stable, so most of the Antarctic biota is very old. Since this isolation has lasted for millions of years, many taxonomic groups are characterized by high degrees (up to 50 per cent) of endemism (Marques & Pena Cantero, 2010).

18.4.3 Plausible Explanations for the Difference in Diversity between the Polar Regions

Since the benthic assemblages on the Arctic continental shelf are only around 6,000–9,000 years old, recruitment is still an important ongoing process (Sirenko & Piepenburg, 1994). Compared to the Arctic, the smaller Antarctic continental shelf has experienced a relatively more stable environment since its isolation 20 million years ago (Ma) and therefore the Antarctic has accumulated species over a much longer time. This extreme difference in age is probably the main reason for the larger species number sampled in Antarctica (reviewed by Dayton et al., 1994; Arntz et al., 1997; Clarke & Crame, 1997). The importance of age has been recognized in terrestrial systems; for example, Clarke and Crame (1997) suggested that many tropical molluscan clades are so species rich because the tropics are biologically old compared to temperate regions. More generally, the likely strong importance of age in underpinning polar diversity indicates that the SAR approach will be limited when comparing assemblages from both the Arctic and Antarctic.

18.5 Example 4: Diatoms

Diatoms (Bacillariophyta) constitute one of the most diverse and eco-logically important groups of phytoplankton. It is estimated that the global number of diatom species is in the range of 30,000–100,000 species (Mann & Vanormelingen, 2013). Diatoms are widely distributed in almost all aquatic habitats, except the warmest and most hypersaline environments. They occur as attached communities, for example on sea grasses and seaweeds, but also form a significant component of the pelagic food web. Having high reproduction rates (up to a doubling a day in culture experiments) and being a component of the plankton, they can also be dispersed passively over great distances by ocean currents. Diatoms also have a 'mix-and-match genome' that provides them with a range of potentially useful attributes (Armbrust, 2009), such as a rigid silicified cell wall, the presence of vacuoles for nutrient storage, fast responses to changes in ambient light, resting stage formation, ice-binding proteins and a urea cycle. These physiological properties, in addition to the possibility of extremely long-distance dispersal in the oceans, are the main factors explaining why diatoms are so widely distributed. However, as we shall see this does not mean that they cannot be spatially structured.

Hubert et al. (2009) reported an interesting example of microorgan-isms being transported passively over great distances. In the cold Arctic seabed they found thermophilic bacteria that could not be metabolically active. A stable supply of these bacteria into the permanently cold Arctic marine sediment was estimated to occur at a rate exceeding 10^8 spores per square metre of the sediments per year. Thus, dispersal is very powerful and can have a global impact. Cermeno and Falkowski (2009) extracted fossil assemblages of marine diatoms from the Neptune database, a global record of microfossil occurrences reported by the Deep Sea Drilling Project and Ocean Drilling Program. They obtained a dataset consisting of 225 assemblages containing 307 morphologically defined species (seventy genera) from the Pleistocene [~1.5 Ma] to the present. Nonmetric multidimensional scaling revealed remarkable simi-larities between north temperate regions of the Atlantic and Pacific Oceans (sharing more than 75 per cent of their total species pool). Further, there was a clear biogeographic differentiation between these two regions in the Northern Hemisphere and the high-latitude environ-ments in the Southern Ocean. This proximity in the ordination space of fossil assemblages in the Atlantic and Pacific Oceans, and their separation from the Southern Ocean communities, indicates that environmental conditions primarily control the global biogeography of marine diatom assemblages.

Unfortunately, assessing the biodiversity of diatoms at a global scale is a difficult task due to different types of biases. First, sampling effort is simply much larger in some regions of the world (e.g. the Mediterranean, North Sea and North Atlantic) than in other regions. Secondly, for almost the entire twentieth century, biodiversity assessments of plankton were carried out by manual phytoplankton counts of bottle or net samples, serial dilutions or culture studies. However, the advent of molecular techniques has revolutionized the level of detail with which such assessments can be made, providing new insights but also generating bias that makes it difficult to compare data collected with different methods (Ward et al., 1990; Stern et al., 2018).

Overall, it has been very difficult to assess diatom SARs, particularly at larger scales, as observations were simply not detailed enough. In fact, individual diatom species have often been considered to be cosmopolitan in their distribution, with only a small number of taxa restricted to the Southern Ocean (e.g. *Fragilariopsis kerguelensis*; Kloster et al., 2018). However, with molecular methods combined with detailed microscopy studies, this picture is now changing. For example, at the genus level *Skeletonema* is truly cosmopolitan. Previously *Skeletonema costatum* was considered to be mono-specific, whereas it is now known that *Skeletonema* comprises thirty-four species (WORMS, 2019). Many of these species seem to have a relatively restricted distribution (Kooistra et al., 2008). This is only one example of taxa that have now been split into several species that are not only geographically, but in some cases even seasonally, distinct species. The seemingly high prevalence of cryptic species in the phytoplankton clearly hampers comparisons of biodiversity and the identification of large-scale patterns.

Malviya et al. (2016) performed taxonomic profiling of 293 samples derived from forty-six globally distributed sampling sites along the Tara Oceans circumnavigation. A total of 63,371 diatom-assigned ribotypes (based on metabarcoding using the V9 hypervariable region of 18S rDNA) were selected for global diatom distribution and diversity analyses. Species richness varied between the oceanic provinces: the Southern Ocean (SO) stations consistently showed the highest values of species richness, with the South Pacific Ocean (SPO) somewhat lower. Four provinces (North Atlantic Ocean [NAO], Mediterranean Sea [MS], Red Sea (RS) and South Atlantic Ocean [SAO]) had intermediate species richness, while the Indian Ocean (IO) had slightly lower richness than these four.

Malviya et al. (2016) further observed that among the twenty most abundant genera, some had lower diversity in the tropics (e.g. *Fragilariopsis*, *Proboscia* and *Eucampia*), some had lower diversity at high latitudes (e.g. *Attheya* and *Guinardia*) and others displayed a more uniform diversity (e.g. *Thalassosira*). The most diverse, as well as the most abundant, genus was *Chaetoceros*, which had its highest diversity in the Southern Ocean. The diversity of each genus (expressed as the number of ribotypes) was also found to be strikingly variable across ocean provinces. Two major groups of genera were identified: one occurring most frequently in stations from the South Atlantic Ocean, South Pacific Ocean and Southern Ocean and another with maximum occurrence in the Mediterranean Sea (one cluster). However, a substantial degree of sharing was seen among stations from the Southern Ocean, Pacific Ocean and Mediterranean Sea, with no obvious large-scale biodiversity patterns for the group as a whole.

Finally, Malviya et al. (2016) remarked that their data are helpful for addressing the Baas-Becking hypothesis: 'everything is everywhere, but the environment selects'. First, they observed that the twenty most abundant diatom genera were seen in all oceanic provinces, although their abundance patterns were highly variable. However, the diversity of ribotypes from the same genera varied greatly across stations. Many taxa (ribotypes) were only found at a few stations and only a handful of diatom sequences were found everywhere. This again indicates that which patterns are identified across large spatial scales depends to some extent on the taxonomic resolution used.

18.6 Discussion

18.6.1 The Importance of Spatial Scale

Establishing a detailed understanding of species distribution patterns and the mechanisms that shape them is of paramount importance, not only from a direct ecological perspective but also in a management context. Species–area relationships (here we are referring to species accumulation curves; see Chapter 1) have become an important tool in this regard. Species richness estimation facilitates an estimate of species in unsampled areas and is also used in support of decision-making processes with respect to the set-up of protected areas (Neigel, 2003). However, the above four examples show that at increasingly large, macroecological scales it becomes ever more difficult in the marine realm to interpret these relationships in a concrete manner and in terms of a formal assessment of SARs. This is due to the fact that, at these large scales, many different processes, including speciation and large scale and slow-acting geological impacts (e.g. Gutt, 2001), interact to determine contemporary richness. It may be almost impossible to disentangle the roles of these different processes. Moreover, statistical evaluations of such patterns become difficult if only three or four biogeographic regions are compared (see Ellison et al., 1999).

To make valid comparisons between studies and to extrapolate to larger areas it is also important to define the baseline and scope of the investigation and, again, examples 1–4 highlight the difficulties associated with this process when investigating large spatial scales. Too broad a focus, for instance, on 'all benthic species', is likely to generate uninformative curves (in marine as well as terrestrial areas), as many potentially differently structured habitats with different environmental forcings would be lumped together; in all likelihood masking potential drivers of the observed relationships (Gray et al., 2004). In the sea, in particular, it is vital that depth ranges are provided when assessing SARs.

Too broad/large a scale of observation is also problematic from a pure data science point of view. Studies such as the large-scale investigations of latitudinal patterns in gastropods by Roy et al. (1998) had to rely not on discrete samples but on descriptions from the literature and this will necessarily lead to a loss of information compared to the original sampling methodologies. Hence, the validity of interpretations of any SARs constructed using these types of data will be limited.

The worldwide pattern of fish biodiversity in estuaries supports Rosenzweig's (1995) viewpoint of different processes governing SARs at different spatial scales, because a statistical analysis of 786 samples of fish species in 430 estuaries indicated that richness is regulated by multiple factors that are spatially hierarchical (Vasconcelos et al., 2015). Biologically, estuarine species richness is governed first by the number of species that can potentially colonize the area from the wider region and then second by ecological factors that influence the colonization of the estuary. In the global model, estuary area explained a very low amount of variance in the data (<1 per cent). The relatively low explanatory power of area is thus clearly evident for fish in estuaries, molluscs on the American continental shelf and macrobenthos in the polar areas. In all these communities, the diversity is governed by the number of species that can potentially colonize the community and some important factors that influence the colonization process so strongly that they override the area effect, such as geological age (history) and temperature (energy input).

18.6.2 Accurately Measuring Area and Richness in Marine Systems

Generally, marine ecosystems have a much larger geographic extent than terrestrial systems and in particular can be extremely deep, making it important to define area precisely. This was evident in the example of polar macrobenthos communities where an increase in sampling effort changed the notion of an impoverished benthic Arctic and whereby assessment of area based on depth structure led to a re-assessment of the area available for the communities.

Obviously, the identification of a SAR requires an adequate measure of species number (and of area) as well as a concise definition of what is being examined. Since no inventory is so exhaustive that all species are recorded, species richness estimators play a key role in the measurement of biodiversity. In general, estimation of total species richness of large heterogeneous benthic communities faces two difficult problems (Ugland et al., 2003). First, even with exceptionally large international sampling collaborations, it is unlikely that the samples will cover more than a tiny fraction of about 10^{-9} of the total area (as, for example, 4,000 grab samples of $0.25\ m^2$, which would require a formidable identification effort, on a continental shelf of 1 million km^2). In brief, even a large sampling effort will cover only a minute section of an enormous sea area. Second, the multitude of factors affecting species distributions at very large scales is likely to increasingly affect species accumulation curves as samples cover more and more area.

A heavily biased richness estimate not only leads to erroneous conclusions, but also to the wrong questions being asked. Similar sampling problems occur when examining plankton communities. Traditionally, phytoplankton were collected by net and analysed microscopically. Since these methods were time consuming, sampling effort was low. Many taxa were unidentified, while others were often only identified to higher taxonomic levels such as family (with no knowledge of the actual species numbers). Assessments of larger geographic areas were therefore based on few samples, low taxonomic resolution and varying taxonomic expertise. This can easily give the impression that 'everything is everywhere'. However, this might be an artifact of observational biases. Modern molecular methods, as applied as part of large-scale observational efforts such as the Tara Oceans Circumnavigation Project and the FRAM Project, as well as numerous smaller studies, are now starting to revise our understanding of species distributions in the plankton. Ultimately, these efforts will facilitate a better assessment of SARs at different spatial scales in the marine realm (Malviya et al., 2016).

18.7 Conclusions

Investigations of SARs have long been used for describing biodiversity patterns in terrestrial and marine systems. They have also developed into a management tool used, for example, in the design of conservation areas. However, here we have described some examples from the marine realm highlighting some of the problems and limitations of constructing and interpreting SARs at very large spatial scales, especially where the datasets underlying the analyses are based on heterogeneous data sources and/or literature data, which limits the scope of any investigation and risks the introduction of biases. This does not mean of course that the questions being asked are not important. However, to analyse SARs at very large scales in the marine realm, access to methodologically and taxonomically comparable sample data is required in order to facilitate coherent analyses. Molecular tools will probably provide the best scope for more detailed and uniform approaches to assessing sample biodiversity in the future, particularly in the microbial realm, but this is not guaranteed. It will require a great deal of standardization in methods and procedures and more detailed reporting of these procedures than is commonly the case today.

References

Armbrust, E. V. (2009) The life of diatoms in the world's oceans. *Nature*, **459**, 186–192.

Arntz, W. E. & Gallardo, V. A. (1994) Antarctic benthos: Present position and future prospects. *Antarctic science* (ed. by G. Hempel), pp. 243–277. Berlin: Springer Verlag.

Arntz, W. E., Gutt, J. & Klages, M. (1997) Antarctic marine biodiversity: An overview. *Antarctic communities: Species structure and survival* (ed. by B. Battaglia, J. Valencia and D. W. H. Walton), pp. 3–14. Cambridge: Cambridge University Press.

Arrhenius, O. (1921) Species and area. *Journal of Ecology*, **9**, 95–99.

Bluhm, B. A., Gebruk, A. V., Gradinger, R., Hopcroft, R. R., Huettmann, F., Kosobokova, K. N., Sirenko, B. & Weslawski, J. M. (2011) Arctic marine biodiversity: An update of species richness and examples of biodiversity change. *Oceanography*, **24**, 232–248.

Cermeno, P. & Falkowski, P. (2009) Controls on diatom biogeography in the ocean. *Science*, **325**, 1539–1541.

Clarke, A. (2008) Antarctic marine benthic diversity: Patterns and processes. *Journal of Experimental Marine Biology and Ecology*, **366**, 48–55.

Clarke, A. & Crame, J. A. (1997) Diversity, latitude and time: Patterns in shallow seas. *Marine biodiversity: Patterns and processes* (ed. by R. F. G. Ormond, J. D. Gage and M. V. Angel), pp. 122–147. Cambridge: Cambridge University Press.

Clarke, A. & Johnston, N. M. (2003) Antarctic marine benthic diversity. *Oceanography and Marine Biology: An Annual Review*, **41**, 47–114.

Dayton, P. K. (1990) Polar benthos. *Polar oceanography Part B: Chemistry, biology and geology* (ed. by W. O. Smith, Jr.), pp. 631–686. San Diego, CA: Academic Press.

Dayton, P. K., Mordida, B. J. & Bacon, F. (1994) Polar marine communities. *American Zoologist*, **34**, 90–99.

Dell, R. K. (1972) Antarctic benthos. *Advances in Marine Biology*, **10**, 1–216.

Denton, G. H., Anderson, R. F., Toggweiler, J. R., Edwards, R. L., Schaefer, J. M. & Putnam, A. E. (2010) The last glacial termination. *Science*, **328**, 1652–1656.

Drakare, S., Lennon, J. J. & Hillebrand, H. (2006) The imprint of the geographical, evolutionary and ecological context on species–area relationships. *Ecology Letters*, **9**, 215–227.

Dunton, K. (1992) Arctic biogeography: The paradox of the marine benthic fauna and flora. *Trends in Ecology & Evolution*, **7**, 183–189.

Ellison, A. M., Farnsworth, E. J. & Merkt, R. E. (1999) Origins of mangrove ecosystems and the mangrove biodiversity anomaly. *Global Ecology & Biogeography*, **8**, 95–115.

Foster, N. L., Foggo, A. & Howell, K. L. (2013) Using species–area relationships to inform baseline conservation targets for the deep North East Atlantic. *PLoS ONE*, **8**, e58941.

Gray, J. S. (2000) The measurement of marine species diversity, with an application to the benthic fauna of the Norwegian continental shelf. *Journal of Experimental Marine Biology and Ecology*, **250**, 23–49.

Gray, J. S., Ugland, K. I. & Lambshead, J. (2004) Species accumulation and species–area curves – a comment on Scheiner (2003). *Global Ecology & Biogeography*, **13**, 473–476.

Grebmeier, J. M. & Barry, J. P. (1991) The influence of oceanographic processes on pelagic benthic coupling in polar regions: A benthic perspective. *Journal of Marine Systems*, **2**, 498–518.

Griffiths, H. J. (2010) Antarctic marine biodiversity – what do we know about the distribution of life in the Southern Ocean? *PLoS One*, **5**, e11683.

Gutt, J. (2001) On the direct impact of ice on marine benthic communities, a review. *Polar Biology*, **24**, 553–564.

Gutt, J., Griffiths, H. J. & Jones, C. D. (2013) Circumpolar overview and spatial heterogeneity of Antarctic macrobenthic communities. *Marine Biodiversity*, **43**, 481–487.

Hawkins, S. J. & Hartnoll, R. G. (1980) A study of the small-scale relationship between species number and area on a rocky shore. *Estuarine and Coastal Marine Science*, **10**, 201–214.

Howell, K. I., Billet, D. S. M. & Tyler, P. A. (2002) Depth-related distribution and abundance of seastars (Echinodermata: Asteroidea) in the Porcupine Seabight and Porcupine Abyssal Plain, N.E. Atlantic. *Deep-Sea Research Part I: Oceanographic Research Papers*, **49**, 1901–1920.

Hubert, C., Loy, A., Nickel, M., Arnosti, C., Baranyi, C., Brüchert, V., Ferdelman, T., Finster, K., Christensen, F. M., de Rezende, J. R., Vandieken, V. & Jörgensen, B. B. (2009) A constant flux of diverse thermophilic bacteria into the cold Arctic seabed. *Science*, **325**, 1541–1544.

Kendall, M. A. (1996) Are Arctic soft-sediment macrobenthic communities impoverished? *Polar Biology*, **16**, 393–399.

Kloster, M., Kauer, G., Esper, O., Fuchs, N. & Beszteri, B. (2018) Morphometry of the diatom *Fragilariopsis kerguelensis* from Southern Ocean sediment: High-throughput measurements show second morphotype occurring during glacials. *Marine Micropaleontology*, **143**, 70–79.

Kooistra, H. C. F., Sarno, D., Balzano, S., Gu, H., Anderson, R. A. & Zingone, A. (2008) Global diversity and biogeography of *Skeletonema* species (Bacillariophyta). *Protist*, **159**, 177–193.

Lambeck, K., Rouby, H., Purcell, A., Sun, Y. & Sambridge, M. (2014) Sea level and global ice volumes from the last glacial maximum to the holocene. *Proceedings of the National Academy of Sciences USA*, **11**, 15296–15303.

Leibold, M. A., Holyoak, M., Mouquet, N., Amarasekare, P., Chase, J. M., Hoopes, M. F., Holt, R. D., Shurin, J. B., Law, R., Tilman, D., Loreau, M. & Gonzalez, A. (2004) The metacommunity concept: A framework for multi-scale community ecology. *Ecology Letters*, **7**, 601–613.

Lomolino, M. V. (2001) The species–area relationship: New challenges for an old pattern. *Progress in Physical Geography*, **25**, 1–21.

Lowe, J. J. & Walker, M. J. C. (1997) *Reconstructing quaternary environments*. London: Routledge.

MacArthur, R. H. & Wilson, E. O. (1967) *The theory of island biogeography*. Princeton, NJ: Princeton University Press.

Malviya, S., Scalco, E., Audic, S., Vincent, F., Veluchamy, A., Poulain, J., Wincker, P., Iudicone, D., de Vargas, C., Bittner, L., Zingone, A. & Bowler, D. E. (2016) Insights into global diatom distribution and diversity in the world's ocean. *Proceedings of the National Academy of Sciences USA*, **113**, E1516–E1525.

Mann, D. G. & Vanormelingen, P. (2013) An inordinate fondness? The number, distributions and origins of diatom species. *Eukaryotic Microbiology*, **60**, 414–420.

Marques, A. C. & Pena Cantero, A. L. (2010) Areas of endemism in the Antarctic – a case study of the benthic hydrozoan genus *Oswaldella* (Cnidaria, Kirchenpaueriidae). *Journal of Biogeography*, **37**, 617–623.

McCallum, H. I., Kuris, A., Harvell, C. D., Lafferty, K. D., Smith, W. O. & Porter, J. (2004) Does terrestrial epidemiology apply to marine systems? *Trends in Ecology & Evolution*, **19**, 586–591.

McClain, C. R. & Barry, J. P. (2010) Habitat heterogeneity, biogenic disturbance, and resource availability work in concert to regulate biodiversity in deep submarine canyons. *Ecology*, **91**, 964–976.

McManus, M. A. & Woodson, C. B. (2012) Plankton distribution and ocean dispersal. *Journal of Experimental Biology*, **215**, 1008–1016.

Neigel, J. E. (2003) Species–area relationships and marine conservation. *Ecological Applications*, **13**, S138–S145.

Piepenburg, D. (2005) Recent research on Arctic benthos: Common notions need to be revised. *Polar Biology*, **28**, 733–755.

Preston, F. W. (1948) The commonness, and rarity, of species. *Ecology*, **29**, 254–283.

Preston, F. W. (1962) The canonical distribution of commonness and rarity: Part I. *Ecology*, **43**, 185–215.

Rosenzweig, M. L. (1995) *Species diversity in space and time*. New York: Cambridge University Press.

Rosenzweig, M. L. (1999) Heeding the warning in biodiversity's basic law. *Science*, **284**, 276–277.

Roy, K., Jablonski, D., Valentine, J. W. & Rosenberg, G. (1998) Marine latitudinal diversity gradients: Tests of causal hypotheses. *Proceedings of the National Academy of Sciences USA*, **95**, 3699–3702.

Sirenko, B. I. (2001) List of species of free-living invertebrates of Eurasian Arctic seas and adjacent deep waters. *Exploration of the Fauna of the Seas*, 51, pp. 1–129. St. Petersburg: Russian Academy of Sciences.

Sirenko, B. I. & Piepenburg, D. (1994) Current knowledge on biodiversity and benthic zonation patterns of Eurasian Arctic shelf seas with special reference to the Laptev Sea. *Russian–German cooperation in the Siberian shelf seas: Geo-system Laptev Sea. Berichte zur Polarforschung, 144* (ed. by H. Kassens, H. W. Hubberten, S. M. Prymikov and R. Stein), pp. 69–77. Bremerhaven: Alfred Wegener Institute for Polar and Marine Research.

Sirenko, B. I., Clarke, C., Hopcroft, R. R., Huettmann, F., Bluhm, B. A. & Gradinger, R. (eds.) (2020) *The Arctic Register of Marine Species (ARMS) compiled by the Arctic Ocean Diversity (ArcOD)*. www.marinespecies.org/arms

Spalding, M. D., Fox, H. E., Allen, G. R., Davidson, N., Ferdana, Z. A., Finlayson, M., Halpern, B. S., Jorge, M. A., Lombana, A., Lourie, S. A., Martin, K. D., McManus, E., Molnar, J., Recchia, C. A. & Robertson, J. (2007) Marine ecoregions of the world: A bioregionalization of coastal and shelf areas. *BioScience*, **57**, 573–583.

Stern, R., Kraberg, A. C., Bresnan, E., Kooistra, H. C. F., Lovejoy, C., Montresor, M., Morán, X. A., Not, F., Salas, R., Siano, R., Vaulot, D., Amaral-Zettler, L., Zingone, A. & Metfies, K. (2018) Molecular analyses of protists in long-term observation programmes – current status and future perspectives. *Journal of Plankton Research*, **40**, 519–536.

Ugland, K. I., Gray, J. S. & Ellingsen, K. E. (2003) The species accumulation curve and estimation of species richness. *Journal of Animal Ecology*, **72**, 888–897.

Vasconcelos, R. P., Henriques, S., França, S., Pasquaud, S., Cardoso, I., Laborde, M. & Cabral, H. N. (2015) Global patterns and predictors of fish species richness in estuaries. *Journal of Animal Ecology*, **84**, 1331–1341.

Ward, D. M., Weller, R. & Bateson, M. M. (1990) 16S ribosomal RNA sequences reveal numerous uncultured microorganisms in a natural community. *Nature*, **345**, 63–65.

Webb, T. J. (2012) Marine and terrestrial ecology: Unifying concepts, revealing differences. *Trends in Ecology & Evolution*, **27**, 535–541.

WORMS (2019) *The World Register of Marine Species*. http://marinespecies.org.

Part V

Future Directions in Species–Area Relationship Research

19 · *The Island Species–Area Relationship: Rosenzweig's Dinosaur Is Still Alive*

KOSTAS A. TRIANTIS

I am not aware of any other book that has shed so much light on species diversity patterns and generalities across time and space as has M. L. Rosenzweig's (1995) *Species diversity in space and time*. In the preface of his seminal contribution, Rosenzweig compared the study of species diversity patterns with a dinosaur that has come alive and is challenging us. He then (pp. 378–381) listed 10 major challenges that we as ecologists and biogeographers have to address. Some of these are focused on arguably the most general pattern, and definitely one of the oldest known patterns, in nature, that is, the species–area relationship (SAR). Rosenzweig, 25 years ago, was signifying that, regardless of the vast number of studies on SARs, we were still lacking, at his time of writing, a complete understanding of the pattern. All of the preceding chapters in this book highlight the amount of progress that has been made since Rosenzweig's book, the range of new methods and patterns that have been brought to light and the many new and insightful questions that have been asked; some have been answered, others still challenge us.

In this final chapter, I will discuss issues related to the SAR that still require our attention, some of which have been noted in the preceding chapters. I will focus solely on the island species–area relationship (ISAR) and won't discuss species accumulation curves or nested species–area relationships that, due to their mode of construction, are arguably more conveniently studied and better understood (see Chapter 1 and especially Chapters 8 and 9). ISARs are built using independent units of analysis, both in terms of area and species richness – however, spatial autocorrelation has always to be considered. A larger island can have fewer species than a smaller one; a situation that is not permittable in species accumulation curves. Also, in the extreme case, an ISAR can be built with islands for which, for a specific taxon such as land snails, all species included in the analysis are endemic to each single island and, thus, there is no species overlap between islands. In contrast, as species accumulation curves are based on nested sampling, there is always a degree of overlap between samples of increasing size. It is the ISAR, the archetype[1] of the SAR, which challenges our understanding of the natural world and also provides a lens through which the highly fragmented world of the Anthropocene can be viewed.

19.1 Re-doing the Theory of Species–Area Curves

The original theory of species–area curves depended on a fixed lognormal distribution getting unveiled as area grows. This may be appropriate for islands of different size. But I am guessing it is not suitable for provinces (Rosenzweig, 1995, p. 379). Every species in a sample or on an island is represented by individuals and, thus, density of individuals, relative abundance and spatial aggregation should be critical components of a theory explaining the increase of species richness with area across all spatial scales. However, while considerable theoretical progress has been made for nested SARs (e.g. Martín & Goldenfeld, 2006; Storch et al., 2007; see also Chapters 8–11), with regard to the ISAR, such a theory has not been proposed since Preston (1962), who assumed that species abundance distributions (SAD) are log-normal. We now know from numerous empirical SAD studies that they are not always best represented by log-normal-like distribution shapes, with log-series and even multimodal curve shapes being frequently observed (e.g. Ulrich et al., 2010; Matthews et al., 2019a). Arguably, we also need to gain a better understanding of how the SAD and the ISAR interact more generally. Thus, we still lack a theory that connects the SAD with the ISAR and that culminates in a mathematical model for the ISAR whose parameters have biological significance.

In general, after the inflation of (mostly statistical) SAR functions in recent decades and the increasing number of studies comparing SAR models (see Flather, 1996; Tjørve, 2003, 2009; Drakare et al., 2006;

---

[1] As noted in Triantis et al. (2012), the first known description of the species–area pattern comes from G. Forster (1777, Book I, Chapter VIII, p. 156): 'The small size of the island, together with its vast distance from either the eastern or western continent, did not admit of a great variety of animals'.

Williams et al., 2009, Triantis et al., 2012; Matthews et al., 2019b, c; Chapter 7), we are in need of more functions supported by theory.

19.2 Developing the Species–Time Curve

Just as we need a working theory for species–area curves, we also need one for species–time curves (Rosenzweig, 1995, p. 379). Preston (1960) proposed that there is a temporal analogue of the SAR: the species–time relationship (STR). The STR describes how the number of species recorded at a site increases as that site is observed for longer periods of time. Since Preston's original proposition many significant steps forward have been made, but the pattern has received much less attention than the SAR and only a few papers have been published on the pattern using actual data (see discussions in e.g. Rosenzweig, 1995, 1998; Adler & Lauenroth, 2003; Adler et al., 2005; White et al., 2006; White, 2007; McGlinn & Palmer, 2009; Song et al., 2018).

As Rosenzweig (1998, p. 312) stated, 'If Preston was correct, the SPAR–SPTI[2] combination would form a powerful and unique ecological rule. It would say, in some deep sense, that time and space are interchangeable in their effect on diversity'. Adler and Lauenroth (2003) and Adler et al. (2005) have since suggested that a unified, general species–time–area relationship (STAR) holds, with comparable, non-independent scaling of richness across space and time. Indeed, both SARs and STRs have been found to have similar power model z-values (i.e. between 0.2 and 0.4; White, 2007).

In regards to the form of the STR, Preston suggested that, because the processes underpinning the STR should overlap with those underpinning the SAR, the two patterns should have the same functional form (i.e. a power function; Preston, 1960). Other studies have suggested and tested other functional forms, including the logarithmic function and even triphasic STRs (Fisher et al., 1943; White, 2007), and there is currently no consensus as to the 'best' or most general STR model (Rosenzweig, 1995; Adler & Lauenroth; 2003; White et al., 2006). Studies comparing the fit of multiple models (including a range of different functional forms, e.g. models including an asymptote; see White, 2007) using a range of STR datasets, as has been done with ISAR

---

[2] SPAR: species–area relationship; SPTI: species–time relationship.

data (e.g. Triantis et al., 2012; Matthews et al., 2016), would be an interesting avenue for future research.

Most STRs are generally constructed using a temporal sliding window approach which results in nested samples (White, 2007). In these cases, they are more similar to nested SARs than ISARs. As with nested SARs, there are statistical issues in that the data points are not independent. STRs equivalent to ISARs have been constructed, but rarely; primarily using fossils sampled from different strata within a site (Rosenzweig, 1995; see also Song et al., 2018).

Going forward, there are at least two points to consider in future STR studies. First, STRs might vary according to the temporal scale. Rosenzweig (1998) discussed STRs with respect to the sampling period, distinguishing three categories: i) short term, ii) in ecological time and iii) in evolutionary time. The sampling period has been argued to affect both the general shape of the STR and its slope. Second, it is critical that we identify how the two major categories of STRs (i.e. nested and non-nested STRs; e.g. Song et al., 2018), analogous with SARs (i.e. nested SARs and ISARs), differ in terms of general patterns. We could also take another step forward and build STARs for habitat fragments of varying age of formation, islands created by the construction of dams (Chapter 17) and even, at the evolutionary time scale, volcanic islands of multiple archipelagos (e.g. Whittaker et al., 2008; Cameron et al., 2013). More generally, further theoretical work is needed to derive a unified theory of STRs that links with the ISAR.

19.3 Island Species–Area Architectural Constraints

"The apparent linearity of the relationship between species number and area may be the result of sampling a narrow range of areas" (Connor & McCoy, 1979, p. 796). Although the constraints limiting and canalizing the nested species–area pattern have been discussed in depth (Preston, 1962; Connor & McCoy, 1979; Chapters 8, 9 and 18), I am not aware of an extensive analytical study, using for example simulation (e.g. Borregaard et al., 2016) or mechanistic models (Cabral et al., 2017), of how different factors affect the ISAR in terms of goodness of fit, shape, best fitting model and model parameters values. These different factors could include i) overall species richness, ii) overall number of islands, iii) total area, iv) species richness range, v) island area range, vi) species scale (i.e. the ratio of maximum richness to minimum richness) and vii) area scale (see Connor & McCoy, 1979; Santos et al., 2010; Triantis et al., 2012;

Matthews et al., 2016, 2019b for partial approaches; and Chapters 3, 4 and 7 for reviews). What is for example the range of z-values allowed if the overall number of species of an island group (i.e. gamma diversity) is eight and thus the number of species across islands can vary from zero to eight? Is the slope value further limited if the island areas do not vary much and particularly if the full island area range is very small? Is the linear model always the best fitting model when a narrow range of island areas is considered?

I do believe that if we study in depth these ISAR architectural constraints using actual data and simulations, we will be further informed on the biological importance of the pattern, which in turn will increase the ISAR's usefulness for conservation biology.

19.4 (Island) Biogeography Is Not a Single-Variable Discipline

"Our ultimate theory of species diversity may not mention area, because area seldom exerts a direct effect on a species' presence. More often area allows a large enough sample of habitats, which in turn control species occurrence. However, in the absence of good information on diversity of habitats, we first turn to island areas" (MacArthur & Wilson, 1967, p. 8). We know that, as area increases, species diversity usually does also. Area has proven to be the best-single proxy for the available resource space for species to establish and thus the capacity of a given geographical unit for species richness (e.g. Triantis et al., 2012; Matthews et al., 2016). Triantis et al. (2012) showed that, when applying the log–log version of the Arrhenius model to 449 datasets, the explained variance as measured by the coefficient of determination R^2 ranged from 0.060 to 0.993, with a mean $\pm$ SD value of 0.640 ± 0.204. Thus, on average area solely can explain more than 60 per cent of the variance in species richness in many datasets, despite the vast differences between datasets in other aspects, such as taxon, number of islands and total island area. However, this still means that more than 30 per cent (on average) of the variation in species richness can potentially be explained by other factors. Quantifying these factors (e.g. time, energy, topographic relief, climate, elevation, isolation, past archipelago configuration, environmental heterogeneity, human impact and sampling artefacts) may improve our ability to predict species numbers, both across a mainland and within insular (or isolate) systems (e.g. Preston, 1960; Wright, 1983; Shmida & Wilson, 1985; Williamson,

1988; Triantis et al., 2003; Hortal et al., 2009; Weigelt & Kreft, 2013; Stein et al., 2014; Norder et al., 2019). Towards this aim, a method for disentangling the ecological mechanisms underpinning ISARs was recently published by Chase et al. (2019; see also McGlinn et al., 2019 and Liu et al., 2020).

The definition, quantification and inclusion within models of environmental heterogeneity remains our major challenge. Although the ecological Tower of Babel is increasing in size and height, with immense opportunities for describing our world in detail (e.g. Antonelli et al. 2018), the vocabulary we use for describing the heterogeneity of the environment is not organized, precise or simple. It lacks two guiding principles: utility and standardization (see Udvardy, 1959; Looijen, 1995; Hall et al., 1997; Triantis et al., 2003; Stein et al., 2014; Sfenthourakis & Triantis, 2017). The diversity of habitats, envisioned by MacArthur and Wilson (1967) to be a critical component of our ultimate theory, is the place to start from, so as to enable global comparisons and syntheses of a wide range of study systems.

19.5 The Small-Island Effect

When the total number of species per islet is plotted logarithmically against islet size two linear relationships are evident (Fig. 32). Those islets less than 3.5 acres fall along one line with little variation in number of species. The other relationship includes those islets 3.5 acres and over and shows more strikingly the direct relationship between islet size and number of species (Niering, 1956, p. 4)[3]. The influential paper by Lomolino and Weiser (2001) reminded us of an almost forgotten phenomenon, a phenomenon that both Preston (1962) and MacArthur and Wilson (1967) were aware of: the *small-island effect*. Since 2001, the study of the *small-island effect* (SIE) has been gaining increasing space in the literature (e.g. Triantis et al., 2006; Burns et al., 2009; Dengler, 2010; Morrison, 2011; Tjørve & Tjørve, 2011; Schrader et al., 2019; Wang et al., 2016, 2018). According to Google Scholar,[4] until the year 2000 the term appeared only fifty-two times in publications and was often used in the context of other phenomena. However, from 2001 to 2019 the term was used in 460 publications, with the rate of usage still increasing.

---

[3] Although we usually assign the first graphical representation of the small-island effect to Niering (1963), it was in fact eight years earlier (1956) by the same author.

[4] Search on 11 August 2019.

Currently, the SIE is defined as the independent variation of species richness with island area on small islands (Lomolino & Weiser, 2001) or, more generally, as an 'anomalous' feature of species richness on smaller islands in comparison with larger ones (e.g. Triantis & Sfenthourakis, 2012). The SIE has been attributed to a range of factors that have a larger proportional effect on small compared to large islands, such as limited habitat diversity, higher levels of disturbance and greater stochastic species turnover (Whittaker & Fernández-Palacios, 2007; Triantis & Sfenthourakis, 2012; Wang et al., 2016, 2018; Schrader et al., 2019). Multiple methodological frameworks have been introduced for the detection of the SIE and there is a continuous debate about which is preferable and, more generally, on the nature of the SIE itself (e.g. Dengler, 2010; Tjørve & Tjørve, 2011; Triantis & Sfenthourakis, 2012).

At present the principal approach for the detection of the SIE involves fitting a variety of piecewise and linear regression models, using exclusively island area. However, in my opinion, the use of island area by itself is insufficient and we need to complement our models with measures of environmental heterogeneity. Additionally, in studies of the SIE thus far, we have only considered species' taxonomic identity as the descriptor of island communities. This approach assumes all species are evolutionarily independent and ecologically equivalent; an assumption we know to be false in island communities. As such, future studies of the SIE using phylogenetic and functional diversity-based metrics might provide novel insights into the mechanisms shaping the pattern.

19.6 The Evolutionary ISAR

... *and it is found that, other things being equal, the numbers, variety and importance of the forms of animal and vegetable life, do bear some approximate relation to extent of area* (A. R. Wallace, 1876; Chapter IV, On Zoological Regions, p. 54). Wallace (1876, p. 52), discussing 'the most natural primary divisions of the Earth as regards its forms of animal life', that is, the zoogeographic regions of the world (see Holt et al., 2013), recognized that differences in species richness across these broad and evolutionary distinct regions of the world had to be related to differences in the size of the regions. This is probably one of the first descriptions of the evolutionary (interprovincial) species–area relationship. Later, Williams (1943) and Preston (1960) showed that the slope of the species–area relationship changes with geographical scale (although they were using nested SARs). This was codified in Rosenzweig's (1995) scale-structured model of species–area relationships, which however included both ISARs and nested SARs (see Scheiner, 2003; Whittaker & Fernández-Palacios, 2007), with the interprovincial ISAR exhibiting a slope always higher than 0.6 and generally close to unity (see Rosenzweig, 1995, 1998, 2001).

As speciation becomes the dominant process of species regulation, the slope of the SAR tends to rise: 'the slope of the evolutionary SAR should be higher than that resulting when immigration is the sole source of new species' (Losos & Schluter, 2000, p. 848) (see also Rosenzweig, 1995; Triantis et al., 2008, 2015; Kisel et al., 2011; Whittaker et al., 2017). Recent meta-analyses have shown that ISAR slope increases from habitat, to continental shelf, to oceanic islands (e.g. Triantis et al., 2012; Matthews et al., 2019b). This reasoning is further supported by the fact that inter-archipelago species–area relationships (ASARs) for oceanic archipelagos are systematically steeper than the constituent ISARs (Triantis et al., 2015; Whittaker et al., 2017). Remote oceanic archipelagos typically host a high proportion of species arising from in situ speciation (Whittaker & Fernández-Palacios, 2007) and this qualifies them as biotic provinces (Triantis et al., 2015; Whittaker et al., 2017), that is, independent from an evolutionary perspective (Rosenzweig, 1995). Moreover, if within an archipelago we consider only the single-island endemic species (SIES) for each island and construct an ISAR based on the SIES, slope values resemble those of the interprovincial ISAR (Triantis et al., 2008) (see Figure 19.1). Thus, we can generate a scale-structured model with the three types of evolutionary ISARs (Figure 19.1).

Additional tests of the above generalities are required and particularly comparisons of the three evolutionary ISARs (Figure 19.1). Finally, we still face the challenge of fully understanding why, when the dominant process of species addition changes from immigration to speciation, the slope of the ISAR is higher (Heaney, 2000; Losos & Schluter, 2001; Pigolotti & Cencini, 2009; Rabosky & Glor, 2010; Kisel et al., 2011).

19.7 Ecological Space and Carrying Capacity Instead of Area

"It is first obvious that the processes of evolution of communities must be under various sorts of external control, and that in some cases such control limits the

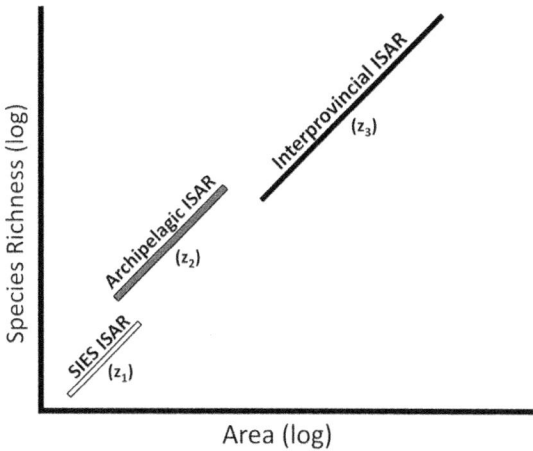

Figure 19.1 Three different types of evolutionary island species–area relationship. SIES ISAR: between islands in the same archipelago, considering only single island endemic species (e.g. Hawaiian snails). Archipelagic ISAR: between different archipelagos, all with high endemism for the focal taxon (e.g. oceanic archipelagos). Interprovincial ISAR: between different biotic provinces. Note that the spatial extent of interprovincial ISARs can be greater, stretching downwards to the size of archipelagos or even single islands depending on the taxon used. Here, I refer to the spatial scale of biogeographic regions. Empirical data suggest that the slopes of the three evolutionary ISARs are steep, close to unity and similar (i.e. $z_1 \approx z_2 \approx z_3$).

possible diversity" (Hutchinson, 1959, p. 150). Brown et al. (2001, p. 328) defined the carrying capacity for species richness as a 'steady-state level of richness specific to a particular site or local ecosystem, that is set by resource availability and other local conditions and is maintained despite changes in species composition'. This definition incorporates two distinct ideas. First, the idea of an equilibrium where species richness remains constant, but with species composition changing as a result of the balanced contribution of processes adding (immigration and speciation) and removing (extinction) species (MacArthur & Wilson, 1967), without an upper limit restricting diversity. Second, the idea of a theoretical maximum species carrying capacity of a region, that is, the maximum species diversity that a region can support. The maximum capacity may also be equal to the biomass and/or the total number of individual species that can conceivably fit within a region (Whittaker et al., 2008; Rabosky, 2013). The conceptual analogy between the maximum species carrying capacity and the species equilibrium is the theoretical (fundamental) and the realized niche. In practice, the species richness at equilibrium is the portion of the maximum capacity realized, given the constraining effects of biological interactions, such as competition, the dispersal ability of the taxon and the time available for colonization and diversification etc. (see also Borregaard et al., 2016; Valente et al., 2017).

The species carrying capacity should depend upon the available ecological space (see Gillespie, 2007). Ecological space thus encompasses the combination of abiotic environmental conditions and biotic conditions (including the historically determined species pool, the prevailing propagule rain, the ecological requirements of the taxa, their dispersal abilities, typical population sizes, average spatial extent of intraspecific gene flow and their propensity to form new species) that constrain actual levels of island diversity (e.g. Rosenzweig, 1995, Whittaker & Fernández-Palacios, 2007; Losos & Ricklefs, 2009; see also the recent debate on 'ecological limits'; Rabosky & Hurlbert, 2015; Harmon & Harrison, 2015) (see Figure 19.2).

A promising approach for describing ecological space is that of 'environmental space or volume' as defined by Weigelt et al. (2013) for 17,883 islands globally, based on ten bioclimatic and physical variables (see also Cabral et al., 2014). However, the biotic component is still missing; considering general descriptors of the ecology of the focal taxon (e.g. body size) might be the way forward. Either way, building species–ecological space relationships instead of traditional ISARs might prove to be insightful in terms of describing and understanding patterns that have gone unnoticed. For example, we might observe similar slope values for different taxa from the same island group, if instead of area we use taxon-standardized ecological space.

As stated in Chapter 1, and in many of the intervening chapters, the ISAR is one of ecology's few laws. Although more than a century of research has been undertaken on the ISAR, revealing many of the pattern's mysteries, the pattern does not yet qualify as a dinosaur. Indeed, the SAR more generally still forms a core part of contemporary theoretical development in ecology and biogeography, such as maximum entropy theory (Chapter 10), neutral theory (Chapter 11) and network analysis (Chapter 15). The chapters in this volume show how far we have come from the early days of the first descriptions of ISAR curve shape (Chapter 2) to more complex model fitting and parameter understanding (Chapters 3, 4 and 7), the use of alternative measures of diversity, such as functional and phylogenetic (Chapter 5), and the expansion of the SAR and SAR concepts to other areas, such as invasion biology (Chapter 6),

Figure 19.2 The theoretical species richness carrying capacity for the same island, centre, for four different taxonomic groups (i.e. mammals, spiders, lizards and land snails), based on the available ecological space, which varies among taxa and is not equal to the island's area. Expected order of carrying capacity size according to taxa ecological requirements, dispersal abilities, typical population sizes, average spatial extent of intraspecific gene flow and propensity to form new species: snails>spiders>lizards>mammals. Pictograms courtesy of PhyloPic (www.phylopic .org): mammal: Zimices, spider: Birgit Lang, lizard: Oliver Griffith, and land snail: Gareth Monger.

the marine realm (Chapter 18) and trophic biogeography (Chapter 12). It is also reassuring to see how far applied SAR research has progressed in recent years (e.g. Chapters 13, 14, 16 and 17). However, the research gaps and future questions discussed in all chapters in this volume also highlight how far we have yet to go.

The SAR remains a fascinating area of research and, although fancy modern tools and substantial datasets are now available to us, it is always fascinating reading and listening to old stories, with or without dinosaurs. For example, see the correspondence between A. R. Wallace (1877) and A. Newton (1877) where, amongst other things, they discuss how the bird species of Madagascar, Comoros, Mauritius, Bourbon and Rodriguez islands vary with island area. Even earlier, Charles Darwin (1859, p. 379) was stating 'The species of all kinds which inhabit oceanic islands are few in number compared with those on equal continental areas'.

Acknowledgements

I thank Rob Whittaker, François Guilhaumon, Mark Lomolino, Robert Cameron, Larry Heaney, Sakis Mylonas, Tom Matthews and Spyros Sfenthourakis for all those discussions on the species–area relationship. Special thanks to Tom Matthews and Rob Whittaker for commenting on a previous version of the chapter.

References

Adler, P. B. & Lauenroth, W. K. (2003) The power of time: Spatiotemporal scaling of species diversity. *Ecology Letters*, **6**, 749–756.

Adler, P. B., White, E. P., Lauenroth, W. K., Kaufman, D. M., Rassweiler, A. & Rasak, J. A. (2005) Evidence for a general species–time–area relationship. *Ecology*, **86**, 2032–2039.

Antonelli, A., Kissling, W. D., Flantua, S. G. A., Bermúdez, M. A., Mulch, A., Muellner-Riehl, A. N., Kreft, H., Linder, H. P., Badgley, C., Fjeldså, J., Fritz, S. A., Rahbek, C., Herman, F., Hooghiemstra, H. & Hoorn, C. (2018) Geological and climatic influences on mountain biodiversity. *Nature Geoscience*, **11**, 718–725.

Borregaard, M. K., Matthews, T. J. & Whittaker, R. J. (2016) The general dynamic model: Towards a unified theory of island biogeography? *Global Ecology & Biogeography*, **25**, 805–816.

Brown, J. H., Ernest, S. K. M., Parody, J. M. & Haskell, J. P. (2001) Regulation of diversity: Maintenance of species richness in changing environments. *Oecologia*, **126**, 321–332.

Burns, K. C., McHardy, R. P. & Pledger, S. (2009) The small island effect: Fact or artefact? *Ecography*, **32**, 269–276.

Cabral, J. S., Valente, L. & Hartig, F. (2017) Mechanistic models in macroecology and biogeography: State-of-art and prospects. *Ecography*, **40**, 267–280.

Cabral, J. S., Weigelt, P., Kissling, W. D. & Kreft, H. (2014) Biogeographic, climatic and spatial drivers differentially affect alpha, beta and gamma diversities on oceanic archipelagos. *Proceedings of the Royal Society B: Biological Sciences*, **281**, 20133246.

Cameron, R. A. D., Triantis, K. A., Parent, C. E., Guilhaumon, F., Alonso, M. R., Ibáñez, M., Martins, A. M. F., Ladle, R. J. & Whittaker, R. J. (2013) Snails on oceanic islands: Testing the general dynamic model of oceanic island biogeography using linear mixed effect models. *Journal of Biogeography*, **40**, 117–130.

Chase, J. M., Gooriah, L., May, F., Ryberg, W. A., Schuler, M. S., Craven, D. & Knight, T. M. (2019) A framework for disentangling ecological mechanisms underlying the island species–area relationship. *Frontiers of Biogeography*, **11**, e40844.

Connor, E. F. & McCoy, E. D. (1979) Statistics and biology of the species–area relationship. *The American Naturalist*, **113**, 791–833.

Darwin, C. R. (1859) *On the origin of species by means of natural selection, or the preservation of favoured races in the struggle for life.* London: John Murray.

Dengler, J. (2010) Robust methods for detecting a small island effect. *Diversity and Distributions*, **16**, 256–266.

Drakare, S., Lennon, J. J. & Hillebrand, H. (2006) The imprint of geographical, evolutionary and ecological context on species–area relationships. *Ecology Letters*, **9**, 215–227.

Fisher, R. A., Corbet, A. S. & Williams, C. B. (1943) The relation between the number of species and the number of individuals in a random sample of an animal population. *Journal of Animal Ecology*, **12**, 42–58.

Flather, C. H. (1996) Fitting species-accumulation functions and assessing regional land use impacts on avian diversity. *Journal of Biogeography*, **23**, 155–168.

Gillespie, R. G. (2007) Oceanic islands: Models of diversity. *Encyclopedia of biodiversity* (ed. by S. A. Levin), pp. 1–13. Oxford: Elsevier Ltd.

Hall, L. S., Krausman, P. R. & Morrison, M. L. (1997) The habitat concept and a plea for standard terminology. *Wildlife Society Bulletin*, **25**, 173–182.

Harmon, L. J. & Harrison, S. (2015) Species diversity is dynamic and unbounded at local and continental scales. *The American Naturalist*, **185**, 584–593.

Heaney, L. R. (2000) Dynamic equilibrium: A long-term, large-scale perspective on the equilibrium model of island biogeography. *Global Ecology & Biogeography*, **9**, 59–74.

Holt, B. G., Lessard, J.-P., Borregaard, M. K., Fritz, S. A., Araújo M. B., Dimitrov, D., Fabre, P.-H., Graham C. H., Graves G. R., Jønsson, K. A., Nogués-Bravo, D., Wang, Z., Whittaker, R. J., Fjeldså, J. & Rahbek, C. (2013) An update of Wallace's zoogeographic regions of the world. *Science*, **339**, 74–78.

Hortal, J., Triantis, K. A., Meiri, S., Thébault, E. & Sfenthourakis, S. (2009) Island species richness increases with habitat diversity. *The American Naturalist*, **174**, E205–E217.

Hutchinson, G. E. (1959) Homage to Santa Rosalia or why are there so many kinds of animals? *The American Naturalist*, **93**, 145–159.

Kisel, Y., McInnes, L., Toomey, N. H. & Orme, C. D. L. (2011) How diversification rates and diversity limits combine to create large-scale species–area relationships. *Philosophical Transactions of the Royal Society B: Biological Sciences*, **366**, 2514–2525.

Liu, J., Matthews, T. J., Zhong, L., Liu, J., Wu, D. & Yu, M. (2020) Environmental filtering underpins the island species–area relationship in a subtropical anthropogenic archipelago. *Journal of Ecology*, **108**, 424–432.

Lomolino, M. V. & Weiser, M. D. (2001) Towards a more general species–area relationship: Diversity on all islands, great and small. *Journal of Biogeography*, **28**, 431–445.

Looijen, R. C. (1995) On the distinction between habitat and niche, and some implications for species' differentiation. *Poznań Studies in the Philosophy of the Sciences and the Humanities*, **45**, 87–108.

Losos, J. B. & Ricklefs, R. E. (2009) Adaptation and diversification on islands. *Nature*, **457**, 830–836.

Losos, J. B. & Schluter, D. (2000) Analysis of an evolutionary species–area relationship. *Nature*, **408**, 847–850.

MacArthur, R. H. & Wilson, E. O. (1967) *The theory of island biogeography*. Princeton, NJ: Princeton University Press.

Martín, H. G. & Goldenfeld, N. (2006) On the origin and robustness of power-law species–area relationships in ecology. *Proceedings of the National Academy of Sciences USA*, **103**, 10310–10315.

Matthews, T. J., Borregaard, M. K., Guilhaumon, F., Triantis, K. A. & Whittaker, R. J. (2016) On the form of species–area relationships in habitat islands and true islands. *Global Ecology & Biogeography*, **25**, 847–858.

Matthews, T. J., Rigal, F., Triantis, K. A. & Whittaker, R. J. (2019b) A global model of island species–area relationships. *Proceedings of the National Academy of Sciences USA*, **25**, 12337–12342.

Matthews, T. J., Sadler, J. P., Kubota, Y., Woodall, C. W. & Pugh, T. A. M. (2019a) Systematic variation in North American tree species abundance distributions along macroecological climatic gradients. *Global Ecology & Biogeography*, **28**, 601–611.

Matthews, T. J., Triantis, K., Whittaker, R. J. & Guilhaumon, F. (2019c) sars: An R package for fitting, evaluating and comparing species–area relationship models. *Ecography*, **42**, 1446–1455.

McGlinn, D. J. & Palmer, M. W. (2009) Modeling the sampling effect in the species–time–area relationship. *Ecology*, **90**, 836–846.

McGlinn, D. J., Xiao, X., May, F., Gotelli, N. J., Engel, T., Blowes, S. A., Knight, T. M., Purschke, O., Chase, O. J. M. & McGill, B. J. (2019) Measurement of Biodiversity (MoB): A method to separate the scale-dependent effects of species abundance distribution, density, and aggregation on diversity change. *Methods in Ecology and Evolution*, **10**, 258–269.

Morrison, L. W. (2011) Why do some small islands lack vegetation? Evidence from long-term observations and introduction experiments. *Ecography*, **34**, 384–391.

Newton, A. (1877) Hartlaub's 'Birds of Madagascar'. *Nature*, **17**, 9–10.

Niering, W. A. (1956) Bioecology of Kapingamarangi Atoll. Caroline Islands: Terrestrial aspects. *Atoll Research Bulletin*, **49**, 1–32.

Niering, W. A. (1963) Terrestrial ecology of Kapingamarangi Atoll, Caroline Islands. *Ecological Monographs*, **33**, 131–160.

Norder, S. J., Proios, K. V., Whittaker, R. J., Alonso, M. R., Borges, P. A. V., Borregaard, M. K., Cowie, R. H., Florens, F. B. V., de Frias Martins, A. M., Ibáñez, M., Kissling, W. D., de Nascimento, L., Otto, R., Parent, C. E., Rigal, F., Warren, B. H., Fernández-Palacios, J. M., van Loon, E. E., Triantis, K. A. & Rijsdijk, K. F. (2019) Beyond the Last Glacial Maximum: Island endemism is best explained by long-lasting archipelago configurations. *Global Ecology & Biogeography*, **28**, 184–197.

Pigolotti, S. & Cencini, M. (2009) Speciation-rate dependence in species–area relationships. *Journal of Theoretical Biology*, **260**, 83–89.

Preston, F. W. (1960) Time and space and the variation of species. *Ecology*, **41**, 611–627.

Preston, F. W. (1962) The canonical distribution of commonness and rarity: Part I. *Ecology*, **43**, 185–215.

Rabosky, D. L. (2013) Diversity-dependence, ecological speciation, and the role of competition in macroevolution. *Annual Review of Ecology, Evolution, and Systematics*, **44**, 481–502.

Rabosky, D. L. & Glor, R. E. (2010) Equilibrium speciation dynamics in a model adaptive radiation of island lizards. *Proceedings of the National Academy of Sciences USA*, **107**, 22178–22183.

Rabosky, D. L. & Hurlbert, A. H. (2015) Species richness at continental scales is dominated by ecological limits. *The American Naturalist*, **185**, 572–583.

Rosenzweig, M. L. (1995) *Species diversity in space and time*. New York: Cambridge University Press.

Rosenzweig, M. L. (1998) Preston's ergodic conjecture: The accumulation of species in space and time. *Biodiversity dynamics: Turnover of populations, taxa, and communities* (ed. by M. L. McKinney and J. A. Drake), pp. 311–348. New York: Columbia University Press.

Rosenzweig, M. L. (2001) Loss of speciation rate will impoverish future diversity. *Proceedings of the National Academy of Sciences USA*, **98**, 5404–5410.

Santos, A. M. C., Whittaker, R. J., Triantis, K. A., Jones, O. R., Borges, P. A. V., Quicke, D. L. J. & Hortal, J. (2010) Are species–area relationships from entire archipelagos congruent with those of their constituent islands? *Global Ecology & Biogeography*, **19**, 527–540.

Scheiner, S. M. (2003) Six types of species-area curves. *Global Ecology & Biogeography*, **12**, 441–447.

Schrader, J., Moeljono, S., Keppel, G. & Kreft, H. (2019) Plants on small islands revisited: The effects of spatial scale and habitat quality on the species–area relationship. *Ecography*, **42**, 1405–1414.

Sfenthourakis, S. & Triantis, K. A. (2017) The Aegean archipelago: A natural laboratory of evolution, ecology and civilizations. *Journal of Biological Research-Thessaloniki*, **24**, 1–13.

Shmida, A. & Wilson, M. V. (1985) Biological determinants of species diversity. *Journal of Biogeography*, **12**, 1–20.

Song, X., Holt, R. D., Si, X., Christman, M. C. & Ding, P. (2018) When the species–time–area relationship meets island biogeography: Diversity patterns of avian communities over time and space in a subtropical archipelago. *Journal of Biogeography*, **45**, 664–675.

Stein, A., Gerstner, K. & Kreft, H. (2014) Environmental heterogeneity as a universal driver of species richness across taxa, biomes and spatial scales. *Ecology Letters*, **17**, 866–880.

Storch, D., Marquet, P. A. & Brown, J. H. (eds.) (2007) *Scaling biodiversity*. Cambridge: Cambridge University Press.

Tjørve, E. (2003) Shapes and functions of species–area curves: A review of possible models. *Journal of Biogeography*, **30**, 827–835.

Tjørve, E. (2009) Shapes and functions of species–area curves (II): A review of new models and parameterizations. *Journal of Biogeography*, **36**, 1435–1445.

Tjørve, E. & Tjørve, K. M. C. (2011) Subjecting the theory of the small-island effect to Ockham's razor. *Journal of Biogeography*, **38**, 1836–1839.

Triantis, K. A. & Sfenthourakis, S. (2012) Island biogeography is not a single-variable discipline: The small island effect debate. *Diversity and Distributions*, **18**, 92–96.

Triantis, K. A., Guilhaumon, F. & Whittaker, R. J. (2012) The island species–area relationship: Biology and statistics. *Journal of Biogeography*, **39**, 215–231.

Triantis, K. A., Mylonas, M., Lika, K. & Vardinoyannis, K. (2003) A model for the species–area–habitat relationship. *Journal of Biogeography*, **30**, 19–27.

Triantis, K. A., Mylonas, M. & Whittaker, R. J. (2008) Evolutionary species–area curves as revealed by single-island endemics: Insights for the interprovincial species–area relationship. *Ecography*, **31**, 401–407.

Triantis, K. A., Economo, E. P., Guilhaumon, F. & Ricklefs, R. E. (2015) Diversity regulation at macro-scales: Species richness on oceanic archipelagos. *Global Ecology & Biogeography*, **24**, 594–605.

Triantis, K. A., Vardinoyannis, K., Tsolaki, E. P., Botsaris, I., Lika, K. & Mylonas, M. (2006) Re-approaching the small island effect. *Journal of Biogeography*, **33**, 914–923.

Udvardy, M. F. D. (1959) Notes on the ecological concepts of habitat, biotope and niche. *Ecology*, **40**, 725–728.

Ulrich, W., Ollik, M. & Ugland, K. I. (2010) A meta-analysis of species–abundance distributions. *Oikos*, **119**, 1149–1155.

Valente, L., Illera, J. C., Havenstein, K., Pallien, T., Etienne, R. S. & Tiedemann, R. (2017) Equilibrium bird species diversity in Atlantic islands. *Current Biology*, **27**, 1660–1666.

Wallace, A. R. (1876) *The geographical distribution of animals: With a study of the relations of living and extinct faunas Volume 1.* Cambridge: Cambridge University Press.

Wallace, A. R. (1877) The comparative richness of faunas and floras tested numerically. *Nature*, **17**, 100–101.

Wang, Y., Chen, C. & Millien, V. (2018) A global synthesis of the small-island effect in habitat islands. *Proceedings of the Royal Society B: Biological Sciences*, **285**, 20181868.

Wang, Y., Millien, V. & Ding, P. (2016) On empty islands and the small-island effect. *Global Ecology & Biogeography*, **25**, 1333–1345.

Weigelt, P. & Kreft, H. (2013) Quantifying island isolation – insights from global patterns of insular plant species richness. *Ecography*, **36**, 417–429.

Weigelt, P., Jetz, W. & Kreft, H. (2013) Bioclimatic and physical characterization of the World's islands. *Proceedings of the National Academy of Sciences USA*, **110**, 15307–15312.

White, E. P. (2007) Spatiotemporal scaling of species richness: Patterns, processes, and implications. *Scaling biodiversity* (ed. by D. Storch, P. A. Marquet and J. H. Brown), pp. 325–346. Cambridge: Cambridge University Press.

White, E. P., Adler, P. B., Lauenroth, W. K., Gill, R. A., Greenberg, D., Kaufman, D. M., Rassweiler, A., Rusak, J. A., Smith, M. D., Steinbeck, J. R., Waide, R. B. & Yao, J. (2006) A comparison of the species–time relationship across ecosystems and taxonomic groups. *Oikos*, **112**, 185–195.

Whittaker, R. J. & Fernández-Palacios, J. M. (2007) *Island biogeography: Ecology, evolution, and conservation*, 2nd ed. Oxford: Oxford University Press.

Whittaker, R. J., Triantis, K. A. & Ladle, R. J. (2008) A general dynamic theory of oceanic island biogeography. *Journal of Biogeography*, **35**, 977–994.

Whittaker, R. J., Fernández-Palacios, J. M., Matthews, T. J., Borregaard, M. K. & Triantis, K. A. (2017) Island biogeography: Taking the long view of nature's laboratories. *Science*, **357**, eaam8326.

Williams, C. B. (1943) Area and number of species. *Nature*, **152**, 264–267.

Williams, M. R., Lamont, B. B. & Henstridge, J. D. (2009) Species–area functions revisited. *Journal of Biogeography*, **36**, 1994–2004.

Williamson, M. (1988) Relationship of species number to area, distance and other variables. *Analytical biogeography* (ed. by A. A. Myers and P. S. Giller), pp. 91–115. New York: Chapman & Hall.

Wright, D. H. (1983) Species–energy theory: An extension of species–area theory. *Oikos*, **41**, 496–506.

Index